Lecture Notes in Mathematics

Volume 2333

This series reports on new developments in all areas of mathematics and their applications - quickly, informally and at a high level. Mathematical texts analysing new developments in modelling and numerical simulation are welcome. The type of material considered for publication includes:

1. Research monographs
2. Lectures on a new field or presentations of a new angle in a classical field
3. Summer schools and intensive courses on topics of current research.

Texts which are out of print but still in demand may also be considered if they fall within these categories. The timeliness of a manuscript is sometimes more important than its form, which may be preliminary or tentative.

Titles from this series are indexed by Scopus, Web of Science, Mathematical Reviews, and zbMATH.

Andrew D. Lewis

Geometric Analysis on Real Analytic Manifolds

Springer

Andrew D. Lewis
Department of Mathematics and Statistics
Queens University
Kingston, ON, Canada

ISSN 0075-8434 ISSN 1617-9692 (electronic)
Lecture Notes in Mathematics
ISBN 978-3-031-37912-3 ISBN 978-3-031-37913-0 (eBook)
https://doi.org/10.1007/978-3-031-37913-0

Mathematics Subject Classification: 32C05, 46E10, 46T20, 47H30, 53B05, 58A07, 58A20

This work was supported by Natural Sciences and Engineering Research Council of Canada

This Springer imprint is published by the registered company Springer Nature Switzerland AG
The registered company address is: Gewerbestrasse 11, 6330 Cham, Switzerland

Dedicated to the memory of Winston

Preface

This book arose from my own efforts over a number of years to be able to work with things that are real analytic in differential geometry. The difficulties of working with real analytic objects defined on real analytic manifolds are numerous, compared to working with smooth objects. For example, in smooth differential geometry, it is elementary to extend local constructions to global ones using smooth cut-off functions. In the real analytic case, local constructions can typically *not* be extended to global ones, and one must work with sheaf theoretic constructions or embedding theorems arising from the seminal work of Cartan [13] and Grauert [23]. Similarly, for differential topology, there are various well-understood and classical topologies that one can use in the smooth case, depending on what one wishes to achieve, e.g., [32, 50]. In the real analytic case, the topologies are more rigid as is shown in the work of Martineau [48]. Moreover, usable descriptions of these topologies were not available until fairly recently; as far as we are aware, the first clear and rigorous presentation of seminorms for the real analytic topology on the space of real analytic functions is found in the work of Vogt [66]. It is these seminorms that allow a (mere) user of real analytic topologies to have a chance of proving new theorems, an example being the work of Jafarpour and Lewis [35] on flows of time-varying vector fields.

The focus in this book is

1. A careful development of the topology for the space of real analytic functions on real analytic manifolds and, more generally, for the space of real analytic sections of a real analytic vector bundle and real analytic mappings of real analytic manifolds.
2. An illustration of how our characterisations of the real analytic topology can be used to prove basic and useful facts about geometric operations on real analytic manifolds.
3. Do the previous two things in a global, intrinsic, and geometric manner.

The Approach

Let us describe in a little more detail the approach we take to real analytic analysis and at the same time prepare the reader to navigate the rather detailed constructions that we undergo.

A barrier to our objectives right at the start is that the appropriate topology for real analytic functions (functions, for simplicity) is not so easily envisaged, in contrast to the smooth case. While a suitable real analytic topology has been around since at least the work of Martineau [48]—who provided two descriptions of such a topology and showed that they agree—there has not been a "user-friendly" description of the real analytic topology, i.e., a description using seminorms, until quite recently. Some useful initial formulae are provided in [52], and seminorms are provided in the lecture notes of [16]. However, as far as we are aware, it is only in the technical note of Vogt [66] that we see a clear proof of the suitability of these seminorms. These were adapted to the geometric setting for sections of a real analytic vector bundle by Jafarpour and Lewis [35]. Part of this development was a decomposition of jet bundles using connections. The initial developments of that monograph are the starting point for our approach here.

Another complicating facet of the real analytic theory arises when one considers lifts from the base space to the total space of a real analytic vector bundle $\pi_{\mathsf{E}} \colon \mathsf{E} \to \mathsf{M}$, e.g., vertical lift of a section of E or horizontal lift of a vector field on M. The first of these operations requires no additional structure, but the second requires a connection. However, both require (in our approach) connections to study their real analytic continuity, because one needs to provide bounds for the jets on the codomain (i.e., on E) in terms of jets on the domain (i.e., on M). To provide seminorms, one also needs Riemannian and vector bundle metrics, and all of this data has to fit together nicely to provide the bounds required. For instance, one has a natural Riemannian metric on the manifold E arising from (1) a Riemannian metric on M, (2) a fibre metric on E, (3) an affine connection on M, and (4) a linear connection in E. Using an adaptation of a construction of Sasaki [61], this data gives rise to a Riemannian metric on E, and so its Levi-Civita affine connection. Moreover, the resulting Riemannian metric makes $\pi_{\mathsf{E}} \colon \mathsf{E} \to \mathsf{M}$ a Riemannian submersion, and enables some useful constructions of O'Neill [56].

In order to illustrate the nature of the difficulties one encounters, let us consider a specific and illustrative instance of the sort of argument that one must piece together to prove continuity in the real analytic case. The reader should regard the discussion immediately following as a sort of "thought experiment."

Suppose that we have a real analytic vector bundle $\pi_{\mathsf{E}} \colon \mathsf{E} \to \mathsf{M}$ with $\nabla^{\pi_{\mathsf{E}}}$ a real analytic linear connection in E. Let X be a real analytic vector field on M which we horizontally lift to a real analytic vector field X^{h} on E. To assess the continuity of the map $X \mapsto X^{\mathrm{h}}$ in the real analytic topology, one needs to compute jets of X^{h} and relate these to jets of X. Thus, one needs to differentiate X^{h} arbitrarily many times. This differentiation must be done on E, as this is the base on which X^{h} is defined. Trying this directly in local coordinates is, in principle, possible, but it is pretty unlikely that one will be able to produce the refined estimates required in this way.

Thus, in our approach, one needs an affine connection on E (thinking of E as just a manifold now). One can now see that there will be a complicated intermingling of the linear connection $\nabla^{\pi_{\mathsf{E}}}$, an affine connection ∇^{M} on M (to compute jets of X), and a fabricated affine connection ∇^{E} on E (to compute jets of X^{h}). This is only the beginning of the difficulties one faces. One also needs, not only formulae for the derivatives of X^{h}, but also recursive formulae relating how a derivative of X^{h} of order, say, k is related to the derivatives of X of orders $0, 1, \dots, k$. These recursive formulae are essential for being able to obtain growth estimates for the derivatives needed to relate the seminorms applied to X^{h} to those applied to X. Moreover, since the mapping $X \mapsto X^{\mathrm{h}}$ is injective, one might hope that the mapping is not just continuous, but is an homeomorphism onto its image. To prove this, one now needs to get estimates for the jets of X from formulae involving the jets of X^{h}. Thus, one needs estimates that go "both ways."[1] It is also worth mentioning that the estimates one needs from these recursive formulae are quite unforgiving, and so their form has to be very precisely managed. This requires extensive bookkeeping. This bookkeeping occupies us for a substantial portion of the book. This is contrasted with the smooth case, where very sloppy bookkeeping suffices; we shall say a few words about this contrast at illustrative places in the book.

Another difficulty is that the use of connections to compute derivatives for jets forces one to address the matter of whether the topologies defined by the seminorms used for jets, and derived from the use of connections, are actually not dependent on the chosen connection. Thus, one must compare iterated covariant derivatives with respect to different connections and show that these are related to one another in such a way that the resulting real analytic topology is well defined. This, in itself, is a substantial undertaking. It is done in an ad hoc way in [35, Lemma 2.5]; here we do this in a systematic and geometric way that offers many benefits towards the objectives of this book, apart from rendering more attractive the computations of [35].

We mention that the idea of obtaining recursive formulae for derivatives is given in a local setting by Thilliez [64] during the course of the proof of his Proposition 2.5, and can be applied to the mapping $\mathrm{C}^{\omega}(\mathsf{N}) \ni f \mapsto \Phi^* f \in \mathrm{C}^{\omega}(\mathsf{M})$ of pull-back by a real analytic mapping $\Phi \in \mathrm{C}^{\omega}(\mathsf{M}; \mathsf{N})$. We are able to extend the ideas in Thilliez' computations to general classes of geometric operations. For example, as we mention above, a local working out of the estimates for the horizontal lift operation seems like it will be very difficult. However, once one *does* get these things to work out, it is relatively straightforward to prove the main results of the book, which are the continuity of the fundamental geometric operations mentioned above.

[1] Note that this is an open mapping argument, and so raises the question of whether the open mapping theorem holds for the real analytic topology. It does, in fact, using the properties of this topology that we prove in Proposition 2.14. We shall, at times, use the resulting Open Mapping Theorem, but at other times we shall explicitly prove the openness using seminorms since the seminorm approach is the *raison d'être* of our methodology.

One of the features of the book is that almost all constructions are done intrinsically. While this may seem to unnecessarily complicate things, this is not, in fact, so. Even were one to work locally, there would still arise two difficult problems that we overcome in our approach, but that still must be overcome in a local approach: (1) the difficulty of lifts as described in detail above; (2) the verification that the topologies do not depend on various choices made (charts in the local calculations, and metrics and connections in the intrinsic calculations). Thus, while the intrinsic calculations are sometimes complicated, they are only a little more complicated than the necessarily already complicated local calculations. And we believe that the intrinsic approach is ultimately easier to use, once one understands how to use it. An objective of this book is to do a lot of tedious hard work required to produce methods and results that are themselves more or less straightforward.

As a side-benefit to our approach, we also are able to easily provide proofs in the finitely differentiable and smooth cases. We point out the relevant places where modifications can be made to the real analytic proofs to give the results in the finitely differentiable and smooth cases.

An Outline of the Book and Tips for Reading It

We open in Chap. 1 by providing an overview of the background required and/or made use of in the book.

In Chap. 2, we review real analytic differential geometry and the definition of the real analytic topology as per [48], give some of the properties of this topology, and define in geometric language the seminorms for this topology as constructed in [35]. We give a proof that these seminorms do indeed define the real analytic topology, following [66]. In Chap. 2, we also consider at length the topology of the space of real analytic mappings between real analytic manifolds. We provide a number of descriptions of this topology, which give as a consequence other characterisations of the topology for the set of sections of a real analytic vector bundle. To our knowledge, the topology for the space of real analytic mappings is presented here for the first time.

In Chap. 3, we provide a host of geometric constructions whose bearing on the main goal of the book will be difficult to glean on a first reading. Some sketchy motivation for the constructions of this chapter is outlined above in our discussion of the difficulties one will encounter trying to prove continuity of the horizontal lift mapping $X \mapsto X^{\mathrm{h}}$. In Sect. 3.1 we perform constructions with functions, vector fields, and tensors on the total space of a vector bundle. These form the basis for derivative computations done in Sect. 3.2. Particularly, in Sect. 3.2.1, we give $\pi_{\mathsf{E}} \colon \mathsf{E} \to \mathsf{M}$ the structure of a Riemannian submersion, following [56]. This allows us to relate, in a natural way, constructions on E with those on M. In Sect. 3.3, we provide the crucial recursive formulae that relate derivatives on E with those on M. We do this for a few of the standard geometric lifts one has for a vector bundle with a linear connection. Some of these we do because they are intrinsically interesting. Some we do because they are required for our general approach, even if one is not interested in them per se.

While Chap. 3 is focussed on the geometric constructions we need for our main results, the results of Chap. 4 focus on analytical results, specifically providing norms estimates relating the various operations from Sect. 3.3. In Sect. 4.1, we give fibre norms for various jet bundles that are used to define seminorms corresponding to the geometric constructions of interest. In Sect. 4.2, we put all of our work from Chap. 3 to use to prove Lemma 4.17, the technical lemma that makes everything work. The lemma gives a very precise estimate for the fibre norms of derivatives of coefficients that arise in the recursive constructions of Sect. 3.3. There is no wiggle room in the form of the required estimate, and this is one of the reasons why the computations of Chap. 3 are so laboriously carried out; these computations need to be understood at a high resolution. Once we have these estimates, however, in Sect. 4.3 we show that the fibre norms for jet bundles obtained in Sect. 4.1 behave in the proper way as to make the topologies we construct independent of our choices of connections and metrics. This is stated as Lemma 4.19. The actual proving of the independence of the topologies is carried out by proving in Theorem 4.24 that the topologies are each the same as a topology described using local forms of the seminorms. This device of using a local description carries two benefits.

1. It provides the local description of the seminorms. While our approach is intrinsic as much as this is possible, sometimes in practice one must work locally, and having the explicit local formulae for the fibre norms is beneficial.
2. While we have tried to make our treatment intrinsic, there is a crucial point where a *local* estimate for the growth of derivatives becomes unavoidable, resting as it does on the Cauchy estimates for holomorphic functions. In our proof of Theorem 4.24 is where this seemingly unavoidable local estimate is not avoided.

In Chap. 5, we prove continuity of some representative and some important geometric constructions. There is a long list of these constructions and we only give a few of the more obvious ones; we hope that the tools we develop in the book, and put to use in Chap. 5, will make it easy for researchers down the road to prove some important results in the real analytic setting where continuity is crucial.

Acknowledgements The research reported in this book was financially supported by the Natural Sciences and Engineering Research Council of Canada.

Kingston, ON, Canada
May 2023

Andrew D. Lewis

Contents

Chapter 1
Notation and Background

In this chapter we review some of the notation and machinery we use in the subsequent chapters. We provide references to aid a reader who may lack some of the prerequisites. While most of the material we overview should be regarded as elementary, it is possible that a reader without a substantial background in functional analysis will benefit from probing more deeply, following the references we give in Sect. 1.8.

1.1 Basic Terminology and Notation

When A is a subset of a set X, we write $A \subseteq X$. If we wish to exclude the possibility that $A = X$, we write $A \subsetneq X$. For a family of sets $(X_i)_{i \in I}$, we denote by $\prod_{i \in I} X_i$ the product of these sets. By $\mathrm{pr}_j \colon \prod_{i \in I} X_i \to X_j$ we denote the projection onto the jth factor. For a family of sets $(X_i)_{i \in I}$, their disjoint union is denoted by

$$\mathop{\dot{\bigcup}}_{i \in I} \{(x, i) \mid x \in X_i\}.$$

The identity map on a set X is denoted by id_X. If $f \colon X \to Y$ and if $A \subseteq X$, $f|A$ denotes the restriction of f to A. Sometimes we may wish to indicate a mapping without giving it a name, typically in a framework where there is a great deal of existing notation and we do not want to introduce yet another symbol. To do this, we will write a mapping as $(x \mapsto f(x))$, thereby indicating what the mapping does to an element x of its domain.

The cardinality of a set X is denoted by $\mathrm{card}(X)$.

By $\mathbb{Z}$ we denote the set of integers. We use the notation $\mathbb{Z}_{>0}$ and $\mathbb{Z}_{\geq 0}$ to denote the subsets of positive and nonnegative integers. By $\mathbb{R}$ we denote the set of real

A. D. Lewis, *Geometric Analysis on Real Analytic Manifolds*, Lecture Notes in Mathematics 2333, https://doi.org/10.1007/978-3-031-37913-0_1

numbers and by $\mathbb{C}$ we denote the set of complex numbers. We denote $\mathrm{i} = \sqrt{-1}$. By $\mathbb{R}_{>0}$ we denote the subset of positive real numbers.

We denote by $\mathbb{R}^n$ the n-fold Cartesian product of $\mathbb{R}$. A point in $\mathbb{R}^n$ will typically be denoted in a bold font, e.g., $\boldsymbol{x} = (x_1, \dots, x_n)$. We denote the standard basis for $\mathbb{R}^n$ by $(\boldsymbol{e}_1, \dots, \boldsymbol{e}_n)$.

1.2 Algebra and Linear Algebra

Any intermediate level linear algebra text suffices to provide the concepts we need, e.g., [4]. Tensor algebra at the level we use it is covered in [1, §6.1].

By $\mathfrak{S}_k$ we denote the permutation group of $\{1, \dots, k\}$. For $k, l \in \mathbb{Z}_{\geq 0}$, we denote by $\mathfrak{S}_{k,l}$ the subset of $\mathfrak{S}_{k+l}$ consisting of permutations σ satisfying

$$\sigma(1) < \cdots < \sigma(k), \quad \sigma(k+1) < \cdots < \sigma(k+l).$$

We also denote by $\mathfrak{S}_{k|l}$ the subgroup of $\mathfrak{S}_{k+l}$ consisting of permutations having the form

$$\begin{pmatrix} 1 & \cdots & k & k+1 & \cdots & k+l \\ \sigma_1(1) & \cdots & \sigma_1(k) & k+\sigma_2(1) & \cdots & k+\sigma_2(l) \end{pmatrix}$$

for $\sigma_1 \in \mathfrak{S}_k$ and $\sigma_2 \in \mathfrak{S}_l$. We note that $\mathfrak{S}_{k+l}/\mathfrak{S}_{k|l} \simeq \mathfrak{S}_{k,l}$, so that (1) if $\sigma \in \mathfrak{S}_{k+l}$, then $\sigma = \sigma_1 \circ \sigma_2$ for $\sigma_1 \in \mathfrak{S}_{k,l}$ and $\sigma_2 \in \mathfrak{S}_{k|l}$ and (2) $\operatorname{card}(\mathfrak{S}_{k,l}) = \frac{(k+l)!}{k!l!}$. Note that, similarly, $\mathfrak{S}_{k|l} \backslash \mathfrak{S}_{k+1} \simeq \mathfrak{S}_{k,l}$, and so, if $\sigma \in \mathfrak{S}_{k+l}$, then $\sigma = \sigma_1 \circ \sigma_2$ for $\sigma_1 \in \mathfrak{S}_{k|l}$ and $\sigma_2 \in \mathfrak{S}_{k,l}$.

For $\mathbb{R}$-vector spaces U and V, we denote by $\operatorname{Hom}_{\mathbb{R}}(\mathsf{U}; \mathsf{V})$ the set of $\mathbb{R}$-linear mappings from U to V. We shall also adapt this notation to bilinear mappings. Thus, adding a $\mathbb{R}$-vector space W to the conversation, by $\operatorname{Hom}_{\mathbb{R}}(\mathsf{U}, \mathsf{V}; \mathsf{W})$ we denote the set of bilinear mappings from $\mathsf{U} \times \mathsf{V}$ to W. We denote $\operatorname{End}_{\mathbb{R}}(\mathsf{V}) = \operatorname{Hom}_{\mathbb{R}}(\mathsf{V}; \mathsf{V})$. We denote by $\mathsf{V}^* = \operatorname{Hom}_{\mathbb{R}}(\mathsf{V}; \mathbb{R})$ the algebraic dual. If $v \in \mathsf{V}$ and $\alpha \in \mathsf{V}^*$, we will denote the evaluation of α on v at various points by $\alpha(v)$, $\alpha \cdot v$, or $\langle \alpha; v \rangle$, whichever seems most pleasing to us at the moment. If $A \in \operatorname{Hom}_{\mathbb{R}}(\mathsf{U}; \mathsf{V})$, we denote by $A^* \in \operatorname{Hom}_{\mathbb{R}}(\mathsf{V}^*; \mathsf{U}^*)$ the dual of A. If $S \subseteq \mathsf{V}$, then we denote by

$$\operatorname{ann}(S) = \{\alpha \in \mathsf{V}^* \mid \alpha(v) = 0, \ v \in S\}$$

the annihilator subspace.

For a $\mathbb{R}$-vector space V, $\mathrm{T}^k(\mathsf{V})$ is the k-fold tensor product of V with itself. For $r, s \in \mathbb{Z}_{>0}$, we denote

$$\mathrm{T}^r_s(\mathsf{V}) = \underbrace{\mathsf{V} \otimes \cdots \otimes \mathsf{V}}_{r \text{ times}} \otimes \underbrace{\mathsf{V}^* \otimes \cdots \otimes \mathsf{V}^*}_{s \text{ times}}.$$

By $\mathsf{S}^k(\mathsf{V})$ we denote the k-fold symmetric tensor product of V with itself, and we think of this as a subset of $\mathrm{T}^k(\mathsf{V})$. In like manner, by $\bigwedge^k(\mathsf{V})$ we denote the k-fold alternating tensors of V. For $A \in \mathsf{S}^k(\mathsf{V})$ and $B \in \mathsf{S}^l(\mathsf{V})$, we define the symmetric tensor product of A and B to be

$$A \odot B = \sum_{\sigma \in \mathfrak{S}_{k,l}} \sigma(A \otimes B). \tag{1.1}$$

We define $\mathrm{Sym}_k\colon \mathrm{T}^k(\mathsf{V}) \to \mathsf{S}^k(\mathsf{V})$ by

$$\mathrm{Sym}_k(v_1 \otimes \cdots \otimes v_k) = \frac{1}{k!} \sum_{\sigma \in \mathfrak{S}_k} v_{\sigma(1)} \otimes \cdots \otimes v_{\sigma(k)}.$$

We note that we have the alternative formula

$$A \odot B = \frac{(k+l)!}{k!l!} \mathrm{Sym}_{k+l}(A \otimes B) \tag{1.2}$$

for the product of $A \in \mathsf{S}^k(\mathsf{V})$ and $B \in \mathsf{S}^l(\mathsf{V})$ [10, Proposition IV.3.3]. We recall that

$$\dim_{\mathbb{R}}(\mathsf{S}^k(\mathsf{V})) = \binom{\dim_{\mathbb{R}}(\mathsf{V}) + k - 1}{k}, \tag{1.3}$$

when V is finite-dimensional.

For a $\mathbb{R}$-vector space V, let us denote

$$\mathrm{T}^{\leq m}(\mathsf{V}) = \bigoplus_{j=0}^{m} \mathrm{T}^j(\mathsf{V}), \quad \mathsf{S}^{\leq m}(\mathsf{V}) = \bigoplus_{j=0}^{m} \mathsf{S}^j(\mathsf{V}),$$

and define

$$\begin{aligned} \mathrm{Sym}_{\leq m}\colon \mathrm{T}^{\leq m}(\mathsf{V}) &\to \mathsf{S}^{\leq m}(\mathsf{V}) \\ (A_0, A_1, \ldots, A_m) &\mapsto (A_0, \mathrm{Sym}_1(A_1), \ldots, \mathrm{Sym}_m(A_m)). \end{aligned}$$

For $\mathbb{R}$-inner product spaces $(\mathsf{U}, \mathbb{G}_{\mathsf{U}})$ and $(\mathsf{V}, \mathbb{G}_{\mathsf{V}})$, we denote the transpose of $L \in \mathrm{Hom}_{\mathbb{R}}(\mathsf{U}; \mathsf{V})$ as the linear map $L^T \in \mathrm{Hom}_{\mathbb{R}}(\mathsf{V}; \mathsf{U})$ defined by

$$\mathbb{G}_{\mathsf{V}}(L(u), v) = \mathbb{G}_{\mathsf{U}}(u, L^T(v)), \qquad u \in \mathsf{U},\ v \in \mathsf{V}.$$

By $\mathbb{G}_{\mathsf{V}}^\flat\colon \mathsf{V} \to \mathsf{V}^*$ we denote the isomorphism induced by the inner product $\mathbb{G}_{\mathsf{V}}$ on V, and by $\mathbb{G}_{\mathsf{V}}^\sharp$ we denote the inverse of $\mathbb{G}_{\mathsf{V}}^\flat$.

1.3 Real Analysis and Differential Calculus

A standard reference for real analysis suffices for our purposes, e.g., [62]. Our notation for differential calculus follows [1, Chapter 2].

We shall not use any particular notation for the Euclidean norm for $\mathbb{R}^n$, and so will just denote this norm by

$$\|\boldsymbol{x}\| = \left(\sum_{j=1}^{n} |x_j|^2\right)^{1/2}.$$

It is sometimes convenient to use other norms for $\mathbb{R}^n$, particularly the 1- and ∞-norms defined, as usual, by

$$\|\boldsymbol{x}\|_1 = \sum_{j=1}^{n} |x_j|, \quad \|\boldsymbol{x}\|_\infty = \sup\left\{ |x_j| \;\middle|\; j \in \{1, \dots, n\}\right\}.$$

The following relationships between these norms are useful:

$$\|\boldsymbol{x}\| \leq \|\boldsymbol{x}\|_1 \leq \sqrt{n}\|\boldsymbol{x}\|, \quad \|\boldsymbol{x}\|_\infty \leq \|\boldsymbol{x}\| \leq \sqrt{n}\|\boldsymbol{x}\|_\infty,$$
$$\|\boldsymbol{x}\|_\infty \leq \|\boldsymbol{x}\|_1 \leq n\|\boldsymbol{x}\|_\infty. \tag{1.4}$$

If we are using a norm whose definition is evident from context, we will simply denote it by $\|\cdot\|$, expecting that context will ensure that there is no confusion.

For $\boldsymbol{x} \in \mathbb{R}^n$ and $r \in \mathbb{R}_{>0}$, we denote by

$$\mathsf{B}(r, \boldsymbol{x}) = \left\{ \boldsymbol{y} \in \mathbb{R}^n \;\middle|\; \|\boldsymbol{y} - \boldsymbol{x}\| < r\right\}$$

and

$$\overline{\mathsf{B}}(r, \boldsymbol{x}) = \left\{ \boldsymbol{y} \in \mathbb{R}^n \;\middle|\; \|\boldsymbol{y} - \boldsymbol{x}\| \leq r\right\}$$

the ***open ball*** and the ***closed ball*** of radius r centred at $\boldsymbol{x}$. As with the notation for norms, we shall often use the preceding notation for balls in settings different from $\mathbb{R}^n$, and accept the abuse of notation. For $\boldsymbol{r} \in \mathbb{R}^n_{>0}$ and for $\boldsymbol{x} \in \mathbb{R}^n$, we denote by

$$\mathsf{D}^n(\boldsymbol{r}, \boldsymbol{x}) = \prod_{j=1}^{n} (x_j - r_j, x_j + r_j)$$

and

$$\overline{\mathsf{D}}^n(\boldsymbol{r}, \boldsymbol{x}) = \prod_{j=1}^{n} [x_j - r_j, x_j + r_j]$$

the ***open polydisk*** and the ***closed polydisk*** of radius r and centre $\boldsymbol{x}$. We will use the same notation for the analogous polydisks in $\mathbb{C}^n$.

If $\mathcal{U} \subseteq \mathbb{R}^n$ is open and if $\boldsymbol{\Phi}: \mathcal{U} \to \mathbb{R}^m$ is differentiable at $\boldsymbol{x} \in \mathcal{U}$, we denote its derivative by $\boldsymbol{D}\boldsymbol{\Phi}(\boldsymbol{x}) \in \mathrm{Hom}_{\mathbb{R}}(\mathbb{R}^n; \mathbb{R}^m)$. Higher-order derivatives, when they exist, are denoted by $\boldsymbol{D}^k\boldsymbol{\Phi}(\boldsymbol{x})$, k being the order of differentiation. We recall that, if $\boldsymbol{\Phi}: \mathcal{U} \to \mathbb{R}^m$ is of class C^k, $k \in \mathbb{Z}_{>0}$, then $\boldsymbol{D}^k\boldsymbol{\Phi}(\boldsymbol{x})$ is symmetric [1, Proposition 2.4.14].

If $\mathcal{U} \subseteq \mathbb{R}^{n_1} \times \mathbb{R}^{n_2}$ is open and if $\boldsymbol{\Phi}: \mathcal{U} \to \mathbb{R}^m$ is differentiable at $(\boldsymbol{x}_{01}, \boldsymbol{x}_{02}) \in \mathcal{U}$, then the partial derivatives $\boldsymbol{D}_a\boldsymbol{\Phi}(\boldsymbol{x}_{01}, \boldsymbol{x}_{02}) \in \mathrm{Hom}_{\mathbb{R}}(\mathbb{R}^{n_a}; \mathbb{R}^m)$, $a \in \{1, 2\}$, are the derivatives at $\boldsymbol{x}_{0a}$ of the mappings

$$\boldsymbol{x}_1 \mapsto \boldsymbol{\Phi}(\boldsymbol{x}_1, \boldsymbol{x}_{02}), \quad \boldsymbol{x}_2 \mapsto \boldsymbol{\Phi}(\boldsymbol{x}_{01}, \boldsymbol{x}_2).$$

We shall sometimes find it convenient to use multi-index notation for derivatives. A ***multi-index*** with length n is an element of $\mathbb{Z}^n_{\geq 0}$, i.e., an n-tuple $I = (i_1, \dots, i_n)$ of nonnegative integers. If $\boldsymbol{\Phi}: \mathcal{U} \to \mathbb{R}^m$ is a smooth function, then we denote

$$\boldsymbol{D}^I\boldsymbol{\Phi}(\boldsymbol{x}) = \boldsymbol{D}_1^{i_1} \cdots \boldsymbol{D}_n^{i_n}\boldsymbol{\Phi}(\boldsymbol{x}),$$

identifying

$$\mathbb{R}^n = \underbrace{\mathbb{R} \times \cdots \times \mathbb{R}}_{n \text{ times}}$$

and using partial derivative notation as above. We will use the symbol $|I| = i_1 + \cdots + i_n$ to denote the order of the derivative. Another piece of multi-index notation we shall use is

$$\boldsymbol{a}^I = a_1^{i_1} \cdots a_n^{i_n},$$

for $\boldsymbol{a} \in \mathbb{R}^n$ and $I \in \mathbb{Z}^n_{\geq 0}$. Also, we denote $I! = i_1! \cdots i_n!$.

1.4 Differential Geometry

We shall adopt the notation and conventions of smooth differential geometry of [1]. We shall also make use of real analytic differential geometry, of course. In Sect. 2.1 we shall give an overview of real analytic differential geometry with an emphasis on the ways in which it differs from smooth differential geometry. There are no comprehensive textbook references dedicated to real analytic differential geometry, but the book of [14] contains some of what we shall need.

Throughout the book, unless otherwise stated, manifolds are connected, second countable, Hausdorff manifolds. The assumption of connectedness can be dispensed with but is convenient as it allows one to not have to worry about manifolds with components of different dimensions and vector bundles with fibres of different dimensions.

We shall work with regularity classes $r \in \{\infty, \omega\}$, "∞" meaning smooth, "ω" meaning real analytic. If the reader sees the symbol "C^r" without any specific indication of what r is, it is intended that r be either ∞ or ω. The word "smooth" will always mean "infinitely differentiable." Sometimes we do not require infinite differentiability, but will hypothesise it anyway. Other times we will precisely specify the regularity needed; but we will be a little sloppy with this as (1) it is not crucial to the purposes of this book and (2) it is typically easy to know when infinite differentiability is hypothesised but not required. We shall make some use of holomorphic regularity, but will introduce our notation for this when we make use of it.

The tangent bundle of a manifold M is denoted by $\pi_{\mathsf{TM}}: \mathsf{TM} \to \mathsf{M}$ and the cotangent bundle by $\pi_{\mathsf{T}^*\mathsf{M}}: \mathsf{T}^*\mathsf{M} \to \mathsf{M}$.

We denote by $\mathrm{C}^r(\mathsf{M}; \mathsf{N})$ the set of mappings from a manifold M to a manifold N of class C^r. When $\mathsf{N} = \mathbb{R}$, we denote by $\mathrm{C}^r(\mathsf{M}) = \mathrm{C}^r(\mathsf{M}; \mathbb{R})$ the set of scalar-valued functions of class C^r. For $\Phi \in \mathrm{C}^1(\mathsf{M}; \mathsf{N})$, $T\Phi: \mathsf{TM} \to \mathsf{TN}$ denotes the derivative of Φ, and $T_x\Phi = T\Phi|\mathsf{T}_x\mathsf{M}$. A mapping $\Phi \in \mathrm{C}^1(\mathsf{M}; \mathsf{N})$ is a ***submersion*** if $T_x\Phi$ is surjective for each $x \in \mathsf{M}$ and is an ***immersion*** if $T_x\Phi$ is injective for each $x \in \mathsf{M}$. An immersion is an ***embedding*** if it is an homeomorphism onto its image.

For $f \in \mathrm{C}^\nu(\mathsf{M})$, $\nu \in \mathbb{Z}_{>0} \cup \{\infty, \omega\}$, we denote by $\mathrm{d}f \in \Gamma^{\nu-1}(\mathsf{T}^*\mathsf{M})$ the ***differential*** of f, defined by

$$T_x f(v_x) = (f(x), \langle \mathrm{d}f(x); v_x\rangle), \qquad v_x \in \mathsf{T}_x\mathsf{M}.$$

We denote by $T_x^*\Phi$ the dual of $T_x\Phi$.

Let $\pi_{\mathsf{E}}: \mathsf{E} \to \mathsf{M}$ be a vector bundle of class C^r. We shall sometimes denote the fibre over $x \in \mathsf{M}$ by E_x, noting that this has the structure of a $\mathbb{R}$-vector space. If $A \subseteq \mathsf{M}$, we denote $\mathsf{E}|A = \pi_{\mathsf{E}}^{-1}(A)$. By $\Gamma^r(\mathsf{E})$ we denote the set of sections of E of class C^r. This space has the structure of a $\mathbb{R}$-vector space with the vector space operations

$$(\xi + \eta)(x) = \xi(x) + \eta(x), \quad (a\xi)(x) = a(\xi(x)), \qquad x \in \mathsf{M},$$

and of a $\mathrm{C}^r(\mathsf{M})$-module with the additional operation of multiplication

$$(f\xi)(x) = f(x)\xi(x), \qquad x \in \mathsf{M},$$

for $f \in \mathrm{C}^r(\mathsf{M})$, $\xi, \eta \in \Gamma^r(\mathsf{E})$, and $a \in \mathbb{R}$. By $\mathbb{R}^k_{\mathsf{M}}$ we denote the trivial bundle $\mathbb{R}^k_{\mathsf{M}} = \mathsf{M} \times \mathbb{R}^k$ with vector bundle projection being projection onto the first factor. The dual bundle E^* of a vector bundle E is the set of vector bundle mappings from E to $\mathbb{R}_{\mathsf{M}}$ over id_{M}. We note that there is a natural identification of $\Gamma^r(\mathbb{R}_{\mathsf{M}})$ with $\mathrm{C}^r(\mathsf{M})$. Given a C^r-vector bundle $\pi_{\mathsf{E}}: \mathsf{E} \to \mathsf{M}$ and a mapping $\Phi \in \mathrm{C}^r(\mathsf{N}; \mathsf{M})$, we denote by $\Phi^*\pi_{\mathsf{E}}: \Phi^*\mathsf{E} \to \mathsf{N}$ the pull-back bundle. For C^r-vector bundles $\pi_{\mathsf{E}}: \mathsf{E} \to \mathsf{M}$

and $\pi_{\mathsf{F}}\colon \mathsf{F} \to \mathsf{M}$ over the same base, we denote by $\mathrm{VB}^r(\mathsf{E}; \mathsf{F})$ the set of C^r-vector bundle mappings from E to F over id_{M}. We denote by $\mathrm{T}^r_s(\mathsf{E})$ the tensor bundle over M whose fibre over x is $\mathrm{T}^r_s(\mathsf{E}_x)$. We shall also denote by $\mathrm{S}^k(\mathsf{E})$ the symmetric tensor bundle and by $\bigwedge^k(\mathsf{E})$ the alternating tensor bundle.

For a vector field X and a differentiable function f, $\mathscr{L}_X f$ denotes the Lie derivative of f with respect to X. We might also write $Xf = \mathscr{L}_X f$. For smooth vector fields X and Y, we denote by $[X, Y]$ the Lie bracket of these vector fields, this being defined by the formula

$$\mathscr{L}_{[X,Y]} f = \mathscr{L}_X \mathscr{L}_Y f - \mathscr{L}_Y \mathscr{L}_X f, \qquad f \in \mathrm{C}^r(\mathsf{M}).$$

For $X \in \Gamma^r(\mathsf{TM})$, the flow of X is denoted by Φ^X_t, meaning that, for $x \in \mathsf{M}$, we have

$$\frac{\mathrm{d}}{\mathrm{d}t}\Phi^X_t(x) = X \circ \Phi^X_t(x), \quad \Phi^X_0(x) = x.$$

Said otherwise, $t \mapsto \Phi^X_t(x)$ is the integral curve of X through x at time 0.

The Lie derivative for vector fields extends to a derivation of the tensor algebra for a manifold. Specifically, for $X \in \Gamma^\infty(\mathsf{TM})$, we denote

$$\mathscr{L}_X f = \langle \mathrm{d}f; X\rangle, \quad \mathscr{L}_X Y = [X, Y], \qquad f \in \mathrm{C}^\infty(\mathsf{M}),\ X \in \Gamma^\infty(\mathsf{TM}).$$

For $\alpha \in \Gamma^\infty(\mathsf{T}^*\mathsf{M})$, we can then define its Lie derivative with respect to X by

$$\langle \mathscr{L}_X \alpha; Y\rangle = \mathscr{L}_X \langle \alpha; Y\rangle - \langle \alpha; \mathscr{L}_X Y\rangle, \qquad Y \in \Gamma^\infty(\mathsf{TM}).$$

The Lie derivative of a tensor field $A \in \Gamma^\infty(\mathrm{T}^r_s(\mathsf{TM}))$ is then defined by

$$\begin{aligned}\mathscr{L}_X A(\alpha^1, \dots, \alpha^r, X_1, \dots, X_s) &= \mathscr{L}_X(A(\alpha^1, \dots, \alpha^r, X_1, \dots, X_s)) \\ &\quad - \sum_{j=1}^{r} A(\alpha^1, \dots, \mathscr{L}_X \alpha^j, \dots, \alpha^r, X_1, \dots, X_s) \\ &\quad - \sum_{j=1}^{s} A(\alpha^1, \dots, \alpha^r, X_1, \dots, \mathscr{L}_X X_j, \dots, X_s). \qquad (1.5)\end{aligned}$$

Of course, these constructions make sense for tensor fields and vector fields that are less regular than smooth.

We shall make occasional use of notation and methods from sheaf theory. A readable introduction to the theory and its uses in differential geometry is the book [60]. By $\mathscr{G}^r_{\mathsf{E}}$ we denote the sheaf of C^r-sections of E. Thus

$$\mathscr{G}^r_{\mathsf{E}}(\mathcal{U}) = \Gamma^r(\mathsf{E}|\mathcal{U})$$

when $\mathcal{U} \subseteq \mathsf{M}$ is open. By $\mathscr{C}^r_{\mathsf{M}}$ we denote the sheaf of C^r-functions on M. Thus

$$\mathscr{C}^r_{\mathsf{M}}(\mathcal{U}) = \mathrm{C}^r(\mathcal{U})$$

when $\mathcal{U} \subseteq \mathsf{M}$ is open.

1.5 Riemannian Geometry and Connections

We shall make use of basic constructions from Riemannian geometry. We also work a great deal with connections, both affine connections and linear connections in vector bundles. We refer to [42] as a standard reference, and [43] is also a useful reference.

First suppose that $r \in \{\infty, \omega\}$. A ***C^r-fibre metric*** on a C^r-vector bundle $\pi_{\mathsf{E}}\colon \mathsf{E} \to \mathsf{M}$ is $\mathbb{G}_{\pi_{\mathsf{E}}} \in \Gamma^r(\mathrm{S}^2(\mathsf{E}^*))$ such that $\mathbb{G}_{\pi_{\mathsf{E}}}(x)$ is an inner product on E_x for each $x \in \mathsf{M}$. The associated norm on fibres we denote by $\|\cdot\|_{\mathbb{G}}$. In case E is the tangent bundle of M, then a fibre metric is a ***Riemannian metric***, and we will use the notation $\mathbb{G}_{\mathsf{M}}$ in this case.

A ***C^r-linear connection*** in a C^r-vector bundle $\pi_{\mathsf{E}}\colon \mathsf{E} \to \mathsf{M}$ will be denoted by $\nabla^{\pi_{\mathsf{E}}}$; this defines a mapping

$$\Gamma^r(\mathsf{TM}) \times \Gamma^r(\mathsf{E}) \ni (X, \xi) \mapsto \nabla^{\pi_{\mathsf{E}}}_X \xi \in \Gamma^r(\mathsf{E})$$

that is $\mathbb{R}$-bilinear, $\mathrm{C}^r(\mathsf{M})$-linear in X, and that satisfies the following derivation property in ξ:

$$\nabla^{\pi_{\mathsf{E}}}_X (f\xi) = f\nabla^{\pi_{\mathsf{E}}}_X \xi + (\mathscr{L}_X f)\xi.$$

There are many equivalent ways to define a linear connection, all being equivalent to the way we give above. Another way to make the definition, and one we will find useful, is via a ***connector*** for $\nabla^{\pi_{\mathsf{E}}}$. This is a mapping $K_{\nabla^{\pi_{\mathsf{E}}}}\colon \mathsf{TE} \to \mathsf{E}$ such that the two diagrams

$$\begin{array}{ccc} \mathsf{TE} & \xrightarrow{K_{\nabla^{\pi_{\mathsf{E}}}}} & \mathsf{E} \\ {\scriptstyle T\pi_{\mathsf{E}}}\downarrow & & \downarrow{\scriptstyle \pi_{\mathsf{E}}} \\ \mathsf{TM} & \xrightarrow[\pi_{\mathsf{TM}}]{} & \mathsf{M} \end{array} \qquad \begin{array}{ccc} \mathsf{TE} & \xrightarrow{K_{\nabla^{\pi_{\mathsf{E}}}}} & \mathsf{E} \\ {\scriptstyle \pi_{\mathsf{TE}}}\downarrow & & \downarrow{\scriptstyle \pi_{\mathsf{E}}} \\ \mathsf{E} & \xrightarrow[\pi_{\mathsf{E}}]{} & \mathsf{M} \end{array} \tag{1.6}$$

define vector bundle mappings [43, §11.11]. The relationship between the connection and the connector is

$$\nabla^{\pi_{\mathsf{E}}}_X \xi = K_{\nabla^{\pi_{\mathsf{E}}}} \circ T\xi \circ X. \tag{1.7}$$

Given a linear connection $\nabla^{\pi_{\mathsf{E}}}$ in $\pi\colon \mathsf{E} \to \mathsf{M}$, one can define the notion of parallel transport along a curve. If $x \in \mathsf{M}$, if $e \in \mathsf{E}_x$, and if $\gamma\colon [0, T] \to \mathsf{M}$ is a smooth curve with $\gamma(0) = x$, then we have the initial value problem

$$\nabla^{\pi_{\mathsf{E}}}_{\gamma'(t)} \xi(t) = 0, \quad \xi(0) = e,$$

in E. The solution defines a mapping $\tau^{\gamma}_t\colon \mathsf{E}_x \to \mathsf{E}_{\gamma(t)}$ called ***parallel transport*** along γ. Because $\nabla^{\pi_{\mathsf{E}}}$ is a *linear* connection, parallel transport is a linear isomorphism of fibres [42, page 114].

In case E is the tangent bundle of M, then a linear connection is called an ***affine connection***, and we will denote it by ∇^{M}. Given a Riemannian metric $\mathbb{G}_{\mathsf{M}}$, there is a distinguished affine connection ∇^{M} defined by the following requirements:

1. $\mathscr{L}_X(\mathbb{G}_{\mathsf{M}}(Y, Z)) = \mathbb{G}_{\mathsf{M}}(\nabla^{\mathsf{M}}_X Y, Z) + \mathbb{G}_{\mathsf{M}}(Y, \nabla^{\mathsf{M}}_X Z)$, $X, Y, Z \in \Gamma^\infty(\mathsf{TM})$;
2. $\nabla^{\mathsf{M}}_X Y - \nabla^{\mathsf{M}}_Y X = [X, Y]$, $X, Y \in \Gamma^\infty(\mathsf{TM})$

[42, Theorem IV.2.2]. The unique affine connection ∇^{M} with these properties is called the ***Levi-Civita connection***. We will sometimes, but not always, require that the affine connection on M with which we work be the Levi-Civita affine connection for a Riemannian metric.

For an affine connection ∇^{M} on M, a C^2-curve $\gamma\colon I \to \mathsf{M}$ ($I \subseteq \mathbb{R}$ is an interval) is a ***geodesic*** if it satisfies $\nabla^{\mathsf{M}}_{\gamma'(t)} \gamma'(t) = 0$.

A Riemannian manifold, i.e., a manifold M with a Riemannian metric $\mathbb{G}_{\mathsf{M}}$, has a natural metric space structure as follows. For a piecewise C^1-curve $\gamma\colon [a, b] \to \mathsf{M}$,[1] define its ***length*** by

$$\ell_{\mathbb{G}_{\mathsf{M}}}(\gamma) = \int_a^b \sqrt{\mathbb{G}_{\mathsf{M}}(\gamma'(t), \gamma'(t))}\, \mathrm{d}t.$$

One can easily show that the length is independent of parameterisation, and so one can then consider curves defined on $[0, 1]$. This being the case, the ***distance*** between

[1] Thus γ is continuous and is of class C^1 on a finite number of disjoint open subintervals for which the closure of their union is $[a, b]$, and the derivative has well-defined limits at the endpoints of the intervals.

$x_1, x_2 \in \mathsf{M}$ is

$$\begin{aligned} d_{\mathbb{G}_\mathsf{M}}(x_1, x_2) = \inf\bigl\{\ell_{\mathbb{G}_\mathsf{M}}(\gamma) \bigm| \\ \gamma\colon [0, 1] \to \mathsf{M} \text{ is piecewise differentiable and } \gamma(0) = x_1 \text{ and } \gamma(1) = x_2\bigr\}. \end{aligned}$$

This distance function can be verified to have the properties of a metric [1, Proposition 5.5.10].

A linear connection in a vector bundle $\pi_\mathsf{E}\colon \mathsf{E} \to \mathsf{M}$ induces a splitting of the short exact sequence

$$0 \longrightarrow \ker(T\pi_\mathsf{E}) \longrightarrow \mathsf{TE} \xrightarrow{T\pi_\mathsf{E}} \mathsf{TM} \longrightarrow 0$$

[43, §11.10]. For $e \in \mathsf{E}$, we thus have a splitting of the tangent space $\mathsf{T}_e\mathsf{E} \simeq \mathsf{T}_{\pi_\mathsf{E}(e)}\mathsf{M} \oplus \mathsf{E}_{\pi_\mathsf{E}(e)}$. The first component in this splitting we call ***horizontal*** and denote by $\mathsf{H}_e\mathsf{E}$, and the second we call ***vertical*** and denote by $\mathsf{V}_e\mathsf{E}$. By hor and ver we denote the projections onto the horizontal and vertical subspaces, respectively.

We note that covariant differentiation with respect to a vector field X of sections of E, along with Lie differentiation of functions, gives rise to covariant differentiation of tensors, just as we saw above for $\mathscr{L}_X$. A little more generally, if we have vector bundles $\pi_\mathsf{E}\colon \mathsf{E} \to \mathsf{M}$ and $\pi_\mathsf{F}\colon \mathsf{F} \to \mathsf{E}$, and linear connections ∇^{π_E} and ∇^{π_F}, then we have a connection in $\mathsf{E} \otimes \mathsf{F}$ denoted by $\nabla^{\pi_\mathsf{E} \otimes \pi_\mathsf{F}}$ and defined by

$$\nabla_X^{\pi_\mathsf{E} \otimes \pi_\mathsf{F}}(\xi \otimes \eta) = (\nabla_X^{\pi_\mathsf{E}}\xi) \otimes \eta + \xi \otimes (\nabla_X^{\pi_\mathsf{F}}\eta). \tag{1.8}$$

We will also use the ***covariant differential*** for tensor fields. This is defined for an affine connection ∇^M and $A \in \Gamma^\infty(\mathsf{T}^k(\mathsf{T}^*\mathsf{M}))$ by

$$\nabla^\mathsf{M} A(X_0, X_1, \dots, X_k) = (\nabla_{X_0}^\mathsf{M} A)(X_1, \dots, X_k),$$

giving $\nabla^\mathsf{M} A \in \Gamma^\infty(\mathsf{T}^{k+1}(\mathsf{T}^*\mathsf{M}))$. We will especially consider this in the case of tensors $A \in \Gamma^\infty(\mathsf{T}^k(\mathsf{T}^*\mathsf{M}) \otimes \mathsf{E})$, where the above definition is applied to the first component of the tensor product.

If ∇^{π_E} is C^r-linear connection in a C^r-vector bundle $\pi_\mathsf{E}\colon \mathsf{E} \to \mathsf{M}$, its ***curvature tensor*** is $R_{\nabla^{\pi_\mathsf{E}}} \in \Gamma^r(\bigwedge^2(\mathsf{T}^*\mathsf{M}) \otimes \mathsf{E}^* \otimes \mathsf{E})$ defined by

$$\begin{aligned} R_{\nabla^{\pi_\mathsf{E}}}(X, Y)(\xi) = \nabla_X^{\pi_\mathsf{E}}\nabla_Y^{\pi_\mathsf{E}}\xi - \nabla_Y^{\pi_\mathsf{E}}\nabla_X^{\pi_\mathsf{E}}\xi - \nabla_{[X,Y]}^{\pi_\mathsf{E}}\xi, \\ X, Y \in \Gamma^r(\mathsf{TM}),\ \xi \in \Gamma^r(\mathsf{E}). \end{aligned}$$

For a C^r-affine connection ∇^M, its ***torsion tensor*** is $T_{\nabla^\mathsf{M}} \in \Gamma^r(\bigwedge^2(\mathsf{T}^*\mathsf{M}) \otimes \mathsf{TM})$ defined by

$$T_{\nabla^\mathsf{M}}(X, Y) = \nabla_X^\mathsf{M} Y - \nabla_Y^\mathsf{M} X - [X, Y], \qquad X, Y \in \Gamma^r(\mathsf{TM}).$$

1.6 Jet Bundles

We shall make extensive use of jet bundles of various sorts. We can recommend [64] and [43, §12] as useful references.

Let M be a C^r-manifold and let $m \in \mathbb{Z}_{\geq 0}$. In general, an m-jet at $x \in \mathsf{M}$ of a function, section, or mapping is an equivalence class of functions, sections, or mappings, defined locally around x, where the equivalence is agreement (say in local coordinates) of all objects to order m.

Let us consider the various special cases.

For $x \in \mathsf{M}$ and $a \in \mathbb{R}$, by $\mathsf{J}^m_{(x,a)}(\mathsf{M}; \mathbb{R})$ we denote the m-jets of functions at x taking value a at x. For a C^r-function f defined in a neighbourhood of x, we denote by $j_m f(x) \in \mathsf{J}^m_{(x,f(x))}\mathsf{M}$ the m-ket of f. Of particular interest is the set $\mathsf{T}^{*m}_x\mathsf{M} = \mathsf{J}^m_{(x,0)}(\mathsf{M}; \mathbb{R})$ of jets of functions taking the value 0 at x. This has the structure of a $\mathbb{R}$-algebra with the algebra structure defined by the three operations

$$
\begin{aligned}
j_m f(x) + j_m g(x) &= j_m(f+g)(x),\\
(j_m f(x))(j_m g(x)) &= j_m(fg)(x),\\
a(j_m f(x)) &= j_m(af)(x),
\end{aligned}
$$

for functions f and g and for $a \in \mathbb{R}$. We denote $\mathsf{T}^{*m}\mathsf{M} = \mathring{\cup}_{x\in\mathsf{M}}\mathsf{T}^{*m}_x\mathsf{M}$. For $m, l \in \mathbb{Z}_{\geq 0}$ with $m \geq l$, we have projections $\rho^m_l\colon \mathsf{T}^{*m}\mathsf{M} \to \mathsf{T}^{*l}\mathsf{M}$. Note that $\mathsf{T}^{*0}\mathsf{M} \simeq \mathsf{M}$ and that $\mathsf{T}^{*1}\mathsf{M} \simeq \mathsf{T}^*\mathsf{M}$. We abbreviate $\rho_m \triangleq \rho^m_0\colon \mathsf{T}^{*m}\mathsf{M} \to \mathsf{M}$, which has the structure of a vector bundle.

Let $\pi_{\mathsf{E}}\colon \mathsf{E} \to \mathsf{M}$ be a C^r-vector bundle. For $x \in \mathsf{M}$ and $m \in \mathbb{Z}_{\geq 0}$, $\mathsf{J}^m_x\mathsf{E}$ denotes the set of m-jets of sections of E at x. For a C^r-section ξ defined in some neighbourhood of x, $j_m\xi(x) \in \mathsf{J}^m_x\mathsf{E}$ denotes the m-jet of ξ. We denote by $\mathsf{J}^m\mathsf{E} = \mathring{\cup}_{x\in\mathsf{M}}\mathsf{J}^m_x\mathsf{E}$ the bundle of m-jets. For $m, l \in \mathbb{Z}_{\geq 0}$ with $m \geq l$, we denote by $\pi^m_{\mathsf{E},l}\colon \mathsf{J}^m\mathsf{E} \to \mathsf{J}^l\mathsf{E}$ the projection. Note that $\mathsf{J}^0\mathsf{E} \simeq \mathsf{E}$. We abbreviate $\pi_{\mathsf{E},m} \triangleq \pi_{\mathsf{E}} \circ \pi^m_{\mathsf{E},0}\colon \mathsf{J}^m\mathsf{E} \to \mathsf{M}$, and note that $\mathsf{J}^m\mathsf{E}$ has the structure of a vector bundle over M, with addition and scalar multiplication defined by

$$
j_m\xi(x) + j_m\eta(x) = j_m(\xi+\eta)(x), \quad a(j_m\xi(x)) = j_m(a\xi)(x)
$$

for sections ξ and η and for $a \in \mathbb{R}$. One can show that

$$\mathsf{J}^m\mathsf{E} \simeq (\mathbb{R}_\mathsf{M} \oplus \mathsf{T}^{*m}\mathsf{M}) \otimes \mathsf{E}. \tag{1.9}$$

1.7 Topology

We shall make extensive use of elementary notions of general topology; the book [73] is a reference.

1.7.1 Basics

We shall denote by $\operatorname{int}(A)$ and $\operatorname{cl}(A)$ the interior and closure of a subset A.

A subset $A \subseteq \mathcal{X}$ of a topological space $(\mathcal{X}, \mathscr{O})$ is ***precompact*** if $\operatorname{cl}(A)$ is compact. Often this property is called "relative compactness."

A ***base*** for a topology $\mathscr{O}$ for $\mathcal{X}$ is a collection $\mathscr{B} \subseteq \mathscr{O}$ such that every open set is a union of sets from $\mathscr{B}$. A ***subbase*** for a topology is a collection of subsets for which the collection of finite intersections is a base for the topology. A ***neighbourhood base*** at $x \in \mathcal{X}$ is a collection $\mathscr{B}(x)$ of neighbourhoods of x such that, if $\mathcal{O}$ is a neighbourhood of x, then there exists $\mathcal{B} \in \mathscr{B}(x)$ such that $\mathcal{B} \subseteq \mathcal{O}$.

An Hausdorff topological space $(\mathcal{X}, \mathscr{O})$ is ***locally compact*** if every point possesses a base of neighbourhoods that are precompact.

We denote by $\mathrm{C}^0(\mathcal{X}; \mathcal{Y})$ the set of continuous mappings from a topological space $(\mathcal{X}, \mathscr{O}_\mathcal{X})$ to a topological space $(\mathcal{Y}, \mathscr{O}_\mathcal{Y})$. A mapping $\Phi \in \mathrm{C}^0(\mathcal{X}; \mathcal{Y})$ is ***proper*** if $\Phi^{-1}(\mathcal{K})$ is compact for every compact $\mathcal{K} \subseteq \mathcal{Y}$.

We shall make use of the notion of a ***compact exhaustion*** for a topological space $(\mathcal{X}, \mathscr{O})$, by which we mean a countable family $(\mathcal{K}_j)_{j\in\mathbb{Z}_{>0}}$ of compact subsets of $\mathcal{X}$ with the following properties:

1. $\mathcal{K}_j \subseteq \operatorname{int}(\mathcal{K}_{j+1})$, $j \in \mathbb{Z}_{>0}$;
2. $\cup_{j\in\mathbb{Z}_{>0}}\mathcal{K}_j = \mathcal{X}$.

A second countable, locally compact, Hausdorff topological space always possesses a compact exhaustion [2, Lemma 2.76].

This is a book about function space topologies, particularly for functions that are real analytic. At times we will make reference to the "compact-open" topology. In general, this topology is defined as follows. Let $(\mathcal{X}, \mathscr{O}_\mathcal{X})$ and $(\mathcal{Y}, \mathscr{O}_\mathcal{Y})$ be topological spaces. The ***compact-open topology*** for $\mathrm{C}^0(\mathcal{X}; \mathcal{Y})$ is the topology with the sets

$$\mathcal{B}(\mathcal{K}, \mathcal{V}) = \{\Phi \in \mathrm{C}^0(\mathcal{X}; \mathcal{Y}) \mid \Phi(\mathcal{K}) \subseteq \mathcal{V}\}, \qquad \mathcal{K} \subseteq \mathcal{X} \text{ compact}, \ \mathcal{V} \in \mathscr{O}_\mathcal{Y},$$

as a subbase. Thus open sets in the compact-open topology are unions of finite intersections of subsets of these subbasic sets. We shall provide rather more concrete characterisations of the compact-open topology in cases of interest to us.

1.7.2 Metric and Uniform Spaces

We shall, at times, make use of facts about metric spaces, and use [11] as a reference. If $(\mathcal{M}, \mathrm{d})$ is a metric space, if $x_0 \in \mathcal{M}$, and if $r \in \mathbb{R}_{>0}$, then we denote by $\mathsf{B}_{\mathrm{d}}(r, x_0)$ and $\overline{\mathsf{B}}_{\mathrm{d}}(r, x_0)$ the open ball and closed ball, respectively, of radius r and centre x_0. The ***metric topology*** for a metric space $(\mathcal{M}, \mathrm{d})$ is the topology with the set of open balls as a base.

We shall also make use of the idea of a uniform space. There are multiple ways of defining the notion of a uniform space, but we choose the most concrete of these, consistent with the theme of the book. Let $\mathcal{X}$ be a set. A ***semimetric***[2] for $\mathcal{X}$ is the same as a metric, absent the property that the only pairs of points with distance zero between them are the pairs of identical points. A ***uniform structure*** for $\mathcal{X}$ is defined by a family $(\mathrm{d}_i)_{i \in I}$ of semimetrics. Specifically, one defines a topology for $\mathcal{X}$ by the requirement that the sets

$$\bigcap_{j=1}^{k} \{x \in \mathcal{X} \mid \mathrm{d}_{i_j}(x, x_0) < r_j\}, \qquad r_1, \dots, r_k \in \mathbb{R}_{>0},\ i_1, \dots, i_k \in I,\ k \in \mathbb{Z}_{>0},$$

comprise a neighbourhood base at $x_0 \in \mathcal{X}$. A topology of this form is special because one has the required structure to define notions such as Cauchy sequences (and so completeness) and uniform continuity normally associated with metric spaces. The semimetric point of view is often undeveloped in modern approaches to uniform spaces. However, [40] gives a good development of the semimetric point of view.

We shall also make mention of the notion of Polish spaces and their brethren. A ***Polish space*** is a topological space $(\mathcal{X}, \mathscr{O})$ whose topology is the metric topology for some metric whose metric uniformity is complete and whose metric topology is separable. There may be many metrics which give the topology of a Polish space, and one must take care to understand that a "natural" metric on a Polish space may not be the one that makes it a Polish space.[3] A ***Lusin space*** is an Hausdorff topological space $(\mathcal{X}, \mathscr{O})$ such that there is a Polish space $(\mathcal{X}', \mathscr{O}')$ and a bijective continuous mapping $\Phi\colon \mathcal{X}' \to \mathcal{X}$. A ***Suslin space*** is an Hausdorff topological space

[2] It is not uncommon to call this a "pseudometric." We shall use "semimetric," consistent with our use of "seminorm."

[3] For example, $(0, 1)$ is a Polish space, but the restriction of the standard metric on $\mathbb{R}$ is not the right metric to make this assertion.

$(\mathcal{X}, \mathscr{O})$ such that there exists a Polish space $(\mathcal{X}', \mathscr{O}')$ and a surjective continuous map $\Phi\colon \mathcal{X}' \to \mathcal{X}$. We note that both Lusin and Suslin spaces are separable since the image of a separable space under a continuous map is separable [73, Theorem 16.4(a)].

Another important attribute of topological spaces is that of being sequential [22]. This attribute has a host of equivalent definitions, but perhaps the easiest to understand is the following: a topological space $(\mathcal{X}, \mathscr{O})$ is ***sequential*** if a subset $\mathcal{C} \subseteq \mathcal{X}$ is closed if and only if every convergent sequence $(x_j)_{j\in\mathbb{Z}_{>0}}$ in $\mathcal{C}$ has its limit in $\mathcal{C}$.

1.7.3 Initial and Final Topologies, Inverse and Direct Limits

We shall make frequent use of the notion of initial and final topologies. We let $((\mathcal{X}_i, \mathscr{O}_i))_{i\in I}$ be a family of topological spaces, let $\mathcal{Y}$ be a set, and let $\Phi_i : \mathcal{Y} \to \mathcal{X}_i$ and $\Psi_i : \mathcal{X}_i \to \mathcal{Y}$, $i \in I$, be families of mappings.

The ***initial topology*** for $\mathcal{Y}$ defined by the mappings Φ_i, $i \in I$, is the coarsest topology for $\mathcal{Y}$ such that each of the mappings Φ_i, $i \in I$, is continuous. The subsets $\Phi_i^{-1}(\mathcal{O}_i)$, $\mathcal{O}_i \in \mathscr{O}_i$, $i \in I$, are a base for the initial topology. The initial topology is characterised by the following fact. If $(\mathcal{Z}, \mathscr{O})$ is a topological space, a mapping $\Psi\colon \mathcal{Z} \to \mathcal{Y}$ is continuous if and only if the diagram

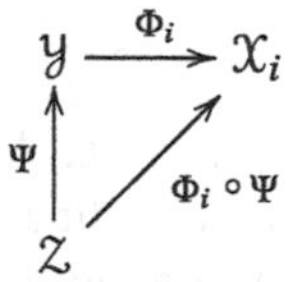

is a commutative diagram of topological spaces for each $i \in I$.

The ***final topology*** for $\mathcal{Y}$ defined by the mappings Ψ_i, $i \in I$, is the finest topology for $\mathcal{Y}$ such that each of the mappings Ψ_i, $i \in I$, is continuous. A subset $\mathcal{U} \subseteq \mathcal{Y}$ is open for the final topology if and only if $\Psi_i^{-1}(\mathcal{U}) \in \mathscr{O}_i$, $i \in I$. The final topology is characterised by the following fact. If $(\mathcal{Z}, \mathscr{O})$ is a topological space, a mapping $\Phi\colon \mathcal{Y} \to \mathcal{Z}$ is continuous if and only if the diagram

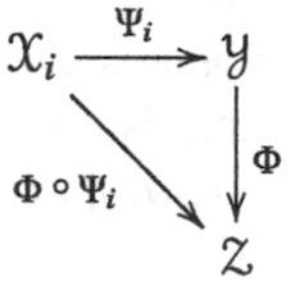

is a commutative diagram of topological spaces for each $i \in I$. We shall refer to these as the "universal properties" of the initial and final topologies, although they are not strictly universal in the category theoretic sense.

An important example of an initial topology is the topology induced on a subset of a topological space. Precisely, let $(\mathcal{X}, \mathscr{O})$ be a topological space and let $A \subseteq \mathcal{X}$ with $\iota_A : A \to \mathcal{X}$ the inclusion. Then the ***induced topology*** of A from $\mathscr{O}$ is topology

$$\mathscr{O}_A = \{\mathcal{U} \cap A \mid \mathcal{U} \in \mathscr{O}\},$$

and is easily seen to be the initial topology for the family of mappings (ι_A).

An important example of a final topology is the quotient topology. Let $(\mathcal{X}, \mathscr{O})$ be a topological space, let $\mathcal{Y}$ be a set and let $\Phi: \mathcal{X} \to \mathcal{Y}$ be a surjective map. The ***quotient topology*** for $\mathcal{Y}$ is the finest topology for which Φ is continuous. One readily verifies that the quotient topology is the final topology associated with the family of mappings (Φ).

Other important classes of initial and final topologies arise from inverse and direct limits of topological spaces. Since we routinely use these in not entirely trivial ways, we give the detailed definitions.

Definition 1.1 (Inverse Limit of Topological Spaces) Let $(I, \preceq)$ be a directed set. An ***inverse system*** of topological spaces is a pair $(((\mathcal{X}_i, \mathscr{O}_i))_{i\in I}, (\Phi_{ii'})_{i\preceq i'})$ where

(i) $(\mathcal{X}_i, \mathscr{O}_i)$, $i \in I$, is a topological space and
(ii) $\Phi_{ii'} \in \mathrm{C}^0(\mathcal{X}_{i'}; \mathcal{X}_i)$, $i, i' \in I$, $i \preceq i'$, are continuous maps satisfying

 (a) $\Phi_{ii} = \mathrm{id}_{\mathcal{X}_i}$, $i \in I$, and
 (b) $\Phi_{ii''} = \Phi_{ii'} \circ \Phi_{i'i''}$ for all $i, i', i'' \in I$ such that $i \preceq i' \preceq i''$.

An ***inverse limit*** of an inverse system $(((\mathcal{X}_i, \mathscr{O}_i))_{i\in I}, (\Phi_{ii'})_{i\preceq i'})$ is a topological space $(\mathcal{X}, \mathscr{O}_{\mathcal{X}})$ and a family $\pi_i \in \mathrm{C}^0(\mathcal{X}; \mathcal{X}_i)$, $i \in I$, of continuous maps such that

(iii) the diagram

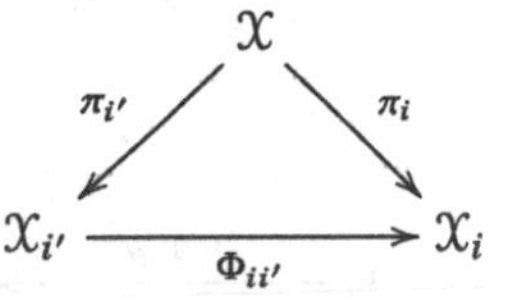

(1.10)

 commutes for every $i, i' \in I$ such that $i \preceq i'$ and
(iv) if $(\mathcal{Y}, \mathscr{O}_{\mathcal{Y}})$ is a topological space and if $\Psi_i \in \mathrm{C}^0(\mathcal{Y}; \mathcal{X}_i)$, $i \in I$, are continuous maps such that the diagram

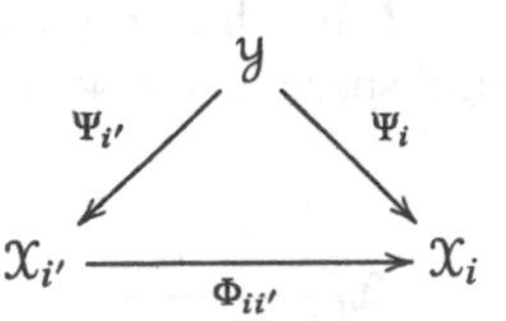

commutes for every $i, i' \in I$ such that $i \preceq i'$, then there exists a unique $\Theta \in \mathrm{C}^0(\mathcal{Y}; \mathcal{X})$ such that the diagram

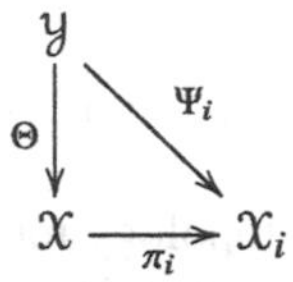

commutes for every $i \in I$.

We may sometimes denote the inverse limit by $\varprojlim_{i \in I}(\mathcal{X}_i, \mathscr{O}_i)$. ∘

The topology of the inverse limit $(\mathcal{X}, \mathscr{O})$ in the definition is the initial topology defined by the mappings π_i, $i \in I$. Frequently one sees "projective limit" where we choose to use "inverse limit."

Next we consider direct limits. In the literature, these are typically known as "inductive limits," but we use the terminology "direct limit" since it agrees with usage in the rest of category theory.

Definition 1.2 (Direct Limit of Topological Spaces) Let $(I, \preceq)$ be a directed set. A ***directed system*** of topological spaces is a pair $(((\mathcal{X}_i, \mathscr{O}_i))_{i \in I}, (\Phi_{ii'})_{i \preceq i'})$ where

(i) $(\mathcal{X}_i, \mathscr{O}_i)$, $i \in I$, is a topological space and
(ii) $\Phi_{ii'} \in \mathrm{C}^0(\mathcal{X}_i; \mathcal{X}_{i'})$, $i, i' \in I$, $i \preceq i'$, are continuous maps satisfying
    (a) $\Phi_{ii} = \mathrm{id}_{\mathcal{X}_i}$, $i \in I$;
    (b) $\Phi_{ii''} = \Phi_{i'i''} \circ \Phi_{ii'}$ for $i, i', i'' \in I$ satisfying $i \preceq i'$ and $i' \preceq i''$.

A ***direct limit*** of a directed system $(((\mathcal{X}_i, \mathscr{O}_i))_{i \in I}, (\Phi_{ii'})_{i \preceq i'})$ of topological spaces is a topological space $(\mathcal{X}, \mathscr{O}_{\mathcal{X}})$ and a family $\kappa_i \in \mathrm{C}^0(\mathcal{X}_i; \mathcal{X})$, $i \in I$, of continuous maps such that

(iii) the diagram

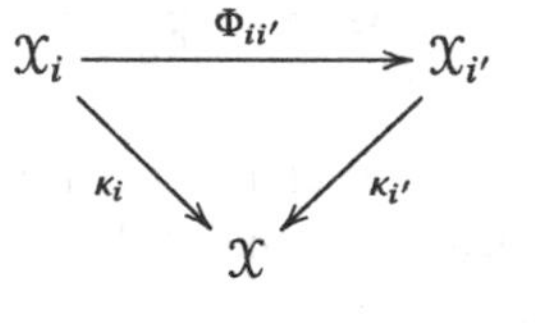

commutes for every $i, i' \in I$ for which $i \preceq i'$ and
(iv) if $(\mathcal{Y}, \mathscr{O}_{\mathcal{Y}})$ is a topological space and if $\Psi_i \in \mathrm{C}^0(\mathcal{X}_i; \mathcal{Y})$, $i \in I$, are such that the diagram

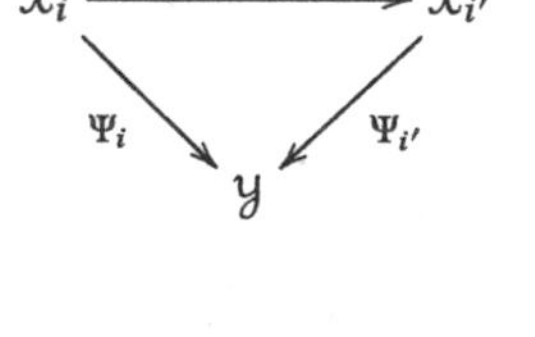

commutes for every $i, i' \in I$ for which $i \preceq i'$, then there exists a unique $\Theta \in \mathrm{C}^0(\mathcal{X}; \mathcal{Y})$ such that the diagram

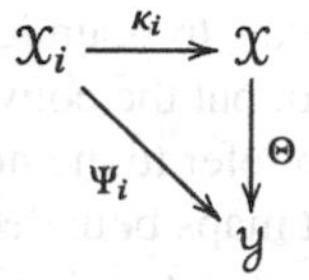

commutes for every $i \in I$.

We may sometimes denote the direct limit by $\varinjlim_{i \in I}(\mathcal{X}_i, \mathscr{O}_i)$. ∘

The topology of the direct limit $(\mathcal{X}, \mathscr{O})$ in the definition is the final topology defined by the mappings κ_i, $i \in I$.

1.8 Locally Convex Topological Vector Spaces

We shall make occasional use of ideas from the theory of locally convex topological vector spaces. We refer to [37] as a comprehensive reference source for the basic theory.

Let V be a $\mathbb{R}$-vector space. A ***seminorm*** on V is the same as a norm, absent the property that the only vector with zero "norm" is the zero vector. Unlike for normed vector spaces where the topology is defined by a single norm p, for a ***locally convex topological vector space*** the topology is defined by a family of seminorms $(p_i)_{i \in I}$. By "defined by" we mean that there is a neighbourhood base at $v \in \mathsf{V}$ given by

$$\bigcap_{j=1}^{k}(v + p_{i_j}^{-1}([0, r_j))), \qquad r_1, \dots, r_k \in \mathbb{R}_{>0},\ i_1, \dots, i_k \in I,\ k \in \mathbb{Z}_{>0}.$$

A locally convex topological vector space is evidently a uniform space, and is quite often characterised by means other than by seminorms, just as a uniform space often has a less concrete characterisation than by semimetrics. The uniformity of a locally convex topology means that one has the notion of completeness in a locally convex topological vector space.

For linear maps, one has the usual notion of continuity, and we denote by $\mathrm{L}(\mathsf{U}; \mathsf{V})$ the set of continuous linear mappings from the locally convex topological vector space $(\mathsf{U}, \mathscr{O}_{\mathsf{U}})$ to the locally convex topological vector space $(\mathsf{V}, \mathscr{O}_{\mathsf{V}})$. One can show that $L \in \mathrm{L}(\mathsf{U}; \mathsf{V})$ if and only if, for any continuous seminorm q for $(\mathsf{V}, \mathscr{O}_{\mathsf{V}})$, there exists a continuous seminorm p for $(\mathsf{U}, \mathscr{O}_{\mathsf{U}})$ such that $q \circ L(u) \leq p(u)$ for every $u \in \mathsf{U}$. We denote $\mathsf{V}' = \mathrm{L}(\mathsf{V}; \mathbb{R})$, which is the ***topological dual*** of V. A difference with the theory of normed vector spaces is the absence of the equivalence of the notions of continuity and boundedness. To explain, let $(\mathsf{V}, \mathscr{O})$ be a locally convex topological vector space and let $(p_i)_{i \in I}$ be a family of seminorms defining the topology. A subset $\mathcal{B} \subseteq \mathsf{V}$ is ***bounded*** if $p_i|\mathcal{B}$ is bounded for every $i \in I$. A linear map $L \in \mathrm{Hom}_{\mathbb{R}}(\mathsf{U}; \mathsf{V})$ between locally convex topological vector spaces is ***bounded*** if it maps bounded subsets to bounded subsets. One can then show that a continuous linear map is bounded, but the converse may not hold.

We shall also have occasion to refer to the notion of a ***compact*** continuous linear map, by which we mean one that maps bounded sets to precompact sets.

When working with locally convex topological vector spaces, it is advantageous to understand the rôle of specific properties of the topology in ways that do not arise organically in the theory of normed vector spaces. A systematic exploration of the various properties that are possible and what are the benefits of each is precisely the theory of locally convex topological vector spaces, and requires a lengthy book to understand with any level of comprehensiveness. Therefore, we shall limit ourselves here to cursory explanations of those properties of which we shall make use.

1.8.1 Inverse and Direct Limits of Locally Convex Topological Vector Spaces

Locally convex limits are an important way of constructing locally convex topologies. We shall give a rapid overview here.

We start with inverse limits. In the literature, these are typically known as "projective limits," but we use the terminology "inverse limit" since it agrees with usage in the rest of category theory.

Definition 1.3 (Inverse Limit of Locally Convex Topological Vector Spaces) Let $(I, \preceq)$ be a directed set. An ***inverse system*** of locally convex topological vector spaces is a pair $(((\mathsf{V}_i, \mathscr{O}_i))_{i \in I}, (\phi_{ii'})_{i \preceq i'})$ where

(i) $(\mathsf{V}_i, \mathscr{O}_i)$, $i \in I$, is a locally convex topological vector space and
(ii) $\phi_{ii'} \in \mathrm{L}(\mathsf{V}_{i'}; \mathsf{V}_i)$, $i, i' \in I$, $i \preceq i'$, are continuous linear maps satisfying
   (a) $\phi_{ii} = \mathrm{id}_{\mathsf{V}_i}$, $i \in I$, and
   (b) $\phi_{ii''} = \phi_{ii'} \circ \phi_{i'i''}$ for all $i, i', i'' \in I$ such that $i \preceq i' \preceq i''$.

An ***inverse limit*** of an inverse system $(((\mathsf{V}_i, \mathscr{O}_i))_{i \in I}, (\phi_{ii'})_{i \preceq i'})$ is a locally convex topological vector space $(\mathsf{V}, \mathscr{O}_{\mathsf{V}})$ and a family $\pi_i \in \mathrm{L}(\mathsf{V}; \mathsf{V}_i)$, $i \in I$, of continuous linear maps such that

(iii) the diagram

$$\begin{array}{ccc} & \mathsf{V} & \\ {}^{\pi_{i'}}\swarrow & & \searrow^{\pi_i} \\ \mathsf{V}_{i'} & \xrightarrow[\phi_{ii'}]{} & \mathsf{V}_i \end{array} \tag{1.11}$$

commutes for every $i, i' \in I$ such that $i \preceq i'$ and

(iv) if $(\mathsf{U}, \mathscr{O}_{\mathsf{U}})$ is a locally convex topological vector space and if $\psi_i \in \mathrm{L}(\mathsf{U}; \mathsf{V}_i)$, $i \in I$, are continuous linear maps such that the diagram

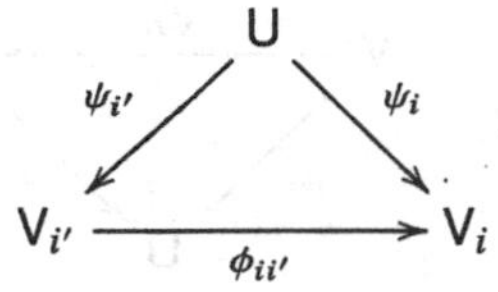

commutes for every $i, i' \in I$ such that $i \preceq i'$, then there exists a unique $\theta \in \mathrm{L}(\mathsf{U}; \mathsf{V})$ such that the diagram

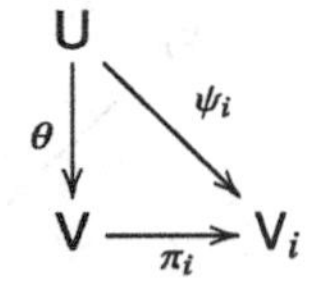

commutes for every $i \in I$.

We may sometimes denote the inverse limit by $\varprojlim_{i \in I}(\mathsf{V}_i, \mathscr{O}_i)$. $\circ$

The topology of the inverse limit $(\mathsf{V}, \mathscr{O})$ in the definition is the initial topology defined by the mappings π_i, $i \in I$.

Next we consider direct limits. In the literature, these are typically known as "inductive limits," but we use the terminology "direct limit."

Definition 1.4 (Direct Limit of Locally Convex Topological Vector Spaces) Let $(I, \preceq)$ be a directed set. A ***directed system*** of locally convex topological vector spaces is a pair $(((\mathsf{V}_i, \mathscr{O}_i))_{i \in I}, (\phi_{ii'})_{i \preceq i'})$ where

(i) $(\mathsf{V}_i, \mathscr{O}_i)$, $i \in I$, is a locally convex topological vector space and
(ii) $\phi_{ii'} \in \mathrm{L}(\mathsf{V}_i; \mathsf{V}_{i'})$, $i, i' \in I$, $i \preceq i'$, are continuous linear maps satisfying

(a) $\phi_{ii} = \mathrm{id}_{\mathsf{V}_i}$, $i \in I$;
(b) $\phi_{ii''} = \phi_{i'i''} \circ \phi_{ii'}$ for $i, i', i'' \in I$ satisfying $i \preceq i'$ and $i' \preceq i''$.

A ***locally convex direct limit*** of a directed system $(((\mathsf{V}_i, \mathscr{O}_i))_{i \in I}, (\phi_{ii'})_{i \preceq i'})$ of locally convex topological vector spaces is a locally convex topological vector space $(\mathsf{V}, \mathscr{O}_{\mathsf{V}})$ and a family $\kappa_i \in \mathrm{L}(\mathsf{V}_i; \mathsf{V})$, $i \in I$, of continuous linear maps such that

(iii) the diagram

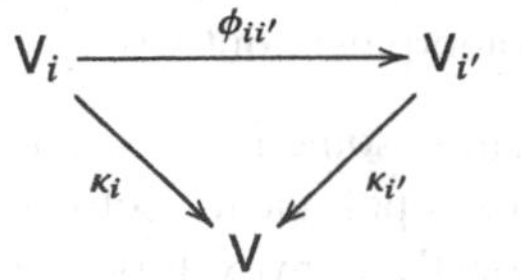

commutes for every $i, i' \in I$ for which $i \preceq i'$ and

(iv) if $(\mathsf{U}, \mathscr{O}_{\mathsf{U}})$ is a locally convex topological vector space and if $\psi_i \in \mathrm{L}(\mathsf{V}_i; \mathsf{U})$, $i \in I$, are such that the diagram

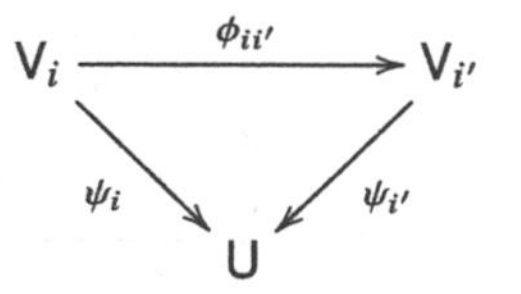

commutes for every $i, i' \in I$ for which $i \preceq i'$, then there exists a unique $\theta \in \mathrm{L}(\mathsf{V}; \mathsf{U})$ such that the diagram

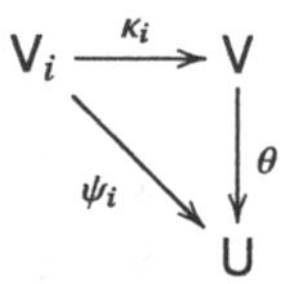

commutes for every $i \in I$.

We may sometimes denote the direct limit by $\varinjlim_{i \in I}(\mathsf{V}_i, \mathscr{O}_i)$. ○

The topology of the direct limit $(\mathsf{V}, \mathscr{O})$ in the definition is *not* generally the final topology defined by the mappings κ_i, $i \in I$. It is, instead, the final *locally convex* topology defined by these mappings. Precisely, this means that, in the definition of final topology above, one requires that the final topology be locally convex, e.g., by taking convex hulls of the open sets from the topological final topology.

For direct limits, one typically needs to endow them with additional attributes in order for the direct limit to have useful properties. The following is a list of some such attributes that will be useful for us.

Definition 1.5 (Attributes of Direct Limits of Locally Convex Topological Vector Spaces) Let $(I, \preceq)$ be a directed set, and let $((\mathsf{V}_i)_{i \in I}, (\phi_{ii'})_{i \preceq i'})$ be a directed system in the category of locally convex topological vector spaces with locally convex direct limit V and induced mappings $\phi_i \in \mathrm{L}(\mathsf{V}_i; \mathsf{V})$, $i \in I$. The directed system is:

(i) ***compact*** if the connecting maps $\phi_{ii'}$, $i \preceq i'$, are compact;
(ii) ***regular*** if, given a bounded set $\mathcal{B} \subseteq \mathsf{V}$, there exists $i \in I$ and a bounded set $\mathcal{B}_i \subseteq \mathsf{V}_i$ such that $\mathcal{B} \subseteq \phi_i(\mathcal{B}_i)$;
(iii) ***boundedly retractive*** if the connecting maps $\phi_{ii'}$, $i \preceq i'$, are injective, and if, given a bounded set $\mathcal{B} \subseteq \mathsf{V}$, there exists $i \in I$ and a bounded set $\mathcal{B}_i \subseteq \mathsf{V}_i$ such that $\phi_i|\mathcal{B}_i$ is an homeomorphism onto $\mathcal{B}$. ○

As inverse and direct limits, sometimes in combination, are a common means of arriving at locally convex topological vector spaces in practice, it is useful to know what attributes of locally convex topological vector spaces persist under inverse and/or direct limits. For instance, countable inverse limits of Fréchet spaces (see below) are Fréchet spaces, while countable direct limits of Fréchet spaces are generally not. We shall make substantial use of permanence properties of direct limits when we give our construction of the real analytic topology, and we shall reference the appropriate places in the literature at that time.

1.8.2 Metrisable Locally Convex Topological Vector Spaces

A locally convex topological vector space $(\mathsf{V}, \mathscr{O})$ is ***metrisable*** if it is Hausdorff and if its topology is the metric topology for a translation-invariant metric on V. One can show that $(\mathsf{V}, \mathscr{O})$ is metrisable if and only if its topology can be defined by a countable or finite family of seminorms. A complete metrisable locally convex topological vector space is a ***Fréchet space***.

Typically, Fréchet spaces are to be regarded as somewhat friendly. However, the real analytic topology we define in this book for the space of sections of a real analytic vector bundle is not metrisable. For this reason, one should establish other useful properties for this topology in order for it to be manageable. A useful generalisation of separable Fréchet spaces is to make use of the generalisations of complete separable metric spaces above to Lusin and Suslin spaces. Indeed, we shall be able to make use of the fact that certain Hausdorff countable direct limits of Suslin spaces are Suslin spaces and that all countable inverse limits of Suslin spaces are Suslin spaces. It is difficult to piece together these facts from the literature, so let us illustrate how this can be done.

Proposition 1.6 (Direct and Inverse Limits of Suslin Locally Convex Topological Vector Spaces) *Let $(\mathsf{V}_j, \mathscr{O}_j)$, $j \in \mathbb{Z}_{>0}$, be a countable family of Suslin locally convex topological vector spaces. Then the following statements hold:*

(i) if $\mathsf{V}_j \subseteq \mathsf{V}_{j+1}$ and if the inclusion of V_j in V_{j+1} is continuous (i.e., $\mathscr{O}_j$ is finer than the induced topology for V_j from $\mathscr{O}_{j+1}$), then $\varinjlim_{j\to\infty}(\mathsf{V}_j, \mathscr{O}_j)$ is Suslin if it is Hausdorff;

(ii) if $\phi_j \in \mathrm{L}(\mathsf{V}_{j+1}; \mathsf{V}_j)$, $j \in \mathbb{Z}_{>0}$ and if we consider the obvious inverse system associated to this collection of continuous linear maps,[4] then $\varprojlim_{j\to\infty}(\mathsf{V}_j, \mathscr{O}_j)$ is Suslin.

Proof

(i) Let $(\mathsf{V}, \mathscr{O}) = \varinjlim_{j\to\infty}(\mathsf{V}_j, \mathscr{O}_j)$ be the locally convex direct limit. Let us fix $j \in \mathbb{Z}_{>0}$ for a moment. Let $\mathscr{O}'_j$ be the topology for V_j induced by $\mathscr{O}$, noting that V_j is identified with a linear subspace of V. Since the definition of direct limit implies that the inclusion of $(\mathsf{V}_j, \mathscr{O}_j)$ in $(\mathsf{V}, \mathscr{O})$ is continuous, we have $\mathscr{O}'_j \subseteq \mathscr{O}_j$. Thus the identity map from $(\mathsf{V}_j, \mathscr{O}_j)$ to $(\mathsf{V}_j, \mathscr{O}'_j)$ is continuous, and so $(\mathsf{V}_j, \mathscr{O}'_j)$ is the

[4] That is to say, the inverse system where, if $j_1 \leq j_2$, then take the linear map in $\mathrm{L}(\mathsf{V}_{j_2}; \mathsf{V}_{j_1})$ to be $\phi_{j_1} \circ \cdots \circ \phi_{j_2-1}$.

continuous image of the Suslin space $(\mathsf{V}_j, \mathscr{O}_j)$, and so is itself Suslin (carefully noting that $\mathscr{O}'_j$ is assumed to be Hausdorff).

Now we make use of a lemma.

Lemma 1 *Let $(\mathcal{X}, \mathscr{O})$ be an Hausdorff topological space and let $\mathcal{S}_j \subseteq \mathcal{X}$, $j \in \mathbb{Z}_{>0}$, be topological subspaces for which*

(i) the induced topology for $\mathcal{S}_j$ from $(\mathcal{X}, \mathscr{O})$ is Suslin and
(ii) $\mathcal{X} = \cup_{j\in\mathbb{Z}_{>0}} \mathcal{S}_j$.

Then $(\mathcal{X}, \mathscr{O})$ is a Suslin space.

Proof For each $j \in \mathbb{Z}_{>0}$, let $(\mathcal{Y}_j, \mathscr{O}_j)$ be a Polish space and let $\Phi_j \in \mathrm{C}^0(\mathcal{Y}_j; \mathcal{S}_j)$ be a continuous surjective mapping. Since the topology on $\mathcal{S}_j$ is the induced topology, we can regard $\Phi_j \in \mathrm{C}^0(\mathcal{Y}_j; \mathcal{X})$. It is easy to verify that the disjoint union $\mathring{\cup}_{j\in\mathbb{Z}_{>0}}(\mathcal{Y}_j, \mathscr{O}_j)$ is itself a Polish space.[5] Now define

$$\begin{aligned}\Phi\colon \mathring{\bigcup_{j\in\mathbb{Z}_{>0}}} \mathcal{Y}_j &\to \mathcal{X}\\ (y_j, j) &\mapsto \Phi_j(y_j),\end{aligned}$$

which is readily verified to be continuous. Moreover, $\mathrm{image}(\Phi) = \cup_{j\in\mathbb{Z}_{>0}} \mathcal{S}_j = \mathcal{X}$, which gives the result. ▽

This part of the proposition follows immediately from the lemma and what we proved before the lemma.

(ii) Here we note some more or less well known facts about Suslin spaces:

1. a countable product of Suslin spaces is a Suslin space [8, Lemma 6.6.5(iii)];
2. a closed subspace of a Suslin space is a Suslin space [8, Lemma 6.6.5(ii)].

Now, since the inverse limit of an inverse system of Hausdorff locally convex topological vector spaces is a closed subspace of the product of the spaces from the inverse family [37, Proposition 2.6.1], this part of the proposition follows. □

1.8.3 Open Mapping Theorems for Locally Convex Topological Vector Spaces

An important theorem in the linear analysis of normed vector spaces is the Open Mapping Theorem which asserts that a surjective continuous linear mapping

[5] Countable disjoint unions of separable spaces are pretty clearly themselves separable. Moreover, disjoint unions of metric spaces are also metric spaces if one defines the distance between points in disjoint components of the disjoint union to be distance 1 from one another. If each of the metrics is complete, it is easy to see that the resulting metric for the disjoint union is also complete.

between Banach spaces is open. In general, to assert an analogous result for locally convex topological vector spaces requires various sorts of hypotheses on both the domain and codomain. We shall report an approach to this developed by De Wilde [16].

First of all, let us introduce the standard terminology for continuous, open linear mappings between locally convex topological vector spaces.

Definition 1.7 (Topological Homomorphism, Topological Monomorphism, Topological Epimorphism) Let $(\mathsf{U}, \mathscr{O}_\mathsf{U})$ and $(\mathsf{V}, \mathscr{O}_\mathsf{V})$ be locally convex topological vector spaces and let $L \in \mathrm{L}(\mathsf{U}; \mathsf{V})$. Then L is:

(i) a ***topological homomorphism*** if it is an open mapping onto image(L) (with the induced topology);
(ii) a ***topological monomorphism*** if it is an injective topological homomorphism;
(iii) a ***topological epimorphism*** if it is a surjective topological homomorphism. ○

Frequently, one uses the simpler terms "homomorphism," "monomorphism," and "epimorphism." However, this can lead to confusion since, for example, linear monomorphisms are not generally topological monomorphisms, and so it seems best to make the category explicit.

For the next piece of terminology, a locally convex topological vector space is ***ultrabornological*** if it is an Hausdorff direct limit of Banach spaces. Next, a ***web*** in a locally convex topological vector space $(\mathsf{V}, \mathscr{O})$ is a family $C_{j_1,\dots,j_k}$, $j_1, \dots, j_k \in \mathbb{Z}_{>0}$, $k \in \mathbb{Z}_{>0}$, of balanced[6] convex subsets satisfying the following:

1. $\cup_{j\in\mathbb{Z}_{>0}} C_j = \mathsf{V}$;
2. $\cup_{j\in\mathbb{Z}_{>0}} C_{j_1,\dots,j_k,j} = C_{j_1,\dots,j_k}$ for all $j_1, \dots, j_k \in \mathbb{Z}_{>0}$, $k \in \mathbb{Z}_{>0}$;
3. for each sequence $(j_k)_{k\in\mathbb{Z}_{>0}}$ in $\mathbb{Z}_{>0}$, there exists a sequence $(r_k)_{k\in\mathbb{Z}_{>0}}$ in $\mathbb{R}_{>0}$ such that, for every sequence $(v_k)_{k\in\mathbb{Z}_{>0}}$ for which $v_k \in C_{j_1,\dots,j_k}$, $k \in \mathbb{Z}_{>0}$, the series $\sum_{k=1}^{\infty} r_k v_k$ converges.

A locally convex topological vector space with a web is called ***webbed***. If the notion of a webbed topological vector space seems a little contrived, it is: specifically, it is contrived to give rise to an Open Mapping Theorem. The notion of a web is also useful because it is persistent under countable inverse limits and well-behaved countable direct limits [37, Corollary 5.3.3].

The De Wilde Open Mapping Theorem states that a surjective continuous linear mapping from a webbed locally convex topological vector space onto an ultrabornological locally convex topological vector space is open. A fairly succinct proof can be found in [51, Theorem 24.30].

---

[6] A subset B of a $\mathbb{R}$-vector space is ***balanced*** if $v \in B$ implies that $-v \in B$.

1.8.4 Nuclear Locally Convex Topological Spaces

Nuclear locally convex topological vector spaces form a special, yet interesting, class of locally convex topological vector spaces with useful properties resembling finite-dimensional vector spaces. For example, a subset of a nuclear locally convex topological vector space is compact if and only if it is closed and bounded. This implies, as a consequence, that the only normed nuclear spaces are those that are finite-dimensional.

To define this class of locally convex topological vector spaces, we first define the notion of a nuclear mapping between Banach spaces. Thus let $(\mathsf{U}, p_{\mathsf{U}})$ and $(\mathsf{V}, p_{\mathsf{V}})$ be Banach spaces. A mapping $\phi \in \mathrm{L}(\mathsf{U}; \mathsf{V})$ is ***nuclear*** if there exists a sequence $(\lambda_j)_{j\in\mathbb{Z}_{>0}}$ in the closed unit ball of U' (giving U' the operator norm), a sequence $(v_j)_{j\in\mathbb{Z}_{>0}}$ in the closed unit ball of V, and a sequence $(c_j)_{j\in\mathbb{Z}_{>0}}$ in $\ell^1(\mathbb{Z}_{>0}; \mathbb{R})$ for which

$$\phi(u) = \sum_{j=1}^{\infty} c_j \langle \lambda_j; u\rangle v_j.$$

One should think of nuclear linear maps between Banach spaces as having a "small" image.

To make use of this Banach space construction to define the notion of a nuclear locally convex topological vector space, we let $(\mathsf{V}, \mathscr{O})$ be a locally convex topological vector space and let p be a continuous seminorm for V. We can then define the normed vector space $(\mathsf{V}_p, \hat{p})$ by

$$\mathsf{V}_p = \mathsf{V}/p^{-1}(0), \quad \hat{p}(v + p^{-1}(0)) = \inf\{p(v + v') \mid p(v') = 0\},$$

noting that $p^{-1}(0)$ is a subspace. We denote by $(\overline{\mathsf{V}}_p, \hat{p})$ the Banach space completion of $(\mathsf{V}_p, \hat{p})$. We then say that $(\mathsf{V}, \mathscr{O})$ is ***nuclear*** if, for every continuous seminorm p for V, there exists a continuous seminorm p' for V with $p' \geq p$ and such that the natural projection $\pi_{p,p'} \colon \overline{\mathsf{V}}_p \to \overline{\mathsf{V}}_{p'}$ is a nuclear mapping between Banach spaces.

1.8.5 Tensor Products of Locally Convex Topological Vector Spaces

The subject of tensor products of locally convex topological vector spaces is enormously rich and owes its initial development to the early work of Grothendieck [28]. We require only the most elementary parts of the theory, namely the so-called projective tensor product [37, Chapter 15].

Let $(\mathsf{U}, \mathscr{O}_{\mathsf{U}})$ and $(\mathsf{V}, \mathscr{O}_{\mathsf{V}})$ be locally convex topological vector spaces. Note that we have the bilinear and linear mappings

$$\mathsf{U} \times \mathsf{V} \to \mathsf{U} \otimes \mathsf{V}, \quad \mathsf{U} \otimes \mathsf{V} \to \mathrm{Hom}_{\mathbb{R}}(\mathsf{U}^*; \mathsf{V}),$$

respectively, defined by

$$(u, v) \mapsto u \otimes v, \quad u \otimes v \mapsto (\alpha \mapsto \langle \alpha; u \rangle v).$$

These algebraic constructions can be used to induce topologies on $\mathsf{U} \otimes \mathsf{V}$. We shall focus on the first of these mappings, this giving the projective tensor topology. Thus the projective tensor topology can be seen as the topological version of the usual definition of the tensor product in the algebraic setting. The second mapping gives, after some appropriate constructions, the injective tensor topology.

The ***projective tensor topology*** is the locally convex final topology for $\mathsf{U} \otimes \mathsf{V}$ associated with the bilinear mapping $\mathsf{U} \times \mathsf{V}$ to $\mathsf{U} \otimes \mathsf{V}$. One typically denotes by $\mathsf{U} \otimes_\pi \mathsf{V}$ the vector space $\mathsf{U} \otimes \mathsf{V}$ equipped with the projective tensor topology. One way to describe the projective tensor topology is to define a family of seminorms that describes the topology. If q is a continuous seminorm for $(\mathsf{U}, \mathscr{O}_{\mathsf{U}})$ and if p is a continuous seminorm for $(\mathsf{V}, \mathscr{O}_{\mathsf{V}})$, then define $q \otimes_\pi p \colon \mathsf{U} \otimes \mathsf{V} \to \mathbb{R}$ by

$$q \otimes_\pi p(A) = \inf \left\{ \sum_{j=1}^k q(u_j) p(v_j) \,\middle|\, A = \sum_{j=1}^k u_j \otimes v_j,\ u_j \in \mathsf{U},\ v_j \in \mathsf{V},\ j \in \{1, \dots, k\},\ k \in \mathbb{Z}_{>0} \right\}.$$

One can show that $q \otimes_\pi p$ is indeed a seminorm, and that $q \otimes_\pi p(u \otimes v) = q(u)p(v)$. Moreover, the family of seminorms $q \otimes p$, as q and p run over families of seminorms defining the topologies $\mathscr{O}_{\mathsf{U}}$ and $\mathscr{O}_{\mathsf{V}}$, gives the projective tensor topology for $\mathsf{U} \otimes \mathsf{V}$.

There is also a characterisation of the projective tensor product that is akin, for locally convex topological vector spaces, to the usual universal property of the algebraic tensor product. To give this characterisation, we add a locally convex topological vector space $(\mathsf{W}, \mathscr{O}_{\mathsf{W}})$ into the discussion. We denote by $\mathrm{L}(\mathsf{U}, \mathsf{V}; \mathsf{W})$ the set of continuous bilinear mappings from $\mathsf{U} \times \mathsf{V}$ to W. Then the projective tensor topology is the unique topology for which the mapping $\mathsf{U} \times \mathsf{V} \mapsto \mathsf{U} \otimes \mathsf{V}$ is continuous and for which, for any locally convex topological vector $(\mathsf{W}, \mathscr{O}_{\mathsf{W}})$ and

$\beta \in \mathsf{L}(\mathsf{U}, \mathsf{V}; \mathsf{W})$, there exists a unique $\phi_\beta \in \mathsf{L}(\mathsf{U} \otimes_\pi \mathsf{V}; \mathsf{W})$ such that the diagram

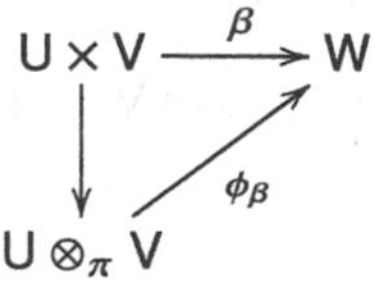

commutes.

1.9 Real Analytic Real Analysis

In order to initiate an uninitiated reader to the subject of real analytic differential geometry, we give a rapid overview of real analytic analysis in Euclidean spaces. A thorough presentation can be found in the book of Krantz and Parks [46].

The notion of real analyticity typically has its origin in its connection to convergent power series. Let us first, therefore, consider convergent power series. Let $n, m \in \mathbb{Z}_{>0}$. A ***real power series*** in n variables with values in $\mathbb{R}^m$ is a formal power series

$$\sum_{j=1}^{m} \sum_{I \in \mathbb{Z}^n_{\geq 0}} \alpha_I^a \boldsymbol{X}^I \boldsymbol{e}_a,$$

for coefficients $\alpha_I^a \in \mathbb{R}$, $a \in \{1, \dots, m\}$, $I \in \mathbb{Z}^n_{\geq 0}$ and indeterminants $\boldsymbol{X} = (X_1, \dots, X_n)$. We recall that, if $I = (i_1, \dots, i_n) \in \mathbb{Z}^n_{\geq 0}$, then $\boldsymbol{X}^I = X_1^{i_1} \cdots X_n^{i_n}$. If one wishes to be a little more precise, one can define the series by the mapping

$$\{1, \dots, m\} \times \mathbb{Z}^n_{\geq 0} \ni (a, I) \mapsto \alpha_I^a \in \mathbb{R},$$

which determines the coefficients of the formal series. Writing these coefficients in their rôle of a series with indeterminates connects the precise definition with how one actually thinks about it. We call such a formal power series ***convergent*** if there exists some $\boldsymbol{r} \in \mathbb{R}^n_{>0}$ such that the series

$$\sum_{j=1}^{m} \sum_{I \in \mathbb{Z}^n_{\geq 0}} \alpha_I^a \boldsymbol{r}^I \boldsymbol{e}_a$$

converges absolutely. In this case, we say that the power series ***converges for r***. If a formal power series converges for $\boldsymbol{r}$, then the series

$$\sum_{j=1}^{m} \sum_{I \in \mathbb{Z}_{\geq 0}^{n}} \alpha_I^a \boldsymbol{x}^I \boldsymbol{e}_a$$

converges absolutely for any $\boldsymbol{x} \in \overline{\mathsf{D}}^n(\boldsymbol{r}, \boldsymbol{0})$ and uniformly on compact subsets of $\mathsf{D}^n(\boldsymbol{r}, \boldsymbol{0})$. Since the series obtained by term-by-term differentiation of the original series is itself a power series, it can be subjected to the same arguments. To this end, one can show that the power series obtained by taking any single partial derivative converges for $\boldsymbol{r}$ if the original series converges for $\boldsymbol{r}$. Thus we can see that a formal power series converges to an infinitely differentiable function whose derivatives can be obtained by iterated term-by-term differentiation of the original power series.

Additionally, the same arguments as the preceding results also give absolute convergence of the complex power series

$$\sum_{j=1}^{m} \sum_{I \in \mathbb{Z}_{\geq 0}^{n}} \alpha_I^a \boldsymbol{z}^I \boldsymbol{e}_a$$

in the *complex* polydisk

$$\{z \in \mathbb{C}^n \mid |z_j| \leq r_j\}$$

and uniform convergence on compact subsets of the corresponding open complex disk. Thus the $\mathbb{R}^m$-valued functions obtained from real power series are actually convergent complex power series, and so define an holomorphic function from the open complex polydisk to $\mathbb{C}^m$. Thus the functions obtained from convergent real power series extend to complex power series with an analogous domain of convergence.

Thus power series. Let us turn now to functions.

For $\mathcal{U} \subseteq \mathbb{R}^n$ open, for $\boldsymbol{f} \in \mathrm{C}^\infty(\mathcal{U}; \mathbb{R}^m)$, and for $\boldsymbol{x}_0 \in \mathcal{U}$, the ***Taylor series*** for $\boldsymbol{f}$ at $\boldsymbol{x}_0$ is the real power series

$$\sum_{I \in \mathbb{Z}_{\geq 0}^{n}} \sum_{a=1}^{m} \frac{\boldsymbol{D}^I f_a(\boldsymbol{x}_0)}{I!} \boldsymbol{X} \boldsymbol{e}_a.$$

This series will generally not converge. Indeed, it is a theorem of Borel [9] that any real power series—convergent or not—is the Taylor series for some C^∞-mapping defined in some neighbourhood of $\boldsymbol{x}_0$.

This leads to the following definition.

Definition 1.8 (Real Analytic Mapping) Let $\mathcal{U} \subseteq \mathbb{R}^n$ be open. A mapping $\boldsymbol{f} \colon \mathcal{U} \to \mathbb{R}^m$ is ***real analytic*** if, for each $\boldsymbol{x}_0 \in \mathcal{U}$, the Taylor series for $\boldsymbol{f}$ at $\boldsymbol{x}_0$

converges and if, in some neighbourhood $\mathcal{V}$ of $\boldsymbol{x}_0$,

$$f(\boldsymbol{x}) = \sum_{I \in \mathbb{Z}^n_{\geq 0}} \frac{\boldsymbol{D}^I f(\boldsymbol{x}_0)}{I!} (\boldsymbol{x} - \boldsymbol{x}_0)^I. \qquad \circ$$

In brief, a mapping is real analytic when it is equal to its own (convergent) Taylor series in some neighbourhood of every point. Note that, while a function may be real analytic on some domain, its Taylor series at a single point may not converge on the entire domain. To see this, consider the two real analytic functions

$$\begin{aligned} f \colon \mathbb{R} &\to \mathbb{R} \\ x &\mapsto \mathrm{e}^x, \end{aligned} \qquad \begin{aligned} g \colon \mathbb{R} &\to \mathbb{R} \\ x &\mapsto \frac{1}{1+x^2}. \end{aligned}$$

For f, we have

$$f(x) = \sum_{j=0}^{\infty} \frac{\boldsymbol{D}^j f(0)}{j!} x^j = \sum_{j=0}^{\infty} \frac{x^j}{j!}$$

pointwise on $\mathbb{R}$, and uniformly on compact subsets of $\mathbb{R}$. On the other hand, the Taylor series of g about 0 is

$$\sum_{j=0}^{\infty} \frac{\boldsymbol{D}^j g(0)}{j!} = \sum_{j=0}^{\infty} (-1)^j x^{2j},$$

and this series converges only on $(-1, 1)$. This does not contradict real analyticity since, for example, g possesses a (different) Taylor series convergent to g in some neighbourhood of ± 1.

A question of some importance, and of whose answer we will make essential use, is that of how one can recognise real analytic mappings in the class of C^∞-mappings.

Theorem 1.9 (Characterisation of Real Analytic Functions) *If $\mathcal{U} \subseteq \mathbb{R}^n$ is open and if $f \colon \mathcal{U} \to \mathbb{R}$ is infinitely differentiable, then the following statements are equivalent:*

(i) f is real analytic;
(ii) for each $\boldsymbol{x}_0 \in \mathcal{U}$, there exists a neighbourhood $\mathcal{V} \subseteq \mathcal{U}$ of $\boldsymbol{x}_0$ and $C, r \in \mathbb{R}_{>0}$ such that

$$|\boldsymbol{D}^I f(\boldsymbol{x})| \leq C I! r^{-|I|}$$

for all $\boldsymbol{x} \in \mathcal{V}$ and $I \in \mathbb{Z}^n_{\geq 0}$.

Proof It is easy to prove the theorem using holomorphic extensions and Cauchy estimates for holomorphic functions [45, Lemma 2.3.9]. A "real" proof is more difficult, and one such is given in [46, Proposition 2.2.10]. □

As we have previously indicated, we shall denote by $\mathrm{C}^{\omega}(\mathcal{U}; \mathbb{R}^m)$ the set of real analytic mappings from $\mathcal{U}$ to $\mathbb{R}^m$. This set of real analytic mappings is blessed with the algebraic properties one expects. To wit,

1. sums of real analytic mappings are real analytic,
2. compositions of real analytic mappings are real analytic, and
3. for scalar-valued functions, products of real analytic functions are real analytic and the quotient of a real analytic function by a nowhere zero real analytic function is real analytic.

If $\mathcal{U}, \mathcal{V} \subseteq \mathbb{R}^n$ are open, a ***real analytic diffeomorphism*** from $\mathcal{U}$ to $\mathcal{V}$ is a real analytic bijection $\boldsymbol{\Phi}\colon \mathcal{U} \rightarrow \mathcal{V}$ with a real analytic inverse. The Inverse Function Theorem is valid for real analytic functions: if $\boldsymbol{\Phi}\colon \mathcal{U} \rightarrow \mathbb{R}^n$ is such that $\boldsymbol{D\Phi}(\boldsymbol{x})$ is invertible, then there is a neighbourhood $\mathcal{U}' \subseteq \mathcal{U}$ of $\boldsymbol{x}$ such that $\boldsymbol{\Phi}|\mathcal{U}'\colon \mathcal{U}' \rightarrow \boldsymbol{\Phi}(\mathcal{U}')$ is a real analytic diffeomorphism. The real analytic Inverse Function Theorem is essential for real analytic differential geometry, just as the smooth version is essential for smooth differential geometry.

Chapter 2
Topology for Spaces of Real Analytic Sections and Mappings

In this chapter we introduce the main characters of the book, the topologies for the space of real analytic sections of a real analytic vector bundle and for the space of real analytic mappings between real analytic manifolds. We first give an overview of real analytic differential geometry. After this, we will give four descriptions of the real analytic topology for the space of sections of a real analytic vector bundle, two due to Martineau [50] and two given by Jafarpour and Lewis [36], based on the note of Vogt [68]. Finally, in Sect. 2.5, we give a variety of descriptions for topologies for the space of real analytic mappings between real analytic manifolds, and we show that these all agree. In particular, we show how the topology for the space of sections of a vector bundle allows us to introduce a uniform structure for the space of real analytic mappings between real analytic mappings. We note that, as far as we know, these descriptions of the topology of the space of real analytic mappings are being given here for the first time.

As can be seen from the above discussion, in this chapter we will arrive at a variety of topologies for spaces of real analytic sections and real analytic mappings. Part of the development will be to show that all of these various topologies agree. This will all contribute to the idea that there is essentially only one meaningful topology for spaces of real analytic sections and mappings. This is a featured shared by the holomorphic analogues, and in contrast to the smooth analogues. We shall have a few words to say about this at the end of the chapter.

2.1 Real Analytic Differential Geometry

In this section we consider some of the basic constructions of real analytic differential geometry. We shall also indicate some relationships between real analytic and holomorphic differential geometry. Correspondingly, we also shed some light on some of the existential problems that arise in real analytic differential geometry.

A. D. Lewis, *Geometric Analysis on Real Analytic Manifolds*, Lecture Notes in Mathematics 2333, https://doi.org/10.1007/978-3-031-37913-0_2

2.1.1 The Fundamental Objects of Real Analytic Differential Geometry

We assume that the reader is thoroughly familiar with the basic constructions of smooth differential geometry. In particular, we assume that the reader will understand that the basic properties of real analytic mappings as enumerated near the end of Sect. 1.9 allow all of the basic constructions of smooth differential geometry to be applied to the real analytic case with a mere substitution of "C^ω" for "C^∞." Thus we immediately have following constructions:

1. a real analytic manifold has an atlas of charts whose overlap maps are real analytic diffeomorphisms;
2. a real analytic submanifold is one admitting real analytic submanifold charts;
3. a real analytic vector bundle has an atlas of vector bundle charts for which the overlap maps are real analytic local vector bundle isomorphisms;
4. a mapping between real analytic manifolds is real analytic, or ***class*** $\mathbf{C}^\omega$, if it has real analytic local representatives;
5. a section of a real analytic vector bundle is real analytic if it is a real analytic mapping in the previous sense;
6. the tangent bundle of a real analytic manifold is itself a real analytic manifold;
7. the various tensor bundles associated with a real analytic vector bundle are real analytic vector bundles;
8. the notion of a real analytic linear connection on a vector bundle makes sense.

As we mentioned in Sect. 1.4, we shall assume that all real analytic manifolds are Hausdorff, connected, and second countable.

We also point out that all of the previous constructions apply with "real analytic" replaced by "holomorphic," with the initial assumption that charts take values in complex Euclidean spaces. We shall say that functions, mappings, or sections are of ***class*** $\mathbf{C}^{\mathbf{hol}}$ if they are holomorphic.

2.1.2 Existential Constructions in Real Analytic Differential Geometry

Given the breezy way in which we indicate that real analytic and holomorphic differential geometry are "the same as" smooth differential geometry, the uninitiated reader can be forgiven for wondering whether there are any differences between the smooth, and the real analytic and holomorphic cases. The differences arise when one uses constructions in smooth differential geometry that make use of partitions of unity. Indeed, there is no such thing as a real analytic or holomorphic partition of unity. This is a result of the so-called Identity Theorem, a version of which says that, if M and N are real analytic or holomorphic manifolds with M connected,[1] if $\Phi, \Psi: \mathsf{M} \to \mathsf{N}$ are real analytic or holomorphic mappings, and if there is an open subset $\mathcal{U} \subseteq \mathsf{M}$ such that $\Phi|\mathcal{U} = \Psi|\mathcal{U}$, then $\Phi = \Psi$. In particular, if the Taylor series of Φ and Ψ agree at a point, then the two mappings are equal everywhere! We do not know of a statement in the literature of this theorem in the real analytic case (Krantz and Parks [46, §1.2] give a one-dimensional version); this is not necessarily problematic, however, since a proof can be given that follows along the lines of the holomorphic proof. The latter is given in, e.g., [30, Theorem A.3]. What the Identity Theorem indicates is that real analytic and holomorphic differential geometry are far more rigid than smooth differential geometry. This manifests itself in the difficulty of proving basic existential results in the real analytic and holomorphic case that are trivial in the smooth case.

There are two principal (and related) tools for solving existential problems in real analytic and holomorphic differential geometry: sheaf theory and embedding theorems. We shall briefly discuss each of these, and give an example of how they can be used. We will work primarily in the real analytic case, but will consider the holomorphic case in a few places.

First we consider sheaf theory. We shall not take the time to develop the theory in any substantial way... or in any other way, really... but rather we illustrate how the theory can be used. We let $r \in \{\infty, \omega\}$, let M be a C^r-manifold, and let $\pi_{\mathsf{E}}: \mathsf{E} \to \mathsf{M}$ be a C^r-vector bundle. By $\mathscr{C}^r_{\mathsf{M}}$ we denote the sheaf of C^r-functions on M and by $\mathscr{G}^r_{\mathsf{E}}$ we denote the sheaf of C^r-sections of E. A fundamental problem is whether there exist C^r-sections of E subject to certain constraints. A tool for ascertaining whether there is a solution to such a problem are the so-called Cartan's Theorems A and B. These are originally proved by Cartan [12] in the holomorphic case and also by Cartan [13] in the real analytic case.

We do not state Cartan's Theorems, as this is the realm of a different book. However, the following lemma illustrates a use of the real analytic version of Cartan's Theorem B. We state the result also in the smooth case and we give a sheaf theoretic version of the proof in this case as well. A direct solution using cut-off functions is, of course, elementary.

Lemma 2.1 (Extension of Sections with Prescribed Jets) *Let $r \in \{\infty, \omega\}$, let $\pi_{\mathsf{E}}: \mathsf{E} \to \mathsf{M}$ be a C^r-vector bundle, and let $\mathsf{S} \subseteq \mathsf{M}$ be a closed C^r-submanifold. Let $k \in \mathbb{Z}_{\geq 0}$. Let $\mathcal{V}$ be a neighbourhood of S and let $\xi_{\mathsf{S}} \in \Gamma^r(\mathsf{E}|\mathcal{V})$. Then there exists $\xi \in \Gamma^r(\mathsf{E})$ such that $j_k\xi(x) = j_k\xi_{\mathsf{S}}(x)$.*

Proof We begin with a sublemma. Simpler versions of this result are called Hadamard's Lemma, but we could not find a reference to the form we require.

[1] Our blanket assumption is that all manifolds are connected. But connectedness is fundamental here, so we reiterate it.

Sublemma 1 *Let $r \in \{\infty, \omega\}$. Let $\mathcal{U} \subseteq \mathbb{R}^n$ be a neighbourhood of $\mathbf{0}$, let $\mathsf{S} \subseteq \mathbb{R}^n$ be the subspace*

$$\mathsf{S} = \{(x^1, \dots, x^n) \in \mathbb{R}^n \mid x^1 = \cdots = x^s = 0\},$$

let $k \in \mathbb{Z}_{\geq 0}$, and let $f \in \mathrm{C}^r(\mathcal{U})$ satisfy $\boldsymbol{D}^j f(\boldsymbol{x}) = 0$ for all $j \in \{0, 1, \dots, k\}$ and $\boldsymbol{x} \in \mathsf{S} \cap \mathcal{U}$. Let $\mathrm{pr}_{\mathsf{S}} : \mathbb{R}^n \to \mathsf{S}$ be the natural projection onto the first s-components. Then there exist a neighbourhood $\mathcal{V} \subseteq \mathcal{U}$ of $\mathbf{0}$ and functions $g_I \in \mathrm{C}^r(\mathcal{V})$, $I \in \mathbb{Z}^s_{>0}$, $|I| = k+1$, such that

$$f(\boldsymbol{x}) = \sum_{\substack{I \in \mathbb{Z}^s_{>0} \\ |I| = k+1}} g_I(\boldsymbol{x}) \, \mathrm{pr}_{\mathsf{S}}(\boldsymbol{x})^I, \qquad \boldsymbol{x} \in \mathcal{V}.$$

Proof We prove the sublemma by induction on k. For $k = 0$, the hypothesis is that f vanishes on $\mathsf{S} \cap \mathcal{U}$. Let $\mathcal{W} \subseteq \mathsf{S}$ be a neighbourhood of $\mathbf{0}$ and let $\epsilon \in \mathbb{R}_{>0}$ be such that $\mathsf{B}(\epsilon, \boldsymbol{y}) \subseteq \mathcal{U}$ for all $\boldsymbol{y} \in \mathcal{W}$, possibly after shrinking $\mathcal{W}$. Let

$$\mathcal{V} = \bigcup_{\boldsymbol{x} \in \mathcal{W}} \mathsf{B}(\epsilon, \boldsymbol{x}).$$

Let $\boldsymbol{x} = (\boldsymbol{x}_1, \boldsymbol{x}_2) \in \mathcal{V}$ (with $(\mathbf{0}, \boldsymbol{x}_2) \in \mathsf{S}$) and define

$$\begin{aligned} \gamma_{\boldsymbol{x}} : [0, 1] &\to \mathbb{R} \\ t &\mapsto f(t\boldsymbol{x}_1, \boldsymbol{x}_2). \end{aligned}$$

We calculate

$$\begin{aligned} f(\boldsymbol{x}) = f(\boldsymbol{x}_1, \boldsymbol{x}_2) &= f(\boldsymbol{x}_1, \boldsymbol{x}_2) - f(\mathbf{0}, \boldsymbol{x}_2) \\ &= \gamma_{\boldsymbol{x}}(1) - \gamma_{\boldsymbol{x}}(0) = \int_0^1 \gamma'_{\boldsymbol{x}}(t) \, \mathrm{d}t \\ &= \int_0^1 \sum_{j=1}^s x^j \frac{\partial f}{\partial x^j}((t\boldsymbol{x}_1, \boldsymbol{x}_2)) \, \mathrm{d}t = \sum_{j=1}^s x^j g_j(\boldsymbol{x}), \end{aligned}$$

where

$$g_j(\boldsymbol{x}) = \int_0^1 \frac{\partial f}{\partial x^j}(t\boldsymbol{x}_1, \boldsymbol{x}_2) \, \mathrm{d}t, \qquad j \in \{1, \dots, s\}.$$

It remains to show that the functions $g_1, \dots, g_s$ are of class C^r. By standard theorems on interchanging derivatives and integrals [38, Theorem 16.11], we can conclude that $g_1, \dots, g_m$ are smooth when f is smooth. If the data are holomorphic, swapping integrals and derivatives allows us to conclude that $g_1, \dots, g_s$ are holomorphic when f is holomorphic, by verifying the Cauchy–Riemann equations. In the real analytic case, we can complexify to a complex neighbourhood of $\mathbf{0}$, and so conclude real analyticity by holomorphicity of the complexification.

As a standin for a full proof by induction, let us see how the case $k = 1$ follows from the case $k = 0$. The general inductive argument is the same, only with more notation.

We note that, for $\boldsymbol{x} \in \mathcal{V}$, we have

$$\frac{\partial f}{\partial x^k}(\boldsymbol{x}) = \begin{cases} g_k(\boldsymbol{x}) + \sum_{j=1}^{s} x^j \frac{\partial g_j}{\partial x^k}(\boldsymbol{x}), & k \in \{1, \dots, s\}, \\ \sum_{j=1}^{s} x^j \frac{\partial g_j}{\partial x^k}(\boldsymbol{x}), & k \in \{s+1, \dots, n\}. \end{cases}$$

Thus $\boldsymbol{D}f(\boldsymbol{x}) = 0$ for $\boldsymbol{x} \in \mathcal{S} \cap \mathcal{V}$ if and only if $g_1(\boldsymbol{x}) = \cdots = g_s(\boldsymbol{x}) = 0$. Thus one can apply the arguments from the first part of the proof to write

$$g_k(\boldsymbol{x}) = \sum_{j=1}^{s} x^j g_{kj}(\boldsymbol{x})$$

on a neighbourhood of $\mathbf{0}$. Thus

$$f(\boldsymbol{x}) = \sum_{j,k=1}^{s} x^k x^j g_{kj}(\boldsymbol{x}),$$

giving the desired form of f in this case. ▽

To prove the lemma, let $\mathscr{Z}^k_{\mathsf{S}}$ be the sheaf of C^r-sections of E whose k-jet vanishes on S. We have the exact sequence

$$0 \longrightarrow \mathscr{Z}^k_{\mathsf{S}} \longrightarrow \mathscr{G}^r_{\mathsf{E}} \xrightarrow{\Psi} \mathscr{G}^r_{\mathsf{E}}/\mathscr{Z}^k_{\mathsf{S}} \longrightarrow 0$$

Note that the stalk of the quotient sheaf at $x \in \mathsf{S}$ consists of germs of sections whose k-jets agree on S.

Now, if $x \notin \mathsf{S}$, then there is a neighbourhood $\mathcal{U}$ of x such that $\mathcal{U} \cap \mathsf{S} = \emptyset$, and so $\mathscr{Z}^k_{\mathsf{S}}(\mathcal{U}) = \mathscr{G}^r_{\mathsf{E}}(\mathcal{U})$. That $\mathscr{Z}^k_{\mathsf{S}}$ is locally finitely generated at x then follows since $\mathscr{G}^r_{\mathsf{E}}$ is locally finitely generated. If $x \in \mathsf{S}$, choose a submanifold chart $(\mathcal{U}, \chi)$ for S about x so that

$$\mathsf{S} \cap \mathcal{U} = \{y \in \mathcal{U} \mid \phi(y) = (0, \dots, 0, x^{s+1}, \dots, x^n)\}.$$

Then the k-jet of a function f on $\mathcal{U}$ vanishes on S if and only if it is a $\mathrm{C}^r(\mathcal{U})$-linear combination of polynomial functions in $x^1, \dots, x^s$ of degree $k + 1$; this follows by the above sublemma. Thus, if $\xi_1, \dots, \xi_m$ is a local basis of sections of E about x, then the (finite) set of products of these sections with the polynomial functions in $x^1, \dots, x^s$ of degree at least $k + 1$ generates $\Gamma^r(\mathsf{E}|\mathcal{U})$ as a $\mathrm{C}^r(\mathcal{U})$-module. This shows that $\mathscr{Z}^k_{\mathsf{S}}$ is locally finitely generated about x. This then shows that $\mathscr{Z}^k_{\mathsf{S}}$ is coherent in the case $r = \omega$, by virtue of [25, Consequence A.4.2.2].

Cartan's Theorem B [13, Proposition 6] shows, in the case $r = \omega$, that Ψ is surjective on global sections. The case of $r = \infty$ follows in a similar way, using the fact that positive cohomology groups for sheave of modules of smooth functions vanish ([69, Proposition 2.3.11], along with [69, Examples 2.3.4(d,e)] and [69, Proposition 2.3.5]). For the degree 1 cohomology group, this implies that there exists $\xi \in \Gamma^r(\mathsf{E})$ such that, for each $x \in \mathsf{S}$, $[\xi]_x = [\xi_{\mathsf{S}}]_x$. This, however, means precisely that $j_k\xi(x) = j_k\xi_{\mathsf{S}}(x)$ for each $x \in \mathsf{S}$. □

Remarks 2.2 (On the Extension of Sections) Let us make two comments on this lemma, as they illustrate the subtle correspondences—and lack of correspondences—between smooth, real analytic, and holomorphic differential geometry.

1. The previous lemma is rather trivial using partitions of unity in the smooth case, and does not require tools like sheaf cohomology. Indeed, one can sometimes regard partitions of unity and sheaf theory as being interchangeable in the smooth case. This is not so in the real analytic case, where sheaf theory provides powerful tools that play the rôle of partitions of unity in the solution of certain kinds of problems.
2. The previous lemma is generally false in the holomorphic case. Indeed, holomorphic differential geometry is far more rigid that real analytic differential geometry. This is most succinctly illustrated by the following fact: the set of sections of an holomorphic vector bundle with a compact base space is finite-dimensional.[2] ∘

Next we consider the matter of embedding theorems. For smooth manifolds, Whitney's Embedding Theorem asserts that an Hausdorff, connected, second countable, smooth manifold[3] can be properly embedded in a sufficiently high-dimensional copy of real Euclidean space [71]. The real analytic version is the following result, which we enunciate clearly for future reference.

Theorem 2.3 (Grauert's Embedding Theorem) *If* M *is an Hausdorff, connected, second countable, real analytic manifold, then there exist* $N \in \mathbb{Z}_{>0}$ *and a proper real analytic embedding* $\iota_{\mathsf{M}} \in \mathrm{C}^{\omega}(\mathsf{M}; \mathbb{R}^N)$.

Proof [24, Theorem 3]. □

We note that there is generally no such embedding theorem for holomorphic manifolds, e.g., by the Maximum Modulus Theorem, there exists no nonconstant holomorphic mapping from a compact connected holomorphic manifold into com-

[2] An interesting way to prove this is to note that the space of sections in this case is a normed nuclear space, and so must be finite-dimensional.

[3] We recall that the our standing geometric assumptions in Sect. 1.4 are that manifolds are always Hausdorff, connected, and second countable. We reiterate this here since it is essential for the validity of the embedding theorem.

plex Euclidean space. Holomorphic manifolds that *do* admit proper holomorphic embeddings into complex Euclidean space are important, and are called ***Stein manifolds*** [26]. The embedding theorem for Stein manifolds was proved by Remmert [61].

As an application of Grauert's real analytic embedding theorem, we prove the following existential result concerning metrics and connections.

Lemma 2.4 (Existence of Real Analytic Connections and Fibre Metrics) *If* $\pi_{\mathsf{E}} : \mathsf{E} \to \mathsf{M}$ *is a real analytic vector bundle, then there exist*

(i) a real analytic linear connection on E,
(ii) a real analytic affine connection on M,
(iii) a real analytic fibre metric on E, *and*
(iv) a real analytic Riemannian metric on M.

Proof By Theorem 2.3, there exists a proper real analytic embedding ι_{E} of E in $\mathbb{R}^N$ for some $N \in \mathbb{Z}_{>0}$. There is then an induced proper real analytic embedding ι_{M} of M in $\mathbb{R}^N$ by restricting ι_{E} to the zero section of E. Let us take the subbundle $\hat{\mathsf{E}}$ of $\mathsf{T}\mathbb{R}^N|\iota_{\mathsf{M}}(\mathsf{M})$ whose fibre at $\iota_{\mathsf{M}}(x) \in \iota_{\mathsf{M}}(\mathsf{M})$ is

$$\hat{\mathsf{E}}_{\iota_{\mathsf{M}}(x)} = T_{0_x}\iota_{\mathsf{E}}(\mathsf{V}_{0_x}\mathsf{E}).$$

Now recall that $\mathsf{E} \simeq \zeta^*\mathsf{VE}$, where $\zeta : \mathsf{M} \to \mathsf{E}$ is the zero section [43, page 55]. Let us abbreviate $\hat{\iota}_{\mathsf{E}} = T\iota_{\mathsf{E}}|\zeta^*\mathsf{VE}$. We then have the following diagram

$$\begin{array}{ccc} \mathsf{E} \simeq \zeta^*\mathsf{VE} & \xrightarrow{\hat{\iota}_{\mathsf{E}}} & \mathbb{R}^N \times \mathbb{R}^N \\ {\scriptstyle \pi_{\mathsf{E}}}\downarrow & & \downarrow{\scriptstyle \mathrm{pr}_2} \\ \mathsf{M} & \xrightarrow[\iota_{\mathsf{M}}]{} & \mathbb{R}^N \end{array} \tag{2.1}$$

describing a monomorphism of real analytic vector bundles over the proper embedding ι_{M}, with the image of $\hat{\iota}_{\mathsf{E}}$ being $\hat{\mathsf{E}}$.

To prescribe a linear connection in the vector bundle E, we will take the prescription whereby one defines a connector $K : \mathsf{TE} \to \mathsf{E}$ such that the diagrams (1.6) are vector bundle mappings. We define K as follows. For $e_x \in \mathsf{E}_x$ and $X_{e_x} \in \mathsf{T}_{e_x}\mathsf{E}$ we have

$$T_{e_x}\hat{\iota}_{\mathsf{E}}(X_{e_x}) \in \mathsf{T}_{\hat{\iota}_{\mathsf{E}}(e_x)}(\mathbb{R}^N \times \mathbb{R}^N) \simeq \mathbb{R}^N \oplus \mathbb{R}^N,$$

and we define K so that

$$\hat{\iota}_{\mathsf{E}} \circ K(X_{e_x}) = \mathrm{pr}_2 \circ T_{e_x}\hat{\iota}_{\mathsf{E}}(X_{e_x});$$

this uniquely defines K by injectivity of $\hat{\iota}_{\mathsf{E}}$, and amounts to using on E the connection induced on image$(\hat{\iota}_{\mathsf{E}})$ by the trivial connection on $\mathbb{R}^N \times \mathbb{R}^N$. In particular, this means that we think of $\hat{\iota}_{\mathsf{E}} \circ K(X_{e_x})$ as being an element of the fibre of the trivial bundle $\mathbb{R}^N \times \mathbb{R}^N$ at $\iota_{\mathsf{M}}(x)$.

If $v_x \in \mathsf{TM}$, if $e, e' \in \mathsf{E}$, and if $X \in \mathsf{T}_e\mathsf{E}$ and $X' \in T_{e'}\mathsf{E}$ satisfy $X, X' \in T\pi_{\mathsf{E}}^{-1}(v_x)$, then note that

$$\begin{aligned} T_e\pi_{\mathsf{E}}(X) = T_{e'}\pi_{\mathsf{E}}(X') &\implies T_e(\iota_{\mathsf{M}} \circ \pi_{\mathsf{E}})(X) = T_{e'}(\iota_{\mathsf{M}} \circ \pi_{\mathsf{E}})(X') \\ &\implies T_e(\mathrm{pr}_2 \circ \hat{\iota}_{\mathsf{E}})(X) = T_{e'}(\mathrm{pr}_2 \circ \hat{\iota}_{\mathsf{E}})(X') \\ &\implies T_{\iota_{\mathsf{M}}(x)}\,\mathrm{pr}_2 \circ T_e\hat{\iota}_{\mathsf{E}}(X) = T_{\iota_{\mathsf{M}}(x)}\,\mathrm{pr}_2 \circ T_{e'}\hat{\iota}_{\mathsf{E}}(X'). \end{aligned}$$

Thus we can write

$$T_e\hat{\iota}_{\mathsf{E}}(X) = (\boldsymbol{x}, \boldsymbol{e}, \boldsymbol{u}, \boldsymbol{v}), \quad T_{e'}\hat{\iota}_{\mathsf{E}}(X) = (\boldsymbol{x}, \boldsymbol{e}', \boldsymbol{u}, \boldsymbol{v}')$$

for suitable $\boldsymbol{x}, \boldsymbol{u}, \boldsymbol{e}, \boldsymbol{e}', \boldsymbol{v}, \boldsymbol{v}' \in \mathbb{R}^N$. Therefore,

$$\hat{\iota}_{\mathsf{E}} \circ K(X) = (\boldsymbol{x}, \boldsymbol{v}), \quad \hat{\iota}_{\mathsf{E}} \circ K(X') = (\boldsymbol{x}, \boldsymbol{v}'), \quad \hat{\iota}_{\mathsf{E}} \circ K(X + X') = (\boldsymbol{x}, \boldsymbol{v} + \boldsymbol{v}'),$$

from which we immediately conclude that, for addition in the vector bundle $T\pi_{\mathsf{E}} \colon \mathsf{TE} \to \mathsf{TM}$, we have

$$\hat{\iota}_{\mathsf{E}} \circ K(X + X') = \hat{\iota}_{\mathsf{E}} \circ K(X) + \hat{\iota}_{\mathsf{E}} \circ K(X'),$$

showing that the diagram on the left in (1.6) makes K a vector bundle mapping.

On the other hand, if $e_x \in \mathsf{E}$ and if $X, X' \in \mathsf{T}_{e_x}\mathsf{E}$, then we have, using vector bundle addition in $\pi_{\mathsf{TE}} \colon \mathsf{TE} \to \mathsf{E}$,

$$\begin{aligned} \hat{\iota}_{\mathsf{E}} \circ K(X + X') &= \mathrm{pr}_2 \circ T_{e_x}\hat{\iota}_{\mathsf{E}}(X + X') \\ &= \mathrm{pr}_2 \circ T_{e_x}\hat{\iota}_{\mathsf{E}}(X) + \mathrm{pr}_2 \circ T_{e_x}\hat{\iota}_{\mathsf{E}}(X') \\ &= \hat{\iota}_{\mathsf{E}} \circ K(X) + \hat{\iota}_{\mathsf{E}} \circ K(X'), \end{aligned}$$

giving that the diagram on the right in (1.6) makes K a vector bundle mapping. Since K is real analytic, this defines a real analytic linear connection $\nabla^{\pi_{\mathsf{E}}}$ on E as in [43, §11.11].

The existence of a fibre metric $\mathbb{G}_{\pi_{\mathsf{E}}}$, a Riemannian metric $\mathbb{G}_{\mathsf{M}}$, and an affine connection ∇^{M} are straightforward. Indeed, we let $\mathbb{G}_{\mathbb{R}^N}$ be the Euclidean metric on $\mathbb{R}^N$, and define $\mathbb{G}_{\pi_{\mathsf{E}}}$ and $\mathbb{G}_{\mathsf{M}}$ by

$$\mathbb{G}_{\pi_{\mathsf{E}}}(e_x, e'_x) = \mathbb{G}_{\mathbb{R}^N}(\hat{\iota}_{\mathsf{E}}(e_x), \hat{\iota}_{\mathsf{E}}(e'_x))$$

and

$$\mathbb{G}_{\mathsf{M}}(v_x, v_x') = \mathbb{G}_{\mathbb{R}^N}(T_x\iota_{\mathsf{M}}(v_x), T_x\iota_{\mathsf{M}}(v_x')).$$

An affine connection ∇^{M} can be taken to be the Levi-Civita connection of $\mathbb{G}_{\mathsf{M}}$. □

The upshot of the lemma is that we can always define the following data:

1. a linear connection $\nabla^{\pi_{\mathsf{E}}}$ in E;
2. an affine connection ∇^{M} on M;
3. a fibre metric $\mathbb{G}_{\pi_{\mathsf{E}}}$ on E;
4. a Riemannian metric $\mathbb{G}_{\mathsf{M}}$ on M.

We shall frequently assume this data without mention.

We note that the constructions of the lemma give the following result of independent interest. We comment that this result is an example of something that is true in real analytic differential geometry, but not generally true in holomorphic differential geometry.

Corollary 2.5 (Real Analytic Vector Bundles Are Subbundles of Trivial Bundles) *If* $\pi \colon \mathsf{E} \to \mathsf{M}$ *is a real analytic vector bundle, then* E *is a* C^ω*-subbundle of a trivial bundle* $\mathrm{pr}_1 \colon \mathbb{R}^N_{\mathsf{M}} = \mathsf{M} \times \mathbb{R}^N \to \mathsf{M}$ *for suitable* $N \in \mathbb{Z}_{>0}$.

Proof This follows directly from the constructions that gave rise to the diagram (2.1). □

2.1.3 Complexification of Real Analytic Manifolds and Vector Bundles

It is not too surprising that, to study real analytic differential geometry, one approach is to "complexify" to holomorphic differential geometry. In this section we describe how one does this.

We let $\pi_{\mathsf{E}} \colon \mathsf{E} \to \mathsf{M}$ be a real analytic vector bundle. We assume the data required to make the diagram (2.1) giving $\pi_{\mathsf{E}} \colon \mathsf{E} \to \mathsf{M}$ as the image of a real analytic vector bundle monomorphism in the trivial vector bundle $\mathbb{R}^N \times \mathbb{R}^N$ for some suitable $N \in \mathbb{Z}_{>0}$.

Now we complexify. Recall that, if V is a $\mathbb{C}$-vector space, then multiplication by $\sqrt{-1}$ induces a $\mathbb{R}$-linear map $J \in \mathrm{End}_{\mathbb{R}}(\mathsf{V})$.[4] A $\mathbb{R}$-subspace U of V is ***totally real*** if $\mathsf{U} \cap J(\mathsf{U}) = \{0\}$. A submanifold of an holomorphic manifold, thinking of the latter as a smooth manifold, is ***totally real*** if its tangent spaces are totally real subspaces. By [72, Proposition 1], for a real analytic manifold M there exists a complexification $\overline{\mathsf{M}}$ of M, i.e., an holomorphic manifold having M as a totally real

[4] Thus J satisfying $J(v) = \mathrm{i}v$ for $v \in \mathsf{V}$.

submanifold and where $\overline{\mathsf{M}}$ has the same $\mathbb{C}$-dimension as the $\mathbb{R}$-dimension of M. As shown by Grauert [24, §3.4], for any neighbourhood $\overline{\mathcal{U}}$ of M in $\overline{\mathsf{M}}$, there exists a Stein neighbourhood $\overline{\mathcal{S}}$ of M contained in $\overline{\mathcal{U}}$. By arguments involving extending convergent real power series to convergent complex power series (the conditions on coefficients for convergence are the same for both real and complex power series, as we indicated in Sect. 1.9), one can show that there is an holomorphic extension of ι_{M} to $\iota_{\overline{\mathsf{M}}}\colon \overline{\mathsf{M}} \to \mathbb{C}^N$, possibly after shrinking $\overline{\mathsf{M}}$ [14, Lemma 5.40]. By applying similar reasoning to the transition maps for the real analytic vector bundle E, one obtains an holomorphic vector bundle $\pi_{\overline{\mathsf{E}}}\colon \overline{\mathsf{E}} \to \overline{\mathsf{M}}$ for which the diagram

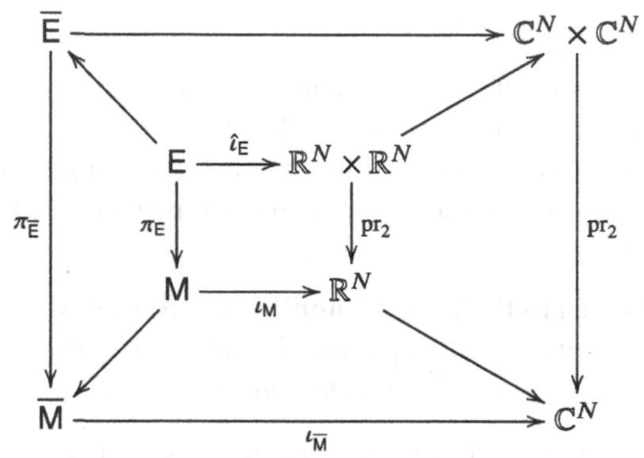

commutes, where all diagonal arrows are complexification and where the inner diagram is as defined in the proof of Lemma 2.4. One can then define a real analytic Hermitian fibre metric $\mathbb{G}_{\pi_{\overline{\mathsf{E}}}}$ on $\overline{\mathsf{E}}$ induced from the standard Hermitian metric on the fibres of the vector bundle $\mathbb{C}^N \times \mathbb{C}^N$ and an Hermitian metric $\mathbb{G}_{\overline{\mathsf{M}}}$ on $\overline{\mathsf{M}}$ induced from the standard Hermitian metric on $\mathbb{C}^N$.

2.2 Martineau's Descriptions of the Real Analytic Topology

We shall give a description of two topologies for the space $\Gamma^\omega(\mathsf{E})$ of real analytic sections of a real analytic vector bundle $\pi_{\mathsf{E}}\colon \mathsf{E} \to \mathsf{M}$. The original work of Martineau [50] describes these topologies for the space of real analytic functions, but it is evident that the same considerations apply to sections of a general vector bundle. Each description offers benefits in terms of providing immediately some useful properties of the topology, although showing that they agree is something of an undertaking, and we shall make some comments about this.

That all being said, let us turn to Martineau's characterisations of the topology for $\Gamma^\omega(\mathsf{E})$.

2.2.1 *Germs of Holomorphic Sections Over Subsets of a Real Analytic Manifold*

In two different places, we will need to consider germs of holomorphic sections. In this section we organise the methodology for doing this to unify the notation. Throughout this section we let $\pi_{\mathsf{E}}\colon \mathsf{E} \to \mathsf{M}$ be a real analytic vector bundle and we assume that we have the constructions in place to make use of the complexification constructions of Sect. 2.1.3.

Let $A \subseteq \mathsf{M}$ and let $\overline{\mathscr{N}}_A$ be the set of neighbourhoods of A in the complexification $\overline{\mathsf{M}}$. For $\overline{\mathcal{U}}, \overline{\mathcal{V}} \in \overline{\mathscr{N}}_A$, and for $\overline{\xi} \in \Gamma^{\mathrm{hol}}(\overline{\mathsf{E}}|\overline{\mathcal{U}})$ and $\overline{\eta} \in \Gamma^{\mathrm{hol}}(\overline{\mathsf{E}}|\overline{\mathcal{V}})$, we say that $\overline{\xi}$ is ***equivalent*** to $\overline{\eta}$ if there exist $\overline{\mathcal{W}} \in \overline{\mathscr{N}}_A$ and $\overline{\zeta} \in \Gamma^{\mathrm{hol}}(\overline{\mathsf{E}}|\overline{\mathcal{W}})$ such that $\overline{\mathcal{W}} \subseteq \overline{\mathcal{U}} \cap \overline{\mathcal{V}}$ and such that

$$\overline{\xi}|\overline{\mathcal{W}} = \overline{\eta}|\overline{\mathcal{W}} = \overline{\zeta}.$$

By $\mathscr{G}^{\mathrm{hol}}_{A,\overline{\mathsf{E}}}$ we denote the set of equivalence classes, which we call the set of ***germs of sections*** of $\overline{\mathsf{E}}$ over A. By $[\overline{\xi}]_A$ we denote the equivalence class of $\overline{\xi} \in \Gamma^{\mathrm{hol}}(\overline{\mathsf{E}}|\overline{\mathcal{U}})$ for some $\overline{\mathcal{U}} \in \overline{\mathscr{N}}_A$. For the particular case of functions, i.e., sections of the vector bundle $\mathbb{R}_{\mathsf{M}}$, we denote the set of germs by $\mathscr{C}^{\mathrm{hol}}_{A,\overline{\mathsf{M}}}$.

Now, for $x \in \mathsf{M}$, E_x is a totally real subspace of $\overline{\mathsf{E}}_x$ with half the real dimension, and so it follows that

$$\overline{\mathsf{E}}_x = \mathsf{E}_x \oplus J(\mathsf{E}_x),$$

where J is the complex structure on the fibres of $\overline{\mathsf{E}}$. For $\overline{\mathcal{U}} \in \overline{\mathscr{N}}_A$, denote by $\Gamma^{\mathrm{hol},\mathbb{R}}(\overline{\mathsf{E}}|\overline{\mathcal{U}})$ those holomorphic sections $\overline{\xi}$ of $\overline{\mathsf{E}}|\overline{\mathcal{U}}$ such that $\overline{\xi}(x) \in \mathsf{E}_x$ for $x \in \overline{\mathcal{U}} \cap \mathsf{M}$. We think of this as being a locally convex topological $\mathbb{R}$-vector space with the seminorms $p^{\mathrm{hol}}_{\overline{\mathcal{K}}}$, $\overline{\mathcal{K}} \subseteq \overline{\mathcal{U}}$ compact, defined by

$$p^{\mathrm{hol}}_{\overline{\mathcal{K}}}(\overline{\xi}) = \sup\left\{ \|\overline{\xi}(\overline{x})\|_{\overline{\mathbb{G}}_{\overline{\mathsf{E}}}} \;\middle|\; \overline{x} \in \overline{\mathcal{K}} \right\}, \tag{2.2}$$

i.e., we use the locally convex structure induced from the usual compact-open topology on $\Gamma^{\mathrm{hol}}(\overline{\mathsf{E}}|\overline{\mathcal{U}})$.

Remark 2.6 (Closedness of "Real" Sections) We note that $\Gamma^{\mathrm{hol},\mathbb{R}}(\overline{\mathsf{E}}|\overline{\mathcal{U}})$ is a closed $\mathbb{R}$-subspace of $\Gamma^{\mathrm{hol}}(\overline{\mathsf{E}})$ in the compact-open topology, i.e., the restriction of requiring "realness" on M is a closed condition. This is easily shown, and we often assume it often without mention. ○

Denote by $\mathscr{G}^{\mathrm{hol},\mathbb{R}}_{A,\overline{\mathsf{E}}}$ the set of germs of sections from $\Gamma^{\mathrm{hol},\mathbb{R}}(\overline{\mathsf{E}}|\overline{\mathcal{U}})$, $\overline{\mathcal{U}} \in \overline{\mathscr{N}}_A$. If $\overline{\mathcal{U}}_1, \overline{\mathcal{U}}_2 \in \overline{\mathscr{N}}_A$ satisfy $\overline{\mathcal{U}}_1 \subseteq \overline{\mathcal{U}}_2$, then we have the restriction mapping

$$\begin{aligned} r_{\overline{\mathcal{U}}_2,\overline{\mathcal{U}}_1} \colon \Gamma^{\mathrm{hol},\mathbb{R}}(\overline{\mathsf{E}}|\overline{\mathcal{U}}_2) &\to \Gamma^{\mathrm{hol},\mathbb{R}}(\overline{\mathsf{E}}|\overline{\mathcal{U}}_1) \\ \overline{\xi} &\mapsto \overline{\xi}|\overline{\mathcal{U}}_1. \end{aligned}$$

This restriction is continuous since, for any compact set $\overline{\mathcal{K}} \subseteq \overline{\mathcal{U}}_1 \subseteq \overline{\mathcal{U}}_2$ and any $\overline{\xi} \in \Gamma^{\mathrm{hol},\mathbb{R}}(\overline{\mathsf{E}}|\overline{\mathcal{U}}_2)$, we have $p^{\mathrm{hol}}_{\overline{\mathcal{K}}}(r_{\overline{\mathcal{U}}_2,\overline{\mathcal{U}}_1}(\overline{\xi})) \leq p^{\mathrm{hol}}_{\overline{\mathcal{K}}}(\overline{\xi})$. We also have maps

$$\begin{aligned} r_{\overline{\mathcal{U}},A} \colon \Gamma^{\mathrm{hol},\mathbb{R}}(\overline{\mathsf{E}}|\overline{\mathcal{U}}) &\to \mathscr{G}^{\mathrm{hol},\mathbb{R}}_{A,\overline{\mathsf{E}}} \\ \overline{\xi} &\mapsto [\overline{\xi}]_A. \end{aligned}$$

Note that $\overline{\mathscr{N}}_A$ is a directed set by reverse inclusion; that is, $\overline{\mathcal{U}}_2 \preceq \overline{\mathcal{U}}_1$ if $\overline{\mathcal{U}}_1 \subseteq \overline{\mathcal{U}}_2$. Thus we have the directed system $(\Gamma^{\mathrm{hol},\mathbb{R}}(\overline{\mathsf{E}}|\overline{\mathcal{U}}))_{\overline{\mathcal{U}} \in \overline{\mathscr{N}}_A}$, along with the mappings $r_{\overline{\mathcal{U}}_2,\overline{\mathcal{U}}_1}$, in the category of locally convex topological $\mathbb{R}$-vector spaces. This then gives rise to the locally convex direct limit topology for the direct limit $\mathscr{G}^{\mathrm{hol},\mathbb{R}}_{A,\overline{\mathsf{E}}}$, as in Definition 1.4.

We shall use this general development in two different ways, one of which we turn our attention to now.

2.2.2 A Natural Direct Limit Topology for the Space of Real Analytic Sections

We let $\pi_{\mathsf{E}} \colon \mathsf{E} \to \mathsf{M}$ be a real analytic vector bundle. As in Sect. 2.1.3, we shall extend E to an holomorphic vector bundle $\pi_{\overline{\mathsf{E}}} \colon \overline{\mathsf{E}} \to \overline{\mathsf{M}}$ that will serve as an important device for all of our constructions.

The following lemma is key.

Lemma 2.7 (Real Analytic Sections as Holomorphic Germs) *There is a natural $\mathbb{R}$-vector space isomorphism between $\Gamma^{\omega}(\mathsf{E})$ and $\mathscr{G}^{\mathrm{hol},\mathbb{R}}_{\mathsf{M},\overline{\mathsf{E}}}$.*

Proof Let $\xi \in \Gamma^{\omega}(\mathsf{E})$. As in [14, Lemma 5.40], there is an extension of ξ to a section $\overline{\xi} \in \Gamma^{\mathrm{hol},\mathbb{R}}(\overline{\mathsf{E}}|\overline{\mathcal{U}})$ for some $\overline{\mathcal{U}} \in \overline{\mathscr{N}}_{\mathsf{M}}$. We claim that the map $i_{\mathsf{M}} \colon \Gamma^{\omega}(\mathsf{E}) \to \mathscr{G}^{\mathrm{hol},\mathbb{R}}_{\mathsf{M},\overline{\mathsf{E}}}$ defined by $i_{\mathsf{M}}(\xi) = [\overline{\xi}]_{\mathsf{M}}$ is the desired isomorphism. That i_{M} is independent of the choice of extension $\overline{\xi}$ is a consequence of the fact that the extension to $\overline{\xi}$ is unique inasmuch as any two such extensions agree on some neighbourhood contained in their intersection; this is the uniqueness assertion of [14, Lemma 5.40]. This fact also ensures that i_{M} is injective. For surjectivity, let $[\overline{\xi}]_{\mathsf{M}} \in \mathscr{G}^{\mathrm{hol},\mathbb{R}}_{\mathsf{M},\overline{\mathsf{E}}}$ and let us define $\xi \colon \mathsf{M} \to \mathsf{E}$ by $\xi(x) = \overline{\xi}(x)$ for $x \in \mathsf{M}$. Note that the restriction of $\overline{\xi}$ to

M is real analytic because the values of $\overline{\xi}|\mathsf{M}$ at points in a neighbourhood of $x \in \mathsf{M}$ are given by the restriction of the (necessarily convergent) complex Taylor series of $\overline{\xi}$ to M. Obviously, $i_{\mathsf{M}}(\xi) = [\overline{\xi}]_{\mathsf{M}}$. □

Now we use the locally convex direct limit topology on $\mathscr{G}^{\mathrm{hol},\mathbb{R}}_{\mathsf{M},\overline{\mathsf{E}}}$ described in Sect. 2.2.1, along with the preceding lemma, to immediately give a locally convex topology for $\Gamma^{\omega}(\mathsf{E})$ that we refer to as the ***direct limit topology***.

Let us make an important observation about the direct limit topology for $\Gamma^{\omega}(\mathsf{M})$. Let us denote by $\mathscr{S}_{\mathsf{M}}$ the set of all Stein neighbourhoods of M in $\overline{\mathsf{M}}$. As shown by Grauert [24, §3.4], if $\overline{\mathcal{U}} \in \overline{\mathscr{N}}_{\mathsf{M}}$, then there exists $\overline{\mathcal{S}} \in \mathscr{S}_{\mathsf{M}}$ with $\overline{\mathcal{S}} \subseteq \overline{\mathcal{U}}$. Therefore, $\mathscr{S}_{\mathsf{M}}$ is cofinal[5] in $\overline{\mathscr{N}}_{\mathsf{M}}$ and so the directed systems $(\Gamma^{\mathrm{hol},\mathbb{R}}(\mathsf{E}|\overline{\mathcal{U}}))_{\overline{\mathcal{U}}\in\overline{\mathscr{N}}_{\mathsf{M}}}$ and $(\Gamma^{\mathrm{hol},\mathbb{R}}(\mathsf{E}|\overline{\mathcal{S}}))_{\overline{\mathcal{S}}\in\mathscr{S}_{\mathsf{M}}}$ induce the same final topology on $\Gamma^{\omega}(\mathsf{E})$ [29, page 137].

2.2.3 The Topology of Holomorphic Germs About a Compact Set

We now turn to the second of Martineau's topologies for $\Gamma^{\omega}(\mathsf{M})$. This description first makes use of the constructions from Sect. 2.2.1, applied to the case when $\mathcal{K}$ is a compact subset of M.

We continue with the notation from Sect. 2.2.1. For $\mathcal{K} \subseteq \mathsf{M}$ compact, we have the locally convex direct limit topology, described above for general subsets $A \subseteq \mathsf{M}$, on $\mathscr{G}^{\mathrm{hol},\mathbb{R}}_{\mathcal{K},\overline{\mathsf{E}}}$. We seem to have gained nothing, since we have yet another direct limit topology. However, the direct limit can be shown to be of a friendly sort, namely a *countable* direct limit. Key to doing this is the following construction.

Lemma 2.8 (Compact Sets as Countable Intersections of Open Sets) *If* $(\mathcal{M}, d)$ *is a locally compact metric space,*[6] *if* $\mathcal{K} \subseteq \mathcal{S}$ *is compact, and if* $\mathscr{N}_{\mathcal{K}}$ *is the set of neighbourhoods of* $\mathcal{K}$*, then there exists a sequence* $(\mathcal{U}_j)_{j\in\mathbb{Z}_{>0}}$ *of precompact neighbourhoods of* $\mathcal{K}$ *such that*

(i) $\mathrm{cl}(\mathcal{U}_{j+1}) \subseteq \mathcal{U}_j$, $j \in \mathbb{Z}_{>0}$,
(ii) $\mathcal{K} = \cap_{j\in\mathbb{Z}_{>0}}\mathcal{U}_j$, *and*
(iii) if $\mathcal{U} \in \mathscr{N}_{\mathcal{K}}$*, then there exists* $j_0 \in \mathbb{Z}_{>0}$ *such that* $\mathcal{U}_{j_0} \subseteq \mathcal{U}$.

Proof We first make use of some metric space constructions. For sets $A, B \subseteq \mathcal{M}$, denote

$$\mathrm{dist}(A, B) = \inf\{\mathrm{d}(x, y) \mid x \in A,\ y \in B\}.$$

[5] A subset $J \subseteq I$ is ***cofinal*** if, for every $i \in I$, there exists $j \in J$ with $i \preceq j$.

[6] A manifold, for example.

If $A = \{x\}$, denote $\mathrm{dist}(\{x\}, B) = \mathrm{dist}(x, B)$ and so define a function

$$\begin{aligned}\mathrm{dist}_B \colon \mathcal{M} &\to \mathbb{R}_{\geq 0}\\ x &\mapsto \mathrm{dist}(x, B).\end{aligned}$$

Similarly we denote $\mathrm{dist}(A, \{y\}) = \mathrm{dist}(A, y)$.

The following sublemma then records some useful facts.

Sublemma 1 *If $(\mathcal{M}, d)$ is a metric space, then the following statements hold:*

(i) if $B \subseteq \mathcal{M}$, then the function dist_B *is uniformly continuous in the metric topology;*
(ii) if $A, B \subseteq \mathcal{M}$ are disjoint closed sets, then $\mathrm{dist}(x, B), \mathrm{dist}(A, y) > 0$ *for all $x \in A$ and $y \in B$.*

Proof

(i) Let $\epsilon \in \mathbb{R}_{>0}$ and take $\delta = \frac{\epsilon}{2}$. Let $y \in B$ be such that $\mathrm{d}(x_1, y) - \mathrm{dist}(x_1, B) < \frac{\epsilon}{2}$. Then, if $\mathrm{d}(x_1, x_2) < \delta$,

$$\mathrm{dist}(x_2, B) \leq \mathrm{d}(x_2, y) \leq \mathrm{d}(x_2, x_1) + \mathrm{d}(x_1, y) \leq \mathrm{dist}(x_1, B) + \epsilon.$$

In a symmetric manner one shows that

$$\mathrm{dist}(x_1, B) \leq \mathrm{dist}(x_2, B) + \epsilon,$$

provided that $\mathrm{d}(x_1, x_2) < \delta$. Therefore,

$$|\mathrm{dist}(x_1, B) - \mathrm{dist}(x_2, B)| < \epsilon,$$

provided that $\mathrm{d}(x_1, x_2) < \delta$, giving uniform continuity, as desired.

(ii) Suppose that $\mathrm{dist}(x, B) = 0$. Then there exists a sequence $(y_j)_{j \in \mathbb{Z}_{>0}}$ in B such that $\mathrm{d}(y_j, x) < \frac{1}{j}$ for each $j \in \mathbb{Z}_{>0}$. Thus the sequence $(y_j)_{j \in \mathbb{Z}_{>0}}$ converges to x and so $x \in \mathrm{cl}(B) = B$. Therefore, if $A \cap B = \emptyset$ we can conclude that, if $\mathrm{dist}(x, B) = 0$, then $x \notin A$. That is, $\mathrm{dist}(x, B) > 0$ for every $x \in A$, and similarly $\mathrm{dist}(A, y) > 0$ for every $y \in B$. $\triangledown$

Let $x \in \mathcal{K}$ and let $r_x \in \mathbb{R}_{>0}$ be such that $\overline{\mathsf{B}}_{\mathrm{d}}(r_x, x)$ is compact (by local compactness). Let $\mathcal{V}_x \subseteq \overline{\mathsf{B}}_{\mathrm{d}}(r_x, x)$ be a precompact neighbourhood of x. Since $\mathcal{K}$ is compact, let $x_1, \dots, x_k \in \mathcal{K}$ be such that $\mathcal{K} \subseteq \mathcal{U}_1 \triangleq \cup_{j=1}^{k} \mathcal{V}_{x_j}$. Note that $\mathcal{U}_1$ is precompact, being a finite union of precompact sets. Let $r_1 = \min\{r_{x_1}, \dots, r_{x_k}\}$.

Clear the notation from the preceding paragraph so we can use it again for a different purpose, keeping only $\mathcal{U}_1$ and r_1. Let $x \in \mathcal{K}$ and let $r_x \in \mathbb{R}_{>0}$ be such that $r_x < \frac{r_1}{2}$ and let $\mathcal{V}_x \subseteq \overline{\mathsf{B}}_{\mathrm{d}}(r_x, x)$ be a precompact neighbourhood of x. By compactness of $\mathcal{K}$, let $x_1, \dots, x_k \in \mathcal{K}$ be such that $\mathcal{K} \subseteq \mathcal{U}_2 \triangleq \cup_{j=1}^{k} \mathcal{V}_{x_j}$. Note that $\mathcal{U}_2$ is precompact and that $\mathrm{cl}(\mathcal{U}_2) \subseteq \mathcal{U}_1$. Let $r_2 = \min\{r_{x_1}, \dots, r_{x_k}\}$.

We can continue in this way, each time choosing balls of half the minimum radius of the preceding step, to arrive at a sequence of precompact neighbourhoods having the first property in the statement of the lemma. For the second, we clearly have $\mathcal{K} \subseteq \cap_{j\in\mathbb{Z}_{>0}}\mathcal{U}_j$. If $x \notin \mathcal{K}$, then $x \notin \mathcal{U}_j$ if we choose j sufficiently large that $r_j < \mathrm{dist}(\mathcal{K}, x)$, this being possible by part (ii) of the sublemma. We thus conclude that $\cap_{j\in\mathbb{Z}_{>0}}\mathcal{U}_j \subseteq \mathcal{K}$, giving the first part of the lemma.

Finally, let $\mathcal{U} \in \mathscr{N}_{\mathcal{K}}$ and note that $\mathrm{dist}_{\mathsf{M}\setminus\mathcal{U}}|\mathcal{K}$ is continuous by part (i) of the sublemma and so there exists $r \in \mathbb{R}_{>0}$ such that $\mathrm{dist}_{\mathsf{M}\setminus\mathcal{U}}(x) \geq r$ for every $x \in \mathcal{K}$. Thus, choosing $j_0 \in \mathbb{Z}_{>0}$ such that $r_{j_0} < r$, we have $\mathcal{U}_{j_0} \subseteq \mathcal{U}$. □

By the lemma, there is a *countable* family $(\overline{\mathcal{U}}_{\mathcal{K},j})_{j\in\mathbb{Z}_{>0}}$ from $\overline{\mathscr{N}}_{\mathcal{K}}$ with the property that $\mathrm{cl}(\overline{\mathcal{U}}_{\mathcal{K},j+1}) \subseteq \overline{\mathcal{U}}_{\mathcal{K},j}$ and $\mathcal{K} = \cap_{j\in\mathbb{Z}_{>0}}\overline{\mathcal{U}}_{\mathcal{K},j}$. Moreover, the sequence $(\overline{\mathcal{U}}_{\mathcal{K},j})_{j\in\mathbb{Z}_{>0}}$ is cofinal in $\overline{\mathscr{N}}_{\mathcal{K}}$, i.e., if $\overline{\mathcal{U}} \in \overline{\mathscr{N}}_{\mathcal{K}}$, then there exists $j \in \mathbb{Z}_{>0}$ with $\overline{\mathcal{U}}_{\mathcal{K},j} \subseteq \overline{\mathcal{U}}$. Let us fix such a family of neighbourhoods. Let us fix $j \in \mathbb{Z}_{>0}$ for a moment. Let $\Gamma^{\mathrm{hol},\mathbb{R}}_{\mathrm{bdd}}(\overline{\mathsf{E}}|\overline{\mathcal{U}}_{\mathcal{K},j})$ be the set of bounded sections from $\Gamma^{\mathrm{hol},\mathbb{R}}(\overline{\mathsf{E}}|\overline{\mathcal{U}}_{\mathcal{K},j})$, boundedness being taken relative to an Hermitian fibre metric $\mathbb{G}_{\pi_{\overline{\mathsf{E}}}}$ for $\pi_{\overline{\mathsf{E}}}\colon \overline{\mathsf{E}} \to \overline{\mathsf{M}}$. If we define a norm on $\Gamma^{\mathrm{hol},\mathbb{R}}_{\mathrm{bdd}}(\overline{\mathsf{E}}|\overline{\mathcal{U}}_{\mathcal{K},j})$ by

$$p^{\mathrm{hol}}_{\overline{\mathcal{U}}_{\mathcal{K},j},\infty}(\overline{\xi}) = \sup\left\{ \|\overline{\xi}(\overline{x})\|_{\mathbb{G}_{\pi_{\overline{\mathsf{E}}}}} \;\middle|\; \overline{x} \in \overline{\mathcal{U}}_{\mathcal{K},j}\right\},$$

then this makes $\Gamma^{\mathrm{hol},\mathbb{R}}_{\mathrm{bdd}}(\overline{\mathcal{U}}_{\mathcal{K},j})$ into a normed vector space. This gives context for the next purely holomorphic lemma.

Lemma 2.9 (The Topology of $\Gamma^{\mathrm{hol}}_{\mathrm{bdd}}(\mathsf{E})$) *Let $\pi_{\mathsf{E}}\colon \mathsf{E} \to \mathsf{M}$ be an holomorphic vector bundle. Then the following statements hold:*

(i) $\Gamma^{\mathrm{hol}}_{\mathrm{bdd}}(\mathsf{E})$ is a Banach space;
(ii) the inclusion of $\Gamma^{\mathrm{hol}}_{\mathrm{bdd}}(\mathsf{E})$ with the norm topology in $\Gamma^{\mathrm{hol}}(\mathsf{E})$ with the compact-open topology is continuous and compact;
(iii) if $\mathcal{U} \subseteq \mathsf{M}$ is a precompact open set with $\mathrm{cl}(\mathcal{U}) \subseteq \mathsf{M}$, then the restriction map from $\Gamma^{\mathrm{hol}}(\mathsf{E})$ to $\Gamma^{\mathrm{hol}}_{\mathrm{bdd}}(\mathsf{E}|\mathcal{U})$ is continuous.

Proof

(i) By [32, Theorem 7.9], a Cauchy sequence $(\xi_j)_{j\in\mathbb{Z}_{>0}}$ in $\Gamma^{\mathrm{hol}}_{\mathrm{bdd}}(\mathsf{E})$ converges to a bounded continuous section ξ of E. That ξ is also holomorphic follows since uniform limits of holomorphic sections are holomorphic [30, page 5].

(ii) For continuity, it suffices to show that a sequence $(\xi_j)_{j\in\mathbb{Z}_{>0}}$ in $\Gamma^{\mathrm{hol}}_{\mathrm{bdd}}(\mathsf{E})$ converges to $\xi \in \Gamma^{\mathrm{hol}}_{\mathrm{bdd}}(\mathsf{E})$ uniformly on compact subsets of M if it converges in norm. This, however, is obvious.

To show compactness of the inclusion, let $\mathcal{B} \subseteq \Gamma^{\mathrm{hol}}_{\mathrm{bdd}}(\mathsf{E})$ be bounded. Then $p^{\mathrm{hol}}_{\mathsf{M}}|\mathcal{B}$ is bounded. This implies that $p^{\mathrm{hol}}_{\mathcal{K}}|\mathcal{B}$ is bounded for any compact $\mathcal{K} \subseteq \mathsf{M}$. Thus $\mathcal{B}$ is also a bounded subset of $\Gamma^{\mathrm{hol}}(\mathsf{E})$. Since $\Gamma^{\mathrm{hol}}(\mathsf{E})$ is nuclear [47, Theorem II.8.2], we conclude that $\mathcal{B}$ is precompact in $\Gamma^{\nu}(\mathsf{E}|\mathcal{U})$, giving compactness of the inclusion.

(iii) Since the topology of $\Gamma^{\text{hol}}(\mathsf{E})$ is metrisable [47, Theorem II.8.2], it suffices to show that the restriction of a convergent sequence in $\Gamma^{\text{hol}}(\mathsf{E})$ to $\mathcal{U}$ converges uniformly. This, however, follows since $\text{cl}(\mathcal{U})$ is compact. □

Now, no longer fixing j, the lemma ensures that we have a sequence of inclusions

$$\Gamma^{\text{hol},\mathbb{R}}_{\text{bdd}}(\overline{\mathsf{E}}|\overline{\mathcal{U}}_{\mathcal{K},1}) \subseteq \Gamma^{\text{hol},\mathbb{R}}(\overline{\mathsf{E}}|\overline{\mathcal{U}}_{\mathcal{K},1}) \subseteq \Gamma^{\text{hol},\mathbb{R}}_{\text{bdd}}(\overline{\mathsf{E}}|\overline{\mathcal{U}}_{\mathcal{K},2}) \subseteq \cdots \subseteq \Gamma^{\text{hol},\mathbb{R}}(\overline{\mathsf{E}}|\overline{\mathcal{U}}_{\mathcal{K},j}) \subseteq \Gamma^{\text{hol},\mathbb{R}}_{\text{bdd}}(\overline{\mathsf{E}}|\overline{\mathcal{U}}_{\mathcal{K},j+1}) \subseteq \cdots .$$

The inclusion $\Gamma^{\text{hol},\mathbb{R}}(\overline{\mathcal{U}}_{\mathcal{K},j}) \subseteq \Gamma^{\text{hol},\mathbb{R}}_{\text{bdd}}(\overline{\mathcal{U}}_{\mathcal{K},j+1})$, $j \in \mathbb{Z}_{>0}$, is by restriction from $\overline{\mathcal{U}}_{\mathcal{K},j}$ to the smaller $\overline{\mathcal{U}}_{\mathcal{K},j+1}$, keeping in mind that $\text{cl}(\overline{\mathcal{U}}_{\mathcal{K},j+1}) \subseteq \overline{\mathcal{U}}_{\mathcal{K},j}$. Each of the inclusion maps in the preceding sequence is continuous by Lemma 2.9.

For $j \in \mathbb{Z}_{>0}$ define

$$\begin{aligned} r_{\mathcal{K},j} \colon \Gamma^{\text{hol},\mathbb{R}}_{\text{bdd}}(\overline{\mathsf{E}}|\overline{\mathcal{U}}_{\mathcal{K},j}) &\to \mathscr{G}^{\text{hol},\mathbb{R}}_{\mathcal{K},\overline{\mathsf{E}}} \\ \overline{\xi} &\mapsto [\overline{\xi}]_{\mathcal{K}}. \end{aligned} \tag{2.3}$$

Now one can show that the locally convex direct limit topologies induced on $\mathscr{G}^{\text{hol},\mathbb{R}}_{\mathcal{K},\overline{\mathsf{E}}}$ by the directed system $(\Gamma^{\text{hol},\mathbb{R}}(\overline{\mathsf{E}}|\overline{\mathcal{U}}))_{\overline{\mathcal{U}}\in\overline{\mathscr{N}}_{\mathcal{K}}}$ of Fréchet spaces and by the directed system $(\Gamma^{\text{hol},\mathbb{R}}_{\text{bdd}}(\overline{\mathsf{E}}|\overline{\mathcal{U}}_{\mathcal{K},j}))_{j\in\mathbb{Z}_{>0}}$ of Banach spaces agree [47, Theorem 8.4].

Let us give some properties of this topology.

Proposition 2.10 (Properties of $\mathscr{G}^{\text{hol},\mathbb{R}}_{\mathcal{K},\overline{\mathsf{E}}}$) *Let $\pi_{\mathsf{E}} \colon \mathsf{E} \to \mathsf{M}$ be a real analytic vector bundle and let $\mathcal{K} \subseteq \mathsf{M}$ be compact and connected. Then the direct limit topology described above for $\mathscr{G}^{\text{hol},\mathbb{R}}_{\mathcal{K},\overline{\mathsf{E}}}$ has the following properties:*

(i) it is boundedly retractive (meaning that the direct limit is boundedly retractive);
(ii) it is Hausdorff;
(iii) it is complete;
(iv) it is sequential;
(v) it is Suslin;
(vi) it is nuclear;
(vii) it is ultrabornological;
(viii) it is webbed.

Proof

(i) Let $(\mathcal{U}_j)_{j\in\mathbb{Z}_{>0}}$ be a cofinal sequence in $\overline{\mathscr{N}}_{\mathcal{K}}$, as in Lemma 2.8. Since $\mathcal{K}$ is connected, the neighbourhoods $\mathcal{U}_j$ of Lemma 2.8 can be chosen to also be connected. Thus the inclusion mapping

$$\Gamma^r(\mathsf{E}|\mathcal{U}_j) \hookrightarrow \Gamma^r(\mathsf{E}|\mathcal{U}_{j+1})$$

is injective by the Identity Theorem. By Lemma 2.9, the inclusion map

$$\Gamma^r_{\mathrm{bdd}}(\mathsf{E}|\mathcal{U}_j) \hookrightarrow \Gamma^r(\mathsf{E}|\mathcal{U}_j)$$

is compact. Thus, by Jarchow [37, Proposition 17.1.1], the inclusion

$$\Gamma^r_{\mathrm{bdd}}(\mathsf{E}|\mathcal{U}_j) \hookrightarrow \Gamma^r_{\mathrm{bdd}}(\mathsf{E}|\mathcal{U}_{j+1})$$

is compact. Thus the direct limit topology is compact, and so is boundedly retractive as follows from [44, Theorem 6'].

(ii) By definition, boundedly retractive direct limits are regular. Thus Hausdorffness of the direct limit topology follows from [37, Proposition 4.5.3].

(iii) As we saw in the proof of part (i), $\mathscr{G}^{\mathrm{hol},\mathbb{R}}_{\mathcal{K},\overline{\mathsf{E}}}$ is a locally convex direct limit with compact connecting maps. Completeness then follows from [44, Theorem 6'].

(iv) Fréchet spaces are sequential and Hausdorff direct limits of sequential spaces are sequential [22, Corollary 1.7].

(v) This follows from Proposition 1.6(i).

(vi) First we note that the space of sections of an holomorphic vector bundle with the compact-open topology is nuclear, essentially by [47, Theorem 8.2]. Thus $\mathscr{G}^{\mathrm{hol},\mathbb{R}}_{\mathcal{K},\overline{\mathsf{E}}}$ is a countable Hausdorff locally convex direct limit of nuclear spaces. Its nuclearity then follows from [37, Corollary 21.2.3].

(vii) First we note that Fréchet spaces are ultrabornological [37, Corollary 13.1.4]. The space of sections of an holomorphic vector bundle is a Fréchet space by [47, Theorem 8.2], and so is ultrabornological. Thus $\mathscr{G}^{\mathrm{hol},\mathbb{R}}_{\mathcal{K},\overline{\mathsf{E}}}$ is ultrabornological by [37, Corollary 13.1.6] as it is a countable Hausdorff locally convex direct limit.

(viii) The space of sections of an holomorphic vector bundle is a Fréchet space by [47, Theorem 8.2], and so a webbed space by [37, Proposition 5.2.2]. Thus $\mathscr{G}^{\mathrm{hol},\mathbb{R}}_{\mathcal{K},\overline{\mathsf{E}}}$ is webbed by [37, Corollary 5.3.3(b)] as it is a countable Hausdorff locally convex direct limit. □

2.2.4 *An Inverse Limit Topology for the Space of Real Analytic Sections*

Now we shall use the constructions from the preceding section to easily arrive at a topology on $\Gamma^\omega(\mathsf{E})$ induced by the locally convex topologies on the spaces $\mathscr{G}^{\mathrm{hol},\mathbb{R}}_{\mathcal{K},\overline{\mathsf{E}}}$, $\mathcal{K} \subseteq \mathsf{M}$ compact.

For a compact set $\mathcal{K} \subseteq \mathsf{M}$ we have an inclusion $i_{\mathcal{K}} : \Gamma^{\omega}(\mathsf{E}) \to \mathscr{G}^{\mathrm{hol},\mathbb{R}}_{\mathcal{K},\overline{\mathsf{E}}}$ defined as follows. If $\xi \in \Gamma^{\omega}(\mathsf{E})$, then ξ admits an holomorphic extension $\overline{\xi}$ defined on a neighbourhood $\overline{\mathcal{U}} \subseteq \overline{\mathsf{M}}$ of M [14, Lemma 5.40]. Since $\overline{\mathcal{U}} \in \overline{\mathscr{N}}_{\mathcal{K}}$ we define $i_{\mathcal{K}}(\xi) = [\overline{\xi}]_{\mathcal{K}}$. We can consider the set $\mathscr{K}_{\mathsf{M}}$ of compact subsets of M to be a directed set by inclusion: if $\mathcal{K}_1, \mathcal{K}_2 \in \mathscr{K}_{\mathsf{M}}$, we take $\mathcal{K}_1 \preceq \mathcal{K}_2$ if $\mathcal{K}_1 \subseteq \mathcal{K}_2$. We then have the inverse system

$$\left((\mathscr{G}^{\mathrm{hol},\mathbb{R}}_{\mathcal{K},\overline{\mathsf{E}}})_{\mathcal{K}\in\mathscr{K}_{\mathsf{M}}}, (\pi_{\mathcal{K}_2,\mathcal{K}_1})_{\mathcal{K}_1\subseteq\mathcal{K}_2}\right), \tag{2.4}$$

where we have

$$\begin{aligned} \pi_{\mathcal{K}_2,\mathcal{K}_1} : \mathscr{G}^{\mathrm{hol},\mathbb{R}}_{\mathcal{K}_2,\overline{\mathsf{E}}} &\to \mathscr{G}^{\mathrm{hol},\mathbb{R}}_{\mathcal{K}_1,\overline{\mathsf{E}}} \\ [\overline{\xi}]_{\mathcal{K}_2} &\mapsto [\overline{\xi}]_{\mathcal{K}_1}, \end{aligned}$$

bearing in mind that $\overline{\mathscr{N}}_{\mathcal{K}_2} \subseteq \overline{\mathscr{N}}_{\mathcal{K}_1}$ if $\mathcal{K}_1 \subseteq \mathcal{K}_2$. We shall denote by $\varprojlim_{\mathcal{K}\in\mathscr{K}_{\mathsf{M}}} \mathscr{G}^{\mathrm{hol},\mathbb{R}}_{\mathcal{K},\overline{\mathsf{E}}}$ the corresponding inverse limit in the category of locally convex topological vector spaces. The corresponding induced mappings from the direct limit we denote by

$$\pi_{\mathcal{K}} : \varprojlim_{\mathcal{K}'\in\mathscr{K}_{\mathsf{M}}} \mathscr{G}^{\mathrm{hol},\mathbb{R}}_{\mathcal{K}',\overline{\mathsf{E}}} \to \mathscr{G}^{\mathrm{hol},\mathbb{R}}_{\mathcal{K},\overline{\mathsf{E}}}.$$

We then have the following result.

Lemma 2.11 (Real Analytic Sections as Inverse Limits) *There is a natural* $\mathbb{R}$-*vector space isomorphism between* $\Gamma^{\omega}(\mathsf{E})$ *and* $\varprojlim_{\mathcal{K}\in\mathscr{K}_{\mathsf{M}}} \mathscr{G}^{\mathrm{hol},\mathbb{R}}_{\mathcal{K},\overline{\mathsf{E}}}$.

Proof By Lemma 2.7, we have a vector space isomorphism $\mathscr{G}^{\mathrm{hol},\mathbb{R}}_{\mathsf{M},\overline{\mathsf{E}}} \simeq \Gamma^{\omega}(\mathsf{M})$. Thus it suffices to establish a vector space isomorphism

$$\mathscr{G}^{\mathrm{hol},\mathbb{R}}_{\mathsf{M},\overline{\mathsf{E}}} \simeq \varprojlim_{\mathcal{K}\in\mathscr{K}_{\mathsf{M}}} \mathscr{G}^{\mathrm{hol},\mathbb{R}}_{\mathcal{K},\overline{\mathsf{E}}}.$$

We define a linear mapping

$$\begin{aligned} i_{\mathcal{K}} : \mathscr{G}^{\mathrm{hol},\mathbb{R}}_{\mathsf{M},\overline{\mathsf{E}}} &\to \mathscr{G}^{\mathrm{hol},\mathbb{R}}_{\mathcal{K},\overline{\mathsf{E}}} \\ [\overline{\xi}]_{\mathsf{M}} &\mapsto [\overline{\xi}]_{\mathcal{K}}. \end{aligned}$$

We claim that this mapping is injective. Indeed, knowledge of a germ $[\overline{\xi}]_{\mathcal{K}}$ implies, in particular, knowledge of the germ of $\overline{\xi}$ at every $x \in \mathcal{K}$. This, then, uniquely prescribes the germ $[\overline{\xi}]_{\mathsf{M}}$ by the Identity Theorem (here we use explicitly our assumption that M is connected). Now we note that the universal property of inverse limits (used here in the category of $\mathbb{R}$-vector spaces) gives the unique vertical linear mapping in the following diagram:

$$\begin{array}{ccc} \varprojlim_{\mathcal{K}'\in\mathscr{K}_{\mathsf{M}}} \mathscr{G}^{\mathrm{hol},\mathbb{R}}_{\mathcal{K}',\overline{\mathsf{E}}} & \xrightarrow{\pi_{\mathcal{K}}} & \mathscr{G}^{\mathrm{hol},\mathbb{R}}_{\mathcal{K},\overline{\mathsf{E}}} \\ {\scriptstyle \iota_{\mathsf{M}}}\uparrow & \nearrow {\scriptstyle i_{\mathcal{K}}} & \\ \mathscr{G}^{\mathrm{hol},\mathbb{R}}_{\mathsf{M},\overline{\mathsf{E}}} & & \end{array}$$

Since $i_{\mathcal{K}}$ is injective, so too is $\pi_{\mathcal{K}} \circ \iota_{\mathsf{M}}$. This directly gives injectivity of ι_{M}. To show that ι_{M} is surjective, we note that an element of the inverse limit is an element of the product [37, Proposition 2.6.1], and so we can write an element of the inverse limit as $([\overline{\xi}_{\mathcal{K}}]_{\mathcal{K}})_{\mathcal{K}\in\mathscr{K}_{\mathsf{M}}}$, where $[\overline{\xi}_{\mathcal{K}}]_{\mathcal{K}} \in \mathscr{G}^{\mathrm{hol},\mathbb{R}}_{\mathcal{K},\overline{\mathsf{E}}}$. We note that the germ $[\overline{\xi}_{\mathcal{K}}]_{\mathcal{K}}$ specifies the germ at every $x \in \mathcal{K}$. Moreover, the definition of the inverse limit ensures that the germs of $[\overline{\xi}_{\mathcal{K}_1}]_{\mathcal{K}_1}$ and $[\overline{\xi}_{\mathcal{K}_2}]_{\mathcal{K}_2}$ agree at each $x \in \mathcal{K}_1 \cap \mathcal{K}_2$. Since M is covered by the union of its compact subsets, this shows that an element of the inverse limit uniquely prescribes a germ of a section at every $x \in \mathsf{M}$; thus this uniquely prescribes the germ of a global section (here we use the fact that the presheaf of real analytic sections is a sheaf). □

Thus the inverse limit topology for $\varprojlim_{\mathcal{K}\in\mathscr{K}_{\mathsf{M}}} \mathscr{G}^{\mathrm{hol},\mathbb{R}}_{\mathcal{K},\overline{\mathsf{E}}}$ induces a locally convex topology for $\Gamma^{\omega}(\mathsf{M})$ that we call the ***inverse limit topology***, naturally enough.

Suppose now that we have a compact exhaustion $(\mathcal{K}_j)_{j\in\mathbb{Z}_{>0}}$ of M. Since $\overline{\mathscr{N}_{\mathcal{K}_{j+1}}} \subseteq \overline{\mathscr{N}_{\mathcal{K}_j}}$ we have a projection

$$\begin{aligned} \pi_j \colon \mathscr{G}^{\mathrm{hol},\mathbb{R}}_{\mathcal{K}_{j+1},\overline{\mathsf{E}}} &\to \mathscr{G}^{\mathrm{hol},\mathbb{R}}_{\mathcal{K}_j,\overline{\mathsf{E}}} \\ [\overline{\xi}]_{\mathcal{K}_{j+1}} &\mapsto [\overline{\xi}]_{\mathcal{K}_j}. \end{aligned}$$

One can check that, as $\mathbb{R}$-vector spaces, the inverse limit of the inverse system $(\mathscr{G}^{\mathrm{hol},\mathbb{R}}_{\mathcal{K}_j,\overline{\mathsf{E}}})_{j\in\mathbb{Z}_{>0}}$ is isomorphic to that for the inverse system (2.4). This shows that the inverse limit topology can be obtained as a *countable* inverse limit of countable direct limits.

2.2.5 Properties of the C^{ω}-Topology

We now have two topologies for $\Gamma^{\omega}(\mathsf{E})$, the direct limit topology of Sect. 2.2.2 and the inverse limit (of direct limit topologies) topology of Sect. 2.2.4. Let us consider the relationship between these two topologies.

We first show that the mappings $i_{\mathcal{K}}$ from the proof of Lemma 2.11 are continuous.

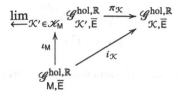

Lemma 2.12 (Continuity of Mappings of Germs) *For $\mathcal{K} \subseteq \mathsf{M}$ compact, the mapping*

$$i_{\mathcal{K}} \colon \mathscr{G}^{\mathrm{hol},\mathbb{R}}_{\mathsf{M},\overline{\mathsf{E}}} \to \mathscr{G}^{\mathrm{hol},\mathbb{R}}_{\mathcal{K},\overline{\mathsf{E}}}$$
$$[\overline{\xi}]_{\mathsf{M}} \mapsto [\overline{\xi}]_{\mathcal{K}}$$

is continuous if the domain and codomain have their direct limit topologies.

Proof By definition of the direct limit topology for $\mathscr{G}^{\mathrm{hol},\mathbb{R}}_{\mathsf{M},\overline{\mathsf{E}}}$, the lemma will be proved if we can show that, for any $\overline{\mathcal{V}} \in \overline{\mathscr{N}}_{\mathsf{M}}$, in the diagram

$$\begin{array}{ccc} \Gamma^{\mathrm{hol},\mathbb{R}}(\overline{\mathsf{E}}|\overline{\mathcal{V}}) & \xrightarrow{r_{\overline{\mathcal{V}},\mathsf{M}}} & \mathscr{G}^{\mathrm{hol},\mathbb{R}}_{\mathsf{M},\overline{\mathsf{E}}} \\ & \searrow^{i_{\mathcal{K}} \circ r_{\overline{\mathcal{V}},\mathsf{M}}} & \downarrow i_{\mathcal{K}} \\ & & \mathscr{G}^{\mathrm{hol},\mathbb{R}}_{\mathcal{K},\overline{\mathsf{E}}} \end{array}$$

the diagonal mapping is continuous. Let $(\overline{\mathcal{U}}_j)_{j\in\mathbb{Z}_{>0}}$ be a sequence in $\overline{\mathscr{N}}_{\mathcal{K}}$ as in Lemma 2.8. By part (iii) of that lemma, let $j_0 \in \mathbb{Z}_{>0}$ be such that $\overline{\mathcal{U}}_{j_0} \subseteq \overline{\mathcal{V}}$. Then we have the diagram

$$\begin{array}{ccc} \Gamma^{\mathrm{hol},\mathbb{R}}(\overline{\mathsf{E}}|\overline{\mathcal{V}}) & \xrightarrow{r_{\overline{\mathcal{V}},\mathsf{M}}} & \mathscr{G}^{\mathrm{hol},\mathbb{R}}_{\mathsf{M},\overline{\mathsf{E}}} \\ {\scriptstyle r_{\overline{\mathcal{V}},\overline{\mathcal{U}}_{j_0}}}\downarrow & & \downarrow i_{\mathcal{K}} \\ \Gamma^{\mathrm{hol},\mathbb{R}}(\overline{\mathsf{E}}|\overline{\mathcal{U}}_{j_0}) & \xrightarrow[r_{\overline{\mathcal{U}}_{j_0},\mathcal{K}}]{} & \mathscr{G}^{\mathrm{hol},\mathbb{R}}_{\mathcal{K},\overline{\mathsf{E}}} \end{array}$$

which is directly verified to commute in the category of $\mathbb{R}$-vector spaces. Moreover, the restriction mapping $r_{\overline{\mathcal{V}},\overline{\mathcal{U}}_{j_0}}$ is continuous in the compact-open topologies, as is directly verified, and the restriction mapping $r_{\overline{\mathcal{U}}_{j_0},\mathcal{K}}$ is continuous by definition of the direct limit topology for $\mathscr{G}^{\mathrm{hol},\mathbb{R}}_{\mathcal{K},\overline{\mathsf{E}}}$. This gives the lemma. □

Now we can prove the following result.

Proposition 2.13 (The Inverse Limit Topology Is Coarser Than the Direct Limit Topology) *The identity map on $\Gamma^{\omega}(\mathsf{E})$ is continuous if the domain has the direct limit topology and the codomain has the inverse limit topology.*

Proof For each compact $\mathcal{K} \subseteq \mathsf{M}$, we have the diagram

$$\begin{array}{ccc} \varprojlim_{\mathcal{K}' \in \mathscr{K}_\mathsf{M}} \mathscr{G}^{\mathrm{hol},\mathbb{R}}_{\mathcal{K}',\overline{\mathsf{E}}} & \xrightarrow{\pi_\mathcal{K}} & \mathscr{G}^{\mathrm{hol},\mathbb{R}}_{\mathcal{K},\overline{\mathsf{E}}} \\ \uparrow \mathrm{id} & \nearrow i_\mathcal{K} & \\ \mathscr{G}^{\mathrm{hol},\mathbb{R}}_{\mathsf{M},\overline{\mathsf{E}}} & & \end{array}$$

in the category of $\mathbb{R}$-vector spaces. If $\mathscr{G}^{\mathrm{hol},\mathbb{R}}_{\mathsf{M},\overline{\mathsf{E}}}$ has the direct limit topology, then the diagonal arrow is continuous, as per the preceding lemma. If $\varprojlim_{\mathcal{K} \in \mathscr{K}_\mathsf{M}} \mathscr{G}^{\mathrm{hol},\mathbb{R}}_{\mathcal{K},\overline{\mathsf{E}}}$ has the inverse limit topology, then this topology is the initial topology associated with the mappings $\pi_\mathcal{K}$. By the universal property of the inverse limit topology, the vertical arrow is continuous. Bearing in mind Lemmata 2.7 and 2.11, we get the proposition. □

One can now wonder whether the identity map on $\Gamma^\omega(\mathsf{E})$ is, in fact, open, which would imply the equality of the direct limit and inverse limit topologies. If there were a suitable Open Mapping Theorem from the direct limit topology to the inverse limit topology, then this would give the result. However, the common Open Mapping Theorems fail to give the result. For example, the De Wilde Open Mapping Theorem would require that the direct limit topology be webbed (something that does not follow easily from the definition of the direct limit topology[7]) and that the inverse limit topology be ultrabornological (something that does not follow easily from the definition of the inverse limit topology[8]). Thus the properties of the domain and codomain are the opposite of what they need to be to apply the De Wilde Open Mapping Theorem. There are various ways in which one can establish the openness of the identity map from Proposition 2.13, and a nice discussion of these ideas can be found in the notes [17]. The original approach in [50, Theorem 1.2(a)] was by showing that $\cup_{j \in \mathbb{Z}_{>0}} (\mathscr{G}^{\mathrm{hol}}_{\mathcal{K}_j,\overline{\mathsf{E}}})^*$ is a dense subspace of the dual of $\Gamma^\omega(\mathsf{E})$ equipped with the direct limit topology, using earlier results in [49] on analytic functionals. A modern approach, using homological methods, equates an inverse limit being ultrabornological with the vanishing of Proj^1, where Proj is a functor on inverse systems devised by Palamodov [59]; see also [70]. In all cases, showing equality of the two topologies is not straightforward and we shall just accept this equality as true. In any case, we call the resulting topology, however it is defined, the ***C^ω-topology***.

The two constructions of the C^ω-topology by Martineau permit some fairly easy conclusions about properties of this topology.

---

[7] The difficulty is that the direct limit is not countable in this case.

[8] The difficulty is that subspaces of ultrabornological spaces, even closed subspaces, are not necessarily ultrabornological.

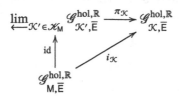

Proposition 2.14 (Properties of $\Gamma^\omega(\mathsf{E})$) *For a real analytic vector bundle $\pi_{\mathsf{E}}\colon \mathsf{E} \to \mathsf{M}$, the C^ω-topology for $\Gamma^\omega(\mathsf{E})$ has the following properties:*

(i) it is Hausdorff;
(ii) it is complete;
(iii) it is not metrisable;
(iv) it is sequential;
(v) it is Suslin;
(vi) it is nuclear;
(vii) it is ultrabornological;
(viii) it is webbed.

Proof We shall implicitly make use of the conclusions of Proposition 2.10 throughout the proof. We will also make use of both of our equivalent constructions of the real analytic topology, one from Sect. 2.2.2 and one from Sect. 2.2.4.

(i) The C^ω-topology is an inverse limit of Hausdorff topologies, and so is Hausdorff by [37, Proposition 2.6.1(a)], noting that a product of Hausdorff topologies is Hausdorff and that a subspace of a Hausdorff space is Hausdorff.
(ii) The C^ω-topology is an inverse limit of Hausdorff complete locally convex topological vector spaces, and so is complete by [37, Corollary 3.2.7].
(iii) This is a nontrivial property of the C^ω-topology, and we refer to [20] and [67, Theorem 10] for details.
(iv) Fréchet spaces are sequential and Hausdorff direct limits of sequential spaces are sequential [22, Corollary 1.7].
(v) The C^ω-topology is a countable inverse limit of Suslin locally convex topological vector spaces, and so is Suslin by Proposition 1.6(ii).
(vi) The C^ω-topology is a countable inverse limit of nuclear locally convex topological vector spaces, and so is nuclear by [37, Corollary 21.2.3].
(vii) The C^ω-topology is an Hausdorff direct limit of ultrabornological spaces, and so is ultrabornological by [37, Corollary 13.1.5].
(viii) The C^ω-topology is a countable inverse limit of webbed spaces, and so is webbed by [37, Corollary 5.3.3(a)]. □

2.3 Constructions with Jet Bundles

In this section we provide some constructions that relate jet bundles and connections; these constructions will feature prominently in the book, starting in Sect. 2.4 where we use them to define seminorms for the C^ω-topology. Later, substantial portions of the text will be dedicated to finding certain uniform estimates for jet bundle norms; these estimates will be phrased for tensors arising from constructions with jet bundles and connections. It is fair to say, therefore, that constructions with jet bundles are at the core of our geometric approach.

2.3.1 Decompositions for Jet Bundles

Throughout this section, we will consider $r \in \{\infty, \omega\}$.

Let $\pi_{\mathsf{E}} \colon \mathsf{E} \to \mathsf{M}$ be a C^r-vector bundle. As per Lemma 2.4, we suppose that we have a linear connection $\nabla^{\pi_{\mathsf{E}}}$ on the vector bundle E and an affine connection ∇^{M} on M, all data being of class C^r. We then have induced connections, that we also denote by $\nabla^{\pi_{\mathsf{E}}}$ and ∇^{M}, in various tensor bundles of E and TM, respectively. As in (1.8), the connections $\nabla^{\pi_{\mathsf{E}}}$ and ∇^{M} extend naturally to connections in various tensor products of TM and E, all of these being denoted by $\nabla^{\mathsf{M},\pi_{\mathsf{E}}}$. Note that, if $\xi \in \Gamma^\infty(\mathsf{E})$, then

$$\nabla^{\mathsf{M},\pi_{\mathsf{E}},m}\xi \triangleq \underbrace{\nabla^{\mathsf{M},\pi_{\mathsf{E}}} \cdots (\nabla^{\mathsf{M},\pi_{\mathsf{E}}}}_{m-1 \text{ times}}(\nabla^{\pi_{\mathsf{E}}}\xi)) \in \Gamma^\infty(\mathsf{T}^m(\mathsf{T}^*\mathsf{M}) \otimes \mathsf{E}). \tag{2.5}$$

Now, given $\xi \in \Gamma^\infty(\mathsf{E})$ and $m \in \mathbb{Z}_{\geq 0}$, we define

$$D^m_{\nabla^{\mathsf{M}},\nabla^{\pi_{\mathsf{E}}}}(\xi) = \mathrm{Sym}_m \otimes \mathrm{id}_{\mathsf{E}}(\nabla^{\mathsf{M},\pi_{\mathsf{E}},m}\xi) \in \Gamma^\infty(\mathsf{S}^m(\mathsf{T}^*\mathsf{M}) \otimes \mathsf{E}), \tag{2.6}$$

We take the convention that $D^0_{\nabla^{\mathsf{M}},\nabla^{\pi_{\mathsf{E}}}}(\xi) = \xi$.

The following lemma is then key for our presentation.

Lemma 2.15 (Decomposition of Jet Bundles) *Let $r \in \{\infty, \omega\}$ and let $\pi_{\mathsf{E}} \colon \mathsf{E} \to \mathsf{M}$ be a C^r-vector bundle. The map*

$$\begin{aligned} S^m_{\nabla^{\mathsf{M}},\nabla^{\pi_{\mathsf{E}}}} \colon \mathsf{J}^m\mathsf{E} &\to \bigoplus_{j=0}^m (\mathsf{S}^j(\mathsf{T}^*\mathsf{M}) \otimes \mathsf{E}) \\ j_m\xi(x) &\mapsto (\xi(x), D^1_{\nabla^{\mathsf{M}},\nabla^{\pi_{\mathsf{E}}}}(\xi)(x), \dots, D^m_{\nabla^{\mathsf{M}},\nabla^{\pi_{\mathsf{E}}}}(\xi)(x)) \end{aligned}$$

is an isomorphism of C^r-vector bundles, and, for each $m \in \mathbb{Z}_{>0}$, the diagram

$$\begin{array}{ccc} \mathsf{J}^{m+1}\mathsf{E} & \xrightarrow{S^{m+1}_{\nabla^{\mathsf{M}},\nabla^{\pi_{\mathsf{E}}}}} & \oplus_{j=0}^{m+1}(\mathsf{S}^j(\mathsf{T}^*\mathsf{M}) \otimes \mathsf{E}) \\ {\scriptstyle \pi^{m+1}_{\mathsf{E},m}} \downarrow & & \downarrow {\scriptstyle \mathrm{pr}^{m+1}_m} \\ \mathsf{J}^m\mathsf{E} & \xrightarrow[S^m_{\nabla^{\mathsf{M}},\nabla^{\pi_{\mathsf{E}}}}]{} & \oplus_{j=0}^{m}(\mathsf{S}^j(\mathsf{T}^*\mathsf{M}) \otimes \mathsf{E}) \end{array}$$

commutes, where pr^{m+1}_m is the obvious projection, stripping off the last component of the direct sum.

Proof We prove the result by induction on m. For $m = 0$ the result is a tautology. For $m = 1$, as in [43, §17.1], we have a vector bundle mapping $S_{\nabla^{\pi_{\mathsf{E}}}} : \mathsf{E} \to \mathsf{J}^1\mathsf{E}$ over id_{M} that determines the connection $\nabla^{\pi_{\mathsf{E}}}$ by

$$\nabla^{\pi_{\mathsf{E}}}\xi(x) = j_1\xi(x) - S_{\nabla^{\pi_{\mathsf{E}}}}(\xi(x)). \tag{2.7}$$

Let us show that $S^1_{\nabla^{\mathsf{M}},\nabla^{\pi\mathsf{E}}}$ is well-defined. Thus let $\xi, \eta \in \Gamma^\infty(\mathsf{E})$ be such that $j_1\xi(x) = j_1\eta(x)$. Then, clearly, $\xi(x) = \eta(x)$, and the formula (2.7) shows that $\nabla^{\pi\mathsf{E}}\xi(x) = \nabla^{\pi\mathsf{E}}\eta(x)$, and so $S^1_{\nabla^{\mathsf{M}},\nabla^{\pi\mathsf{E}}}$ is indeed well defined. It is clearly linear on fibres, so it remains to show that it is an isomorphism. This will follow from dimension counting if it is injective. However, if $S^1_{\nabla^{\mathsf{M}},\nabla^{\pi\mathsf{E}}}(j_1\xi(x)) = 0$ then $j_1\xi(x) = 0$ by (2.7).

For the induction step, we begin with a sublemma.

Sublemma 1 *Let* F *be a field and consider the following commutative diagram of finite-dimensional* F*-vector spaces with exact rows and columns:*

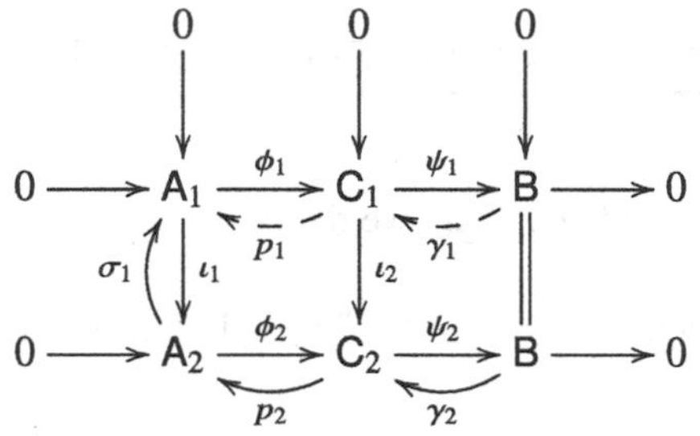

If there exists a mapping $\gamma_2 \in \mathrm{Hom}_{\mathsf{F}}(\mathsf{B}; \mathsf{C}_2)$ *such that* $\psi_2 \circ \gamma_2 = \mathrm{id}_{\mathsf{B}}$ *(with* $p_2 \in \mathrm{Hom}_{\mathsf{F}}(\mathsf{C}_2; \mathsf{A}_2)$ *the corresponding projection), then there exists a unique mapping* $\gamma_1 \in \mathrm{Hom}_{\mathsf{F}}(\mathsf{B}; \mathsf{C}_1)$ *such that* $\psi_1 \circ \gamma_1 = \mathrm{id}_{\mathsf{B}}$ *and such that* $\gamma_2 = \iota_2 \circ \gamma_1$*. There is also induced a projection* $p_1 \in \mathrm{Hom}_{\mathsf{F}}(\mathsf{C}_1; \mathsf{A}_1)$*.*

Moreover, if there additionally exists a mapping $\sigma_1 \in \mathrm{Hom}_{\mathsf{F}}(\mathsf{A}_2; \mathsf{A}_1)$ *such that* $\sigma_1 \circ \iota_1 = \mathrm{id}_{\mathsf{A}_1}$*, then the projection* p_1 *is uniquely determined by the condition* $p_1 = \sigma_1 \circ p_2 \circ \iota_2$*.*

Proof We begin by extending the diagram to one of the form

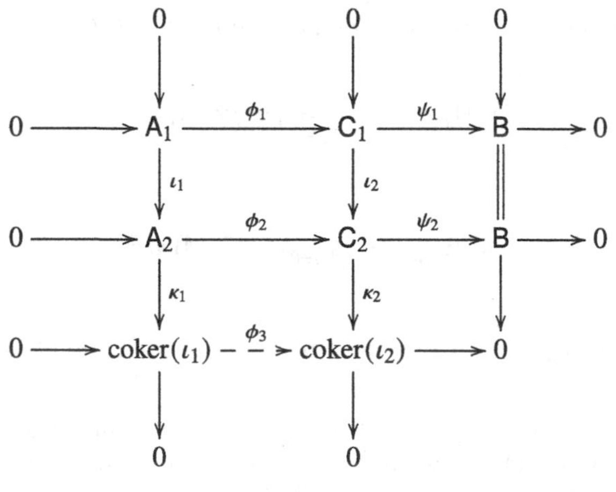

also with exact rows and columns. We claim that there is a natural mapping ϕ_3 between the cokernels, as indicated by the dashed arrow in the diagram, and that ϕ_3 is, moreover, an isomorphism. Suppose that $u_2 \in \text{image}(\iota_1)$ and let $u_1 \in \mathsf{A}_1$ be such that $\iota_1(u_1) = u_2$. By commutativity of the diagram, we have

$$\phi_2(u_2) = \phi_2 \circ \iota_1(u_1) = \iota_2 \circ \phi_1(u_1),$$

showing that $\phi_2(\text{image}(\iota_1)) \subseteq \text{image}(\iota_2)$. We thus have a well-defined homomorphism

$$\begin{aligned} \phi_3 \colon \text{coker}(\iota_1) &\to \text{coker}(\iota_2) \\ u_2 + \text{image}(\iota_1) &\mapsto \phi_2(u_2) + \text{image}(\iota_2). \end{aligned}$$

We now claim that ϕ_3 is injective. Indeed,

$$\phi_3(u_2 + \text{image}(\iota_1)) = 0 \quad \Longrightarrow \quad \phi_2(u_2) \in \text{image}(\iota_2).$$

Thus let $v_1 \in \mathsf{C}_1$ be such that $\phi_2(u_2) = \iota_2(v_1)$. Thus

$$\begin{aligned} 0 = \psi_2 \circ \phi_2(u_2) &= \psi_2 \circ \iota_2(v_1) = \psi_1(v_1) \\ \Longrightarrow \quad v_1 &\in \ker(\psi_1) = \text{image}(\phi_1). \end{aligned}$$

Thus $v_1 = \phi_1(u_1')$ for some $u_1' \in \mathsf{A}_1$. Therefore,

$$\phi_2(u_2) = \iota_2 \circ \phi_1(u_1') = \phi_2 \circ \iota_1(u_1'),$$

and injectivity of ϕ_2 gives $u_2 \in \text{image}(\iota_1)$ and so $u_2 + \text{image}(\iota_1) = 0 + \text{image}(\iota_1)$, giving the desired injectivity of ϕ_3.

Now note that

$$\dim(\text{coker}(\iota_1)) = \dim(\mathsf{A}_2) - \dim(\mathsf{A}_1)$$

by exactness of the left column. Also,

$$\dim(\text{coker}(\iota_2)) = \dim(\mathsf{C}_2) - \dim(\mathsf{C}_1)$$

by exactness of the middle column. By exactness of the top and middle rows, we have

$$\dim(\mathsf{B}) = \dim(\mathsf{C}_2) - \dim(\mathsf{A}_2) = \dim(\mathsf{C}_1) - \dim(\mathsf{A}_1).$$

This proves that

$$\dim(\text{coker}(\iota_1)) = \dim(\text{coker}(\iota_2)).$$

Thus the homomorphism ϕ_3 is an isomorphism, as claimed.

Now we proceed with the proof, using the extended diagram, and identifying the bottom cokernels with the isomorphism ϕ_3. The existence of the stated homomorphism γ_2 means that the middle row in the diagram splits. Therefore, $\mathsf{C}_2 = \text{image}(\phi_2) \oplus \text{image}(\gamma_2)$. Thus there exists a well-defined projection $p_2 \in \text{Hom}_{\mathsf{F}}(\mathsf{C}_2; \mathsf{A}_2)$ such that $p_2 \circ \phi_2 = \text{id}_{\mathsf{A}_2}$ [31, Theorem 41.1].

We will now prove that $\text{image}(\gamma_2) \subseteq \text{image}(\iota_2)$. By commutativity of the diagram and since ψ_1 is surjective, if $w \in \mathsf{B}$ then there exists $v_1 \in \mathsf{C}_1$ such that $\psi_2 \circ \iota_2(v_1) = w$. Since $\psi_2 \circ \gamma_2 = \text{id}_{\mathsf{B}}$, we have

$$\psi_2 \circ \iota_2(v_1) = \psi_2 \circ \gamma_2(w) \quad \Longrightarrow \quad \iota_2(v_1) - \gamma_2(w) \in \ker(\psi_2) = \text{image}(\phi_2).$$

Let $u_2 \in \mathsf{A}_2$ be such that $\phi_2(u_2) = \iota_2(v_1) - \gamma_2(w)$. Since $p_2 \circ \phi_2 = \text{id}_{\mathsf{A}_2}$ we have

$$u_2 = p_2 \circ \iota_2(v_1) - p_2 \circ \gamma_2(w),$$

whence

$$\kappa_1(u_2) = \kappa_1 \circ p_2 \circ \iota_2(v_1) - \kappa_1 \circ p_2 \circ \gamma_2(w) = 0,$$

noting that (1) $\kappa_1 \circ p_2 = \kappa_2$ (by commutativity), (2) $\kappa_2 \circ \iota_2 = 0$ (by exactness), and (3) $p_2 \circ \gamma_2 = 0$ (by exactness). Thus $u_2 \in \ker(\kappa_1) = \text{image}(\iota_1)$. Let $u_1 \in \mathsf{A}_1$ be such that $\iota_1(u_1) = u_2$. We then have

$$\iota_2(v_1) - \gamma_2(w) = \phi_2 \circ \iota_1(u_1) = \iota_2 \circ \phi_1(u_1),$$

which gives $\gamma_2(w) \in \text{image}(\iota_2)$, as claimed.

Now we define $\gamma_1 \in \text{Hom}_{\mathsf{F}}(\mathsf{B}; \mathsf{C}_1)$ by asking that $\gamma_1(w) \in \mathsf{C}_1$ have the property that $\iota_2 \circ \gamma_1(w) = \gamma_2(w)$, this making sense since we just showed that $\text{image}(\gamma_2) \subseteq \text{image}(\iota_2)$. Moreover, since ι_2 is injective, the definition uniquely prescribes γ_1. Finally we note that

$$\psi_1 \circ \gamma_1 = \psi_2 \circ \iota_2 \circ \gamma_1 = \psi_2 \circ \gamma_2 = \text{id}_{\mathsf{B}},$$

as claimed.

To prove the final assertion, let us denote $\hat{p}_1 = \sigma_1 \circ p_2 \circ \iota_2$. We then have

$$\hat{p}_1 \circ \phi_1 = \sigma_1 \circ p_2 \circ \iota_2 \circ \phi_1 = \sigma_1 \circ p_2 \circ \phi_2 \circ \iota_1 = \sigma_1 \circ \iota_1 = \text{id}_{\mathsf{A}_1},$$

using commutativity. We also have

$$\hat{p}_1 \circ \gamma_1 = \sigma_1 \circ p_2 \circ \iota_2 \circ \gamma_1 = \sigma_1 \circ p_2 \circ \gamma_2 = 0.$$

The two preceding conclusions show that $\hat{p}_1$ is the projection defined by the splitting of the top row of the diagram, i.e., $\hat{p}_1 = p_1$. ▽

Now suppose that the lemma is true for $m \in \mathbb{Z}_{>0}$. For any $k \in \mathbb{Z}_{>0}$, we have a short exact sequence

$$0 \longrightarrow \mathsf{S}^k(\mathsf{T}^*\mathsf{M}) \otimes \mathsf{E} \xrightarrow{\epsilon_k} \mathsf{J}^k\mathsf{E} \xrightarrow{\pi^k_{\mathsf{E},k-1}} \mathsf{J}^{k-1}\mathsf{E} \longrightarrow 0$$

for which we refer to [64, Theorem 6.2.9]. Recall from [64, Definition 6.2.25] that we have an inclusion $\iota_{1,m}$ of $\mathsf{J}^{m+1}\mathsf{E}$ in $\mathsf{J}^1(\mathsf{J}^m\mathsf{E})$ by $j_{m+1}\xi(x) \mapsto j_1(j_m\xi(x))$ (see Sect. 3.2.4). We also have an induced injection

$$\hat{\iota}_{1,m} \colon \mathsf{S}^{m+1}(\mathsf{T}^*\mathsf{M}) \otimes \mathsf{E} \to \mathsf{T}^*\mathsf{M} \otimes \mathsf{J}^m\mathsf{E}$$

defined by the composition

$$\mathsf{S}^{m+1}(\mathsf{T}^*\mathsf{M}) \otimes \mathsf{E} \longrightarrow \mathsf{T}^*\mathsf{M} \otimes \mathsf{S}^m(\mathsf{T}^*\mathsf{M}) \otimes \mathsf{E} \xrightarrow{\mathrm{id} \otimes \epsilon_m} \mathsf{T}^*\mathsf{M} \otimes \mathsf{J}^m\mathsf{E}$$

Explicitly, the left arrow is defined by

$$\alpha^1 \odot \cdots \odot \alpha^{m+1} \otimes \xi \mapsto \sum_{j=1}^{m+1} \alpha^j \otimes \alpha^1 \odot \cdots \odot \alpha^{j-1} \odot \alpha^{j+1} \odot \cdots \odot \alpha^{m+1} \otimes \xi,$$

$\odot$ denoting the symmetric tensor product defined in (1.1); see Lemma 3.20. We thus have the following commutative diagram with exact rows and columns:

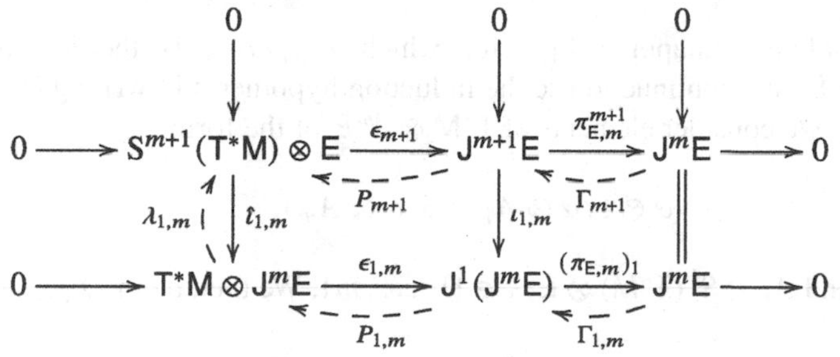

(2.8)

We shall define a connection on $(\pi_{\mathsf{E},m})_1 \colon \mathsf{J}^1(\mathsf{J}^m\mathsf{E}) \to \mathsf{J}^m\mathsf{E}$ which gives a splitting $\Gamma_{1,m}$ and $P_{1,m}$ of the lower row in the diagram. By the sublemma, this will give a splitting Γ_{m+1} and P_{m+1} of the upper row, and so give a projection from $\mathsf{J}^{m+1}\mathsf{E}$ onto $\mathsf{S}^{m+1}(\mathsf{T}^*\mathsf{M}) \otimes \mathsf{E}$, which will allow us to prove the induction step. To compute P_{m+1} from the sublemma, we shall also give a map $\lambda_{1,m}$ as in the diagram so that $\lambda_{1,m} \circ \hat{\iota}_{1,m}$ is the identity on $\mathsf{S}^{m+1}(\mathsf{T}^*\mathsf{M}) \otimes \mathsf{E}$.

We start, under the induction hypothesis, by making the identification

$$\mathsf{J}^m\mathsf{E} \simeq \bigoplus_{j=0}^{m} \mathsf{S}^j(\mathsf{T}^*\mathsf{M}) \otimes \mathsf{E},$$

and consequently writing a section of $\mathsf{J}^m\mathsf{E}$ as

$$x \mapsto (\xi(x), D^1_{\nabla^{\mathsf{M}},\nabla^{\pi_{\mathsf{E}}}}(\xi(x)), \dots, D^m_{\nabla^{\mathsf{M}},\nabla^{\pi_{\mathsf{E}}}}(\xi(x))).$$

We then have a connection $\nabla^{\pi_{\mathsf{E}},m}$ on $\mathsf{J}^m\mathsf{E}$ given by

$$\begin{aligned}\nabla^{\pi_{\mathsf{E}},m}_X&(\xi, D^1_{\nabla^{\mathsf{M}},\nabla^{\pi_{\mathsf{E}}}}(\xi), \dots, D^m_{\nabla^{\mathsf{M}},\nabla^{\pi_{\mathsf{E}}}}(\xi))\\ &= (\nabla^{\pi_{\mathsf{E}}}_X\xi, \nabla^{\mathsf{M},\pi_{\mathsf{E}}}_X D^1_{\nabla^{\mathsf{M}},\nabla^{\pi_{\mathsf{E}}}}(\xi), \dots, \nabla^{\mathsf{M},\pi_{\mathsf{E}}}_X D^m_{\nabla^{\mathsf{M}},\nabla^{\pi_{\mathsf{E}}}}(\xi)),\end{aligned}$$

cf. constructions from Sect. 3.2.4. Thus

$$\begin{aligned}\nabla^{\pi_{\mathsf{E}},m}&(\xi, D^1_{\nabla^{\mathsf{M}},\nabla^{\pi_{\mathsf{E}}}}(\xi), \dots, D^m_{\nabla^{\mathsf{M}},\nabla^{\pi_{\mathsf{E}}}}(\xi))\\ &= (\nabla^{\pi_{\mathsf{E}}}\xi, \nabla^{\mathsf{M},\pi_{\mathsf{E}}} D^1_{\nabla^{\mathsf{M}},\nabla^{\pi_{\mathsf{E}}}}(\xi), \dots, \nabla^{\mathsf{M},\pi_{\mathsf{E}}} D^m_{\nabla^{\mathsf{M}},\nabla^{\pi_{\mathsf{E}}}}(\xi)),\end{aligned}$$

which—according to the jet bundle characterisation of connections from [43, §17.1] and which we have already employed in (2.7)—gives the mapping $P_{1,m}$ in the diagram (2.8) as

$$\begin{aligned}P_{1,m}&(j_1(\xi, D^1_{\nabla^{\mathsf{M}},\nabla^{\pi_{\mathsf{E}}}}(\xi), \dots, D^m_{\nabla^{\mathsf{M}},\nabla^{\pi_{\mathsf{E}}}}(\xi)))\\ &= (\nabla^{\pi_{\mathsf{E}}}\xi, \nabla^{\mathsf{M},\pi_{\mathsf{E}}} D^1_{\nabla^{\mathsf{M}},\nabla^{\pi_{\mathsf{E}}}}(\xi), \dots, \nabla^{\mathsf{M},\pi_{\mathsf{E}}} D^m_{\nabla^{\mathsf{M}},\nabla^{\pi_{\mathsf{E}}}}(\xi)).\end{aligned}$$

Now we define a mapping $\lambda_{1,m}$ for which $\lambda_{1,m} \circ \hat{\iota}_{1,m}$ is the identity on $\mathsf{S}^{m+1}(\mathsf{T}^*\mathsf{M}) \otimes \mathsf{E}$. We continue to use the induction hypothesis in writing elements of $\mathsf{J}^m\mathsf{E}$, so that we consider elements of $\mathsf{T}^*\mathsf{M} \otimes \mathsf{J}^m\mathsf{E}$ of the form

$$(\alpha \otimes \xi, \alpha \otimes A_1, \dots, \alpha \otimes A_m),$$

for $\alpha \in \mathsf{T}^*\mathsf{M}$ and $A_k \in \mathsf{S}^k(\mathsf{T}^*\mathsf{M}) \otimes \mathsf{E}$, $k \in \{1, \dots, m\}$. We then define $\lambda_{1,m}$ by

$$\begin{aligned}\lambda_{1,m}(\alpha_0 \otimes \xi, \alpha_0 \otimes \alpha^1_1 \otimes \xi, \dots, \alpha_0 \otimes \alpha^1_m \odot \cdots \odot \alpha^m_m \otimes \xi)&\\ = \mathrm{Sym}_{m+1} \otimes \mathrm{id}_{\mathsf{E}}(\alpha_0 \otimes \alpha^1_m \odot \cdots \odot \alpha^m_m \otimes \xi).&\end{aligned}$$

Note that, with the form of $\mathsf{J}^m\mathsf{E}$ from the induction hypothesis, we have

$$\begin{aligned}&\hat{\iota}_{1,m}(\alpha^1 \odot \cdots \odot \alpha^{m+1} \otimes \xi)\\ &= \left(0, \dots, 0, \frac{1}{m+1}\sum_{j=1}^{m+1} \alpha^j \otimes \alpha^1 \odot \cdots \odot \alpha^{j-1} \odot \alpha^{j+1} \odot \cdots \odot \alpha^{m+1} \otimes \xi\right).\end{aligned}$$

We then directly verify that $\lambda_{1,m} \circ \hat{\iota}_{1,m}$ is indeed the identity.

We finally claim that

$$P_{m+1}(j_{m+1}\xi(x)) = D^{m+1}_{\nabla^{\mathsf{M}},\nabla^{\pi_{\mathsf{E}}}}(\xi), \tag{2.9}$$

which will establish the lemma. To see this, first note that it suffices to define P_{m+1} on image(ϵ_{m+1}) since

1. $\mathsf{J}^{m+1}\mathsf{E} \simeq (\mathsf{S}^{m+1}(\mathsf{T}^*\mathsf{M}) \otimes \mathsf{E}) \oplus \mathsf{J}^m\mathsf{E}$,
2. P_{m+1} is zero on $\mathsf{J}^m\mathsf{E} \subseteq \mathsf{J}^{m+1}\mathsf{E}$ (thinking of the inclusion arising from the connection-induced isomorphism from the preceding item), and
3. $P_{m+1} \circ \epsilon_{m+1}$ is the identity map on $\mathsf{S}^{m+1}(\mathsf{T}^*\mathsf{M}) \otimes \mathsf{E}$.

In order to connect the algebra and the geometry, let us write elements of $\mathsf{S}^{m+1}(\mathsf{T}^*\mathsf{M}) \otimes \mathsf{E}$ in a particular way. We let $x \in \mathsf{M}$ and let $f^1, \dots, f^{m+1}$ be smooth functions contained in the maximal ideal of $\mathsf{C}^\infty(\mathsf{M})$ at x, i.e., $f^j(x) = 0$, $j \in \{1, \dots, m+1\}$. Let ξ be a smooth section of E. We then can work with elements of $\mathsf{S}^{m+1}(\mathsf{T}^*\mathsf{M}) \otimes \mathsf{E}$ of the form

$$\mathrm{d}f^1(x) \odot \cdots \odot \mathrm{d}f^{m+1}(x) \otimes \xi(x).$$

We then have

$$\epsilon_{m+1}(\mathrm{d}f^1(x) \odot \cdots \odot \mathrm{d}f^{m+1}(x) \otimes \xi(x)) = j_{m+1}(f^1 \cdots f^{m+1}\xi)(x);$$

this is easy to see using the Leibniz Rule, cf. [23, Lemma 2.1]. (See [1, Supplement 2.4A] for a description of the higher-order Leibniz Rule.) Now, using the last part of the sublemma, we compute

$$\begin{aligned}
&P_{m+1}(j_{m+1}(f^1 \cdots f^{m+1}\xi)(x)) \\
&\quad = \lambda_{1,m} \circ P_{1,m} \circ \iota_{1,m}(j_{m+1}(f^1 \cdots f^{m+1}\xi)(x)) \\
&\quad = \lambda_{1,m} \circ P_{1,m}(j_1(f^1 \cdots f^{m+1}\xi, D^1_{\nabla^{\mathsf{M}},\nabla^{\pi_{\mathsf{E}}}}(f^1 \cdots f^{m+1}\xi), \dots, \\
&\qquad D^m_{\nabla^{\mathsf{M}},\nabla^{\pi_{\mathsf{E}}}}(f^1 \cdots f^{m+1}\xi))(x)) \\
&\quad = \lambda_{1,m}(\nabla^{\pi_{\mathsf{E}}}(f^1 \cdots f^{m+1}\xi)(x), \nabla^{\mathsf{M},\pi_{\mathsf{E}}} D^1_{\nabla^{\mathsf{M}},\nabla^{\pi_{\mathsf{E}}}}(f^1 \cdots f^{m+1}\xi)(x), \dots, \\
&\qquad \nabla^{\mathsf{M},\pi_{\mathsf{E}}} D^m_{\nabla^{\mathsf{M}},\nabla^{\pi_{\mathsf{E}}}}(f^1 \cdots f^{m+1}\xi)(x)) \\
&\quad = \mathrm{Sym}_{m+1} \otimes \mathrm{id}_{\mathsf{E}}(\nabla^{\mathsf{M},\pi_{\mathsf{E}}} D^m_{\nabla^{\mathsf{M}},\nabla^{\pi_{\mathsf{E}}}}(f^1 \cdots f^{m+1}\xi)(x)) \\
&\quad = D^{m+1}_{\nabla^{\mathsf{M}},\nabla^{\pi_{\mathsf{E}}}}(f^1 \cdots f^{m+1}\xi)(x),
\end{aligned}$$

which shows that, with P_{m+1} defined as in (2.9), $P_{m+1} \circ \epsilon_{m+1}$ is indeed the identity on $\mathsf{S}^{m+1}(\mathsf{T}^*\mathsf{M}) \otimes \mathsf{E}$.

The commuting of the diagram in the statement of the lemma follows directly from the recursive nature of the constructions. □

There are a couple of special cases of interest.

1. Jets of functions fit into the framework of the lemma by using the trivial line bundle $\mathbb{R}_{\mathsf{M}} = \mathsf{M} \times \mathbb{R}$. The identification of a function with a section of this bundle is specified by $f \mapsto \xi_f$, with $\xi_f(x) = (x, f(x))$; we shall make this identification implicitly. In this case, the bundle has a canonical flat connection defined by $\nabla^{\pi_{\mathsf{E}}} f = \mathrm{d}f$. Therefore, the decomposition of Lemma 2.15 is determined by an affine connection ∇^{M} on M, and so we have a mapping

$$\begin{aligned} S^m_{\nabla^{\mathsf{M}}} \colon \mathsf{J}^m(\mathsf{M}; \mathbb{R}) &\to \bigoplus_{j=0}^{m} \mathsf{S}^j(\mathsf{T}^*\mathsf{M}) \\ f(x) &\mapsto (f(x), \mathrm{d}f(x), \dots, \mathrm{Sym}_m \circ \nabla^{\mathsf{M},m-1}\mathrm{d}f(x)). \end{aligned} \tag{2.10}$$

This can be restricted to $\mathsf{T}^{*m}\mathsf{M}$ to give the mapping

$$\begin{aligned} S^m_{\nabla^{\mathsf{M}}} \colon \mathsf{T}^{*m}\mathsf{M} &\to \bigoplus_{j=1}^{m} \mathsf{S}^j(\mathsf{T}^*\mathsf{M}) \\ f(x) &\mapsto (\mathrm{d}f(x), \dots, \mathrm{Sym}_m \circ \nabla^{\mathsf{M},m-1}\mathrm{d}f(x)), \end{aligned} \tag{2.11}$$

adopting a mild abuse of notation. We recall that $\mathsf{T}^{*m}_x\mathsf{M}$ is an $\mathbb{R}$-algebra, and the induced $\mathbb{R}$-algebra structure on $\oplus_{j=1}^m \mathsf{S}^j(\mathsf{T}^*_x\mathsf{M})$ is that of polynomial functions that vanish at 0 and with degree at most m.

2. Another special case is that of jets of vector fields. In this case, the vector bundle is $\pi_{\mathsf{TM}} \colon \mathsf{TM} \to \mathsf{M}$. We can make use of an affine connection ∇^{M} on M to provide everything we need to define the mapping

$$\begin{aligned} S^m_{\nabla^{\mathsf{M}}} \colon \mathsf{J}^m\mathsf{TM} &\to \bigoplus_{j=0}^{m} (\mathsf{S}^j(\mathsf{T}^*\mathsf{M}) \otimes \mathsf{TM}) \\ X(x) &\mapsto (X(x), \nabla^{\mathsf{M}}X(x), \dots, \mathrm{Sym}_m \circ \nabla^{\mathsf{M},m}X(x)). \end{aligned} \tag{2.12}$$

Of course, this applies equally well to jets of one-forms on M, or any other sections of tensor bundles associated with the tangent bundle.

This case of vector fields is the setting of Jafarpour and Lewis [36] in their study of flows of time-varying vector fields.

2.3.2 *Fibre Norms for Jet Bundles of Vector Bundles*

With the decomposition of jet bundles from the preceding section, we now indicate how to provide norms for jet bundles using a Riemannian metric on M and a fibre metric on E. We let $r \in \{\infty, \omega\}$ and let $\pi_{\mathsf{E}} : \mathsf{E} \to \mathsf{M}$ be a C^r-vector bundle. We shall suppose that we have a C^r-affine connection ∇^{M} on M and a C^r-vector bundle connection $\nabla^{\pi_{\mathsf{E}}}$ in E, as in Sect. 2.3.1. This allows us to give the decomposition of $\mathsf{J}^m\mathsf{E}$ as in Lemma 2.15. We additionally suppose that we have a C^r-Riemannian metric $\mathbb{G}_{\mathsf{M}}$ on M and a C^r-fibre metric $\mathbb{G}_{\pi_{\mathsf{E}}}$ on E. Note that the existence of the metrics and connections is ensured in the real analytic case by Lemma 2.4; we note that this lemma can be applied equally well in the smooth case. In the smooth case, one can also use standard constructions using partitions of unity give existence of metrics [42, Proposition III.1.4], and the existence of connections is ensured by [42, Theorem II.2.1].

The first step in making the constructions of this section is the following result concerning inner products on tensor products.

Lemma 2.16 (Inner Products on Tensor Products) *Let* U *and* V *be finite-dimensional* $\mathbb{R}$*-vector spaces and let* $\mathbb{G}$ *and* $\mathbb{H}$ *be inner products on* U *and* V*, respectively. Then the element* $\mathbb{G} \otimes \mathbb{H}$ *of* $\mathrm{T}^2(\mathsf{U}^* \otimes \mathsf{V}^*)$ *defined by*

$$\mathbb{G} \otimes \mathbb{H}(u_1 \otimes v_1, u_2 \otimes v_2) = \mathbb{G}(u_1, u_2)\mathbb{H}(v_1, v_2)$$

is an inner product on $\mathsf{U} \otimes \mathsf{V}$.

Proof Let $(e_1, \dots, e_m)$ and $(f_1, \dots, f_n)$ be orthonormal bases for U and V, respectively. Then

$$\{e_a \otimes f_j \mid a \in \{1, \dots, m\},\ j \in \{1, \dots, n\}\} \tag{2.13}$$

is a basis for $\mathsf{U} \otimes \mathsf{V}$. Note that

$$\mathbb{G} \otimes \mathbb{H}(e_a \otimes f_j, e_b \otimes f_k) = \mathbb{G}(e_a, e_b)\mathbb{H}(f_j, f_k) = \delta_{ab}\delta_{jk},$$

which shows that $\mathbb{G} \otimes \mathbb{H}$ is indeed an inner product, as (2.13) is an orthonormal basis. □

With $\mathbb{G}_{\pi_{\mathsf{E}}}$ a fibre metric on E and with $\mathbb{G}_{\mathsf{M}}$ a Riemannian metric on M as above, let us denote by $\mathbb{G}_{\mathsf{M}}^{-1}$ the associated fibre metric on $\mathsf{T}^*\mathsf{M}$ defined by

$$\mathbb{G}_{\mathsf{M}}^{-1}(\alpha_x, \beta_x) = \mathbb{G}_{\mathsf{M}}(\mathbb{G}_{\mathsf{M}}^{\sharp}(\alpha_x), \mathbb{G}_{\mathsf{M}}^{\sharp}(\beta_x)).$$

In like manner, one has a fibre metric $\mathbb{G}_{\pi_{\mathsf{E}}}^{-1}$ on E^*. Then, by induction using the preceding lemma, we have a fibre metric in all tensor spaces associated with TM and E and their tensor products. We shall denote by $\mathbb{G}_{\mathsf{M},\pi_{\mathsf{E}}}$ any of these various fibre metrics. In particular, we have a fibre metric $\mathbb{G}_{\mathsf{M},\pi_{\mathsf{E}}}$ on $\mathrm{T}^j(\mathsf{T}^*\mathsf{M}) \otimes \mathsf{E}$ induced by $\mathbb{G}_{\mathsf{M}}^{-1}$ and $\mathbb{G}_{\pi_{\mathsf{E}}}$. By restriction, this gives a fibre metric on $\mathrm{S}^j(\mathsf{T}^*\mathsf{M}) \otimes \mathsf{E}$. We can thus define a fibre metric $\mathbb{G}_{\mathsf{M},\pi_{\mathsf{E}},m}$ on $\mathsf{J}^m\mathsf{E}$ given by

$$\mathbb{G}_{\mathsf{M},\pi_{\mathsf{E}},m}(j_m\xi(x), j_m\eta(x)) = \sum_{j=0}^{m} \mathbb{G}_{\mathsf{M},\pi_{\mathsf{E}}}\left(\frac{1}{j!}D^j_{\nabla^{\mathsf{M}},\nabla^{\pi_{\mathsf{E}}}}(\xi)(x), \frac{1}{j!}D^j_{\nabla^{\mathsf{M}},\nabla^{\pi_{\mathsf{E}}}}(\eta)(x)\right). \tag{2.14}$$

Associated to this inner product on fibres is the norm on fibres, which we denote by $\|\cdot\|_{\mathbb{G}_{\mathsf{M},\pi_{\mathsf{E}},m}}$. We shall use these fibre norms continually in our descriptions of our various topologies for real analytic vector bundles, cf. Sect. 4.1. The appearance of the factorials in the fibre metric (2.14) appears superfluous at this point. However, it is essential in order for the real analytic topology defined by our seminorms to be independent of the choices of ∇^{M}, $\nabla^{\pi_{\mathsf{E}}}$, $\mathbb{G}_{\mathsf{M}}$, and $\mathbb{G}_{\pi_{\mathsf{E}}}$, cf. Theorem 4.24.

One may wonder why we have made our constructions of fibre norms for $\mathsf{J}^m\mathsf{E}$ so complicated. Since $\pi_{\mathsf{E}}\colon \mathsf{E} \to \mathsf{M}$ is a vector bundle, could we not just use this fact to endow it with a fibre metric, and hence a fibre norm? In the smooth case, indeed this is an entirely feasible idea. However, in the real analytic case, one is working, essentially, with infinite jets, and so having fibre norms for $\mathsf{J}^m\mathsf{E}$ that vary arbitrarily with $m \in \mathbb{Z}_{\geq 0}$ will create a theory where the topology is not independent of the choice of these fibre norms. By inducing these fibre norms from data on M, we ensure that the topologies are not dependent on our choices, although this is not easy to prove, as we shall see in Sect. 4.3.

2.4 Seminorms for the C^ω-Topology

We now provide seminorms for the C^ω-topology. To do this, we shall first offer another characterisation of the topology on the space $\mathscr{G}^{\mathrm{hol},\mathbb{R}}_{\mathcal{K},\overline{\mathsf{E}}}$ of germs of holomorphic sections about a compact set $\mathcal{K}$.

2.4.1 A Weighted Direct Limit Topology for Sections of Bundles of Infinite Jets

Here we provide a direct limit topology for a subspace of the space of continuous sections of the infinite jet bundle of a vector bundle. In Proposition 2.18 we shall connect this direct limit topology to the direct limit topology described above for germs of holomorphic sections about a compact set. The topology we give here has the advantage of providing explicit seminorms for the topology of germs, and subsequently for the space of real analytic sections.

For this description, we work with infinite jets, so let us introduce the notation we will use for this, referring to [64, Chapter 7] for details. Let us denote by $\mathsf{J}^\infty\mathsf{E}$ the bundle of infinite jets of a vector bundle $\pi_{\mathsf{E}}\colon \mathsf{E} \to \mathsf{M}$, this being the inverse limit (in the category of sets, for the moment) of the inverse system $(\mathsf{J}^m\mathsf{E})_{m\in\mathbb{Z}_{\geq 0}}$ with mappings $\pi^k_{\mathsf{E},l}$, $k,l \in \mathbb{Z}_{\geq 0}$, $k \geq l$. Precisely,

$$\mathsf{J}^\infty\mathsf{E} = \left\{ \phi \in \prod_{m\in\mathbb{Z}_{\geq 0}} \mathsf{J}^m\mathsf{E} \;\middle|\; \pi^k_{\mathsf{E},l} \circ \phi(k) = \phi(l),\ k,l \in \mathbb{Z}_{\geq 0},\ k \geq l \right\}.$$

We let $\pi^\infty_{\mathsf{E},m} : \mathsf{J}^\infty\mathsf{E} \to \mathsf{J}^m\mathsf{E}$ be the projection defined by $\pi^\infty_{\mathsf{E},m}(\phi) = \phi(m)$; these mappings serve as the mappings induced by the inverse limit. For $\xi \in \Gamma^\infty(\mathsf{E})$, we let $j_\infty\xi : \mathsf{M} \to \mathsf{J}^\infty\mathsf{E}$ be defined by $\pi^\infty_{\mathsf{E},m} \circ j_\infty\xi(x) = j_m\xi(x)$. By a theorem of Borel [9], if $\phi \in \mathsf{J}^\infty\mathsf{E}$, there exist $\xi \in \Gamma^\infty(\mathsf{E})$ and $x \in \mathsf{M}$ such that $j_\infty\xi(x) = \phi$. We can define sections of $\mathsf{J}^\infty\mathsf{E}$ in the usual manner: a section is a map $\Xi : \mathsf{M} \to \mathsf{J}^\infty\mathsf{E}$ satisfying $\pi^\infty_{\mathsf{E},0} \circ \Xi(x) = x$ for every $x \in \mathsf{M}$. We shall equip $\mathsf{J}^\infty\mathsf{E}$ with the initial topology so that a section Ξ is continuous if and only if $\pi^\infty_{\mathsf{E},m} \circ \Xi$ is continuous for every $m \in \mathbb{Z}_{\geq 0}$. We denote the space of continuous sections of $\mathsf{J}^\infty\mathsf{E}$ by $\Gamma^0(\mathsf{J}^\infty\mathsf{E})$. Since we are only dealing with continuous sections, we can talk about sections defined on any subset $A \subseteq \mathsf{M}$, using the induced topology on A. The continuous sections defined on $A \subseteq \mathsf{M}$ will be denoted by $\Gamma^0(\mathsf{J}^\infty\mathsf{E}|A)$.

Now let $\mathcal{K} \subseteq \mathsf{M}$ be compact and, for $j \in \mathbb{Z}_{>0}$, denote

$$\mathscr{E}_j(\mathcal{K}) = \Big\{ \Xi \in \Gamma^0(\mathsf{J}^\infty\mathsf{E}|\mathcal{K}) \;\Big|\; \sup\{ j^{-m} \|\pi^\infty_{\mathsf{E},m} \circ \Xi(x)\|_{\mathbb{G}_{\mathsf{M},\pi_{\mathsf{E}},m}} \mid m \in \mathbb{Z}_{\geq 0},\ x \in \mathcal{K}\} < \infty \Big\},$$

and on $\mathscr{E}_j(\mathcal{K})$ we define a norm $p_{\mathcal{K},j}$ by

$$p_{\mathcal{K},j}(\Xi) = \sup\left\{ j^{-m} \|\pi^\infty_{\mathsf{E},m} \circ \Xi(x)\|_{\mathbb{G}_{\mathsf{M},\pi_{\mathsf{E}},m}} \;\middle|\; m \in \mathbb{Z}_{\geq 0},\ x \in \mathcal{K} \right\}.$$

One readily verifies that, for each $j \in \mathbb{Z}_{>0}$, $(\mathscr{E}_j(\mathcal{K}), p_{\mathcal{K},j})$ is a Banach space. Note that $\mathscr{E}_j(\mathcal{K}) \subseteq \mathscr{E}_{j+1}(\mathcal{K})$ and that $p_{\mathcal{K},j+1}(\Xi) \leq p_{\mathcal{K},j}(\Xi)$ for $\Xi \in \mathscr{E}_j(\mathcal{K})$, and so the inclusion of $\mathscr{E}_j(\mathcal{K})$ in $\mathscr{E}_{j+1}(\mathcal{K})$ is continuous. We let $\mathscr{E}(\mathcal{K})$ be the direct limit of the directed system $(\mathscr{E}_j(\mathcal{K}))_{j\in\mathbb{Z}_{>0}}$ with the mappings being inclusions.

The following attribute of the direct limit topology for $\mathscr{E}(\mathcal{K})$ will be useful.

Lemma 2.17 (**$\mathscr{E}(\mathcal{K})$ Is A Regular Direct Limit**) *The direct limit topology for $\mathscr{E}(\mathcal{K})$ is regular. As a consequence, $\mathscr{E}(\mathcal{K})$ is Hausdorff.*

Proof Let $\overline{\mathsf{B}}_j(1,0) \subseteq \mathscr{E}_j(\mathcal{K})$, $j \in \mathbb{Z}_{>0}$, be the closed unit ball with respect to the norm topology. We claim that $\overline{\mathsf{B}}_j(1,0)$ is closed in the direct limit topology of $\mathscr{E}(\mathcal{K})$. To prove this, we shall prove that $\overline{\mathsf{B}}_j(1,0)$ is closed in a topology that is coarser than the direct limit topology.

The coarser topology we use is the topology induced by the topology of pointwise convergence in $\Gamma^0(\mathsf{J}^\infty\mathsf{E}|\mathcal{K})$. To be precise, let $\widehat{\mathscr{E}}_j(\mathcal{K})$ be the vector space $\mathscr{E}_j(\mathcal{K})$

with the topology defined by the seminorms

$$p_{x,j}(\Xi) = \sup\left\{ j^{-m} \|\pi^{\infty}_{\mathsf{E},m} \circ \Xi(x)\|_{\mathbb{G}_{\mathsf{M},\pi_{\mathsf{E}},m}} \;\middle|\; m \in \mathbb{Z}_{\geq 0}\right\}, \qquad x \in \mathcal{K}.$$

Clearly the identity map from $\mathscr{E}_j(\mathcal{K})$ to $\widehat{\mathscr{E}}_j(\mathcal{K})$ is continuous, and so the topology of $\widehat{\mathscr{E}}_j(\mathcal{K})$ is coarser than the usual topology of $\mathscr{E}(\mathcal{K})$. Now let $\widehat{\mathscr{E}}(\mathcal{K})$ be the direct limit of the directed system $(\widehat{\mathscr{E}}_j(\mathcal{K}))_{j\in\mathbb{Z}_{>0}}$. Note that, algebraically, $\widehat{\mathscr{E}}(\mathcal{K}) = \mathscr{E}(\mathcal{K})$, but the spaces have different topologies, the topology for $\widehat{\mathscr{E}}(\mathcal{K})$ being coarser than that for $\mathscr{E}(\mathcal{K})$.

We will show that $\overline{\mathsf{B}}_j(1,0)$ is closed in $\widehat{\mathscr{E}}(\mathcal{K})$. Let $(I,\preceq)$ be a directed set and let $(\Xi_i)_{i\in I}$ be a convergent net in $\overline{\mathsf{B}}_j(1,0)$ in the topology of $\widehat{\mathscr{E}}(\mathcal{K})$. Thus we have a map $\Xi\colon \mathcal{K} \to \mathsf{J}^{\infty}\mathsf{E}|\mathcal{K}$ such that, for each $x \in \mathcal{K}$, $\lim_{i\in I}\Xi_i(x) = \Xi(x)$. If $\Xi \notin \overline{\mathsf{B}}_j(1,0)$, then there exists $x \in \mathcal{K}$ such that

$$\sup\left\{ j^{-m} \|\pi^{\infty}_{\mathsf{E},m} \circ \Xi(x)\|_{\mathbb{G}_{\mathsf{M},\pi_{\mathsf{E}},m}} \;\middle|\; m \in \mathbb{Z}_{\geq 0}\right\} > 1.$$

Let $\epsilon \in \mathbb{R}_{>0}$ be such that

$$\sup\left\{ j^{-m} \|\pi^{\infty}_{\mathsf{E},m} \circ \Xi(x)\|_{\mathbb{G}_{\mathsf{M},\nabla_{\mathsf{E}},m}} \;\middle|\; m \in \mathbb{Z}_{\geq 0}\right\} > 1+\epsilon$$

and let $i_0 \in I$ be such that

$$\sup\left\{ j^{-m} \|\pi^{\infty}_{\mathsf{E},m} \circ \Xi_i(x) - \pi^{\infty}_{\mathsf{E},m} \circ \Xi(x)\|_{\mathbb{G}_{\mathsf{M},\pi_{\mathsf{E}},m}} \;\middle|\; m \in \mathbb{Z}_{\geq 0}\right\} < \epsilon$$

for $i_0 \preceq i$, this by pointwise convergence. We thus have, for all $i_0 \preceq i$,

$$\begin{aligned}
\epsilon &< \sup\left\{ j^{-m} \|\pi^{\infty}_{\mathsf{E},m} \circ \Xi(x)\|_{\mathbb{G}_{\mathsf{M},\pi_{\mathsf{E}},m}} \;\middle|\; m \in \mathbb{Z}_{\geq 0}\right\} \\
&\quad - \sup\left\{ j^{-m} \|\pi^{\infty}_{\mathsf{E},m} \circ \Xi_i(x)\|_{\mathbb{G}_{\mathsf{M},\pi_{\mathsf{E}},m}} \;\middle|\; m \in \mathbb{Z}_{\geq 0}\right\} \\
&\leq \sup\left\{ j^{-m} \|\pi^{\infty}_{\mathsf{E},m} \circ \Xi_i(x) - \pi^{\infty}_{\mathsf{E},m} \circ \Xi(x)\|_{\mathbb{G}_{\mathsf{M},\pi_{\mathsf{E}},m}} \;\middle|\; m \in \mathbb{Z}_{\geq 0}\right\} < \epsilon,
\end{aligned}$$

which contradiction gives the conclusion that $\Xi \in \overline{\mathsf{B}}_j(1,0)$.

Since $\overline{\mathsf{B}}_j(1,0)$ has been shown to be closed in $\mathscr{E}(\mathcal{K})$, the regularity of the direct limit topology now follows from [7, Corollary 3.7]. Hausdorffness follows from [37, Proposition 4.5.3]. □

The next result explains the relevance of the above constructions.

Proposition 2.18 **($\mathscr{G}^{\mathrm{hol},\mathbb{R}}_{\mathcal{K},\overline{\mathsf{E}}}$ Is a Topological Subspace of $\mathscr{E}(\mathcal{K})$)** *Let $\pi_{\mathsf{E}}\colon \mathsf{E}\to\mathsf{M}$ be a real analytic vector bundle and let $\mathcal{K}\subseteq\mathsf{M}$ be compact. Then the mapping*

$$L_{\mathcal{K}}\colon \mathscr{G}^{\mathrm{hol},\mathbb{R}}_{\mathcal{K},\overline{\mathsf{E}}}\to\mathscr{E}(\mathcal{K})$$
$$[\overline{\xi}]_{\mathcal{K}}\mapsto j_\infty\xi|\mathcal{K}$$

is a continuous open injection, and so is a topological monomorphism.

Proof Throughout the proof, we let $\pi_{\overline{\mathsf{E}}}\colon\overline{\mathsf{E}}\to\overline{\mathsf{M}}$ be a complexification of $\pi_{\mathsf{E}}\colon\mathsf{E}\to\mathsf{M}$ and we let $(\overline{\mathcal{U}}_j)_{j\in\mathbb{Z}_{>0}}$ be a sequence of neighbourhoods of $\mathcal{K}$ as in Lemma 2.8.

Let us first prove that $L_{\mathcal{K}}$ is well-defined, i.e., show that, if $[\overline{\xi}]_{\mathcal{K}}\in\mathscr{G}^{\mathrm{hol},\mathbb{R}}_{\mathcal{K},\overline{\mathsf{E}}}$, then $L_{\mathcal{K}}([\overline{\xi}]_{\mathcal{K}})\in\mathscr{E}_j(\mathcal{K})$ for some $j\in\mathbb{Z}_{>0}$. Let $\overline{\mathcal{U}}$ be a neighbourhood of $\mathcal{K}$ in $\overline{\mathsf{M}}$ on which the section $\overline{\xi}$ is defined, holomorphic, and bounded. Then $\xi|(\mathsf{M}\cap\overline{\mathcal{U}})$ is real analytic and so, by Lemma 2.22, there exist $C,r\in\mathbb{R}_{>0}$ such that

$$\|j_m\xi(x)\|_{\mathbb{G}_{\mathsf{M},\pi_{\mathsf{E}},m}}\le Cr^{-m},\qquad x\in\mathcal{K},\ m\in\mathbb{Z}_{\ge0}.$$

If $j>r^{-1}$, it immediately follows that

$$\sup\left\{j^{-m}\|j_m\xi(x)\|_{\mathbb{G}_{\mathsf{M},\pi_{\mathsf{E}},m}}\;\middle|\;x\in\mathcal{K},\ m\in\mathbb{Z}_{\ge0}\right\}<\infty,$$

i.e., $L_{\mathcal{K}}([\overline{\xi}]_{\mathcal{K}})\in\mathscr{E}_j(\mathcal{K})$.

By the universal property of direct limits, to show that $L_{\mathcal{K}}$ is continuous, it suffices to show that $L_{\mathcal{K}}|\Gamma^{\mathrm{hol},\mathbb{R}}_{\mathrm{bdd}}(\overline{\mathsf{E}}|\overline{\mathcal{U}}_j)$ is continuous for each $j\in\mathbb{Z}_{\ge0}$. We will prove this by showing that, for each $j\in\mathbb{Z}_{>0}$, there exists $j'\in\mathbb{Z}_{>0}$ such that $L_{\mathcal{K}}(\Gamma^{\mathrm{hol}}_{\mathrm{bdd}}(\overline{\mathsf{E}}|\overline{\mathcal{U}}_j))\subseteq\mathscr{E}_{j'}(\mathcal{K})$ and such that $L_{\mathcal{K}}$ is continuous as a map from $\Gamma^{\mathrm{hol}}_{\mathrm{bdd}}(\overline{\mathsf{E}}|\overline{\mathcal{U}}_j)$ to $\mathscr{E}_{j'}(\mathcal{K})$. Since $\mathscr{E}_{j'}(\mathcal{K})$ is continuously included in $\mathscr{E}(\mathcal{K})$, this will give the continuity of $L_{\mathcal{K}}$. First let us show that $L_{\mathcal{K}}(\Gamma^{\mathrm{hol}}_{\mathrm{bdd}}(\overline{\mathsf{E}}|\overline{\mathcal{U}}_j))\subseteq\mathscr{E}_{j'}(\mathcal{K})$ for some $j'\in\mathbb{Z}_{>0}$. By Lemma 2.21, there exist $C,r\in\mathbb{R}_{>0}$ such that

$$\|j_m\xi(x)\|_{\mathbb{G}_{\mathsf{M},\pi_{\mathsf{E}},m}}\le Cr^{-m}p^{\mathrm{hol}}_{\overline{\mathcal{U}}_j,\infty}(\overline{\xi})$$

for every $m\in\mathbb{Z}_{\ge0}$ and $\overline{\xi}\in\Gamma^{\mathrm{hol}}_{\mathrm{bdd}}(\overline{\mathsf{E}}|\overline{\mathcal{U}}_j)$. Taking $j'\in\mathbb{Z}_{>0}$ such that $j'\ge r^{-1}$, we have $L_{\mathcal{K}}(\Gamma^{\mathrm{hol}}_{\mathrm{bdd}}(\overline{\mathsf{E}}|\overline{\mathcal{U}}_j))\subseteq\mathscr{E}_{j'}(\mathcal{K})$, as claimed. To show that $L_{\mathcal{K}}$ is continuous as a map from $\Gamma^{\mathrm{hol}}_{\mathrm{bdd}}(\overline{\mathsf{E}}|\overline{\mathcal{U}}_j)$ to $\mathscr{E}_{j'}(\mathcal{K})$, let $(\overline{\xi}_k)_{k\in\mathbb{Z}_{>0}}$ be a sequence in $\Gamma^{\mathrm{hol}}_{\mathrm{bdd}}(\overline{\mathsf{E}}|\overline{\mathcal{U}}_j)$ converging to zero. We then have

$$\begin{aligned}\lim_{k\to\infty}\sup\left\{(j')^{-m}\|j_m\xi_k(x)\|_{\mathbb{G}_{\mathsf{M},\pi_{\mathsf{E}},m}}\;\middle|\;x\in\mathcal{K},\ m\in\mathbb{Z}_{\ge0}\right\}&\\ \le\lim_{k\to\infty}C\sup\left\{\|\overline{\xi}_k(z)\|_{\mathbb{G}_{\overline{\mathsf{M}},\pi_{\overline{\mathsf{E}}}}}\;\middle|\;z\in\overline{\mathcal{U}}_j\right\}&=0,\end{aligned}$$

giving the desired continuity.

Since germs of holomorphic sections are uniquely determined by their infinite jets by the Identity Theorem, injectivity of $L_{\mathcal{K}}$ follows.

To show that $L_{\mathcal{K}}$ is open, we cannot use the De Wilde Open Mapping Theorem since $L_{\mathcal{K}}$ is not surjective and since subspaces of ultrabornological spaces are not generally ultrabornological. We will instead use an Open Mapping Theorem from §2 of [5]. This theorem refers to the notion of a DF-space, for which we refer the reader to [37, §12.4]. Accepting this notion, the Open Mapping Theorem we use is the following.

Let $(\mathsf{U}, \mathscr{O}_{\mathsf{U}})$ *and* $(\mathsf{V}, \mathscr{O}_{\mathsf{V}})$ *be locally convex topological vector spaces and let* $L \in \mathsf{L}(\mathsf{U}; \mathsf{V})$. *Assume the following*:

(i) $(\mathsf{U}, \mathscr{O}_{\mathsf{V}})$ *is Hausdorff*;
(ii) *closed and bounded subsets of* U *are compact, e.g.*, $(\mathsf{U}, \mathscr{O}_{\mathsf{U}})$ *is nuclear*;
(iii) $(\mathsf{V}, \mathscr{O}_{\mathsf{V}})$ *is a DF-space*;
(iv) $L^{-1}(\mathcal{B})$ *is bounded for any bounded subset* $\mathcal{B} \subseteq \mathsf{V}$.

Then L is open onto its image.

By Proposition 2.10, the direct limit topology of $\mathscr{G}^{\text{hol},\mathbb{R}}_{\mathcal{K},\overline{\mathsf{E}}}$ is Hausdorff and nuclear. Note also that $\mathscr{E}(\mathcal{K})$ is a DF-space since Banach spaces are DF-spaces [37, Corollary 12.4.4] and Hausdorff countable direct limits of DF-spaces are DF-spaces [37, Theorem 12.4.8]. Therefore, to use the preceding Open Mapping Theorem, we must only show that $L_{\mathcal{K}}^{-1}(\mathcal{B})$ is bounded when $\mathcal{B}$ is bounded.

Thus let $\mathcal{B} \subseteq \mathscr{E}(\mathcal{K})$ be bounded. By Lemma 2.17, if $\mathcal{B} \subseteq \mathscr{E}(\mathcal{K})$ is bounded, then $\mathcal{B}$ is contained and bounded in $\mathscr{E}_j(\mathcal{K})$ for some $j \in \mathbb{Z}_{>0}$. Therefore, there exists $C \in \mathbb{R}_{>0}$ such that, if $L_{\mathcal{K}}([\overline{\xi}]_{\mathcal{K}}) \in \mathcal{B}$, then

$$\|j_m\xi(x)\|_{\mathbb{G}_{\mathsf{M},\pi_{\mathsf{E}},m}} \leq Cj^{-m}, \qquad x \in \mathcal{K},\ m \in \mathbb{Z}_{\geq 0}.$$

Let $x \in \mathcal{K}$ and let $(\mathcal{V}_x, \boldsymbol{\nu}_x)$ be a vector bundle chart for E about x with corresponding chart $(\mathcal{U}_x, \boldsymbol{\chi}_x)$ for M. Suppose that the fibre dimension of E over $\mathcal{U}_x$ is k and that $\boldsymbol{\chi}_x$ takes values in $\mathbb{R}^n$. Let $\mathcal{U}'_x \subseteq \mathcal{U}_x$ be a precompact neighbourhood of x such that $\mathrm{cl}(\mathcal{U}'_x) \subseteq \mathcal{U}_x$. Denote $\mathcal{K}_x = \mathcal{K} \cap \mathrm{cl}(\mathcal{U}'_x)$. By Lemma 4.22, there exist $C_x, r_x \in \mathbb{R}_{>0}$ such that, if $L_{\mathcal{K}}([\overline{\xi}]_{\mathcal{K}}) \in \mathcal{B}$, then

$$|\boldsymbol{D}^I\xi^a(\boldsymbol{x})| \leq C_x I! r_x^{-|I|}, \qquad \boldsymbol{x} \in \boldsymbol{\chi}_x(\mathcal{K}_x),\ I \in \mathbb{Z}^n_{\geq 0},\ a \in \{1, \dots, k\},$$

where $\boldsymbol{\xi}$ is the local representative of ξ. We can assume, without loss of generality, that $r_x < 1$. Note that this implies the following for each $[\overline{\xi}]_{\mathcal{K}}$ such that $L_{\mathcal{K}}([\overline{\xi}]_{\mathcal{K}}) \in \mathcal{B}$ and for each $a \in \{1, \dots, k\}$:

1. $\overline{\xi}^a$ admits a convergent power series expansion to an holomorphic function on the polydisk $\mathsf{D}(\boldsymbol{\sigma}_x, \boldsymbol{\chi}_x(x))$ for $0 < \sigma_x < r_x$ (here $\boldsymbol{\sigma}_x = (\sigma_x, \dots, \sigma_x)$);

2. on the polydisk $\mathsf{D}(\boldsymbol{\sigma}_x, \boldsymbol{\chi}_x(x))$, $\overline{\xi}^a$ satisfies $|\overline{\xi}^a| \le C_x(\frac{1}{1-\sigma_x})^n$, recalling that

$$\sum_{k=0}^{\infty} \sigma_x^k = \frac{1}{1-\sigma_x}$$

if $0 < \sigma_x < 1$.

It follows that, if $L_{\mathcal{K}}([\xi]_{\mathcal{K}}) \in \mathcal{B}$ and if $x \in \mathcal{K}$, then $\overline{\xi}$ has a bounded holomorphic extension in some neighbourhood $\overline{\mathcal{V}}_x \subseteq \overline{\mathsf{M}}$ whose image under $\boldsymbol{\chi}_x$ is a polydisk. Let $x_1, \dots, x_k \in \mathcal{K}$ be such that $\mathcal{K} \subseteq \cup_{a=1}^k \overline{\mathcal{V}}_{x_a}$ and then let $j' \in \mathbb{Z}_{>0}$ be such that $\overline{\mathcal{U}}_{j'} \subseteq \cup_{a=1}^k \overline{\mathcal{V}}_{x_a}$. We then have that, for $[\xi]_{\mathcal{K}}$ for which $L_{\mathcal{K}}([\xi]_{\mathcal{K}}) \in \mathcal{B}$, $\overline{\xi} \in \Gamma_{\mathrm{bdd}}^{\mathrm{hol},\mathbb{R}}(\overline{\mathsf{E}}|\overline{\mathcal{U}}_{j'})$. Thus $L_{\mathcal{K}}^{-1}(\mathcal{B})$ is bounded, as claimed. □

2.4.2 Definition of Seminorms

In this section we provide explicit seminorms for Martineau's topologies for $\Gamma^\omega(\mathsf{E})$. Throughout this section, we will work with a real analytic vector bundle $\pi_{\mathsf{E}} : \mathsf{E} \to \mathsf{M}$ and the real analytic data ∇^{M}, $\nabla^{\pi_{\mathsf{E}}}$, $\mathbb{G}_{\mathsf{M}}$, and $\mathbb{G}_{\pi_{\mathsf{E}}}$ that define the fibre metrics for jet bundles as per Sect. 2.3.2. To define seminorms for $\Gamma^\omega(\mathsf{E})$, let $c_0(\mathbb{Z}_{\ge 0}; \mathbb{R}_{>0})$ denote the space of sequences in $\mathbb{R}_{>0}$, indexed by $\mathbb{Z}_{\ge 0}$, and converging to zero. We shall denote a typical element of $c_0(\mathbb{Z}_{\ge 0}; \mathbb{R}_{>0})$ by $\boldsymbol{a} = (a_j)_{j \in \mathbb{Z}_{\ge 0}}$. Now, for $\mathcal{K} \subseteq \mathsf{M}$ and $\boldsymbol{a} \in c_0(\mathbb{Z}_{\ge 0}; \mathbb{R}_{>0})$, we define a seminorm $p_{\mathcal{K},\boldsymbol{a}}^\omega$ for $\Gamma^\omega(\mathsf{E})$ by

$$p_{\mathcal{K},\boldsymbol{a}}^\omega(\xi) = \sup\left\{ a_0 a_1 \cdots a_m \| j_m \xi(x) \|_{\mathbb{G}_{\mathsf{M},\pi_{\mathsf{E}},m}} \;\middle|\; x \in \mathcal{K},\ m \in \mathbb{Z}_{\ge 0} \right\}.$$

The family of seminorms $p_{\mathcal{K},\boldsymbol{a}}^\omega$, $\mathcal{K} \subseteq \mathsf{M}$ compact, $\boldsymbol{a} \in c_0(\mathbb{Z}_{\ge 0}; \mathbb{R}_{>0})$, defines a locally convex topology on $\Gamma^\omega(\mathsf{E})$. Let us prove that this locally convex topology is actually the C^ω-topology.

Theorem 2.19 (Seminorms for $\Gamma^\omega(\mathsf{E})$) *Let $\pi_{\mathsf{E}} : \mathsf{E} \to \mathsf{M}$ be a real analytic vector bundle. Then the family of seminorms*

$$p_{\mathcal{K},\boldsymbol{a}}^\omega, \qquad \mathcal{K} \subseteq \mathsf{M} \text{ compact},\ \boldsymbol{a} \in c_0(\mathbb{Z}_{\ge 0}; \mathbb{R}_{>0}),$$

defines a locally convex topology on $\Gamma^\omega(\mathsf{E})$ agreeing with the C^ω-topology.

Proof First we fix $\mathcal{K}$ and show that the seminorms

$$p_{\mathcal{K},\boldsymbol{a}} = \sup\left\{ a_0 a_1 \cdots a_m \| \pi_{\mathsf{E},m}^\infty \circ \Xi(x) \|_{\mathbb{G}_{\mathsf{M},\pi_{\mathsf{E}},m}} \;\middle|\; m \in \mathbb{Z}_{\ge 0},\ x \in \mathcal{K} \right\},$$

$$\boldsymbol{a} \in c_0(\mathbb{Z}_{\ge 0}; \mathbb{R}_{>0}), \tag{2.15}$$

define the locally convex direct limit topology for $\mathscr{E}(\mathcal{K})$.

To this end, we first show that the seminorms $p_{\mathcal{K},\boldsymbol{a}}$, $\boldsymbol{a} \in c_0(\mathbb{Z}_{\geq 0}; \mathbb{R}_{>0})$, are continuous on $\mathscr{E}(\mathcal{K})$. By the universal property of final topologies, it suffices to show that $p_{\mathcal{K},\boldsymbol{a}}|\mathscr{E}_j(\mathcal{K})$ is continuous for each $j \in \mathbb{Z}_{>0}$. Thus, since $\mathscr{E}_j(\mathcal{K})$ is a Banach space, it suffices to show that, if $(\Xi_k)_{k\in\mathbb{Z}_{>0}}$ is a sequence in $\mathscr{E}_j(\mathcal{K})$ converging to zero in the norm topology, then $\lim_{k\to\infty} p_{\mathcal{K},\boldsymbol{a}}(\Xi_k) = 0$. Let $N \in \mathbb{Z}_{\geq 0}$ be such that $a_m < \frac{1}{j}$ for $m \geq N$. Let $C \geq 1$ be such that

$$a_0 a_1 \cdots a_m \leq C j^{-m}, \qquad m \in \{0, 1, \dots, N\},$$

this being possible since there are only finitely many inequalities to satisfy. Therefore, for any $m \in \mathbb{Z}_{\geq 0}$, we have $a_0 a_1 \cdots a_m \leq C j^{-m}$. Then, for any $\Xi \in \Gamma^0(\mathsf{J}^\infty\mathsf{E}|\mathcal{K})$,

$$a_0 a_1 \cdots a_m \|\pi^\infty_{\mathsf{E},m} \circ \Xi(x)\|_{\mathbb{G}_{\mathsf{M},\pi_{\mathsf{E}},m}} \leq C j^{-m} \|\pi^\infty_{\mathsf{E},m} \circ \Xi(x)\|_{\mathbb{G}_{\mathsf{M},\pi_{\mathsf{E}},m}}$$

for every $x \in \mathcal{K}$ and $m \in \mathbb{Z}_{\geq 0}$. From this we immediately have $\lim_{k\to\infty} p_{\mathcal{K},\boldsymbol{a}}(\Xi_k) = 0$, as desired. This shows that the direct limit topology on $\mathscr{E}(\mathcal{K})$ is finer than the topology defined by the family of seminorms $p_{\mathcal{K},\boldsymbol{a}}$, $\boldsymbol{a} \in c_0(\mathbb{Z}_{\geq 0}; \mathbb{R}_{>0})$.

For the converse, we show that every neighbourhood of $0 \in \mathscr{E}(\mathcal{K})$ in the direct limit topology contains a neighbourhood of zero in the topology defined by the seminorms $p_{\mathcal{K},\boldsymbol{a}}$, $\boldsymbol{a} \in c_0(\mathbb{Z}_{\geq 0}; \mathbb{R}_{>0})$. Let $\mathsf{B}_j(1, 0)$ denote the open unit ball in $\mathscr{E}_j(\mathcal{K})$. A neighbourhood of 0 in the direct limit topology contains a union of balls $\epsilon_j \mathsf{B}_j(1, 0)$ for some $\epsilon_j \in \mathbb{R}_{>0}$, $j \in \mathbb{Z}_{>0}$, [37, Proposition 6.6.5(a)] and we can assume, without loss of generality, that $\epsilon_j \in (0, 1)$ for each $j \in \mathbb{Z}_{>0}$. We define an increasing sequence $(m_j)_{j\in\mathbb{Z}_{>0}}$ in $\mathbb{Z}_{\geq 0}$ as follows. Let $m_1 = 0$. Having defined $m_1, \dots, m_j$, define $m_{j+1} > m_j$ by requiring that $j < \epsilon_{j+1}^{1/m_{j+1}}(j + 1)$. For $m \in \{m_j, \dots, m_{j+1} - 1\}$, define $a_m \in \mathbb{R}_{>0}$ by $a_m^{-1} = \epsilon_j^{1/m_j} j$. Note that, for $m \in \{m_j, \dots, m_{j+1} - 1\}$, we have

$$a_m^{-m} = \epsilon_j^{m/m_j} j^m \leq \epsilon_j j^m.$$

Note that $\lim_{m\to\infty} a_m = 0$. If $\Xi \in \Gamma^0(\mathsf{J}^\infty\mathsf{E}|\mathcal{K})$ satisfies $p_{\mathcal{K},\boldsymbol{a}}(\Xi) \leq 1$, then, for $m \in \{m_j, \dots, m_{j+1} - 1\}$, we have

$$\begin{aligned} j^{-m} \|\pi^\infty_{\mathsf{E},m} \circ \Xi(x)\|_{\mathbb{G}_{\mathsf{M},\pi_{\mathsf{E}},m}} &\leq a_m^m \epsilon_j \|\pi^\infty_{\mathsf{E},m} \circ \Xi(x)\|_{\mathbb{G}_{\mathsf{M},\pi_{\mathsf{E}},m}} \\ &\leq a_0 a_1 \cdots a_m \epsilon_j \|\pi^\infty_{\mathsf{E},m} \circ \Xi(x)\|_{\mathbb{G}_{\mathsf{M},\pi_{\mathsf{E}},m}} \leq \epsilon_j \end{aligned}$$

for $x \in \mathcal{K}$. Thus, if $\Xi \in \Gamma^0(\mathsf{J}^\infty\mathsf{E}|\mathcal{K})$ satisfies $p_{\mathcal{K},\boldsymbol{a}}(\Xi) \leq 1$, then, for $m \in \{m_j, \dots, m_{j+1} - 1\}$, we have $\pi^\infty_{\mathsf{E},m} \circ \Xi \in \epsilon_j \mathsf{B}_j(1, 0)$. Therefore, $\Xi \in$

$\cup_{j\in\mathbb{Z}_{>0}}\epsilon_j \mathsf{B}_j(1,0)$, and this shows that, for $\boldsymbol{a}$ as constructed above,

$$\left\{\Xi \in \Gamma^0(\mathsf{J}^\infty\mathsf{E}|\mathcal{K}) \;\middle|\; p_{\mathcal{K},\boldsymbol{a}}(\Xi) \leq 1\right\} \subseteq \bigcup_{j\in\mathbb{Z}_{>0}} \epsilon_j \mathsf{B}_j(1,0),$$

showing that the seminorms (2.15) give the topology of $\mathscr{E}(\mathcal{K})$.

By Proposition 2.18, we conclude that the topology of $\mathscr{G}^{\mathrm{hol},\mathbb{R}}_{\mathcal{K},\overline{\mathsf{E}}}$ is defined by the seminorms

$$p_{\mathcal{K},\boldsymbol{a}}([\overline{\xi}]_{\mathcal{K}}) = \sup\left\{a_0 a_1 \cdots a_m \| j_m\xi(x)\|_{\mathbb{G}_{\mathsf{M},\pi_{\mathsf{E}},m}} \;\middle|\; m \in \mathbb{Z}_{\geq 0},\ x \in \mathcal{K}\right\},$$
$$\boldsymbol{a} \in \mathrm{c}_0(\mathbb{Z}_{\geq 0}; \mathbb{R}_{>0}).$$

Finally, since $\Gamma^\omega(\mathsf{E})$ is the inverse limit of the spaces $\mathscr{G}^{\mathrm{hol},\mathbb{R}}_{\mathcal{K},\overline{\mathsf{E}}}$ over the directed set of compact subsets $\mathcal{K}$ of M, we conclude that the seminorms

$$p^\omega_{\mathcal{K},\boldsymbol{a}}, \qquad \mathcal{K} \subseteq \mathsf{M} \text{ compact},\ \boldsymbol{a} \in \mathrm{c}_0(\mathbb{Z}_{\geq 0}; \mathbb{R}_{>0}),$$

define the C$^\omega$-topology, as asserted in the statement of the theorem. □

It is sometimes convenient to work with variations of the seminorms $p^\omega_{\mathcal{K},\boldsymbol{a}}$. Specifically, it can be convenient to work with sequences $\boldsymbol{a}$ with particular properties. To this end, we denote

$$\mathrm{c}_0^{\downarrow}(\mathbb{Z}_{\geq 0}; \mathbb{R}_{>0}) = \{\boldsymbol{a} \in \mathrm{c}_0(\mathbb{Z}_{\geq 0}; \mathbb{R}_{>0}) \mid a_j \geq a_{j+1},\ j \in \mathbb{Z}_{\geq 0}\},$$

and, for $\rho \in \mathbb{R}_{>0}$,

$$\mathrm{c}_0(\mathbb{Z}_{\geq 0}; \mathbb{R}_{>0}; \rho) = \{\boldsymbol{a} \in \mathrm{c}_0(\mathbb{Z}_{\geq 0}; \mathbb{R}_{>0}) \mid a_j \leq \rho,\ j \in \mathbb{Z}_{\geq 0}\},$$

and, finally

$$\mathrm{c}_0^{\downarrow}(\mathbb{Z}_{\geq 0}; \mathbb{R}_{>0}; \rho) = \{\boldsymbol{a} \in \mathrm{c}_0^{\downarrow}(\mathbb{Z}_{\geq 0}; \mathbb{R}_{>0}) \mid a_j \leq \rho,\ j \in \mathbb{Z}_{\geq 0}\}.$$

Let us prove that the topology defined by the four flavours of sequences coincide.

Lemma 2.20 (Alternative Seminorms for the Real Analytic Topology) *Let $\pi_{\mathsf{E}}\colon \mathsf{E} \to \mathsf{M}$ be a real analytic vector bundle. The following collections of seminorms define the same locally convex topology for $\Gamma^\omega(\mathsf{E})$:*

(i) $p^\omega_{\mathcal{K},\boldsymbol{a}}$, $\mathcal{K} \subseteq \mathsf{M}$ *compact,* $\boldsymbol{a} \in \mathrm{c}_0(\mathbb{Z}_{\geq 0}; \mathbb{R}_{>0})$*;*
(ii) $p^\omega_{\mathcal{K},\boldsymbol{a}}$, $\mathcal{K} \subseteq \mathsf{M}$ *compact,* $\boldsymbol{a} \in \mathrm{c}_0^{\downarrow}(\mathbb{Z}_{\geq 0}; \mathbb{R}_{>0})$*;*
(iii) $p^\omega_{\mathcal{K},\boldsymbol{a}}$, $\mathcal{K} \subseteq \mathsf{M}$ *compact,* $\boldsymbol{a} \in \mathrm{c}_0(\mathbb{Z}_{\geq 0}; \mathbb{R}_{>0}; \rho)$ *for* $\rho \in \mathbb{R}_{>0}$*;*
(iv) $p^\omega_{\mathcal{K},\boldsymbol{a}}$, $\mathcal{K} \subseteq \mathsf{M}$ *compact,* $\boldsymbol{a} \in \mathrm{c}_0^{\downarrow}(\mathbb{Z}_{\geq 0}; \mathbb{R}_{>0}; \rho)$ *for* $\rho \in \mathbb{R}_{>0}$*.*

Proof Let us denote the topologies defined by the four collections of seminorms by

$$\mathscr{O}^\omega,\ \mathscr{O}^\omega_\downarrow,\ \mathscr{O}^\omega_\rho,\ \mathscr{O}^\omega_{\downarrow,\rho},$$

respectively. Since the seminorms defining $\mathscr{O}^\omega_\downarrow$ and $\mathscr{O}^\omega_\rho$ are a subset of those defining $\mathscr{O}^\omega$, we immediately have

$$\mathscr{O}^\omega \subseteq \mathscr{O}^\omega_\rho \subseteq \mathscr{O}^\omega_{\rho,\downarrow}, \quad \mathscr{O}^\omega \subseteq \mathscr{O}^\omega_\downarrow \subseteq \mathscr{O}^\omega_{\rho,\downarrow}.$$

For $\boldsymbol{a} \in \mathrm{c}_0(\mathbb{Z}_{\geq 0}; \mathbb{R})$ and for $\rho \in \mathbb{R}_{>0}$, we define $\boldsymbol{a}^\rho \in \mathrm{c}_0(\mathbb{Z}_{\geq 0}; \mathbb{R}_{>0}; \rho)$ by $a^\rho_j = \max\{a_j, \rho\}$, $j \in \mathbb{Z}_{\geq 0}$. Given $\boldsymbol{a} \in \mathrm{c}_0(\mathbb{Z}_{\geq 0}; \mathbb{R}_{>0})$, we construct $\boldsymbol{a}^\downarrow \in \mathrm{c}^\downarrow_0(\mathbb{Z}_{\geq 0}; \mathbb{R}_{>0})$ as follows. We take $a^\downarrow_0 = a_0$ and let $m_0 > 0$ be the least integer for which $a_j < a_0$ for $j > m_0$. We then define $a^\downarrow_j = a_0$, $j \in \{0, 1, \dots, m_0\}$. We next let $m_1 > m_0$ be the least integer for which $a_j < a_{m_0+1}$ for $j > m_1$. We then define $a^\downarrow_j = a_{m_0+1}$, $j \in \{m_0+1, \dots, m_1\}$. We carry on in this way to define $a^\downarrow_j$, $j \in \mathbb{Z}_{\geq 0}$. Then we obviously have

$$p^\omega_{\mathcal{K},\boldsymbol{a}^\rho}(\xi) \leq p^\omega_{\mathcal{K},\boldsymbol{a}}(\xi),\ p^\omega_{\mathcal{K},\boldsymbol{a}^\downarrow}(\xi) \leq p^\omega_{\mathcal{K},\boldsymbol{a}}(\xi), \qquad \mathcal{K} \subseteq \mathsf{M} \text{ compact},\ \xi \in \Gamma^\omega(\mathsf{E}).$$

From this we conclude that

$$\mathscr{O}^\omega_{\rho,\downarrow} \subseteq \mathscr{O}^\omega_\rho \subseteq \mathscr{O}^\omega, \quad \mathscr{O}^\omega_{\rho,\downarrow} \subseteq \mathscr{O}^\omega_\downarrow \subseteq \mathscr{O}^\omega,$$

and the lemma follows. □

While we are primarily interested in the difficult real analytic case in this book, it is useful and illustrative to, at times, make comparisons with the other regularity classes, particularly smooth regularity. With this in mind, let us also define seminorms for the sets $\Gamma^\nu(\mathsf{E})$ of sections of class C^ν for $\nu \in \mathbb{Z}_{\geq 0} \cup \{\infty\}$. For $\nu = m \in \mathbb{Z}_{\geq 0}$, we define the seminorms

$$p^m_{\mathcal{K}}(\xi) = \sup\left\{ \|j_m\xi(x)\|_{\mathbb{G}_{\mathsf{M},\pi_{\mathsf{E}},m}} \;\middle|\; x \in \mathcal{K} \right\} \tag{2.16}$$

for $\mathcal{K} \subseteq \mathsf{M}$ compact. For $\nu = \infty$, the appropriate seminorms are

$$p^\infty_{\mathcal{K},m}(\xi) = \sup\left\{ \|j_m\xi(x)\|_{\mathbb{G}_{\mathsf{M},\pi_{\mathsf{E}},m}} \;\middle|\; x \in \mathcal{K} \right\} \tag{2.17}$$

for $\mathcal{K} \subseteq \mathsf{M}$ compact and for $m \in \mathbb{Z}_{\geq 0}$. These seminorms give the topology of uniform convergence of derivatives on compact sets. In particular, the C^0-topology defined in this way is exactly the compact-open topology. A moment's thought will convince oneself that these seminorms define a Polish topology for $\Gamma^\nu(\mathsf{E})$ called the ***C^ν-topology***, $\nu \in \mathbb{Z}_{\geq 0} \cup \{\infty\}$. We note that, for the smooth topology, the seminorms are defined for fixed order jets. As we shall indicate as we go along, it is this fact that leads to simplifications of the results in the book when applied to the smooth case.

Let us close this section with seminorm characterisations of holomorphic and real analytic sections. We note that we have already made use of both of the following lemmata at various points in our constructions above, namely in the proof of Proposition 2.18. Moreover, as we see from the proofs of the lemmata, they make use of Lemma 4.22. Additionally, we have made an independent call to Lemma 4.22 in our proof of Proposition 2.18. One might be led to wonder whether this use of the yet-to-be-proved Lemma 4.22 constitutes circular reasoning. It does not; in the diagram in (4.16) we carefully unwind the logical implications that makes everything consistent.

First we give geometric analogues of the classical Cauchy estimates for holomorphic mappings.

Lemma 2.21 (Cauchy Estimates for Vector Bundles) *Let $\pi_{\mathsf{E}}\colon \mathsf{E} \to \mathsf{M}$ be an holomorphic vector bundle, let $\mathcal{K} \subseteq \mathsf{M}$ be compact, and let $\mathcal{U}$ be a precompact neighbourhood of $\mathcal{K}$. Then there exist $C, r \in \mathbb{R}_{>0}$ such that*

$$p^{\infty}_{\mathcal{K},m}(\xi) \le C r^{-m} p^{\mathrm{hol}}_{\mathcal{U},\infty}(\xi)$$

for every $m \in \mathbb{Z}_{\ge 0}$ and $\xi \in \Gamma^{\mathrm{hol}}_{\mathrm{bdd}}(\mathsf{E}|\mathcal{U})$.

Proof Let $z \in \mathcal{K}$ and let $(\mathcal{W}_z, \boldsymbol{v}_z)$ be an holomorphic vector bundle chart about z with $(\mathcal{U}_z, \boldsymbol{\chi}_z)$ the associated chart for M, supposing that $\mathcal{U}_z \subseteq \mathcal{U}$. Let $k \in \mathbb{Z}_{>0}$ be such that $\boldsymbol{v}_z(\mathcal{W}_z) = \boldsymbol{\chi}_z(\mathcal{U}_z) \times \mathbb{C}^k$. Let $\boldsymbol{z} = \boldsymbol{\chi}_z(z)$ and let $\boldsymbol{\xi}\colon \boldsymbol{\chi}_z(\mathcal{U}_z) \to \mathbb{C}^k$ be the local representative of $\xi \in \Gamma^{\mathrm{hol}}_{\mathrm{bdd}}(\mathsf{E}|\mathcal{U})$. Note that, when taking real derivatives of $\boldsymbol{\xi}$ with respect to coordinates, we can think of taking derivatives with respect to

$$\frac{\partial}{\partial z^j} = \frac{1}{2}\left(\frac{\partial}{\partial x^j} - \mathrm{i}\frac{\partial}{\partial y^j}\right), \quad \frac{\partial}{\partial \bar{z}^j} = \frac{1}{2}\left(\frac{\partial}{\partial x^j} + \mathrm{i}\frac{\partial}{\partial y^j}\right), \qquad j \in \{1, \dots, n\}.$$

Since $\boldsymbol{\xi}$ is holomorphic, the $\frac{\partial}{\partial \bar{z}^j}$ derivatives will vanish [45, page 27]. Thus, for the purposes of the multi-index calculations, we consider multi-indices of length n (not $2n$). In any case, applying the usual Cauchy estimates [45, Lemma 2.3.9], there exists $r \in \mathbb{R}_{>0}$ such that

$$|\boldsymbol{D}^I \xi^a(\boldsymbol{z})| \le I! r^{-|I|} \sup\left\{ |\xi^a(\boldsymbol{\zeta})| \;\middle|\; \boldsymbol{\zeta} \in \overline{\mathsf{D}}(\boldsymbol{r}, \boldsymbol{z})\right\}$$

for every $a \in \{1, \dots, k\}$, $I \in \mathbb{Z}^n_{\ge 0}$, and $\xi \in \Gamma^{\mathrm{hol}}_{\mathrm{bdd}}(\mathsf{E}|\mathcal{U})$. We may choose $r \in (0, 1)$ such that $\overline{\mathsf{D}}(\boldsymbol{r}, \boldsymbol{z})$ is contained in $\boldsymbol{\phi}_z(\mathcal{U}_z)$, where $\boldsymbol{r} = (r, \dots, r)$. Denote $\mathcal{V}_z = \boldsymbol{\chi}_z^{-1}(\mathsf{D}(\boldsymbol{r}, \boldsymbol{z}))$. By continuity, there exists a neighbourhood $\mathcal{V}'_z$ of z such that $\mathrm{cl}(\mathcal{V}'_z) \subseteq \mathcal{V}_z$ and such that

$$|\boldsymbol{D}^I \xi^a(\boldsymbol{z}')| \le 2 I! r^{-|I|} \sup\left\{ |\xi^a(\boldsymbol{\zeta})| \;\middle|\; \boldsymbol{\zeta} \in \overline{\mathsf{D}}(\boldsymbol{r}, \boldsymbol{z})\right\}$$

for every $z' \in \chi_z(\mathcal{V}'_z)$, $\xi \in \Gamma^{\text{hol}}_{\text{bdd}}(\mathsf{E}|\mathcal{U})$, $a \in \{1, \dots, k\}$, and $I \in \mathbb{Z}^n_{\geq 0}$. If $|I| \leq m$, then, since we are assuming that $r < 1$, we have

$$\frac{1}{I!}|\boldsymbol{D}^I \xi^a(\boldsymbol{z}')| \leq 2r^{-m} \sup\left\{ |\xi^a(\boldsymbol{\zeta})| \;\middle|\; \boldsymbol{\zeta} \in \overline{\mathsf{D}}(\boldsymbol{r}, \boldsymbol{z})\right\}$$

for every $a \in \{1, \dots, k\}$, $z' \in \chi_z(\mathcal{V}'_z)$, and $\xi \in \Gamma^{\text{hol}}_{\text{bdd}}(\mathsf{E}|\mathcal{U})$. By Lemma 4.22, it follows that there exist $C_z, r_z \in \mathbb{R}_{>0}$ such that

$$\|j_m\xi(z')\|_{\mathbb{G}_{\mathsf{M},\pi_{\mathsf{E}},m}} \leq C_z r_z^{-m} p^{\text{hol}}_{\mathcal{V}_z,\infty}(\xi)$$

for all $z' \in \mathcal{V}'_z$, $m \in \mathbb{Z}_{\geq 0}$, and $\xi \in \Gamma^{\text{hol}}_{\text{bdd}}(\mathsf{E}|\mathcal{U})$. Let $z_1, \dots, z_k \in \mathcal{K}$ be such that $\mathcal{K} \subseteq \cup_{j=1}^k \mathcal{V}'_{z_j}$, and let $C = \max\{C_{z_1}, \dots, C_{z_k}\}$ and $r = \min\{r_{z_1}, \dots, r_{z_k}\}$. If $z \in \mathcal{K}$, then $z \in \mathcal{V}'_{z_j}$ for some $j \in \{1, \dots, k\}$ and so we have

$$\|j_m\xi(z)\|_{\mathbb{G}_{\mathsf{M},\pi_{\mathsf{E}},m}} \leq C_{z_j} r_{z_j}^{-m} p^{\text{hol}}_{\mathcal{V}_{z_j},\infty}(\xi) \leq C r^{-m} p^{\text{hol}}_{\mathcal{U},\infty}(\xi),$$

and taking supremums over $z \in \mathcal{K}$ on the left gives the result. □

The following lemma, providing bounds for real analytic sections, is a global version of the local result stated as Theorem 1.9.

Lemma 2.22 (Characterisation of Real Analytic Sections) *Let $\pi_{\mathsf{E}} : \mathsf{E} \to \mathsf{M}$ be a real analytic vector bundle and let $\xi \in \Gamma^\infty(\mathsf{E})$. Then the following statements are equivalent:*

(i) $\xi \in \Gamma^\omega(\mathsf{E})$;
(ii) for every compact set $\mathcal{K} \subseteq \mathsf{M}$, there exist $C, r \in \mathbb{R}_{>0}$ such that $p^\infty_{\mathcal{K},m}(\xi) \leq C r^{-m}$ for every $m \in \mathbb{Z}_{\geq 0}$.

Proof (i) $\implies$ (ii) Let $\mathcal{K} \subseteq \mathsf{M}$ be compact, let $x \in \mathcal{K}$, and let $(\mathcal{V}_x, \boldsymbol{\nu}_x)$ be a vector bundle chart for E with $(\mathcal{U}_x, \boldsymbol{\chi}_x)$ the corresponding chart for M. Let $\boldsymbol{\xi} : \boldsymbol{\chi}_x(\mathcal{U}_x) \to \mathbb{R}^k$ be the local representative of ξ. By Theorem 1.9, there exist a neighbourhood $\mathcal{U}'_x \subseteq \mathcal{U}_x$ of x and $B_x, \sigma_x \in \mathbb{R}_{>0}$ such that

$$|\boldsymbol{D}^I \xi^a(\boldsymbol{x}')| \leq B_x I! \sigma_x^{-|I|}$$

for every $a \in \{1, \dots, k\}$, $\boldsymbol{x}' \in \text{cl}(\mathcal{U}'_x)$, and $I \in \mathbb{Z}^n_{\geq 0}$. We can suppose, without loss of generality, that $\sigma_x \in (0, 1)$. In this case, if $|I| \leq m$,

$$\frac{1}{I!}|\boldsymbol{D}^I \xi^a(\boldsymbol{x}')| \leq B_x \sigma_x^{-m}$$

for every $a \in \{1, \dots, k\}$ and $\boldsymbol{x}' \in \mathrm{cl}(\mathcal{U}'_x)$. By Lemma 4.22, there exist $C_x, r_x \in \mathbb{R}_{>0}$ such that

$$\|j_m\xi(x')\|_{\mathbb{G}_{\mathsf{M},\pi_{\mathsf{E}},m}} \leq C_x r_x^{-m}, \qquad x' \in \mathrm{cl}(\mathcal{U}'_x),\ m \in \mathbb{Z}_{\geq 0}.$$

Let $x_1, \dots, x_k \in \mathcal{K}$ be such that $\mathcal{K} \subseteq \cup_{j=1}^k \mathcal{U}'_{x_j}$ and let $C = \max\{C_{x_1}, \dots, C_{x_k}\}$ and $r = \min\{r_{x_1}, \dots, r_{x_k}\}$. Then, if $x \in \mathcal{K}$, we have $x \in \mathcal{U}'_{x_j}$ for some $j \in \{1, \dots, k\}$ and so

$$\|j_m\xi(x)\|_{\mathbb{G}_{\mathsf{M},\pi_{\mathsf{E}},m}} \leq C_{x_j} r_{x_j}^{-m} \leq C r^{-m},$$

as desired.

(ii) $\implies$ (ii) Let $x \in \mathsf{M}$ and let $(\mathcal{V}, \boldsymbol{v})$ be a vector bundle chart for E such that the associated chart $(\mathcal{U}, \boldsymbol{\chi})$ for M is a precompact coordinate chart about x. Let $\boldsymbol{\xi}\colon \boldsymbol{\chi}(\mathcal{U}) \to \mathbb{R}^k$ be the local representative of ξ. By hypothesis, there exist $C, r \in \mathbb{R}_{>0}$ such that $\|j_m\xi(x')\|_{\mathbb{G}_{\mathsf{M},\pi_{\mathsf{E}},m}} \leq C r^{-m}$ for every $m \in \mathbb{Z}_{\geq 0}$ and $x' \in \mathcal{U}$. Let $\mathcal{U}'$ be a precompact neighbourhood of x such that $\mathrm{cl}(\mathcal{U}') \subseteq \mathcal{U}$. By Lemma 4.22, there exist $B, \sigma \in \mathbb{R}_{>0}$ such that

$$|\boldsymbol{D}^I \xi^a(\boldsymbol{x}')| \leq B I! \sigma^{-|I|}$$

for every $a \in \{1, \dots, k\}$, $\boldsymbol{x}' \in \mathrm{cl}(\mathcal{U}')$, and $I \in \mathbb{Z}^n_{\geq 0}$. We conclude real analyticity of ξ in a neighbourhood of x by Theorem 1.9. □

2.5 The Topology for the Space of Real Analytic Mappings

Next we turn to topologising the space of real analytic mappings between real analytic manifolds M and N. We shall provide a few different ways to describe this topology, some of these rather similar to the constructions of Sect. 2.2 for topologising the space of sections of a real analytic vector bundle. Our initial definition of this topology, however, will not be immediately relatable to the vector bundle case. For this initial definition, we shall make a connection to a commonly used topology for finitely differentiable and smooth mappings: the topology of uniform convergence of derivatives on compact sets. After using this as motivation for our definition for the topology for the space of real analytic mappings, we give characterisations of this topology using complexification, in the manner of the two different definitions for a topology for the space of real analytic sections in Sect. 2.2.

2.5.1 Motivation for, and Definition of, the Weak-PB Topology

We first consider a well-known topology for finitely differentiable and smooth mappings. Let us motivate this by recalling the topology for spaces of sections of vector bundles defined by the seminorms (2.16) (in the finitely differentiable case) and the seminorms (2.17) (in the smooth case). We note that these topologies can be regarded as the topology of uniform convergence of finite-order derivatives on compact sets. This idea can be relatively easily adapted to topologies for finitely differentiable and smooth mappings, and we outline this here.

We let M and N be C^∞-manifolds and let $\nu \in \mathbb{Z}_{\geq 0} \cup \{\infty\}$, with $\mathrm{C}^\nu(\mathsf{M}; \mathsf{N})$, therefore, the space of C^ν-mappings from M to N. The topology we consider for this space is called the "weak topology" by [33, §2.1] and the "CO^ν-topology" by Michor [52, §4.3]. It is the adaptation of the standard compact-open topology for spaces of continuous mappings between topological spaces [73, §43]. Somewhat precisely, it is the topology where a sequence $(\Phi_j)_{j\in\mathbb{Z}_{>0}}$ converges to Φ if and only if:

1. $\nu \in \mathbb{Z}_{\geq 0}$: the sequence of all derivatives of $(\Phi_j)_{j\in\mathbb{Z}_{>0}}$ up to and including order ν converge uniformly on compact subsets;
2. $\nu = \infty$: for all $k \in \mathbb{Z}_{\geq 0}$, the sequence of all derivatives of $(\Phi_j)_{j\in\mathbb{Z}_{>0}}$ up to and including order k converge uniformly on compact subsets.

There is another characterisation of this topology, and it is the following.

Theorem 2.23 (Characterisation of "Weak Topology" or "CO^ν-Topology") *For smooth manifolds M and N and for $\nu \in \mathbb{Z}_{\geq 0} \cup \{\infty\}$, the topology for $\mathrm{C}^\nu(\mathsf{M}; \mathsf{N})$ described above is the same as the initial topology associated with the family of mappings*

$$\begin{aligned} \Theta_f \colon \mathrm{C}^\nu(\mathsf{M}; \mathsf{N}) &\to \mathrm{C}^\nu(\mathsf{M}) \\ \Phi &\mapsto \Phi^* f, \end{aligned} \qquad f \in \mathrm{C}^\infty(\mathsf{N}),$$

where $\mathrm{C}^\infty(\mathsf{N})$ has the topology defined by the seminorms (2.17).

By the nature of this characterisation of these topologies, the topologies are the uniform topologies associated with the families of semimetrics

$$d^m_{\mathcal{K},f}(\Phi_1, \Phi_2) = p^m_{\mathcal{K}}(f \circ \Phi_1 - f \circ \Phi_2) \tag{2.18}$$

(in the case $\nu = m \in \mathbb{Z}_{\geq 0}$) and

$$d^\infty_{\mathcal{K},m,f}(\Phi_1, \Phi_2) = p^\infty_{\mathcal{K},m}(f \circ \Phi_1 - f \circ \Phi_2), \qquad m \in \mathbb{Z}_{\geq 0}, \tag{2.19}$$

(in the case $\nu = \infty$) for $f \in \mathrm{C}^\infty(\mathsf{N})$ and $\mathcal{K} \subseteq \mathsf{M}$ compact.

While we were not able to pinpoint a proof of the previous theorem anywhere, it will not come as a surprise to those who understand this topology well; its validity boils down to being able to find globally defined coordinate functions about any point in M and N, these being furnished, for example, by the Whitney Embedding Theorem. Thanks to the Grauert Embedding Theorem, we are similarly able to find globally defined real analytic coordinate functions about any point on a real analytic manifold. We use this as motivation for the following definition.

Definition 2.24 (Weak PB-Topology) Let M and N be real analytic manifolds. The ***weak-PB topology*** for $C^\omega(\mathsf{M};\mathsf{N})$ is the initial topology associated with the family of mappings

$$\begin{aligned} \Theta_f\colon C^\omega(\mathsf{M};\mathsf{N}) &\to C^\omega(\mathsf{M}) \\ \Phi &\mapsto \Phi^* f, \end{aligned} \qquad f \in C^\omega(\mathsf{N}),$$

where $C^\omega(\mathsf{N})$ has the C^ω-topology. ○

The following theorem gives a few alternative means of characterising the weak-PB topology, making use of Grauert's Embedding Theorem. Note that this theorem, applied appropriately in the finitely differentiable and smooth cases, gives a proof of Theorem 2.23.

Theorem 2.25 (Characterisation of Weak-PB Topology for Spaces of Real Analytic Mappings) *Let* M *and* N *be* C^ω*-manifolds and let*

$$\begin{aligned} \iota\colon \mathsf{N} &\to \mathbb{R}^N \\ y &\mapsto (\iota^1(y), \dots, \iota^N(y)) \end{aligned}$$

be a proper C^ω*-embedding. Then the following topologies for* $C^\omega(\mathsf{M};\mathsf{N})$ *agree:*

(i) the initial topology associated with the family of mappings

$$\begin{aligned} \Psi_f\colon C^\omega(\mathsf{M};\mathsf{N}) &\to C^\omega(\mathsf{M}) \\ \Phi &\mapsto \Phi^* f, \end{aligned} \qquad f \in C^\omega(\mathsf{M});$$

(ii) the initial topology associated with the family of mappings

$$\begin{aligned} \Psi_{\iota^j}\colon C^\omega(\mathsf{M};\mathsf{N}) &\to C^\omega(\mathsf{M}) \\ \Phi &\mapsto \Phi^* \iota^j, \end{aligned} \qquad j \in \{1, \dots, N\};$$

(iii) the topology induced on $C^\omega(\mathsf{M};\mathsf{N}) \subseteq C^\omega(\mathsf{M};\mathbb{R}^N)$ *by the weak-PB topology for*

$$C^\omega(\mathsf{M};\mathbb{R}^N) \simeq \bigoplus_{j=1}^N C^\omega(\mathsf{M}).$$

Proof (ii)⊆(i): The topology (ii) is the coarsest topology for which the mappings Ψ_{ι^j}, $j \in \{1, \ldots, N\}$, are continuous. The topology (i) has this property and so (ii)⊆(i).

(iii)⊆(ii): Note that, because the induced topology is the initial topology induced by the inclusion $C^\omega(M; N) \subseteq C^\omega(M; \mathbb{R}^N)$, the topology (iii) is the coarsest topology for which that inclusion is continuous. Since the inclusion is given by

$$\Phi \mapsto (\iota^1 \circ \Phi, \ldots, \iota^N \circ \Phi) = (\Psi_{\iota^1}(\Phi), \ldots, \Psi_{\iota^N}(\Phi)),$$

by Theorem 5.26 the topology (ii) has the property that this inclusion is continuous, and so (iii)⊆(ii).

(i)⊆(iii): The topology (i) is the coarsest topology for which Ψ_f is continuous for every $f \in C^\omega(N)$. We claim that Ψ_f is continuous for all $f \in C^\omega(M)$ for the topology (iii), which will show that (i)⊆(iii).
We first state a lemma.

Lemma 1 *Let M be a C^ω-manifold and let S $\subseteq$ M be a C^ω-embedded submanifold with $\iota_S : S \to M$ the inclusion. Then*

$$\iota_S^* : C^\omega(M) \to C^\omega(S)$$

is an epimorphism, i.e., continuous, surjective, and open.

Proof First note that we can use Lemma 2.1 to show that $\iota_S^* : C^\omega(M) \to C^\omega(S)$ is surjective. It, therefore, remains to show that ι_S^* is continuous and open. Continuity follows from Theorem 5.26. Since $C^\omega(S)$ and $C^\omega(M)$ are ultrabornological webbed spaces (Proposition 2.14), the De Wilde Open Mapping Theorem implies that ι_S^* is open.   ▽

We apply the lemma with $M = \mathbb{R}^N$ and $S = N$. Let $f \in C^\omega(M)$ and, by the lemma, let $\hat{f} \in C^\omega(\mathbb{R}^N)$ be such that $f = \hat{f} \circ \iota$. Now consider the commutative diagram

$$
\begin{array}{ccc}
C^\omega(M; \mathbb{R}^N) & \xrightarrow{\Psi_{\hat{f}}} & C^\omega(M) \\
{\scriptstyle \Phi \mapsto \iota \circ \Phi} \Big\uparrow & \nearrow {\scriptstyle \Psi_f} & \\
C^\omega(M; N) & &
\end{array}
$$

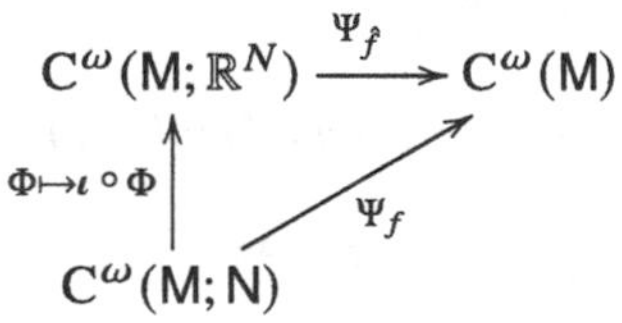

where $C^\omega(M; N)$ has the topology (iii) and (as required by the definition of the topology (iii)) $C^\omega(M; \mathbb{R}^N)$ has the weak-PB topology. Thus the vertical and horizontal arrows are continuous, whence the diagonal arrow is continuous, as desired.

Finally, let us prove the assertion made in part (iii) that the weak-PB topology for $C^\omega(M; \mathbb{R}^N)$ is topologically isomorphic to $\oplus_{j=1}^N C^\omega(M)$. This is equivalent, given what we have already proven, to the assertion that the initial topology for $\mathrm{C}^\omega(\mathsf{M}; \mathbb{R}^N)$ associated with the mappings

$$\begin{aligned} \Psi_j : \mathrm{C}^\omega(\mathsf{M}; \mathbb{R}^N) &\to \mathrm{C}^\omega(\mathsf{M}) \\ \boldsymbol{\Phi} &\mapsto \Phi^j, \end{aligned}$$

$j \in \{1, \dots, N\}$, is isomorphic to $\oplus_{j=1}^N \mathrm{C}^\omega(\mathsf{M})$. Here we write $\boldsymbol{\Phi} = (\Phi^1, \dots, \Phi^N)$. However, this assertion follows from a general assertion regarding initial topologies [73, Theorem 8.12]. Indeed, this general result asserts that the mapping

$$\begin{aligned} \Psi : \mathrm{C}^\omega(\mathsf{M}; \mathbb{R}^N) &\to \bigoplus_{j=1}^N \mathrm{C}^\omega(\mathsf{M}) \\ \boldsymbol{\Phi} &\mapsto (\Phi^1, \dots, \Phi^N) \end{aligned}$$

is an homeomorphism onto its image. Since it is surjective, it is, therefore, an homeomorphism. (Here we also make use of the fact that finite direct sums are topologically isomorphic to products [37, Proposition 4.3.2].)

Our proof, in fact, yields a slightly more general conclusion related to part (iii).

Corollary 2.26 (Weak-PB Topology for Space of Real Analytic Mappings with Values in a Submanifold) *Let* M *and* N *be* C^ω*-manifolds and let* $\mathsf{S} \subseteq \mathsf{N}$ *be an embedded submanifold. Then the weak-PB topology of* $\mathrm{C}^\omega(\mathsf{M}; \mathsf{S})$ *is the topology induced by the weak-PB topology of* $\mathrm{C}^\omega(\mathsf{M}; \mathsf{N})$ *and the inclusion* $\mathrm{C}^\omega(\mathsf{M}; \mathsf{S}) \subseteq \mathrm{C}^\omega(\mathsf{M}; \mathsf{N})$.

The preceding theorem, along with some of our continuity results from Chap. 5, lead to the following result of independent interest.

Proposition 2.27 ("Weak" Characterisations for Topology for Spaces of Real Analytic Sections) *For a* C^ω*-vector bundle* $\pi_{\mathsf{E}} : \mathsf{E} \to \mathsf{M}$*, the following topologies for* $\Gamma^\omega(\mathsf{E})$ *agree:*

(i) the C^ω*-topology;*
(ii) the topology induced by the inclusion $\Gamma^\omega(\mathsf{E}) \subseteq \mathrm{C}^\omega(\mathsf{M}; \mathsf{E})$ *and the weak-PB topology for* $\mathrm{C}^\omega(\mathsf{M}; \mathsf{E})$*;*
(iii) the initial topology associated with the mappings

$$\begin{aligned} \mathrm{ev}_\alpha : \Gamma^\omega(\mathsf{E}) &\to \mathrm{C}^\omega(\mathsf{M}) \\ \xi &\mapsto (x \mapsto \langle \alpha(x); \xi(x) \rangle), \end{aligned} \qquad \alpha \in \Gamma^\omega(\mathsf{E}^*).$$

Proof In the proof, we shall make use of Corollary 2.5 which gives E as a real analytic subbundle of a trivial bundle $\mathrm{pr}_1 : \mathbb{R}^N_{\mathsf{M}} = \mathsf{M} \times \mathbb{R}^N \to \mathsf{M}$. We also define $\mathsf{E}^\perp$

as the subbundle of $\mathbb{R}^N_{\mathsf{M}}$ whose fibres are the orthogonal complements of the fibres of E, using the Euclidean fibre metric. Thus $\mathsf{E}^{\perp}$ is a also a real analytic subbundle.

(i)=(ii) By Theorem 2.25(iii) and Theorem 5.4 below, the obvious vector space isomorphism

$$\mathrm{C}^{\omega}(\mathsf{M}; \mathbb{R}^N) \simeq \Gamma^{\omega}(\mathbb{R}^N_{\mathsf{M}})$$

is an homeomorphism, if $\mathrm{C}^{\omega}(\mathsf{M}; \mathbb{R}^N)$ has the weak-PB topology and $\Gamma^{\omega}(\mathbb{R}^N_{\mathsf{M}})$ has the C^{ω}-topology. Now note that our recollections from the first paragraph of the proof give the diagram

$$\begin{array}{ccc} \mathrm{C}^{\omega}(\mathsf{M}; \mathbb{R}^N) & \xrightarrow{\simeq_{\mathrm{top}}} & \Gamma^{\omega}(\mathbb{R}^N_{\mathsf{M}}) \\ \uparrow & & \uparrow{\scriptstyle \iota_{\mathsf{E}}} \\ \mathrm{C}^{\omega}(\mathsf{M}; \mathsf{E}) & \longleftarrow & \Gamma^{\omega}(\mathsf{E}) \end{array}$$

As indicated, the top arrow is a topological isomorphism. The bottom arrow and the vertical arrows are the natural inclusions. The left inclusion is a topological embedding by Corollary 2.26 (noting that a subbundle is necessarily an embedded submanifold). By Theorem 5.4, we have the topological isomorphism

$$\Gamma^{\omega}(\mathbb{R}^N_{\mathsf{M}}) \simeq \Gamma^{\omega}(\mathsf{E}) \oplus \Gamma^{\omega}(\mathsf{E}^{\perp}).$$

Moreover, from Theorem 5.4 it also follows that the right inclusion in the diagram is a topological monomorphism. Now the bottom inclusion is verified to be continuous and open by elementary arguments with open sets, using the facts just given. This shows that the first two topologies agree.

(ii)=(iii) With E as a subbundle of $\mathbb{R}^N_{\mathsf{M}}$, define

$$\begin{aligned} \iota_j \colon \mathrm{C}^{\omega}(\mathsf{M}; \mathsf{E}) &\to \mathrm{C}^{\omega}(\mathsf{M}) \\ \xi &\mapsto \mathrm{pr}_j \circ \xi, \end{aligned}$$

where

$$\begin{aligned} \mathrm{pr}_j \colon \mathbb{R}^N_{\mathsf{M}} &\to \mathbb{R} \\ (x, (v_1, \dots, v_N)) &\mapsto v_j. \end{aligned}$$

Note that $\iota_j | \Gamma^{\omega}(\mathsf{E}) = \mathrm{ev}_{\alpha_j}$ for $\alpha_j \in \Gamma^{\omega}(\mathsf{E}^*)$ given by

$$\langle \alpha_j(x); e_x \rangle = \mathrm{pr}_j \circ \iota_{\mathsf{E}}(\xi(x)).$$

By Theorem 2.25(iii), the initial topology for $\mathrm{C}^\omega(\mathsf{M};\mathsf{E})$ associated with the mappings ι_j, $j \in \{1,\dots,N\}$, is equal to the weak-PB topology. Thus the initial topology for $\Gamma^\omega(\mathsf{E})$ associated to the mappings $\iota_j|\Gamma^\omega(\mathsf{E}) = \mathrm{ev}_{\alpha_j}$, $j \in \{1,\dots,N\}$, is equal to the topology (ii). □

Note that the weak-PB topology can be characterised as being the uniform topology associated with the family of semimetrics

$$\begin{aligned} \mathrm{d}^\omega_{\mathcal{K},\boldsymbol{a},f}\colon \mathrm{C}^\omega(\mathsf{M};\mathsf{N}) \times \mathrm{C}^\omega(\mathsf{M};\mathsf{N}) &\to \mathbb{R} \\ (\Phi_1,\Phi_2) &\mapsto p^\omega_{\mathcal{K},\boldsymbol{a}}(\Phi_1^*f - \Phi_2^*f), \end{aligned} \tag{2.20}$$

for $\mathcal{K} \subseteq \mathsf{M}$ compact, $\boldsymbol{a} \in c_0(\mathbb{Z}_{\ge 0};\mathbb{R}_{>0})$, and $f \in \mathrm{C}^\omega(\mathsf{N})$. Moreover, the previous theorem ensures that it suffices to restrict attention to the semimetrics of this form for $f \in \{\iota^1,\dots,\iota^N\}$, where $\boldsymbol{\iota}\colon \mathsf{N} \to \mathbb{R}^{\mathsf{N}}$ is a proper C^ω-embedding.

For the remainder of this section, we shall consider descriptions of the weak-PB topology for mappings that mirror the constructions of Sect. 2.2 for the C^ω-topology for the space of real analytic sections of a real analytic vector bundle.

2.5.2 The Topology for the Space of Holomorphic Mappings

Since our constructions will be derived from topologies for spaces of holomorphic mappings, we should first specify and understand this topology. We let M and N be holomorphic manifolds. For the space $\mathrm{C}^{\mathrm{hol}}(\mathsf{M};\mathsf{N})$ of holomorphic mappings, we use the compact-open topology as per [73, §43]. This is the topology with the subbase

$$\mathcal{B}(\mathcal{K},\mathcal{V}) = \{\Phi \in \mathrm{C}^{\mathrm{hol}}(\mathsf{M};\mathsf{N}) \mid \Phi(\mathcal{K}) \subseteq \mathcal{V}\}, \qquad \mathcal{K} \subseteq \mathsf{M} \text{ compact},\ \mathcal{V} \subseteq \mathsf{N} \text{ open}.$$

The topology for the space of holomorphic sections of an holomorphic vector bundle $\pi_{\mathsf{E}}\colon \mathsf{E} \to \mathsf{M}$ with the seminorms

$$p^{\mathrm{hol}}_{\mathcal{K}}(\xi) = \sup\left\{ \|\xi(z)\|_{\mathbb{G}_{\mathsf{E}}} \;\middle|\; z \in \mathcal{K}\right\}, \qquad \mathcal{K} \subseteq \mathsf{M} \text{ compact},$$

(cf. (2.2)) is exactly the compact-open topology, thinking of a section as a mapping from M to E.

In the following result, we give a few different characterisations of the compact-open topology for the space of holomorphic mappings, some valid only for Stein manifolds.

Theorem 2.28 (Characterisation of the Compact-Open Topology for Space of Holomorphic Mappings) *If* M *and* N *are holomorphic manifolds, then the following topologies for* $\mathrm{C}^{\mathrm{hol}}(\mathsf{M}; \mathsf{N})$ *are the same:*

(i) the compact-open topology;
(ii) the topology of uniform convergence on compact subsets of M.

Additionally, if M *and* N *are Stein manifolds and if* $\iota\colon \mathsf{N} \to \mathbb{C}^N$ *is a proper holomorphic embedding, then the preceding two topologies are the same as the following two:*

(iii) the initial topology associated with the family of mappings

$$\begin{aligned}\Theta_f\colon \mathrm{C}^{\mathrm{hol}}(\mathsf{M}; \mathsf{N}) &\to \mathrm{C}^{\mathrm{hol}}(\mathsf{M}) \\ \Phi &\mapsto \Phi^* f,\end{aligned} \qquad f \in \mathrm{C}^{\mathrm{hol}}(\mathsf{N}),$$

where $\mathrm{C}^{\mathrm{hol}}(\mathsf{N})$ *has the compact-open topology;*
(iv) the topology induced on $\mathrm{C}^{\mathrm{hol}}(\mathsf{M}; \mathsf{N}) \subseteq \mathrm{C}^{\mathrm{hol}}(\mathsf{M}; \mathbb{C}^N)$ *by the compact-open topology for*

$$\mathrm{C}^{\mathrm{hol}}(\mathsf{M}; \mathbb{C}^N) \simeq \bigoplus_{j=1}^{N} \mathrm{C}^{\mathrm{hol}}(\mathsf{M}).$$

Proof We shall only sketch the proof since the results are essentially well known or follow along the lines of results we have already proved.

(i)=(ii) This is a standard general result, e.g., [73, Theorem 43.7].

(ii)=(iii) In the same manner as Theorem 2.23 follows from the existence of smooth coordinate functions in a neighbourhood of any point, the assertion in this case follows since, for a Stein manifold, there are holomorphic coordinate functions in a neighbourhood of any point.

(iii)=(iv) This follows in the same manner as the corresponding assertion from Theorem 2.25, after one makes the following observations:

1. for $\Phi \in \mathrm{C}^{\mathrm{hol}}(\mathsf{M}; \mathsf{N})$, the mapping $\Phi^*\colon \mathrm{C}^{\mathrm{hol}}(\mathsf{N}) \to \mathrm{C}^{\mathrm{hol}}(\mathsf{M})$ is continuous;
2. if M is a Stein manifold, then coherent sheaves of $\mathscr{C}^{\mathrm{hol}}_{\mathsf{M}}$-modules have vanishing cohomology [12];
3. as a consequence, the holomorphic analogue of Lemma 2.1 holds when M is a Stein manifold and S is a Stein submanifold;
4. as a consequence, the holomorphic analogue of Lemma 1 from the proof of Theorem 2.25 holds.

The only one of these observations that we shall prove independently is the first. Let $\mathcal{K} \subseteq \mathsf{M}$ be compact and let $(g_j)_{j\in\mathbb{Z}_{>0}}$ be a Cauchy sequence in $\mathrm{C}^{\mathrm{hol}}(\mathsf{N})$. Thus, since

$\Phi(\mathcal{K})$ is compact, for every $\epsilon \in \mathbb{R}_{>0}$, there exists $N \in \mathbb{Z}_{>0}$ such that

$$|g_j(y) - g_k(y)| \leq \epsilon, \qquad j, k \geq N,\ y \in \Phi(\mathcal{K}).$$

Therefore,

$$|\Phi^* g_j(z) - \Phi^* g_k(z)| < \epsilon, \qquad j, k \geq N,\ z \in \mathcal{K},$$

showing that $(\Phi^* g_j)_{j \in \mathbb{Z}_{>0}}$ is a Cauchy sequence in $\mathrm{C}^{\mathrm{hol}}(\mathsf{M})$. This suffices to prove continuity of Φ^*. □

Note that a consequence of the theorem is that the topology of $\mathrm{C}^{\mathrm{hol}}(\mathsf{M}; \mathsf{N})$ is the restriction of the C^0-topology to holomorphic mappings. We also have the corresponding analogues of Corollary 2.26 and Proposition 2.27 for Stein manifolds. The proofs follow in exactly the manner as their real analytic counterparts.

Corollary 2.29 (Weak-PB Topology for Space of Holomorphic Mappings with Values in a Submanifold) *Let* M *and* N *be Stein manifolds and let* $\mathsf{S} \subseteq \mathsf{N}$ *be an embedded Stein submanifold. Then the compact-open topology of* $\mathrm{C}^{\mathrm{hol}}(\mathsf{M}; \mathsf{S})$ *is the topology induced by the compact-open topology of* $\mathrm{C}^{\mathrm{hol}}(\mathsf{M}; \mathsf{N})$ *and the inclusion* $\mathrm{C}^{\mathrm{hol}}(\mathsf{M}; \mathsf{S}) \subseteq \mathrm{C}^{\mathrm{hol}}(\mathsf{M}; \mathsf{N})$.

Proposition 2.30 ("Weak" Characterisations for Topology for Spaces of Holomorphic of Sections) *For an holomorphic vector bundle* $\pi_{\mathsf{E}} \colon \mathsf{E} \to \mathsf{M}$ *with* M *a Stein manifold, the following topologies agree:*

(i) the compact-open topology;
(ii) the topology induced by the inclusion $\Gamma^{\mathrm{hol}}(\mathsf{E}) \subseteq \mathrm{C}^{\mathrm{hol}}(\mathsf{M}; \mathsf{E})$ *and the compact-open topology for* $\mathrm{C}^{\mathrm{hol}}(\mathsf{M}; \mathsf{E})$*;*
(iii) the initial topology associated with the mappings

$$\begin{aligned} \mathrm{ev}_\alpha \colon \Gamma^{\mathrm{hol}}(\mathsf{E}) &\to \mathrm{C}^{\mathrm{hol}}(\mathsf{M}) \\ \xi &\mapsto (z \mapsto \langle \alpha(z); \xi(z) \rangle), \end{aligned} \qquad \alpha \in \Gamma^{\mathrm{hol}}(\mathsf{E}^*).$$

Proof The only missing ingredient in being able to use the same strategy as in the proof of Proposition 2.27 is to note that the total space of an holomorphic vector bundle over a Stein manifold is a Stein manifold [39, Proposition 54.B.4]. □

Also note that the compact-open topology is the uniform topology defined by the family of semimetrics

$$\mathrm{d}^{\mathrm{hol}}_{\mathcal{K}}(\Phi, \Psi) = \sup\{\mathrm{d}(\Phi(z), \Psi(z)) \mid z \in \mathcal{K}\}, \qquad \mathcal{K} \subseteq \mathsf{M} \text{ compact}, \tag{2.21}$$

where d is a metric on M for which the metric topology is the manifold topology, e.g., the metric associated with a Riemannian metric. Since one can consider the semimetrics associated to a compact exhaustion $(\mathcal{K}_j)_{j \in \mathbb{Z}_{>0}}$ of M, this shows that the compact-open topology for $\mathrm{C}^{\mathrm{hol}}(\mathsf{M}; \mathsf{N})$ is metrisable.

2.5.3 *Germs of Holomorphic Mappings Over Subsets of a Real Analytic Manifold*

We next adapt our discussion from Sect. 2.2.1 to mappings rather than sections.

Let M and N be real analytic manifolds with complexifications $\overline{\mathsf{M}}$ and $\overline{\mathsf{N}}$. Let $A \subseteq \mathsf{M}$ and let $\overline{\mathscr{N}}_A$ be the set of neighbourhoods of A in $\overline{\mathsf{M}}$. For $\overline{\mathcal{U}}, \overline{\mathcal{V}} \in \overline{\mathscr{N}}_A$, and for $\overline{\Phi} \in \mathrm{C}^{\mathrm{hol}}(\overline{\mathcal{U}}; \overline{\mathsf{N}})$ and $\overline{\Psi} \in \mathrm{C}^{\mathrm{hol}}(\overline{\mathcal{V}}; \overline{\mathsf{N}})$, we say that $\overline{\Phi}$ is ***equivalent*** to $\overline{\Psi}$ if there exist $\overline{\mathcal{W}} \in \overline{\mathscr{N}}_A$ and $\overline{\Theta} \in \mathrm{C}^{\mathrm{hol}}(\overline{\mathcal{W}}; \overline{\mathsf{N}})$ such that $\overline{\mathcal{W}} \subseteq \overline{\mathcal{U}} \cap \overline{\mathcal{V}}$ and such that

$$\overline{\Phi}|\overline{\mathcal{W}} = \overline{\Psi}|\mathcal{W} = \overline{\Theta}.$$

By $\mathscr{C}^{\mathrm{hol}}_A(\overline{\mathsf{M}}; \overline{\mathsf{N}})$ we denote the set of equivalence classes, which we call the set of ***germs of mappings*** from $\overline{\mathsf{M}}$ to $\overline{\mathsf{N}}$ over A. By $[\overline{\Phi}]_A$ we denote the equivalence class of $\overline{\Phi} \in \mathrm{C}^{\mathrm{hol}}(\overline{\mathcal{U}}; \overline{\mathsf{N}})$ for some $\overline{\mathcal{U}} \in \overline{\mathscr{N}}_A$.

For $\overline{\mathcal{U}} \in \overline{\mathscr{N}}_{\mathsf{M}}$, we let

$$\mathrm{C}^{\mathrm{hol},\mathbb{R}}(\overline{\mathcal{U}}; \overline{\mathsf{N}}) = \{\Phi \in \mathrm{C}^{\mathrm{hol}}(\overline{\mathcal{U}}; \overline{\mathsf{N}}) \mid \Phi(\overline{\mathcal{U}} \cap \mathsf{M}) \subseteq \mathsf{N}\}.$$

Note that, if $\overline{\mathcal{U}} \in \overline{\mathscr{N}}_A$ and if $\overline{\Phi} \in \mathrm{C}^{\mathrm{hol},\mathbb{R}}(\overline{\mathcal{U}}; \overline{\mathsf{N}})$, then, for any mapping $\overline{\Psi} \in \mathrm{C}^{\mathrm{hol}}(\overline{\mathcal{V}}; \overline{\mathsf{N}})$ equivalent to $\overline{\Phi}$, we have $\overline{\Psi} \in \mathrm{C}^{\mathrm{hol},\mathbb{R}}(\overline{\mathcal{V}}; \overline{\mathsf{N}})$. By $\mathscr{C}^{\mathrm{hol},\mathbb{R}}_A(\overline{\mathsf{M}}; \overline{\mathsf{N}})$ we denote the set of equivalence classes of mappings from $\mathrm{C}^{\mathrm{hol},\mathbb{R}}(\overline{\mathcal{U}}; \overline{\mathsf{N}})$.

Now we have mappings

$$\begin{aligned} r_{\overline{\mathcal{U}},A} \colon \mathrm{C}^{\mathrm{hol},\mathbb{R}}(\overline{\mathcal{U}}; \overline{\mathsf{N}}) &\to \mathscr{C}^{\mathrm{hol},\mathbb{R}}_A(\overline{\mathsf{M}}; \overline{\mathsf{N}}) \\ \overline{\Phi} &\mapsto [\overline{\Phi}]_A. \end{aligned}$$

If $\overline{\mathcal{U}}_1, \overline{\mathcal{U}}_2 \in \overline{\mathscr{N}}_A$ satisfy $\overline{\mathcal{U}}_1 \subseteq \overline{\mathcal{U}}_2$, we have the restriction mapping

$$r_{\overline{\mathcal{U}}_2,\overline{\mathcal{U}}_1} \colon \mathrm{C}^{\mathrm{hol},\mathbb{R}}(\overline{\mathcal{U}}_2; \overline{\mathsf{N}}) \to \mathrm{C}^{\mathrm{hol},\mathbb{R}}(\overline{\mathcal{U}}_1; \overline{\mathsf{N}}).$$

We claim that this restriction mapping is continuous. Indeed, if $f \in \mathrm{C}^{\mathrm{hol}}(\overline{\mathsf{N}})$, consider the diagram

$$\begin{array}{ccc} \mathrm{C}^{\mathrm{hol},\mathbb{R}}(\overline{\mathcal{U}}_1; \overline{\mathsf{N}}) & \xrightarrow{\Theta_f} & \mathrm{C}^{\mathrm{hol},\mathbb{R}}(\overline{\mathcal{U}}_1) \\ {\scriptstyle r_{\overline{\mathcal{U}}_2,\overline{\mathcal{U}}_1}}\uparrow & \nearrow{\scriptstyle \Theta_f} & \\ \mathrm{C}^{\mathrm{hol},\mathbb{R}}(\overline{\mathcal{U}}_2; \overline{\mathsf{N}}) & & \end{array}$$

and this gives continuity of the vertical arrow by the universal property of the initial topology. Thus we have the directed system

$$\left((\mathrm{C}^{\mathrm{hol},\mathbb{R}}(\overline{\mathcal{U}};\overline{\mathsf{N}}))_{\overline{\mathcal{U}}\in\overline{\mathscr{N}}_A},(r_{\overline{\mathcal{U}}_2,\overline{\mathcal{U}}_1})_{\overline{\mathcal{U}}_1\subseteq\overline{\mathcal{U}}_2}\right)$$

in the category of topological spaces. This gives the direct limit topology for the direct limit $\mathscr{C}_A^{\mathrm{hol},\mathbb{R}}(\overline{\mathsf{M}};\overline{\mathsf{N}})$.

Let us show that this direct limit topology for the space of germs is independent of complexification. We note that the issue of the independence on complexification also arises in Sect. 2.2, although we did not address it. However, the first part of our (elementary) arguments in the proof of the following lemma apply as well to the case of sections of a vector bundle.

Lemma 2.31 (Independence of Direct Limit Topology on Complexification) *Let* M *and* N *be* C^ω*-manifolds with complexifications* $\overline{\mathsf{M}}$ *and* $\overline{\mathsf{M}}'$, *and* $\overline{\mathsf{N}}$ *and* $\overline{\mathsf{N}}'$, *respectively. Let* $\overline{\mathscr{N}}_A$ *and* $\overline{\mathscr{N}}'_A$ *be the directed sets of neighbourhoods of* $A\subseteq\mathsf{M}$ *in* $\overline{\mathsf{M}}$ *and* $\overline{\mathsf{M}}'$, *respectively. Then the direct limit topologies for the directed systems*

$$\left((\mathrm{C}^{\mathrm{hol},\mathbb{R}}(\overline{\mathcal{U}};\overline{\mathsf{N}}))_{\overline{\mathcal{U}}\in\overline{\mathscr{N}}_A},(r_{\overline{\mathcal{U}}_2,\overline{\mathcal{U}}_1})_{\overline{\mathcal{U}}_1\subseteq\overline{\mathcal{U}}_2}\right),$$
$$\left((\mathrm{C}^{\mathrm{hol},\mathbb{R}}(\overline{\mathcal{U}}';\overline{\mathsf{N}}'))_{\overline{\mathcal{U}}'\in\overline{\mathscr{N}}'_A},(r_{\overline{\mathcal{U}}'_2,\overline{\mathcal{U}}'_1})_{\overline{\mathcal{U}}'_1\subseteq\overline{\mathcal{U}}'_2}\right)$$

are isomorphic in the category of topological spaces.

Proof First we show independence on complexification of M. By uniqueness of complexification, there is a complexification $\overline{\mathsf{M}}''$ which is regarded as an open holomorphic submanifold of both $\overline{\mathsf{M}}$ and $\overline{\mathsf{M}}'$. Since the directed set of neighbourhoods of M in $\overline{\mathsf{M}}''$ is cofinal in the directed sets of neighbourhoods of M in both $\overline{\mathsf{M}}$ and $\overline{\mathsf{M}}'$, the direct limits of the directed systems

$$\left((\mathrm{C}^{\mathrm{hol},\mathbb{R}}(\overline{\mathcal{U}};\overline{\mathsf{N}}))_{\overline{\mathcal{U}}\in\overline{\mathscr{N}}_A},(r_{\overline{\mathcal{U}}_2,\overline{\mathcal{U}}_1})_{\overline{\mathcal{U}}_1\subseteq\overline{\mathcal{U}}_2}\right),$$
$$\left((\mathrm{C}^{\mathrm{hol},\mathbb{R}}(\overline{\mathcal{U}}';\overline{\mathsf{N}}))_{\overline{\mathcal{U}}'\in\overline{\mathscr{N}}'_A},(r_{\overline{\mathcal{U}}'_2,\overline{\mathcal{U}}'_1})_{\overline{\mathcal{U}}'_1\subseteq\overline{\mathcal{U}}'_2}\right)$$

agree in the category of topological spaces.

As concerns the independence on the complexification of N, the reason that there is something to think about is that it is possible that there is a $\overline{\mathcal{U}}\in\overline{\mathscr{N}}_{\mathsf{M}}$ and $\Phi\in\mathrm{C}^{\mathrm{hol},\mathbb{R}}(\overline{\mathcal{U}};\overline{\mathsf{N}})$ such that $\Phi(\overline{\mathcal{U}})\not\subseteq\overline{\mathsf{N}}'$. The way one rectifies this is by choosing a smaller neighbourhood $\overline{\mathcal{U}}'$ on which to represent the germ; one choice is

$$\overline{\mathcal{U}}'=\Phi^{-1}(\overline{\mathsf{N}}\cap\overline{\mathsf{N}}')\cap\overline{\mathcal{U}}.$$

In this case, Φ and $\Phi|\overline{\mathcal{U}}'$ are equivalent. Said otherwise, every germ with a representative in $\mathrm{C}^{\mathrm{hol},\mathbb{R}}(\overline{\mathcal{U}};\overline{\mathsf{N}})$ has a representative in $\mathrm{C}^{\mathrm{hol},\mathbb{R}}(\overline{\mathcal{U}}';\overline{\mathsf{N}}')$ for some $\overline{\mathcal{U}}'$. We need to show that this observation gives rise to an homeomorphism between the direct limits associated to the directed systems

$$\Big((\mathrm{C}^{\mathrm{hol},\mathbb{R}}(\overline{\mathcal{U}};\overline{\mathsf{N}}))_{\overline{\mathcal{U}}\in\mathscr{N}_A},(r_{\overline{\mathcal{U}}_2,\overline{\mathcal{U}}_1})_{\overline{\mathcal{U}}_1\subseteq\overline{\mathcal{U}}_2}\Big),$$
$$\Big((\mathrm{C}^{\mathrm{hol},\mathbb{R}}(\overline{\mathcal{U}};\overline{\mathsf{N}}'))_{\overline{\mathcal{U}}\in\mathscr{N}_A},(r_{\overline{\mathcal{U}}_2,\overline{\mathcal{U}}_1})_{\overline{\mathcal{U}}_1\subseteq\overline{\mathcal{U}}_2}\Big).$$

To do this, we need some notation. By $[(\Phi,\overline{\mathcal{U}})]_A$ and $[(\Phi',\overline{\mathcal{U}})]'_A$ we denote the germs of $\Phi\in\mathrm{C}^{\mathrm{hol},\mathbb{R}}(\overline{\mathcal{U}};\overline{\mathsf{N}})$ and $\Phi'\in\mathrm{C}^{\mathrm{hol},\mathbb{R}}(\overline{\mathcal{U}};\overline{\mathsf{N}}')$; note that we bookkeep the open set on which a representative of a germ is defined, as we will need this added resolution.

We first consider the case that $\overline{\mathsf{N}}'\subseteq\overline{\mathsf{N}}$, and we denote the inclusion by $\iota_{\overline{\mathsf{N}}'}$. We define a mapping

$$\begin{aligned}\kappa_A\colon \varinjlim_{\overline{\mathcal{U}}\in\mathscr{N}_A}\mathrm{C}^{\mathrm{hol},\mathbb{R}}(\overline{\mathcal{U}};\overline{\mathsf{N}})&\to\varinjlim_{\overline{\mathcal{U}}\in\mathscr{N}_A}\mathrm{C}^{\mathrm{hol},\mathbb{R}}(\overline{\mathcal{U}};\overline{\mathsf{N}}')\\ [(\Phi,\overline{\mathcal{U}})]_A&\mapsto[(\Phi|\Phi^{-1}(\overline{\mathsf{N}}')\cap\overline{\mathcal{U}},\Phi^{-1}(\overline{\mathsf{N}}')\cap\overline{\mathcal{U}})]'_A.\end{aligned}$$

First we show that κ_A is well-defined in that the definition does not depend on the representative $(\Phi,\overline{\mathcal{U}})$ of a germ $[(\Phi,\overline{\mathcal{U}})]_A$. Suppose that $[(\Phi_1,\overline{\mathcal{U}}_1)]_A=[(\Phi_2,\overline{\mathcal{U}}_2)]_A$. Then there exists $\overline{\mathcal{V}}\subseteq\overline{\mathcal{U}}_1\cap\overline{\mathcal{U}}_2$ such that $\Phi_1|\overline{\mathcal{V}}=\Phi_2|\overline{\mathcal{V}}$. Note that,

$$\Phi_1^{-1}(\overline{\mathsf{N}}')\cap\overline{\mathcal{V}}\subseteq\Phi_1^{-1}(\overline{\mathsf{N}}')\cap\overline{\mathcal{U}}_1,\quad \Phi_2^{-1}(\overline{\mathsf{N}}')\cap\overline{\mathcal{V}}\subseteq\Phi_2^{-1}(\overline{\mathsf{N}}')\cap\overline{\mathcal{U}}_2$$

and

$$\Phi_1|\Phi_1^{-1}(\overline{\mathsf{N}}')\cap\overline{\mathcal{V}}=\Phi_2|\Phi_2^{-1}(\overline{\mathsf{N}}')\cap\overline{\mathcal{V}},$$

which gives the well-definedness of κ_A. Now we show that κ_A is injective. Suppose that $\kappa_A([\Phi_1,\overline{\mathcal{U}}_1]_A)=\kappa_A([\Phi_2,\overline{\mathcal{U}}_2]_A)$. Then there exists

$$\overline{\mathcal{V}}\subseteq\Phi_1^{-1}(\overline{\mathsf{N}}')\cap\overline{\mathcal{U}}_1\cap\Phi_2^{-1}(\overline{\mathsf{N}}')\cap\overline{\mathcal{U}}_2$$

such that $\Phi_1|\overline{\mathcal{V}}=\Phi_2|\overline{\mathcal{V}}$, and so $[(\Phi_1,\overline{\mathcal{U}}_1)]_A=[(\Phi_2,\overline{\mathcal{U}}_2)]_A$. To show that κ_A is surjective, consider a germ $[(\Phi',\overline{\mathcal{U}})]'_A$. Since $\overline{\mathsf{N}}'\subseteq\overline{\mathsf{N}}$, we immediately have

$$\kappa_A([(\Phi,\overline{\mathcal{U}})]_A)=[(\Phi',\overline{\mathcal{U}})]'_A,$$

where $\Phi=\iota_{\overline{\mathsf{N}}'}\circ\Phi'$.

Note that the mapping

$$\begin{aligned}\lambda_A\colon \varinjlim_{\overline{\mathcal{U}}\in\overline{\mathscr{N}}_A} \mathrm{C}^{\mathrm{hol},\mathbb{R}}(\overline{\mathcal{U}};\overline{\mathsf{N}}') &\to \varinjlim_{\overline{\mathcal{U}}\in\overline{\mathscr{N}}_A} \mathrm{C}^{\mathrm{hol},\mathbb{R}}(\overline{\mathcal{U}};\overline{\mathsf{N}})\\ [(\Phi',\overline{\mathcal{U}})]'_A &\mapsto [(\iota_{\overline{\mathsf{N}}'}\circ\Phi',\overline{\mathcal{U}})]_A\end{aligned}$$

is the inverse of κ_A, as is clear from the preceding arguments. For $\overline{\mathcal{U}}\in\overline{\mathscr{N}}_A$, define

$$\begin{aligned}\lambda_{\overline{\mathcal{U}}}\colon \mathrm{C}^{\mathrm{hol},\mathbb{R}}(\overline{\mathcal{U}};\overline{\mathsf{N}}') &\to \mathrm{C}^{\mathrm{hol},\mathbb{R}}(\overline{\mathcal{U}};\overline{\mathsf{N}})\\ \Phi' &\mapsto \iota_{\overline{\mathsf{N}}'}\circ\Phi',\end{aligned}$$

and note that the diagram

$$\begin{array}{ccc}\mathrm{C}^{\mathrm{hol},\mathbb{R}}(\overline{\mathcal{U}};\overline{\mathsf{N}}') & \xrightarrow{r_{\overline{\mathcal{U}},A}} & \varinjlim_{\overline{\mathcal{U}}\in\overline{\mathscr{N}}_A} \mathrm{C}^{\mathrm{hol},\mathbb{R}}(\overline{\mathcal{U}};\overline{\mathsf{N}}')\\ \Big\downarrow{\scriptstyle \lambda_{\overline{\mathcal{U}}}} & & \Big\downarrow{\scriptstyle \lambda_A}\\ \mathrm{C}^{\mathrm{hol},\mathbb{R}}(\overline{\mathcal{U}};\overline{\mathsf{N}}) & \xrightarrow{r_{\overline{\mathcal{U}},A}} & \varinjlim_{\overline{\mathcal{U}}\in\overline{\mathscr{N}}_A} \mathrm{C}^{\mathrm{hol},\mathbb{R}}(\overline{\mathcal{U}};\overline{\mathsf{N}})\end{array} \tag{2.22}$$

commutes.

To show continuity of κ_A, for $\overline{\mathcal{U}}\in\overline{\mathscr{N}}_A$, consider the diagram

$$\begin{array}{ccc}\mathrm{C}^{\mathrm{hol},\mathbb{R}}(\overline{\mathcal{U}};\overline{\mathsf{N}}) & \xrightarrow{r_{\overline{\mathcal{U}},A}} & \varinjlim_{\overline{\mathcal{U}}'\in\overline{\mathscr{N}}_A} \mathrm{C}^{\mathrm{hol},\mathbb{R}}(\overline{\mathcal{U}}';\overline{\mathsf{N}})\\ & \searrow & \Big\downarrow{\scriptstyle \kappa_A}\\ & & \varinjlim_{\overline{\mathcal{U}}'\in\overline{\mathscr{N}}_A} \mathrm{C}^{\mathrm{hol},\mathbb{R}}(\overline{\mathcal{U}}';\overline{\mathsf{N}}')\end{array}$$

where the diagonal arrow is defined just so the diagram commutes. By the universal property of the direct limit topology, continuity of κ_A will follow if the diagonal arrow is continuous for every $\overline{\mathcal{U}}$. Note that the preimage in $\mathrm{C}^{\mathrm{hol},\mathbb{R}}(\overline{\mathcal{U}};\overline{\mathsf{N}})$ of the diagonal arrow is, by definition of κ_A, the set of mappings $\Phi\in\mathrm{C}^{\mathrm{hol},\mathbb{R}}(\overline{\mathcal{U}};\overline{\mathsf{N}})$ for which $\mathrm{image}(\Phi)\subseteq\overline{\mathsf{N}}'$. We claim that, if $\mathcal{O}\subseteq\varinjlim_{\overline{\mathcal{U}}\in\overline{\mathscr{N}}_A}\mathrm{C}^{\mathrm{hol},\mathbb{R}}(\overline{\mathcal{U}};\overline{\mathsf{N}}')$ is open, then $(\kappa_A\circ r_{\overline{\mathcal{U}},A})^{-1}(\mathcal{O})$ is open. Indeed, observe that, since

$$(\kappa_A\circ r_{\overline{\mathcal{U}},A})^{-1}(\mathcal{O}) = (\lambda_A^{-1}\circ r_{\overline{\mathcal{U}},A})^{-1}(\mathcal{O}) = (r_{\overline{\mathcal{U}},A}\circ\lambda_{\overline{\mathcal{U}}}^{-1})^{-1}(\mathcal{O})$$

and since $r_{\overline{\mathcal{U}},A}^{-1}(\mathcal{O})$ is open in $\mathrm{C}^{\mathrm{hol},\mathbb{R}}(\overline{\mathcal{U}};\overline{\mathsf{N}}')$ since the direct limit topology is the final topology induced by the mappings $r_{\overline{\mathcal{U}},A}$, $\overline{\mathcal{U}}\in\overline{\mathscr{N}}_A$, we need only show that $\lambda_{\overline{\mathcal{U}}}(\mathrm{C}^{\mathrm{hol},\mathbb{R}}(\overline{\mathcal{U}}; \overline{\mathsf{N}}'))$ is open in $\mathrm{C}^{\mathrm{hol},\mathbb{R}}(\overline{\mathcal{U}}; \overline{\mathsf{N}})$. To see that this is true, let $\Phi \in \lambda_{\overline{\mathcal{U}}}(\mathrm{C}^{\mathrm{hol},\mathbb{R}}(\overline{\mathcal{U}}; \overline{\mathsf{N}}'))$ and let $\mathcal{K} \subseteq \overline{\mathcal{U}}$ and $\overline{\mathcal{V}} \subseteq \overline{\mathsf{N}}$ be such that $\Phi(\mathcal{K}) \subseteq \overline{\mathcal{V}}$. Since $\mathrm{image}(\Phi) \subseteq \overline{\mathsf{N}}'$, we can choose the open set $\overline{\mathcal{V}}$ such that $\mathrm{cl}(\overline{\mathcal{V}}) \subseteq \overline{\mathsf{N}}'$ (by local compactness of $\overline{\mathsf{N}}$). Then let $\overline{\mathcal{W}} \subseteq \overline{\mathsf{N}}$ be such that

$$\mathrm{cl}(\overline{\mathcal{V}}) \subseteq \overline{\mathcal{W}} \subseteq \overline{\mathsf{N}}'.$$

Then the subbasic open set $\mathcal{B}(\mathcal{K}, \overline{\mathcal{W}})$ for the compact-open topology of $\mathrm{C}^{\mathrm{hol},\mathbb{R}}(\overline{\mathcal{U}}; \overline{\mathsf{N}})$ is a neighbourhood of Φ contained in

$$\lambda_{\overline{\mathcal{U}}}\left(\varinjlim_{\overline{\mathcal{U}} \in \overline{\mathscr{N}}_A} \mathrm{C}^{\mathrm{hol},\mathbb{R}}(\overline{\mathcal{U}}; \overline{\mathsf{N}}')\right),$$

which shows that this latter set is open.

It thus remains to show that λ_A is continuous. This will follow from (2.22) and the universal property of the direct limit topology if we can show that $\lambda_{\overline{\mathcal{U}}}$ is continuous. Let $\Phi' \in \mathrm{C}^{\mathrm{hol},\mathbb{R}}(\overline{\mathcal{U}}; \overline{\mathsf{N}}')$ and let $\mathcal{K} \subseteq \overline{\mathcal{U}}$ and $\overline{\mathcal{V}} \subseteq \overline{\mathsf{N}}$ be open such that $\mathcal{B}(\mathcal{K}, \overline{\mathcal{V}})$ is a subbasic neighbourhood of $\lambda_{\overline{\mathcal{U}}}(\Phi')$. Then

$$\lambda_{\overline{\mathcal{U}}}^{-1}(\mathcal{B}(\mathcal{K}, \overline{\mathcal{V}})) = \mathcal{B}(\mathcal{K}, \overline{\mathcal{V}} \cap \overline{\mathsf{N}}')$$

is a subbasic neighbourhood of Φ' in $\mathrm{C}^{\mathrm{hol},\mathbb{R}}(\overline{\mathcal{U}}; \overline{\mathsf{N}}')$, giving the desired continuity.

The above proves the lemma for $\overline{\mathsf{N}}' \subseteq \overline{\mathsf{N}}$. In general, we define $\overline{\mathsf{N}}'' = \overline{\mathsf{N}} \cap \overline{\mathsf{N}}'$ and note that the arguments above give homeomorphisms

$$\varinjlim_{\overline{\mathcal{U}} \in \overline{\mathscr{N}}_A} \mathrm{C}^{\mathrm{hol},\mathbb{R}}(\overline{\mathcal{U}}; \overline{\mathsf{N}}) \simeq \varinjlim_{\overline{\mathcal{U}} \in \overline{\mathscr{N}}_A} \mathrm{C}^{\mathrm{hol},\mathbb{R}}(\overline{\mathcal{U}}; \overline{\mathsf{N}}''),$$

$$\varinjlim_{\overline{\mathcal{U}} \in \overline{\mathscr{N}}_A} \mathrm{C}^{\mathrm{hol},\mathbb{R}}(\overline{\mathcal{U}}; \overline{\mathsf{N}}') \simeq \varinjlim_{\overline{\mathcal{U}} \in \overline{\mathscr{N}}_A} \mathrm{C}^{\mathrm{hol},\mathbb{R}}(\overline{\mathcal{U}}; \overline{\mathsf{N}}''),$$

and then the lemma follows by transitivity of the relation "homeomorphic." □

We shall next use the preceding constructions in two ways, mirroring what we did for sections in Sect. 2.2.

2.5.4 *Direct and Inverse Limit Topologies for the Space of Real Analytic Mappings*

Let M and N be real analytic manifolds and let $\overline{\mathsf{M}}$ and $\overline{\mathsf{N}}$ be complexifications of M and N, respectively. Mirroring what we did in Sect. 2.2, in this section we introduce two topologies for $\mathrm{C}^{\omega}(\mathsf{M};\mathsf{N})$ using holomorphic extensions to $\overline{\mathsf{M}}$ and $\overline{\mathsf{N}}$.

The first is a direct application of the constructions from the preceding section to the case $A = \mathsf{M}$, and this gives rise to the directed system

$$\left((\mathrm{C}^{\mathrm{hol},\mathbb{R}}(\overline{\mathcal{U}};\overline{\mathsf{N}}))_{\overline{\mathcal{U}}\in\overline{\mathscr{N}}_{\mathsf{M}}}, (r_{\overline{\mathcal{U}}_2,\overline{\mathcal{U}}_1})_{\overline{\mathcal{U}}_1\subseteq\overline{\mathcal{U}}_2}\right)$$

in the category of topological spaces with direct limit $\mathscr{C}^{\mathrm{hol},\mathbb{R}}_{\mathsf{M}}(\overline{\mathsf{M}};\overline{\mathsf{N}})$. Just as in Lemma 2.7, we have a natural bijection

$$\mathrm{C}^{\omega}(\mathsf{M};\mathsf{N}) \simeq \mathscr{C}^{\mathrm{hol},\mathbb{R}}_{\mathsf{M}}(\overline{\mathsf{M}};\overline{\mathsf{N}}),$$

and so this induces the ***direct limit topology*** for $\mathrm{C}^{\omega}(\mathsf{M};\mathsf{N})$.

Next, for $\mathcal{K} \in \mathscr{K}_{\mathsf{M}}$ a compact set, we have the directed system

$$\left((\mathrm{C}^{\mathrm{hol},\mathbb{R}}(\overline{\mathcal{U}};\overline{\mathsf{N}}))_{\overline{\mathcal{U}}\in\overline{\mathscr{N}}_{\mathcal{K}}}, (r_{\overline{\mathcal{U}}_2,\overline{\mathcal{U}}_1})_{\overline{\mathcal{U}}_1\subseteq\overline{\mathcal{U}}_2}\right)$$

in the category of topological spaces with direct limit $\mathscr{C}^{\mathrm{hol},\mathbb{R}}_{\mathcal{K}}(\overline{\mathsf{M}};\overline{\mathsf{N}})$. This then gives rise to the inverse system

$$\left((\mathscr{C}^{\mathrm{hol},\mathbb{R}}_{\mathcal{K}}(\overline{\mathsf{M}};\overline{\mathsf{N}}))_{\mathcal{K}\in\mathscr{K}_{\mathsf{M}}}, (\pi_{\mathcal{K}_2,\mathcal{K}_1})_{\mathcal{K}_1\subseteq\mathcal{K}_2}\right),$$

where we have

$$\begin{aligned}\pi_{\mathcal{K}_2,\mathcal{K}_1}\colon \mathscr{C}^{\mathrm{hol},\mathbb{R}}_{\mathcal{K}_2}(\overline{\mathsf{M}};\overline{\mathsf{N}}) &\to \mathscr{C}^{\mathrm{hol},\mathbb{R}}_{\mathcal{K}_1}(\overline{\mathsf{M}};\overline{\mathsf{N}})\\ [\overline{\Phi}]_{\mathcal{K}_2} &\mapsto [\overline{\Phi}]_{\mathcal{K}_1}.\end{aligned}$$

The inverse limit is denoted by $\varprojlim_{\mathcal{K}\in\mathscr{K}_{\mathsf{M}}} \mathscr{C}^{\mathrm{hol},\mathbb{R}}_{\mathcal{K}}(\overline{\mathsf{M}};\overline{\mathsf{N}})$. In the same manner as Lemma 2.11, we have a natural bijection

$$\mathrm{C}^{\omega}(\mathsf{M};\mathsf{N}) \simeq \varprojlim_{\mathcal{K}\in\mathscr{K}_{\mathsf{M}}} \mathscr{C}^{\mathrm{hol},\mathbb{R}}_{\mathcal{K}}(\overline{\mathsf{M}};\overline{\mathsf{N}}),$$

and this gives rise to the ***inverse limit topology*** for $\mathrm{C}^{\omega}(\mathsf{M};\mathsf{N})$.

The following theorem establishes the equality of these two holomorphic extension topologies, and also their equality with the weak-PB topology.

Theorem 2.32 (Characterisations of Weak-PB Topology Using Holomorphic Extension) *If* M *and* N *are* C^ω*-manifolds with complexifications* $\overline{\mathsf{M}}$ *and* $\overline{\mathsf{N}}$*, then the following topologies for* $\mathrm{C}^\omega(\mathsf{M};\mathsf{N})$ *are the same:*

(i) the weak-PB topology;
(ii) the direct limit topology;
(iii) the inverse limit topology.

Proof We begin by making a few observations and constructions that will be useful in the proof.

First, as shown in [24, §3.4] and without loss of generality due to Lemma 2.31, we shall assume that the complexification $\overline{\mathsf{N}}$ is a Stein manifold. By the Remmert Embedding Theorem, we let $\hat{\iota}\colon \overline{\mathsf{N}} \to \mathbb{C}^k \simeq \mathbb{R}^{2k}$ be a proper holomorphic embedding. This then gives a proper real analytic embedding $\iota\colon \mathsf{N} \to \mathbb{R}^{2k}$ with holomorphic extension $\bar{\iota}\colon \overline{\mathsf{N}} \to \mathbb{C}^{2k}$. To make our notation more compatible with notation we have already used, let us denote $N = 2k$. We note that we have obvious vector space isomorphisms

$$\Gamma^\omega(\mathbb{R}^N_{\mathsf{M}}) \simeq \mathrm{C}^\omega(\mathsf{M};\mathbb{R}^N), \quad \Gamma^{\mathrm{hol}}(\mathbb{C}^N_{\overline{\mathsf{M}}}) \simeq \mathrm{C}^{\mathrm{hol}}(\overline{\mathsf{M}};\mathbb{C}^N).$$

Moreover, $\mathbb{C}^N_{\overline{\mathsf{M}}}$ is a complexification of the vector bundle $\mathbb{R}^N_{\mathsf{M}}$, as per Sect. 2.1.3. We note that $\mathrm{C}^\omega(\mathsf{M};\mathbb{R}^N)$ has various topologies, all of which agree by Theorem 2.25 and Proposition 2.27. Similarly, $\mathrm{C}^{\mathrm{hol},\mathbb{R}}(\overline{\mathsf{M}};\mathbb{C}^N)$ has multiple topologies, all of which agree, by Theorem 2.28 and Proposition 2.30. We shall, therefore, use whichever description of these topologies that suits our instantaneous needs.

We shall denote by

$$r_{\overline{\mathcal{U}},\mathsf{M}}\colon \mathrm{C}^{\mathrm{hol},\mathbb{R}}(\overline{\mathcal{U}};\overline{\mathsf{N}}) \to \mathscr{C}^{\mathrm{hol},\mathbb{R}}_{\mathsf{M}}(\overline{\mathsf{M}};\overline{\mathsf{N}}),$$
$$\hat{r}_{\overline{\mathcal{U}},\mathsf{M}}\colon \mathrm{C}^{\mathrm{hol},\mathbb{R}}(\overline{\mathcal{U}};\mathbb{C}^N) \to \mathscr{C}^{\mathrm{hol},\mathbb{R}}_{\mathsf{M}}(\overline{\mathsf{M}};\mathbb{C}^N), \qquad \overline{\mathcal{U}} \in \overline{\mathscr{N}}_{\mathsf{M}},$$

the mappings induced by the direct limits and by

$$\pi_{\mathcal{K}}\colon \varprojlim_{\mathcal{K}'\in\mathscr{K}_{\mathsf{M}}} \mathscr{C}^{\mathrm{hol},\mathbb{R}}_{\mathcal{K}'}(\overline{\mathsf{M}};\overline{\mathsf{N}}) \to \mathscr{C}^{\mathrm{hol},\mathbb{R}}_{\mathcal{K}}(\overline{\mathsf{M}};\overline{\mathsf{N}}),$$
$$\hat{\pi}_{\mathcal{K}}\colon \varprojlim_{\mathcal{K}'\in\mathscr{K}_{\mathsf{M}}} \mathscr{C}^{\mathrm{hol},\mathbb{R}}_{\mathcal{K}'}(\overline{\mathsf{M}};\mathbb{C}^N) \to \mathscr{C}^{\mathrm{hol},\mathbb{R}}_{\mathcal{K}}(\overline{\mathsf{M}};\mathbb{C}^N), \qquad \mathcal{K} \in \mathscr{K}_{\mathsf{M}},$$

the mappings induced by the inverse limits.

The following lemma gives a useful characterisation of the direct limit topology.

Lemma 1 *The direct limit topology for* $\mathscr{C}^{\mathrm{hol},\mathbb{R}}_{\mathsf{M}}(\overline{\mathsf{M}};\overline{\mathsf{N}})$ *agrees with the topology induced from* $\mathscr{C}^{\mathrm{hol},\mathbb{R}}_{\mathsf{M}}(\overline{\mathsf{M}};\mathbb{C}^N)$ *by the inclusion*

$$\mathscr{C}^{\mathrm{hol},\mathbb{R}}_{\mathsf{M}}(\overline{\mathsf{M}};\overline{\mathsf{N}}) \ni [\overline{\Phi}]_{\mathsf{M}} \mapsto [\overline{\iota}\circ\overline{\Phi}]_{\mathsf{M}} \in \mathscr{C}^{\mathrm{hol},\mathbb{R}}_{\mathsf{M}}(\overline{\mathsf{M}};\mathbb{C}^N).$$

Proof First let $\mathcal{O} \subseteq \mathscr{C}^{\mathrm{hol},\mathbb{R}}_{\mathsf{M}}(\overline{\mathsf{M}};\overline{\mathsf{N}})$ be open in the direct limit topology. Then, for every $\overline{\mathcal{U}} \in \overline{\mathscr{N}}_{\mathsf{M}}$, $r^{-1}_{\overline{\mathcal{U}},\mathsf{M}}(\mathcal{O})$ is open in $\mathrm{C}^{\mathrm{hol},\mathbb{R}}(\overline{\mathcal{U}};\overline{\mathsf{N}})$. By Theorem 2.28(iv), there exists an open set $\mathcal{O}' \subseteq \mathrm{C}^{\mathrm{hol},\mathbb{R}}(\overline{\mathcal{U}};\mathbb{C}^N)$ such that

$$r^{-1}_{\overline{\mathcal{U}},\mathsf{M}}(\mathcal{O}) = \mathcal{O}' \cap \mathrm{C}^{\mathrm{hol},\mathbb{R}}(\overline{\mathcal{U}};\overline{\mathsf{N}}).$$

Next let us consider open sets in the induced topology. These are subsets $\mathcal{P} \subseteq \mathscr{C}^{\mathrm{hol},\mathbb{R}}_{\mathsf{M}}(\overline{\mathsf{M}};\overline{\mathsf{N}})$ such that there exists an open subset $\mathcal{P}'$ in $\mathscr{C}^{\mathrm{hol},\mathbb{R}}_{\mathsf{M}}(\overline{\mathsf{M}};\mathbb{C}^N)$ such that $\mathcal{P} = \mathcal{P}' \cap \mathscr{C}^{\mathrm{hol},\mathbb{R}}_{\mathsf{M}}(\overline{\mathsf{M}};\overline{\mathsf{N}})$. Openness of $\mathcal{P}'$ in the direct limit topology (making this choice of the many possible descriptions of the topology) means exactly that, for every $\overline{\mathcal{U}} \in \overline{\mathscr{N}}_{\mathsf{M}}$, $\hat{r}^{-1}_{\overline{\mathcal{U}},\mathsf{M}}(\mathcal{P}')$ is open in $\mathrm{C}^{\mathrm{hol},\mathbb{R}}(\overline{\mathcal{U}};\mathbb{C}^N)$. Note that

$$r^{-1}_{\overline{\mathcal{U}},\mathsf{M}}(\mathcal{P}) = \hat{r}^{-1}_{\overline{\mathcal{U}},\mathsf{M}}(\mathcal{P}') \cap \mathrm{C}^{\mathrm{hol},\mathbb{R}}(\overline{\mathcal{U}};\overline{\mathsf{N}}).$$

The above discussion shows that the induced topology is finer than the direct limit topology.

For the opposite inclusion, we consider the diagram

$$\begin{array}{ccc}
\mathscr{C}^{\mathrm{hol},\mathbb{R}}_{\mathsf{M}}(\overline{\mathsf{M}};\overline{\mathsf{N}}) & \longrightarrow & \mathscr{C}^{\mathrm{hol},\mathbb{R}}_{\mathsf{M}}(\overline{\mathsf{M}};\mathbb{C}^N) \\
{\scriptstyle r_{\overline{\mathcal{U}},\mathsf{M}}}\uparrow & \nearrow & \uparrow{\scriptstyle \hat{r}_{\overline{\mathcal{U}},\mathsf{M}}} \\
\mathrm{C}^{\mathrm{hol},\mathbb{R}}(\overline{\mathcal{U}};\overline{\mathsf{N}}) & \longrightarrow & \mathrm{C}^{\mathrm{hol},\mathbb{R}}(\overline{\mathcal{U}};\mathbb{C}^N)
\end{array}$$

for $\overline{\mathcal{U}} \in \overline{\mathscr{N}}_{\mathsf{M}}$, and with the spaces in the top row having the direct limit topologies. Both horizontal arrows are inclusions. The vertical arrows are continuous by definition of the direct limit topologies. By Theorem 2.28(iv), the bottom horizontal arrow is continuous. Thus the dashed diagonal arrow is continuous. By the universal property of direct limits, the top horizontal arrow is continuous. Thus the direct limit topology for $\mathscr{C}^{\mathrm{hol},\mathbb{R}}_{\mathsf{M}}(\overline{\mathsf{M}};\overline{\mathsf{N}})$ is finer than the induced topology from $\mathscr{C}^{\mathrm{hol},\mathbb{R}}_{\mathsf{M}}(\overline{\mathsf{M}};\mathbb{C}^N)$ since the induced topology is the coarsest for which the inclusion is continuous. ▽

(i)$\subseteq$(ii) Let $f \in C^\omega(N)$ and let $\mathcal{V} \in \mathscr{N}_N$ and $\overline{f} \in C^{\mathrm{hol},\mathbb{R}}(\overline{\mathcal{V}})$ be such that $f = \overline{f}|N$ [14, Lemma 5.40]. Note that we have the commutative diagram

$$
\begin{array}{ccc}
C^\omega(M;N) & \xrightarrow{\ \Theta_f\ } & C^\omega(M) \\
\downarrow{\scriptstyle \iota_{M,N}} & & \downarrow{\scriptstyle \iota_M} \\
\mathscr{C}_M^{\mathrm{hol},\mathbb{R}}(\overline{M};\overline{\mathcal{V}}) & \xrightarrow{\ \overline{\Theta}_f\ } & \mathscr{C}_{M,\overline{M}}^{\mathrm{hol},\mathbb{R}} \\
\uparrow{\scriptstyle r_{\overline{\mathcal{U}},M}} & \nearrow & \uparrow{\scriptstyle r_{\overline{\mathcal{U}},M}} \\
C^{\mathrm{hol},\mathbb{R}}(\overline{\mathcal{U}};\overline{\mathcal{V}}) & \xrightarrow{\ \Theta_{\overline{f}}\ } & C^{\mathrm{hol},\mathbb{R}}(\overline{\mathcal{U}})
\end{array}
$$

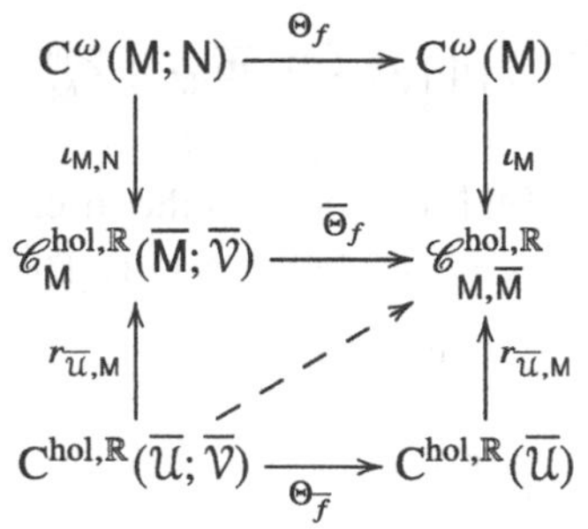

for every $\overline{\mathcal{U}} \in \mathscr{N}_M$. As we have seen, the two upper vertical arrows are homeomorphisms when $C^\omega(M; N)$ has the direct limit topology. Thus $\overline{\Theta}_f$ is defined so that the upper square commutes. By the universal property of the direct limit topology and the continuity of the diagonal arrow, $\overline{\Theta}_f$ is continuous. This shows that Θ_f is continuous if $C^\omega(M; N)$ has the direct limit topology. By definition of the initial topology, this means that the direct limit topology is finer than the weak-PB topology.

(ii)$\subseteq$(i) We consider the diagram

$$
\begin{array}{ccc}
\mathscr{C}_M^{\mathrm{hol},\mathbb{R}}(\overline{M};\overline{N}) & \longrightarrow & \mathscr{C}_M^{\mathrm{hol},\mathbb{R}}(\overline{M};\mathbb{C}^N) \\
\uparrow & \nearrow & \uparrow \\
C^\omega(M;N) & \longrightarrow & C^\omega(M;\mathbb{R}^N)
\end{array}
$$

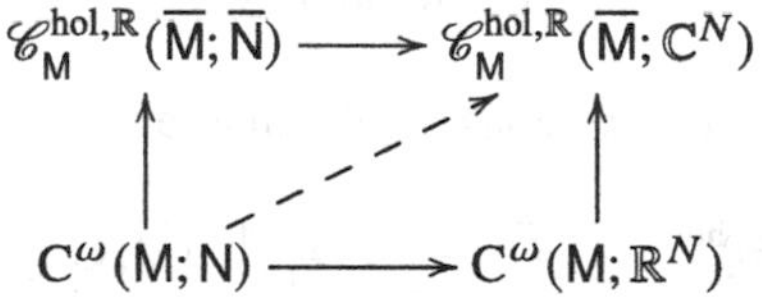

By the lemma above, the direct limit topology for $\mathscr{C}_M^{\mathrm{hol},\mathbb{R}}(\overline{M};\overline{N})$ is the initial topology associated with the top horizontal inclusion, where $\mathscr{C}_M^{\mathrm{hol},\mathbb{R}}(\overline{M};\mathbb{C}^N)$ has the direct limit topology. The bottom horizontal arrow is continuous by Theorem 2.25(iii), if both spaces have the weak-PB topology. If the right vertical arrow is continuous with the topologies just indicated, then the dashed diagonal arrow is continuous, and this would establish the continuity of the left vertical arrow by the universal property of the initial topology.

It thus remains to show that the vector space isomorphism

$$
C^\omega(M;\mathbb{R}^N) \to \mathscr{C}_M^{\mathrm{hol},\mathbb{R}}(\overline{M};\mathbb{C}^N)
$$

is continuous if $C^\omega(M; \mathbb{R}^N)$ has the weak-PB topology and $\mathscr{C}_M^{\mathrm{hol},\mathbb{R}}(\overline{M};\mathbb{C}^N)$ has the direct limit topology. This, however, follows since both topologies are isomorphic to $\oplus_{j=1}^N C^\omega(M)$. For the weak-PB topology, this was shown during the proof of Theorem 2.25. For the direct limit topology, this follows from Theorem 5.4 below.

(ii)⊆(iii) We consider the diagram

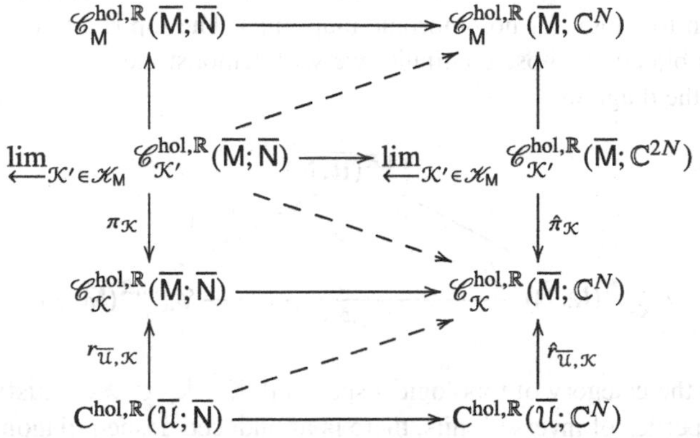

for $\mathcal{K} \in \mathscr{K}_{\mathsf{M}}$ and $\overline{\mathcal{U}} \in \overline{\mathscr{N}}_{\mathcal{K}}$. The spaces in the first row have their direct limit topologies, the spaces in the second row have their inverse limit topologies, the spaces in the third row have their direct limit topologies, and the spaces in the fourth row have the compact-open topology. The first pair of vertical arrows are the canonical mappings and the right of these arrows is continuous [50, Theorem 1.2(a)], i.e., the opposite conclusion of Proposition 2.13, the second pair of vertical arrows are those defining the inverse limits, and the third pair of vertical arrows are those coming from the direct limit topologies. Thus all vertical arrows are continuous with the given topologies, except for the top left arrow whose continuity we will establish. All horizontal arrows are inclusions. The fourth horizontal arrow is continuous by Theorem 2.28(iv). Thus the bottom dashed diagonal arrow is continuous. Thus the third horizontal arrow is continuous by the universal property of the direct limit topology. It follows that the middle dashed diagonal arrow is continuous, and so the second horizontal arrow is continuous by the universal property of inverse limit topologies. Thus the top dashed diagonal arrow is continuous, and we conclude that the first left vertical arrow is continuous since (by the lemma) the first left direct limit topology equals the induced topology, which is the initial topology induced by the inclusion. This all proves this part of the theorem.

(iii)⊆(ii) For this part of the proof, we consider the diagram

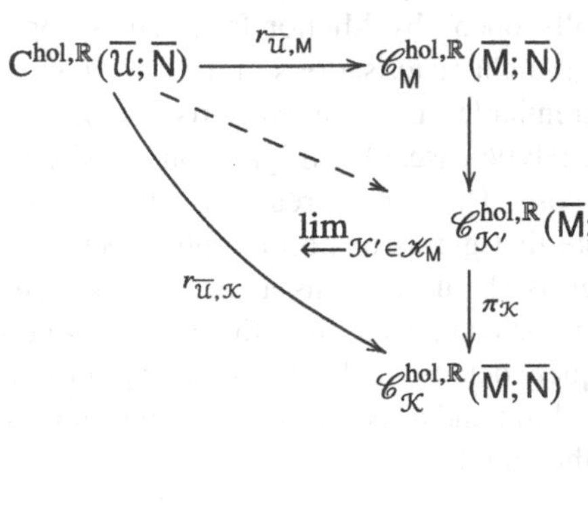

for $\mathcal{K} \in \mathscr{K}_{\mathsf{M}}$ and $\overline{\mathcal{U}} \in \overline{\mathscr{N}_{\mathcal{K}}}$. The topologies are the expected ones: direct limit topologies for spaces of germs, inverse limit topologies for inverse limits, and the compact-open topology for holomorphic mappings. The top right vertical arrow is the canonical bijection whose continuity we will demonstrate.

Note that the diagram

$$\begin{array}{ccccc} & & \mathrm{C}^{\mathrm{hol},\mathbb{R}}(\overline{\mathcal{U}};\overline{\mathsf{N}}) & & \\ & \swarrow^{r_{\overline{\mathcal{U}},\mathcal{K}_2}} & & \searrow^{r_{\overline{\mathcal{U}},\mathcal{K}_1}} & \\ \mathscr{C}^{\mathrm{hol},\mathbb{R}}_{\mathcal{K}_2}(\overline{\mathsf{M}};\overline{\mathsf{N}}) & & \xrightarrow[\pi_{\mathcal{K}_2,\mathcal{K}_2}]{} & & \mathscr{C}^{\mathrm{hol},\mathbb{R}}_{\mathcal{K}_2}(\overline{\mathsf{M}};\overline{\mathsf{N}}) \end{array}$$

commutes in the category of topological spaces if $\mathcal{K}_1, \mathcal{K}_2 \in \mathscr{K}_{\mathsf{M}}$ satisfy $\mathcal{K}_1 \subseteq \mathcal{K}_2$. Thus, by properties of inverse limits, there is an induced dashed diagonal mapping, and this mapping is continuous. Now, by the universal property of the direct limit topology, the top right vertical arrow is continuous, which gives this part of the theorem. □

Our results in this chapter have produced seven characterisations of the topology for the space of real analytic sections of a real analytic vector bundle and four characterisations of the topology for the space of real analytic mappings between real analytic mappings. That all of these characterisations lead to the same topologies speaks to the "uniqueness" of the real analytic topology. Let us say a few words about this.

First of all, for topologies for smooth or finitely differentiable mappings and sections, there are other topologies that one might use than the ones we have introduced; the ones we have introduced are presented mainly for the purpose of comparing them to the real analytic topologies. The topologies we have introduced in the finitely differentiable and smooth cases all have to do with characterising open subsets of mappings by their behaviour on some compact subset of the domain, i.e., by uniform convergence in some way on compact subsets. This seems rather analogous to the inverse limit characterisations of the real analytic topologies. For smooth and finitely differentiable sections and mappings, one might also consider topologies where one wants closeness, not just on compact subsets, but on all of the domain. Such topologies are called "strong" by Hirsch [33], and "Whitney" or "wholly open" by Michor [53]. These topologies seem rather analogous to the direct limit characterisations of the real analytic topologies. In the smooth and finitely differentiable cases, the two sorts of topologies are not the same, but they are in the real analytic case. One might wonder why things are as rigid as this in the real analytic case. The reason, really, is the Identity Theorem, perhaps not surprisingly. We note that germs of holomorphic sections or mappings about compact sets contain germs about points as a special case. The Identity Theorem ensures that, if the domain is connected, then the germ of holomorphic sections or mappings at a point uniquely determine the germs at all points. This, of course, does not *prove* that the inverse limit and direct limit topologies must agree, but perhaps it is suggestive as to why they might.

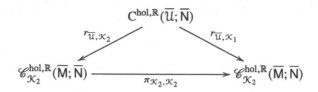

Chapter 3
Geometry: Lifts and Differentiation of Tensors

Many of the geometric constructions we undertake in the book, and estimates associated with these geometric constructions, involve tensors of various sorts defined on the total space of a vector bundle. In this chapter we fairly laboriously carry out the following constructions for a vector bundle $\pi_{\mathsf{E}}\colon \mathsf{E} \to \mathsf{M}$.

1. In Sect. 3.1 we consider various classes of tensors associated with a vector bundle, most defined by some sort of lifting operation from M to E. Some of these lifting constructions can be carried out just using the vector bundle structure, but others involve using a linear connection $\nabla^{\pi_{\mathsf{E}}}$ in E.
2. In Sect. 3.2 we differentiate the tensors we built in Sect. 3.1. The differentiation is carried out by combining an affine connection ∇^{M} on M and a linear connection $\nabla^{\pi_{\mathsf{E}}}$ in E, much in the way that we did in our jet bundle constructions of Sect. 2.3.1.
3. In Sect. 3.3 we provide recursive formulae that relate derivatives of objects on M to their lifts on E, and vice versa. These derivative constructions are essential to our approach for proving the continuity of the various lifting operations.

Many of the constructions we give here will seem, on an initial reading, disconnected from the objectives of the book. This is perhaps especially true of the constructions of Sects. 3.1 and 3.2, at least on a first reading. It is only in the later parts of the book that the relevance of all of these constructions will become apparent. For this reason, perhaps a good strategy would be to skip over these two sections in a first reading, coming back to them when they are subsequently needed.

There is nothing particularly real analytic with the material in this chapter, so the smooth and real analytic cases are considered side-by-side.

A. D. Lewis, *Geometric Analysis on Real Analytic Manifolds*, Lecture Notes in Mathematics 2333, https://doi.org/10.1007/978-3-031-37913-0_3

3.1 Tensors on the Total Space of a Vector Bundle

Let $r \in \{\infty, \omega\}$ and let $\pi_{\mathsf{E}} : \mathsf{E} \to \mathsf{M}$ be a C^r-vector bundle. In this section we define lifts of various kinds of tensors on M to analogous tensors on E. Some of these lifts can be carried out just using the structure of the vector bundle, while others will rely on the introduction of a linear connection in E.

As we mention in the preamble to the chapter, the constructions in this section will seem unmotivated in a first reading, so we suggest the possibility of skipping the section until it is needed later.

3.1.1 Functions on Vector Bundles

Among the geometric constructions we will consider are those associated to a particular set of functions on a vector bundle.

Definition 3.1 (Fibre-Linear Functions) Let $r \in \{\infty, \omega\}$ and let $\pi_{\mathsf{E}} : \mathsf{E} \to \mathsf{M}$ be a vector bundle of class C^r. A function $F \in \mathrm{C}^r(\mathsf{E})$ is ***fibre-linear*** if, for each $x \in \mathsf{M}$, $F|\mathsf{E}_x$ is a linear function. We denote by $\mathrm{Lin}^r(\mathsf{E})$ the set of C^r-fibre-linear functions on E. ∘

Let us give some elementary properties of the sets of fibre-linear functions.

Lemma 3.2 (Properties of Fibre-Linear Functions) *Let* $r \in \{\infty, \omega\}$ *and let* $\pi_{\mathsf{E}} : \mathsf{E} \to \mathsf{M}$ *be a vector bundle of class* C^r. *Then the following statements hold:*

(i) $\mathrm{Lin}^r(\mathsf{E})$ *is a submodule of the* $\mathrm{C}^r(\mathsf{M})$*-module* $\mathrm{C}^r(\mathsf{E})$*;*
(ii) for $F \in \mathrm{Lin}^r(\mathsf{E})$*, there exists* $\lambda_F \in \Gamma^r(\mathsf{E}^*)$ *such that*

$$F(e) = \langle \lambda_F \circ \pi_{\mathsf{E}}(e); e \rangle, \qquad e \in \mathsf{E},$$

and, moreover, the map $F \mapsto \lambda_F$ *is an isomorphism of* $\mathrm{C}^r(\mathsf{M})$*-modules;*

Proof

(i) Let $F \in \mathrm{Lin}^r(\mathsf{E})$ and $f \in \mathrm{C}^r(\mathsf{M})$. Then

$$f \cdot F(e) = (f \circ \pi_{\mathsf{E}}(e))F(e),$$

and so $f \cdot F$ is fibre-linear since a scalar multiple of a linear function is a linear function. Also, since the pointwise sum of linear functions is a linear function, we conclude that $\mathrm{Lin}^r(\mathsf{E})$ is indeed a submodule of $\mathrm{C}^r(\mathsf{E})$.

(ii) This merely follows by definition of the dual bundle E^*. □

In a rather related manner, we can consider other classes of functions on vector bundles.

Definition 3.3 (Lifts and Evaluations of One-Forms and Functions) Let $r \in \{\infty, \omega\}$ and let $\pi_{\mathsf{E}} \colon \mathsf{E} \to \mathsf{M}$ be a C^r-vector bundle.

(i) For $\lambda \in \Gamma^r(\mathsf{E}^*)$, the ***vertical evaluation*** of λ is $\lambda^{\mathrm{e}} \in \mathrm{Lin}^r(\mathsf{E})$ defined by $\lambda^{\mathrm{e}}(e_x) = \langle \lambda(x); e_x \rangle$.

(ii) For $f \in \mathrm{C}^r(\mathsf{M})$, the ***horizontal lift*** of f is the function $f^{\mathrm{h}} \in \mathrm{C}^r(\mathsf{E})$ defined by $f^{\mathrm{h}} = \pi_{\mathsf{E}}^* f$. ○

3.1.2 Vector Fields on Vector Bundles

Next we turn to vector fields on the total space of a vector bundle. As with our consideration of functions in the preceding section, we restrict attention to vector fields that interact nicely with the vector bundle structure.

We begin with the notion of the vertical lift of a section.

Definition 3.4 (Vertical Lift of a Section) Let $r \in \{\infty, \omega\}$ and let $\pi_{\mathsf{E}} \colon \mathsf{E} \to \mathsf{M}$ be a vector bundle of class C^r.

(i) For $e_x, e'_x \in \mathsf{E}_x$, we define the ***vertical lift*** of e'_x to e_x to be

$$\mathrm{vlft}(e_x, e'_x) = \left.\frac{\mathrm{d}}{\mathrm{d}t}\right|_{t=0} (e_x + t e'_x).$$

(ii) Given a section $\xi \in \Gamma^r(\mathsf{E})$, we define the ***vertical lift*** of ξ to E to be the vector field

$$\xi^{\mathrm{v}}(e_x) = \mathrm{vlft}(e_x, \xi(x)).$$ ○

Next we consider another sort of lift, this one requiring a C^r-connection $\nabla^{\pi_{\mathsf{E}}}$ in the vector bundle $\pi_{\mathsf{E}} \colon \mathsf{E} \to \mathsf{B}$. We let $\mathsf{VE} = \ker(T\pi_{\mathsf{E}})$ be the vertical subbundle. As mentioned in Sect. 1.5, the connection $\nabla^{\pi_{\mathsf{E}}}$ defines a complement HE to VE called the horizontal subbundle. We let ver, hor: $\mathsf{TE} \to \mathsf{TE}$ be the projections onto VE and HE, respectively.

Definition 3.5 (Horizontal Lift of a Vector Field) Let $r \in \{\infty, \omega\}$, let $\pi_{\mathsf{E}} \colon \mathsf{E} \to \mathsf{M}$ be a vector bundle of class C^r, and let $\nabla^{\pi_{\mathsf{E}}}$ be a C^r-connection in E.

(i) For $e_x \in \mathsf{E}_x$ and $v_x \in \mathsf{T}_x\mathsf{M}$, the ***horizontal lift*** of v_x to e_x is the unique vector $\mathrm{hlft}(e_x, v_x) \in \mathsf{H}_{e_x}\mathsf{E}$ satisfying

$$T_{e_x}\pi_{\mathsf{E}}(\mathrm{hlft}(e_x, v_x)) = v_x.$$

(ii) For $X \in \Gamma^r(\mathsf{TM})$ on M, we denote by X^{h} the ***horizontal lift*** of X to E, this being the vector field $X^{\mathrm{h}} \in \Gamma^r(\mathsf{TE})$ satisfying

$$X^{\mathrm{h}}(e_x) = \mathrm{hlft}(e_x, X(x)). \qquad \circ$$

Next we provide formulae for differentiating various sorts of functions with respect to various sorts of vector fields.

Lemma 3.6 (Differentiation of Functions on Vector Bundles) *Let* $r \in \{\infty, \omega\}$ *and let* $\pi_{\mathsf{E}} \colon \mathsf{E} \to \mathsf{M}$ *a vector bundle of class* C^r. *Let* $f \in \mathrm{C}^r(\mathsf{M})$, $\lambda \in \Gamma^r(\mathsf{E}^*)$, $X \in \Gamma^r(\mathsf{TM})$, *and* $\xi \in \Gamma^r(\mathsf{E})$.

Then the following statements hold:

(i) $\mathscr{L}_{X^{\mathrm{h}}} f^{\mathrm{h}} = (\mathscr{L}_X f)^{\mathrm{h}}$;
(ii) $\mathscr{L}_{\xi^{\mathrm{v}}} f^{\mathrm{h}} = 0$;
(iii) $\mathscr{L}_{\xi^{\mathrm{v}}} \lambda^{\mathrm{e}} = \langle \lambda; \xi \rangle^{\mathrm{h}}$.

Additionally, let $\nabla^{\pi_{\mathsf{E}}}$ *be a* C^r*-linear connection in* $\pi_{\mathsf{E}} \colon \mathsf{E} \to \mathsf{M}$. *Then*

(iv) $\mathscr{L}_{X^{\mathrm{h}}} \lambda^{\mathrm{e}} = (\nabla^{\pi_{\mathsf{E}}}_X \lambda)^{\mathrm{e}}$.

Proof

(i) We compute

$$\begin{aligned}\mathscr{L}_{X^{\mathrm{h}}} f^{\mathrm{h}}(e) &= \langle \mathrm{d}(\pi_{\mathsf{E}}^* f)(e); X^{\mathrm{h}}(e) \rangle = \langle \mathrm{d}f \circ \pi_{\mathsf{E}}(e); T_e\pi_{\mathsf{E}}(X^{\mathrm{h}}(e)) \rangle \\ &= \langle \mathrm{d}f \circ \pi_{\mathsf{E}}(e); X \circ \pi_{\mathsf{E}}(e) \rangle = (\mathscr{L}_X f)^{\mathrm{h}}(e).\end{aligned}$$

(ii) Since f^{h} is constant on fibres of π_{E} and ξ^{v} is tangent to fibres, we have

$$f^{\mathrm{h}}(e + t\xi \circ \pi_{\mathsf{E}}(e)) = f(e).$$

Differentiating with respect to t at $t = 0$ gives the result.

(iii) Here we compute

$$\begin{aligned}\mathscr{L}_{\xi^{\mathrm{v}}} \lambda^{\mathrm{e}}(e) &= \left.\frac{\mathrm{d}}{\mathrm{d}t}\right|_{t=0} \langle \lambda(e + t\xi \circ \pi_{\mathsf{E}}(e)); e + t\xi \circ \pi_{\mathsf{E}}(e) \rangle \\ &= \langle \lambda \circ \pi_{\mathsf{E}}(e); \xi \circ \pi_{\mathsf{E}}(e) \rangle = \langle \lambda; \xi \rangle^{\mathrm{h}},\end{aligned}$$

so completing the proof.

(iv) Let $e \in \mathsf{E}$ and let $t \mapsto \gamma(t)$ be the integral curve for X satisfying $\gamma(0) = \pi_{\mathsf{E}}(e)$ and let $t \mapsto \gamma^{\mathrm{h}}(t)$ be the integral curve for X^{h} satisfying $\gamma^{\mathrm{h}}(0) = e$. Then $t \mapsto \gamma^{\mathrm{h}}(t)$ is the parallel translation of e along γ, and as such we have $\nabla^{\pi_{\mathsf{E}}}_{\gamma'(t)} \gamma^{\mathrm{h}}(t) =$

0. Then

$$\mathscr{L}_{X^{\mathrm{h}}}\lambda^{\mathrm{e}}(e) = \left.\frac{\mathrm{d}}{\mathrm{d}t}\right|_{t=0} \langle \lambda \circ \gamma(t); \gamma^{\mathrm{h}}(t)\rangle = \langle \nabla_X^{\pi_{\mathsf{E}}} \lambda \circ \pi_{\mathsf{E}}(e); e\rangle,$$

as claimed. □

In Sect. 3.2 we shall have a great deal more to say about differentiation of objects on the total space of a vector bundle when one has more structure present than we use in the preceding result.

3.1.3 Linear Mappings on Vector Bundles

Now we turn to an examination of linear maps associated to a vector bundle $\pi_{\mathsf{E}}\colon \mathsf{E} \to \mathsf{M}$. We shall consider vector bundle mappings of two sorts: (1) with values in the trivial line bundle $\mathbb{R}_{\mathsf{M}}$; (2) with values in E. The first sort of mappings are, of course, simply sections of the dual bundle, or linear functions of the sort studied in Sect. 3.1.1. Our interest here is in lifting such objects to the total space.

First we work with sections of the dual bundle. If we have a connection $\nabla^{\pi_{\mathsf{E}}}$ in a vector bundle $\pi_{\mathsf{E}}\colon \mathsf{E} \to \mathsf{M}$, then this gives us a splitting $\mathsf{TE} = \mathsf{HE} \oplus \mathsf{VE}$, and hence a splitting $\mathsf{T}^*\mathsf{E} = \mathsf{H}^*\mathsf{E} \oplus \mathsf{V}^*\mathsf{E}$ with

$$\mathsf{H}^*\mathsf{E} = \mathrm{ann}(\mathsf{VE}), \quad \mathsf{V}^*\mathsf{E} = \mathrm{ann}(\mathsf{HE}).$$

Note that $\mathsf{H}^*_e\mathsf{E} = \mathrm{image}(T^*_e\pi_{\mathsf{E}})$.

Definition 3.7 (Lifts of One-Forms and Dual Sections) Let $r \in \{\infty, \omega\}$ and let $\pi_{\mathsf{E}}\colon \mathsf{E} \to \mathsf{M}$ be a C^r-vector bundle.

(i) For $\alpha_x \in \mathsf{T}^*_x\mathsf{M}$ and $e_x \in \mathsf{E}_x$, the ***horizontal lift*** of α_x to e_x is $\mathrm{hlft}(e_x, \alpha_x) = T^*_{e_x}\pi_{\mathsf{E}}(\alpha_x)$.

(ii) The ***horizontal lift*** of $\alpha \in \Gamma^r(\mathsf{T}^*\mathsf{M})$ is $\alpha^{\mathrm{h}} = \pi_{\mathsf{E}}^*\alpha \in \Gamma^r(\mathsf{T}^*\mathsf{E})$.

Additionally, let $\nabla^{\pi_{\mathsf{E}}}$ be a connection in E.

(iii) For $\lambda_x \in \mathsf{E}^*_x$ and $e_x \in \mathsf{E}^*_x$, the ***vertical lift*** of λ_x is the unique vector $\mathrm{vlft}(e_x, \lambda_x) \in \mathsf{V}^*_{e_x}\mathsf{E}$ satisfying

$$\langle \mathrm{vlft}(e_x, \lambda_x); \mathrm{vlft}(e_x, u_x)\rangle = \langle \lambda_x; u_x\rangle$$

for every $u_x \in \mathsf{E}_x$.

(iv) The ***vertical lift*** of $\lambda \in \Gamma^r(\mathsf{E}^*)$ is the one-form $\lambda^{\mathrm{v}} \in \Gamma^r(\mathsf{T}^*\mathsf{E})$ satisfying

$$\lambda^{\mathrm{v}}(e_x) = \mathrm{vlft}(e_x, \lambda(x)). \qquad \circ$$

We also have natural ways of lifting homomorphisms of vector bundles.

Definition 3.8 (Vertical Evaluation and Vertical Lift of an Homomorphism) Let $r \in \{\infty, \omega\}$, and let $\pi_{\mathsf{E}} \colon \mathsf{E} \to \mathsf{M}$ and $\pi_{\mathsf{F}} \colon \mathsf{F} \to \mathsf{M}$ be C^r-vector bundles. For $L \in \Gamma^r(\mathsf{F} \otimes \mathsf{E}^*)$,

(i) the ***vertical evaluation*** of L is the section $L^{\mathrm{e}} \in \Gamma^r(\pi_{\mathsf{E}}^*\mathsf{F})$ defined by

$$L^{\mathrm{e}}(e_x) = (e_x, L(e_x)).$$

If, additionally, $\nabla^{\pi_{\mathsf{E}}}$ is a connection in E,

(ii) the ***vertical lift*** of L is the vector bundle homomorphism $L^{\mathrm{v}} \in \Gamma^r(\pi_{\mathsf{E}}^*\mathsf{F} \otimes \mathsf{T}^*\mathsf{E})$ defined by

$$L^{\mathrm{v}}(Z) = (e, L \circ \mathrm{ver}(Z))$$

for $Z \in \mathsf{T}_e\mathsf{E}$, noting that $\mathrm{ver}(Z) \in \mathsf{V}_e\mathsf{E} \simeq \mathsf{E}_{\pi_{\mathsf{E}}(e)}$. $\circ$

We shall be especially interested in two cases of the vector bundle F.

1. $\mathsf{F} = \mathbb{R}_{\mathsf{M}}$: In this case, $\mathsf{F} \otimes \mathsf{E}^* \simeq \mathsf{E}^*$, $\pi_{\mathsf{E}}^*\mathsf{F} \simeq \mathbb{R}_{\mathsf{E}}$, and $\pi_{\mathsf{E}}^*\mathsf{F} \otimes \mathsf{T}^*\mathsf{E} \simeq \mathsf{T}^*\mathsf{E}$. One can easily see that, if $\lambda \in \Gamma^r(\mathsf{E}^*)$, then the vertical evaluation as per Definition 3.8 agrees with that of Definition 3.3, and the vertical lift as per Definition 3.8 agrees with that of Definition 3.7.
2. $\mathsf{F} = \mathsf{E}$: In this case, $\mathsf{F} \otimes \mathsf{E}^* \simeq \mathrm{T}^1_1(\mathsf{E})$, i.e., the bundle of endomorphisms of E. We also have $\pi_{\mathsf{E}}^*\mathsf{F} \simeq \mathsf{VE}$ [42, §6.11]. Thus, for $L \in \Gamma^r(\mathrm{T}^1_1(\mathsf{E}))$, L^{e} is a VE-valued vector field. Also, L^{v} is a VE-valued endomorphism of TE.

Let us perform some analysis of the vertical evaluation and vertical lift of an homomorphism. First of all, for $e_1, e_2 \in \mathsf{E}_x$,

$$L^{\mathrm{e}}(e_1 + e_2) = (e_1, L(e_1)) + (e_2, L(e_2)) = L^{\mathrm{e}}(e_1) + L^{\mathrm{e}}(e_2),$$

where addition is with respect to the vector bundle structure

$$\begin{array}{ccc} \mathsf{E} & \xrightarrow{L^{\mathrm{e}}} & \pi_{\mathsf{E}}^*\mathsf{F} \\ {\scriptstyle \pi_{\mathsf{E}}}\downarrow & & \downarrow{\scriptstyle \pi_{\mathsf{E}}^*\pi_{\mathsf{F}}} \\ \mathsf{M} & \xrightarrow[Z]{} & \mathsf{F} \end{array}$$

where Z is the zero section. Thus L^{e} is a "linear" section over E. We define the vector bundle mapping

$$\begin{aligned} P_{\mathsf{E},\mathsf{F}} \colon \pi_{\mathsf{E}}^*\mathsf{F} \otimes \mathsf{V}^*\mathsf{E} &\to \pi_{\mathsf{E}}^*\mathsf{F} \\ L_e &\mapsto L_e(e) \end{aligned} \tag{3.1}$$

over id_{E}, noting that $e \in \mathsf{E}_{\pi_{\mathsf{E}}(e)} \simeq \mathsf{V}_e\mathsf{E}$. Then, given $A \in \Gamma^r(\pi_{\mathsf{E}}^*\mathsf{F} \otimes \mathsf{V}^*\mathsf{E})$, $P_{\mathsf{E},\mathsf{F}} \circ A$ is a section of $\pi_{\mathsf{E}}^*\mathsf{E}$. Moreover, $P_{\mathsf{E},\mathsf{F}} \circ L^{\mathrm{v}} = L^{\mathrm{e}}$ for $L \in \Gamma^r(\mathsf{F} \otimes \mathsf{E}^*)$.

We shall make use of these observations in Sect. 3.3.

Let us recast the preceding observations in a slightly different way. To start, note that, given $\lambda \in \Gamma^r(\mathsf{E}^*)$ and $\eta \in \Gamma^r(\mathsf{F})$, we have $\eta \otimes \lambda \in \Gamma^r(\mathsf{F} \otimes \mathsf{E}^*)$. The tensor product on the left can be thought of as being of $\mathrm{C}^r(\mathsf{M})$-modules.[1] Moreover, such sections of the bundle of endomorphisms locally generate the sections of the homomorphism bundle. Note that

$$(\eta \otimes \lambda)^{\mathrm{e}} = \xi^{\mathrm{v}} \otimes \lambda^{\mathrm{e}},$$

as is directly verified. In this case, since $\mathrm{C}^r(\mathsf{M})$ is a subring of $\mathrm{C}^r(\mathsf{E})$ (by pull-back), we can regard the tensor product as being of $\mathrm{C}^r(\mathsf{E})$-modules. Therefore,

$$L^{\mathrm{e}} \in \Gamma^r(\Gamma^r(\pi_{\mathsf{E}}^*\mathsf{F}) \otimes \mathrm{Lin}^r(\mathsf{E})).$$

Since $\mathrm{Lin}^r(\mathsf{E}) \subseteq \mathrm{C}^r(\mathsf{E})$, the tensor product is mere multiplication in this case.

A similar sort of analysis can be made for the vertical lift of an homomorphism. In this case, given $\lambda \in \Gamma^r(E^*)$ and $\eta \in \Gamma^r(\mathsf{F})$, we have $\xi \otimes \lambda \in \Gamma^r(\mathsf{F} \otimes \mathsf{E}^*)$, as in the preceding paragraph. In this case, the vertical lift satisfies

$$(\xi \otimes \lambda)^{\mathrm{v}} = \xi^{\mathrm{v}} \otimes \lambda^{\mathrm{v}}.$$

3.1.4 Tensors Fields on Vector Bundles

Next we discuss the extension of our lifts of functions, sections, and vector fields to higher-order tensors. The extension is to tensor powers of the pull-back $\pi_{\mathsf{E}}^*\mathsf{T}^*\mathsf{M}$ of the cotangent bundle to the total space of the vector bundle. Other sorts of lifts are possible, especially in the presence of a connection in the vector bundle. We restrict

[1] This corresponds to the well-known isomorphism

$$\Gamma^r(\mathsf{E}) \otimes_{\mathrm{C}^r(\mathsf{M})} \Gamma^r(\mathsf{F}) \simeq \Gamma^r(\mathsf{E} \otimes \mathsf{F})$$

of $\mathrm{C}^r(\mathsf{M})$-modules that we shall prove as Proposition 5.5 below.

ourselves to the tensor powers of the pull-back of $\mathsf{T}^*\mathsf{M}$ since our interest is in jet bundles, and these tensor powers represent derivatives with respect to the base.

We make the following definitions.

Definition 3.9 (Lifts of Tensors) Let $r \in \{\infty, \omega\}$, and let $\pi_{\mathsf{E}} : \mathsf{E} \to \mathsf{M}$ and $\pi_{\mathsf{F}} : \mathsf{F} \to \mathsf{M}$ be a C^r-vector bundles. Let $k \in \mathbb{Z}_{>0}$.

(i) For $A \in \Gamma^r(\mathrm{T}^k(\mathsf{T}^*\mathsf{M}))$, the ***horizontal lift*** of A is $A^{\mathrm{h}} \in \Gamma^r(\mathrm{T}^k(\mathsf{T}^*\mathsf{E}))$ defined by

$$A^{\mathrm{h}}(Z_1, \ldots, Z_k) = A(T_e\pi_{\mathsf{E}}(Z_1), \ldots, T_e\pi_{\mathsf{E}}(Z_k))$$

for $Z_1, \ldots, Z_k \in \mathsf{T}_e\mathsf{E}$.[2]

(ii) For $A \in \Gamma^r(\mathrm{T}^k(\mathsf{T}^*\mathsf{M}) \otimes \mathsf{E})$, the ***vertical lift*** of A is $A^{\mathrm{v}} \in \Gamma^r(\mathrm{T}^k(\mathsf{T}^*\mathsf{E}) \otimes \mathsf{TE})$ defined by

$$A^{\mathrm{v}}(Z_1, \ldots, Z_k) = \mathrm{vlft}(e, A(T_e\pi_{\mathsf{E}}(Z_1), \ldots, T_e\pi_{\mathsf{E}}(Z_k))),$$

for $Z_1, \ldots, Z_k \in \mathsf{T}_e\mathsf{E}$.

(iii) For $A \in \Gamma^r(\mathrm{T}^k(\mathsf{T}^*\mathsf{M}) \otimes \mathsf{F} \otimes \mathsf{E}^*)$, the ***vertical evaluation*** of A is $A^{\mathrm{e}} \in \Gamma^r(\mathrm{T}^k(\mathsf{T}^*\mathsf{E}) \otimes \pi_{\mathsf{E}}^*\mathsf{F})$ defined by

$$A^{\mathrm{e}}(Z_1, \ldots, Z_k) = (e, A(T_e\pi_{\mathsf{E}}(Z_1), \ldots, T_e\pi_{\mathsf{E}}(Z_k))(e)),$$

for $Z_1, \ldots, Z_k \in \mathsf{T}_e\mathsf{E}$.

Additionally, let $\nabla^{\pi_{\mathsf{E}}}$ be a connection in E.

(iv) For $A \in \Gamma^r(\mathrm{T}^k(\mathsf{T}^*\mathsf{M}) \otimes \mathsf{TM})$, the ***horizontal lift*** of A is $A^{\mathrm{h}} \in \Gamma^r(\mathrm{T}^k(\mathsf{T}^*\mathsf{E}) \otimes \mathsf{TE})$ defined by

$$A^{\mathrm{h}}(Z_1, \ldots, Z_k) = \mathrm{hlft}(e, A(T_e\pi_{\mathsf{E}}(Z_1), \ldots, T_e\pi_{\mathsf{E}}(Z_k)))$$

for $Z_1, \ldots, Z_k \in \mathsf{T}_e\mathsf{E}$.

(v) For $A \in \Gamma^r(\mathrm{T}^k(\mathsf{T}^*\mathsf{M}) \otimes \mathsf{E}^*)$, the ***vertical lift*** of A is $A^{\mathrm{v}} \in \Gamma^r(\mathrm{T}^k(\mathsf{T}^*\mathsf{E}) \otimes \mathsf{T}^*\mathsf{E})$ defined by

$$A^{\mathrm{v}}(Z_1, \ldots, Z_k) = \mathrm{vlft}(e, A(T_e\pi_{\mathsf{E}}(Z_1), \ldots, T_e\pi_{\mathsf{E}}(Z_k)))$$

for $Z_1, \ldots, Z_k \in \mathsf{T}_e\mathsf{E}$.

[2] Of course, this is nothing but the usual definition of pull-back, which we repeat for the sake of symmetry.

(vi) For $A \in \Gamma^r(\mathsf{T}^k(\mathsf{T}^*\mathsf{M}) \otimes \mathsf{F} \otimes \mathsf{E}^*)$, the ***vertical lift*** of A is $A^{\mathrm{v}} \in \Gamma^r(\mathsf{T}^k(\mathsf{T}^*\mathsf{E}) \otimes \pi_{\mathsf{E}}^*\mathsf{F} \otimes \mathsf{T}^*\mathsf{E})$ defined by

$$A^{\mathrm{v}}(Z_1, \dots, Z_k)(Z) = (e, A(T_e\pi_{\mathsf{E}}(Z_1), \dots, T_e\pi_{\mathsf{E}}(Z_k))(\mathrm{ver}(Z))),$$

for $Z_1, \dots, Z_k, Z \in \mathsf{T}_e\mathsf{E}$. $\circ$

3.1.5 Tensor Contractions

In our differentiation results of Sect. 3.2, we shall make use of certain generalisations of the contraction operator on tensors. What we have is a sort of "contraction and insertion" operation. We describe this here in the setting of linear algebra, since this is where it most naturally resides. The constructions can, of course, be extended to vector bundles by performing the vector space constructions on fibres.

Let V be a finite-dimensional $\mathbb{R}$-vector space, let $k \in \mathbb{Z}_{>0}$ and $l \in \mathbb{Z}_{\geq 0}$, and let $\alpha \in \mathrm{T}^k(\mathsf{V}^*)$ and $\beta \in \mathrm{T}^l(\mathsf{V}^*) \otimes \mathsf{V}$. For $j \in \{1, \dots, k\}$, define the ***jth insertion of β in α*** by $\mathrm{Ins}_j(\alpha, \beta) \in \mathrm{T}^{k+l-1}(\mathsf{V}^*)$ by

$$\begin{aligned}
&\mathrm{Ins}_j(\alpha, \beta)(v_1, \dots, v_{k+l-1}) \\
&= \alpha(v_1, \dots, v_{j-1}, \beta(v_j, v_{k+1}, \dots, v_{k+l-1}), v_{j+1}, \dots, v_k).
\end{aligned}$$

To be clear, when $l = 0$ we have

$$\mathrm{Ins}_j(\alpha, v)(v_1, \dots, v_{k-1}) = \alpha(v_1, \dots, v_{j-1}, v, v_j, \dots, v_{k-1}).$$

We will also find it helpful to consider tensor contraction when one of the arguments (the second is the one we care about) is fixed. Thus let $\beta \in \mathrm{T}^l(\mathsf{V}^*) \otimes \mathsf{V}$ and define $\mathrm{Ins}_{j,\beta}(\alpha) = \mathrm{Ins}_j(\alpha, \beta)$.

We shall also need notation for a specific sort of swapping of arguments of a tensor. Let $\alpha \in \mathrm{T}^k(\mathsf{V})$ and let $j_1, j_2 \in \{1, \dots, k\}$. We define

$$\begin{aligned}
&\mathrm{push}_{j_1,j_2}\alpha(v_1, \dots, v_k) \\
&\quad = \begin{cases} \alpha(v_1, \dots, v_{j_1-1}, v_{j_1+1}, \dots, v_{j_2}, v_{j_1}, v_{j_2+1}, \dots, v_k), & j_1 \leq j_2, \\ \alpha(v_1, \dots, v_{j_2-1}, v_{j_1}, v_{j_2}, \dots, v_{j_1-1}, v_{j_1+1}, \dots, v_k), & j_1 > j_2. \end{cases}
\end{aligned}$$

The idea is that push_{j_1,j_2} drops v_{j_1} into the j_2-slot, and shifts the arguments to make room for this. The "insertion" and "push" mappings can be generalised in the obvious way to give $\mathrm{Ins}_j(A, \beta)$ and $\mathrm{push}_{j_1,j_2}(A)$ for $A \in \mathrm{T}^k(\mathsf{V}^*) \otimes \mathsf{U}$ and $\beta \in (\mathrm{T}^l(\mathsf{V}^*) \otimes \mathsf{V}) \otimes \mathsf{U}$ (resp. $A \in \mathsf{U} \otimes \mathrm{T}^k(\mathsf{V}^*)$ and $\beta \in \mathsf{U} \otimes (\mathrm{T}^l(\mathsf{V}^*) \otimes \mathsf{V})$), just by acting on the first (resp. second) component of the tensor product.

The final tensor construction we make is that of a linear tensor derivation. Given $A \in \mathrm{End}_{\mathbb{R}}(\mathsf{V})$, we define a derivation D_A of the tensor algebra $\oplus_{r,s\in\mathbb{Z}_{\geq 0}}\mathrm{T}^r_s(\mathsf{V})$ by $D_A(a) = 0$ for $a \in \mathrm{T}^0_0(\mathsf{V}) \simeq \mathbb{R}$, and $D_A(v) = A(v)$ for $v \in \mathrm{T}^1_0(\mathsf{V}) \simeq \mathsf{V}$. It then follows that $D_A(\alpha) = -A^*(\alpha)$ for $\alpha \in \mathsf{V}^*$. More generally, we have the following result which expresses a well-known formula, e.g., [53, §3.4], in terms of our insertion operation.

Lemma 3.10 (Insertion and Tensor Derivation I) *Let* V *be a finite-dimensional* $\mathbb{R}$*-vector space, let* $A \in \mathrm{End}_{\mathbb{R}}(\mathsf{V})$*, let* $r, s \in \mathbb{Z}_{>0}$*, and let* $T \in \mathrm{T}^r_s(\mathsf{V})$*. Then*

$$D_A(T) = \sum_{j=1}^{r} \mathrm{Ins}_j(T, A^*) - \sum_{j=1}^{s} \mathrm{Ins}_{r+j}(T, A).$$

Proof We have

$$\begin{aligned}
&D_A(T)(\beta^1, \dots, \beta^r, u_1, \dots, u_s)\\
&\quad= \sum_{j=1}^{r} v_1 \otimes \dots \otimes A(v_j) \otimes \dots v_r \otimes \alpha^1 \otimes \dots \otimes \alpha^s(\beta^1, \dots, \beta^r, u_1, \dots, u_s)\\
&\qquad- \sum_{j=1}^{s} v_1 \otimes \dots \otimes v_r \otimes \alpha^1 \otimes \dots \otimes A^*(\alpha^j) \otimes \dots \otimes \alpha^s\\
&\qquad(\beta^1, \dots, \beta^r, u_1, \dots, u_s)\\
&\quad= \sum_{j=1}^{r} \beta^1(v_1) \cdots \beta^j(A(v_j)) \cdots \beta^r(v_r)\alpha^1(u_1) \cdots \alpha^s(u_s)\\
&\qquad- \sum_{j=1}^{s} \beta^1(v_1) \cdots \beta^r(v_r)\alpha^1(u_1) \cdots A^*(\alpha^j)(u_j) \cdots \alpha^s(u_s)\\
&\quad= \sum_{j=1}^{r} \beta^1(v_1) \cdots A^*(\beta^j)(v_j) \cdots \beta^r(v_r)\alpha^1(u_1) \cdots \alpha^s(u_s)\\
&\qquad- \sum_{j=1}^{s} \beta^1(v_1) \cdots \beta^r(v_r)\alpha^1(u_1) \cdots \alpha^j(A(u_j)) \cdots \alpha^s(u_s)\\
&\quad= \sum_{j=1}^{r} T(\beta^1, \dots, A^*(\beta^j), \dots, \beta^r, u_1, \dots, u_s)\\
&\qquad- \sum_{j=1}^{s} T(\beta^1, \dots, \beta^r, u_1, \dots, A(u_j), \dots, u_s),
\end{aligned}$$

as claimed. □

We shall make a minor extension of the preceding notion of a derivation associated to an endomorphism. Let $k, r, s \in \mathbb{Z}_{>0}$. Here we let $T \in \mathrm{T}^r_s(\mathsf{V})$ and $S \in \mathrm{T}^1_k(\mathsf{V})$. For $v_1, \dots, v_{k-1} \in \mathsf{V}$, we define $S_{(v_1,\dots,v_{k-1})} \in \mathrm{End}_{\mathbb{R}}(\mathsf{V})$ by

$$S_{(v_1,\dots,v_{k-1})}(v) = S(v, v_1, \dots, v_{k-1}).$$

Denote $S^* \in \mathrm{T}^1_{k-1}(\mathsf{V}) \otimes \mathsf{V}^*$ by

$$\langle S^*(\beta, v_1, \dots, v_{k-1}); v\rangle = \langle \beta; S(v, v_1, \dots, v_{k-1})\rangle$$

so that

$$S^*(\beta, v_1, \dots, v_{k-1}) = S^*_{(v_1,\dots,v_{k-1})}(\beta).$$

We then define $D_S(T) \in \mathrm{T}^r_{s+k-1}(\mathsf{V})$ by

$$D_S(T)(\beta^1, \dots, \beta^r, u_1, \dots, u_{s+k-1}) = D_{S_{(u_{s+1}\dots u_{s+k-1})}}(T)(\beta^1, \dots, \beta^r, u_1, \dots, u_s). \tag{3.2}$$

The following elementary lemma gives a simpler formula for the previous constructions.

Lemma 3.11 (Insertion and Tensor Derivation II) *Let* V *be a finite-dimensional* $\mathbb{R}$*-vector space, let* $k \in \mathbb{Z}_{>0}$*, let* $S \in \mathrm{T}^1_k(\mathsf{V})$*, let* $r, s \in \mathbb{Z}_{>0}$*, and let* $T \in \mathrm{T}^r_s(\mathsf{V})$*. Then*

$$D_S(T) = \sum_{j=1}^{r} \mathrm{Ins}_j(T, S^*) - \sum_{j=1}^{s} \mathrm{Ins}_{r+j}(T, S).$$

Proof We have

$$\begin{aligned}
D_S(T)&(\beta^1, \dots, \beta^r, u_1, \dots, u_{k+s-1}) \\
&= \sum_{j=1}^{r} \mathrm{Ins}_j(T, S_{(u_{s+1},\dots,u_{s+k-1})}{}^*)(\beta^1, \dots, \beta^r, u_1, \dots, u_s) \\
&\quad - \sum_{j=1}^{s} \mathrm{Ins}_{r+j}(T, S_{(u_{s+1},\dots,u_{s+k-1})})(\beta^1, \dots, \beta^r, u_1, \dots, u_s) \\
&= \sum_{j=1}^{r} T(\beta^1, \dots, S_{(u_{s+1},\dots,u_{s+k-1})}{}^*(\beta_j), \dots, \beta^r, u_1, \dots, u_s) \\
&\quad - \sum_{j=1}^{s} T(\beta^1, \dots, \beta^r, u_1, \dots, S_{(u_{s+1},\dots,u_{s+k-1})}(u_j), \dots, u_s)
\end{aligned}$$

$$= \sum_{j=1}^{r} \mathrm{Ins}_j(T, S^*)(\beta^1, \dots, \beta^r, u_1, \dots, u_{s+k-1})$$
$$- \sum_{j=1}^{r} \mathrm{Ins}_{r+j}(T, S)(\beta^1, \dots, \beta^r, u_1, \dots, u_{s+k-1}),$$

as claimed. □

Let us summarise this in the cases of interest. The cases of interest will be two in number. The first is when $S \in \mathrm{T}^1_2(\mathsf{V})$ and $T = T_0 \otimes v$ for $T_0 \in \mathrm{T}^k(\mathsf{V}^*)$ and $v \in \mathsf{V}$. In this case the preceding lemma gives

$$\begin{aligned}
&D_S(T)(v_1, \dots, v_{k+1}, \beta) \\
&\quad = \mathrm{Ins}_{k+1}(T_0 \otimes v, S^*)(v_1, \dots, v_{k+1}, \beta) - \sum_{j=1}^{k} \mathrm{Ins}_j(T_0 \otimes v, S)(v_1, \dots, v_{k+1}, \beta) \\
&\quad = T_0(v_1, \dots, v_k)\langle \beta; S_{v_{k+1}}(v)\rangle - \langle \beta; v\rangle \sum_{j=1}^{k} \mathrm{Ins}_j(T_0, S)(v_1, \dots, v_{k+1}) \\
&\quad = T_0(v_1, \dots, v_k)\langle \beta; S(v, v_{k+1})\rangle - \langle \beta; v\rangle \sum_{j=1}^{k} \mathrm{Ins}_j(T_0, S)(v_1, \dots, v_{k+1}).
\end{aligned} \tag{3.3}$$

The second case we will consider is when $S \in \mathrm{T}^1_2(\mathsf{V})$ and $T = T_0 \otimes \alpha$ for $T_0 \in \mathrm{T}^k(\mathsf{V}^*)$ and $\alpha \in \mathsf{V}^*$. In this case we have

$$\begin{aligned}
&D_S(T)(v_1, \dots, v_{k+2}) \\
&\quad = -\,\mathrm{Ins}_{k+1}(T_0 \otimes \alpha, S)(v_1, \dots, v_{k+2}) - \sum_{j=1}^{k} \mathrm{Ins}_j(T_0 \otimes \alpha, S)(v_1, \dots, v_{k+2}) \\
&\quad = -\,T_0(v_1, \dots, v_k)\alpha(S(v_{k+1}, v_{k+2})) \\
&\qquad - \langle \alpha; v_{k+2}\rangle \sum_{j=1}^{k} \mathrm{Ins}_j(T_0, S)(v_1, \dots, v_{k+1}).
\end{aligned} \tag{3.4}$$

3.2 Differentiation of Tensors on the Total Space of a Vector Bundle

In this section we establish some technical results for differentiation via connections of various objects—functions, vector fields, tensors—on vector bundles. These results will allow us to intrinsically perform the many calculations required to determine the recursive relations given in Sect. 3.3 between jets on M and jets on E for a vector bundle $\pi_{\mathsf{E}}: \mathsf{E} \to \mathsf{M}$. As with the constructions of the preceding section, the results in this section might seem *non sequitur* to the objectives of the book. And, as with the results of the preceding section, perhaps a good strategy is to hurdle over this section until the results are subsequently needed.

In this section the constructions and results are made and given in both the smooth and real analytic cases.

3.2.1 Vector Bundles as Riemannian Submersions

Let $r \in \{\infty, \omega\}$. Let $\pi_{\mathsf{E}}: \mathsf{E} \to \mathsf{M}$ be a vector bundle with $\pi_{\mathsf{TE}}: \mathsf{TE} \to \mathsf{E}$ its tangent bundle. We suppose that ∇^{M} is an affine connection on M, that $\nabla^{\pi_{\mathsf{E}}}$ is a linear connection on E, that $\mathbb{G}_{\mathsf{M}}$ is a Riemannian metric on M, and that $\mathbb{G}_{\pi_{\mathsf{E}}}$ is a fibre metric for E with all data of class C^r. We shall construct on E a Riemannian metric in a more or less natural way. Not all constructions require that the affine connection on M to be the Levi-Civita connection, but we will only work with the case when it is, since there are useful formulae one can prove in this case. Subsequently we shall show that one can just as well use affine connections other than the Levi-Civita connection.

The Riemannian metric we construct on the total space E is a natural adaptation of the Sasaki metric for tangent bundles [61]. To define the inner product, we use the splitting determined by the connection to give the inner product on $\mathsf{T}_e\mathsf{E}$ by

$$\mathbb{G}_{\mathsf{E}}(w_1, w_2) = \mathbb{G}_{\mathsf{M}}(\mathrm{hor}(w_1), \mathrm{hor}(w_2)) + \mathbb{G}_{\pi_{\mathsf{E}}}(\mathrm{ver}(w_1), \mathrm{ver}(w_2)), \quad w_1, w_2 \in \mathsf{T}_e\mathsf{E}. \tag{3.5}$$

This then turns E into a Riemannian manifold. We denote by ∇^{E} the Levi-Civita connection associated with $\mathbb{G}_{\mathsf{E}}$. Since the connection giving the splitting is of class C^r if $\nabla^{\pi_{\mathsf{E}}}$ is of class C^r, the Riemannian metric $\mathbb{G}_{\mathsf{E}}$ and its Levi-Civita connection are of class C^r if $\mathbb{G}_{\mathsf{M}}$ and $\mathbb{G}_{\pi_{\mathsf{E}}}$ are of class C^r.

We note that the choice of metric $\mathbb{G}_{\mathsf{E}}$ ensures that $\pi_{\mathsf{E}}: \mathsf{E} \to \mathsf{M}$ is a Riemannian submersion if we equip M with its Riemannian metric $\mathbb{G}_{\mathsf{M}}$ used to build $\mathbb{G}_{\mathsf{E}}$. Moreover, the fibres of π_{E} are totally geodesic submanifolds. There are a few constructions involving Riemannian submersions that will be helpful for us, and we review these here, initially in a general setting.

We let $(\mathsf{F}, \mathbb{G}_{\mathsf{F}})$ and $(\mathsf{M}, \mathbb{G}_{\mathsf{M}})$ be Riemannian manifolds and suppose that we have a surjective submersion $\pi \in \mathrm{C}^r(\mathsf{F}; \mathsf{M})$. By ∇^{F} and ∇^{M} we denote the Levi-Civita connections associated with $\mathbb{G}_{\mathsf{F}}$ and $\mathbb{G}_{\mathsf{M}}$, respectively. We let $\mathsf{VF} = \ker(T\pi)$ be the vertical subbundle with HF its $\mathbb{G}_{\mathsf{F}}$-orthogonal complement, which we call the horizontal subbundle HF. We let ver, hor: $\mathsf{TF} \to \mathsf{TF}$ be the projections onto VF and HF, just as we have done for vector bundles. The submersion π is a ***Riemannian submersion*** if, for each $y \in \mathsf{F}$,

$$\mathbb{G}_{\mathsf{M}}(T_y\pi(u), T_y\pi(v)) = \mathbb{G}_{\mathsf{F}}(u, v), \qquad u, v \in \mathsf{H}_y\mathsf{F}.$$

For a vector field X on M, we denote by X^{h} the horizontal lift of X to F. This is the unique HF-valued vector field satisfying $T_y\pi(X^{\mathrm{h}}(y)) = X \circ \pi(y)$ for each $y \in \mathsf{F}$. Thus, for example,

$$\mathbb{G}_{\mathsf{E}}(X^{\mathrm{h}}, Y^{\mathrm{h}}) = \mathbb{G}_{\mathsf{M}}(X, Y), \qquad X, Y \in \Gamma^r(\mathsf{TM}).$$

Following [56], for a C^r-Riemannian submersion $\pi : \mathsf{F} \to \mathsf{N}$, there are two associated tensors that characterise the submersion. Specifically, we define

$$A_\pi, T_\pi \in \Gamma^r(\mathrm{T}^2(\mathsf{T}^*\mathsf{F}) \otimes \mathsf{TF})$$

by

$$\begin{aligned} A_\pi(\xi, \eta) &= \mathrm{ver}(\nabla^{\mathsf{F}}_{\mathrm{hor}(\xi)} \mathrm{hor}(\eta)) + \mathrm{hor}(\nabla^{\mathsf{F}}_{\mathrm{hor}(\xi)} \mathrm{ver}(\eta)), \\ T_\pi(\xi, \eta) &= \mathrm{hor}(\nabla^{\mathsf{F}}_{\mathrm{ver}(\xi)} \mathrm{ver}(\eta)) + \mathrm{ver}(\nabla^{\mathsf{F}}_{\mathrm{ver}(\xi)} \mathrm{hor}(\eta)) \end{aligned} \tag{3.6}$$

for $\xi, \eta \in \Gamma^\infty(\mathsf{TF})$. One can easily verify that A_π and T_π are indeed tensors as claimed. Since the fibres of π are submanifolds, we can define the ***vertical covariant derivative*** as the projection of the covariant derivative onto sections:

$$\nabla^{\mathrm{ver}}_U V = \mathrm{ver}(\nabla^{\mathsf{F}}_U V), \qquad U, V \in \Gamma^r(\mathsf{VF}).$$

Given a submanifold S of a Riemannian manifold $(\mathsf{M}, \mathbb{G}_{\mathsf{M}})$, S inherits the Riemannian metric $\mathbb{G}_{\mathsf{S}}$ obtained by pulling back $\mathbb{G}_{\mathsf{M}}$ by the inclusion $\iota_{\mathsf{S}} : \mathsf{S} \to \mathsf{M}$. The submanifold S is ***totally geodesic*** if every geodesic for $(\mathsf{S}, \mathbb{G}_{\mathsf{S}})$ is also a geodesic for $(\mathsf{M}, \mathbb{G}_{\mathsf{M}})$.

With all this background, we have the following result with tells us how to covariantly differentiate vector fields on the total space of a vector bundle.

Lemma 3.12 (Covariant Derivatives for Riemannian Submersions) *Let* $r \in \{\infty, \omega\}$. *Let* $(\mathsf{F}, \mathbb{G}_{\mathsf{F}})$ *and* $(\mathsf{M}, \mathbb{G}_{\mathsf{M}})$ *be* C^r*-Riemannian manifolds with* ∇^{F} *and* ∇^{M} *the Levi-Civita connections. Let* $\pi : \mathsf{F} \to \mathsf{M}$ *be a* C^r*-Riemannian submersion. Let* $X, Y \in \Gamma^r(\mathsf{TM})$ *and let* $U, V \in \Gamma^r(\mathsf{TF})$ *be vertical vector fields. Then the following statements hold:*

(i) $\mathrm{hor}(\nabla^{\mathsf{F}}_{X^{\mathrm{h}}} Y^{\mathrm{h}}) = (\nabla^{\mathsf{M}}_X Y)^{\mathrm{h}}$*;*
(ii) $A_\pi(X^{\mathrm{h}}, Y^{\mathrm{h}}) = -\frac{1}{2}\,\mathrm{ver}([X^{\mathrm{h}}, Y^{\mathrm{h}}])$*;*
(iii) $\nabla^{\mathsf{F}}_U V = \nabla^{\mathrm{ver}}_U V + T_\pi(U, V)$*;*
(iv) $\nabla^{\mathsf{F}}_V X^{\mathrm{h}} = \mathrm{hor}(\nabla^{\mathsf{F}}_V X^{\mathrm{h}}) + T_\pi(V, X^{\mathrm{h}})$*;*
(v) $\nabla^{\mathsf{F}}_{X^{\mathrm{h}}} V = \mathrm{ver}(\nabla^{\mathsf{F}}_{X^{\mathrm{h}}} V) + A_\pi(X^{\mathrm{h}}, V)$*;*
(vi) $\nabla^{\mathsf{F}}_{X^{\mathrm{h}}} Y^{\mathrm{h}} = (\nabla^{\mathsf{M}}_X Y)^{\mathrm{h}} + A_\pi(X^{\mathrm{h}}, Y^{\mathrm{h}})$*;*
(vii) $\mathbb{G}_{\mathsf{F}}(\nabla^{\mathsf{F}}_V X^{\mathrm{h}}, Y^{\mathrm{h}}) = -\frac{1}{2}\mathbb{G}_{\mathsf{F}}(\mathrm{ver}([X^{\mathrm{h}}, Y^{\mathrm{h}}]), V) = \mathbb{G}_{\mathsf{F}}(\nabla^{\mathsf{F}}_V Y^{\mathrm{h}}, X^{\mathrm{h}})$.

Additionally, if the fibres of π are totally geodesic submanifolds of F, then the following statements hold:

(viii) $T_\pi = 0$*;*
(ix) $\nabla^{\mathrm{ver}}|\mathsf{F}_x$ *is the Levi-Civita connection for the submanifold Riemannian metric on* F_x*;*
(x) $\mathrm{ver}(\nabla^{\mathsf{F}}_{X^{\mathrm{h}}} V) = \mathrm{ver}([X^{\mathrm{h}}, V])$*;*
(xi) $\nabla^{\mathsf{F}}_V X^{\mathrm{h}}$ *is horizontal and* $\nabla^{\mathsf{F}}_V X^{\mathrm{h}} = A_\pi(X^{\mathrm{h}}, V)$*.*

Finally, if $\mathsf{F} = \mathsf{E}$ is the total space of a vector bundle and if $\mathbb{G}_{\mathsf{E}}$ is the Riemannian metric on E defined above, then the following additional statements hold for sections $\xi, \eta \in \Gamma^r(\mathsf{E})$:

(xii) $\nabla^{\mathsf{E}}_{\xi^{\mathrm{v}}} \eta^{\mathrm{v}} = 0$*;*
(xiii) $\mathrm{ver}(\nabla^{\mathsf{E}}_{X^{\mathrm{h}}} \xi^{\mathrm{v}}) = (\nabla^{\pi}_X \xi)^{\mathrm{v}}$*.*

Proof We use the Koszul formula for the Levi-Civita connection:

$$
\begin{aligned}
2\mathbb{G}_{\mathsf{F}}(\nabla^{\mathsf{F}}_\xi \eta, \zeta) = {} & \mathscr{L}_\xi(\mathbb{G}_{\mathsf{F}}(\eta, \zeta)) + \mathscr{L}_\eta(\mathbb{G}_{\mathsf{F}}(\xi, \zeta)) - \mathscr{L}_\zeta(\mathbb{G}(\xi, \eta)) \\
& + \mathbb{G}_{\mathsf{F}}([\xi, \eta], \zeta) - \mathbb{G}_{\mathsf{F}}([\xi, \zeta], \eta) - \mathbb{G}_{\mathsf{F}}([\eta, \zeta], \xi)
\end{aligned} \tag{3.7}
$$

for vector fields ξ, η, and ζ on F [41, Page 160]. We shall also use the formulae

$$
\mathscr{L}_\zeta(\mathbb{G}_{\mathsf{F}}(\xi, \eta)) = \mathbb{G}_{\mathsf{F}}(\nabla^{\mathsf{F}}_\zeta \xi, \eta) + \mathbb{G}_{\mathsf{F}}(\xi, \nabla^{\mathsf{F}}_\zeta \eta) \tag{3.8}
$$

(saying that the Levi-Civita connection is a metric connection) and

$$
\nabla^{\mathsf{F}}_\xi \eta - \nabla^{\mathsf{F}}_\eta \xi = [\xi, \eta] \tag{3.9}
$$

(saying that the Levi-Civita connection is torsion-free).

Let us make some preliminary computations. First, since X^{h} and Y^{h} are π-related to X and Y, we have that $[X^{\mathrm{h}}, Y^{\mathrm{h}}]$ is π-related to $[X, Y]$ [1, Proposition 4.2.25]. Thus

$$
\mathrm{hor}([X^{\mathrm{h}}, Y^{\mathrm{h}}]) = [X, Y]^{\mathrm{h}}. \tag{3.10}
$$

In like manner, since V is π-related to the zero vector field and X^{h} is π-related to X, $[V, X^{\mathrm{h}}]$ is π-related to the zero vector field. That is,

$$\mathrm{hor}([V, X^{\mathrm{h}}]) = 0. \tag{3.11}$$

Next, if f is a function on M, then

$$\mathscr{L}_{X^{\mathrm{h}}}(\pi^* f) = \langle \mathrm{d}(\pi^* f); X^{\mathrm{h}} \rangle = \langle \pi^* \mathrm{d} f; X^{\mathrm{h}} \rangle,$$

from which we deduce

$$\mathscr{L}_{X^{\mathrm{h}}}(\pi^* f)(y) = \langle \mathrm{d} f \circ \pi(y); X \circ \pi(y) \rangle, \qquad y \in \mathsf{F}. \tag{3.12}$$

We trivially have

$$\mathscr{L}_V(\pi^* f) = 0.$$

(i) One can use (3.7) with $\xi = X^{\mathrm{h}}$, $\eta = Y^{\mathrm{h}}$, and $\zeta = Z^{\mathrm{h}}$, and the formulae (3.10) and (3.12) to give

$$\mathbb{G}_{\mathsf{F}}(\nabla^{\mathsf{F}}_{X^{\mathrm{h}}} Y^{\mathrm{h}}, Z^{\mathrm{h}}) = \pi^* \mathbb{G}_{\mathsf{M}}(\nabla^{\mathsf{M}}_X Y, Z).$$

This shows that

$$\mathrm{hor}(\nabla^{\mathsf{F}}_{X^{\mathrm{h}}} Y^{\mathrm{h}}) = (\nabla^{\mathsf{M}}_X Y)^{\mathrm{h}}. \tag{3.13}$$

(ii) Now we use (3.7) with $\xi = X^{\mathrm{h}}$, $\eta = Y^{\mathrm{h}}$, and $\zeta = V$. We immediately have that the first three terms on the right in (3.7) are zero. By (3.11), the last two terms on the right in (3.7) are zero. Thus we have

$$2\mathbb{G}_{\mathsf{F}}(\nabla^{\mathsf{F}}_{X^{\mathrm{h}}} Y^{\mathrm{h}}, V) = \mathbb{G}_{\mathsf{F}}([X, Y]^{\mathrm{h}}, V),$$

and so

$$A_\pi(X^{\mathrm{h}}, Y^{\mathrm{h}}) = \mathrm{ver}(\nabla^{\mathsf{F}}_{X^{\mathrm{h}}} Y^{\mathrm{h}}) = \frac{1}{2}\,\mathrm{ver}([X, Y]^{\mathrm{h}}).$$

(iii) We have

$$\nabla^{\mathsf{F}}_U V = \mathrm{ver}(\nabla^{\mathsf{F}}_U V) + \mathrm{hor}(\nabla^{\mathsf{F}}_U V) = \nabla^{\mathrm{ver}}_U V + T_\pi(U, V),$$

as claimed.

(iv) We have

$$\nabla^{\mathsf{F}}_V X^{\mathrm{h}} = \mathrm{hor}(\nabla^{\mathsf{F}}_V X^{\mathrm{h}}) + \mathrm{ver}(\nabla^{\mathsf{F}}_V X^{\mathrm{h}}) = \mathrm{hor}(\nabla^{\mathsf{F}}_V X^{\mathrm{h}}) + T_\pi(V, X^{\mathrm{h}}),$$

as claimed.

(v) We have

$$\nabla^{\mathsf{F}}_{X^{\mathrm{h}}} V = \mathrm{ver}(\nabla^{\mathsf{F}}_{X^{\mathrm{h}}} V) + \mathrm{hor}(\nabla^{\mathsf{F}}_{X^{\mathrm{h}}} V) = \mathrm{ver}(\nabla^{\mathsf{F}}_{X^{\mathrm{h}}} V) + A_\pi(X^{\mathrm{h}}, V),$$

as claimed.

(vi) We have

$$\nabla^{\mathsf{F}}_{X^{\mathrm{h}}} Y^{\mathrm{h}} = \mathrm{hor}(\nabla^{\mathsf{F}}_{X^{\mathrm{h}}} Y^{\mathrm{h}}) + \mathrm{ver}(\nabla^{\mathsf{F}}_{X^{\mathrm{h}}} Y^{\mathrm{h}}) = (\nabla^{\mathsf{M}}_X Y)^{\mathrm{h}} + A_\pi(X^{\mathrm{h}}, Y^{\mathrm{h}}),$$

using part (i).

(vii) This is a direct computation using (3.9), (3.11), (3.8), and part (i):

$$\begin{aligned} \mathbb{G}_{\mathsf{F}}(\nabla^{\mathsf{F}}_V X^{\mathrm{h}}, Y^{\mathrm{h}}) &= \mathbb{G}_{\mathsf{F}}(\nabla^{\mathsf{F}}_{X^{\mathrm{h}}} V, Y^{\mathrm{h}}) + \mathbb{G}_{\mathsf{F}}([V, X^{\mathrm{h}}], Y^{\mathrm{h}}) \\ &= -\mathbb{G}_{\mathsf{F}}(\nabla^{\mathsf{F}}_{X^{\mathrm{h}}} Y^{\mathrm{h}}, V) \\ &= -\frac{1}{2}\mathbb{G}_{\mathsf{F}}([X^{\mathrm{h}}, Y^{\mathrm{h}}], V) = -\frac{1}{2}\mathbb{G}_{\mathsf{F}}(\mathrm{ver}([X^{\mathrm{h}}, Y^{\mathrm{h}}]), V). \end{aligned} \tag{3.14}$$

(viii) and (ix) These are properties of totally geodesic submanifolds, so we first prove the result for the following situation.

Sublemma 1 *Let* $(\mathsf{M}, \mathbb{G}_{\mathsf{M}})$ *be a Riemannian manifold and let* $\mathsf{S} \subseteq \mathsf{M}$ *be a submanifold. We let* $\mathbb{G}_{\mathsf{S}} = \iota_{\mathsf{S}}^* \mathbb{G}_{\mathsf{M}}$ *be the induced Riemannian metric on* S*. We let* ∇^{M} *and* ∇^{S} *be the Levi-Civita connections. Then* S *is totally geodesic if and only if* $\nabla^{\mathsf{M}}_X Y$ *is tangent to* S *whenever* $X, Y \in \Gamma^\infty(\mathsf{TM})$ *are tangent to* S.

Proof We let $\mathsf{NS} \subseteq \mathsf{TM}|\mathsf{S}$ be the normal bundle. We define the second fundamental form for S to be the section Π_{S} of $\mathrm{T}^2(\mathsf{T}^*\mathsf{S}) \otimes \mathsf{NS}$ defined by

$$\Pi_{\mathsf{S}}(X, Y) = \mathrm{pr}_{\mathsf{NS}}(\nabla^{\mathsf{M}}_X Y)$$

for vector fields X and Y on M that are tangent to S, where $\mathrm{pr}_{\mathsf{NS}}\colon \mathsf{TM}|\mathsf{S} \to \mathsf{NS}$ is the orthogonal projection onto NS.

We claim that Π_{S} is symmetric. Indeed, by (3.9) we have

$$\Pi_{\mathsf{S}}(X, Y) - \Pi_{\mathsf{S}}(Y, X) = \mathrm{pr}_{\mathsf{NS}}([X, Y]) = 0,$$

since $[X, Y]$ is tangent to S if X and Y are tangent to S.

Next we claim that $\mathrm{pr}_{\mathsf{TS}}(\nabla^{\mathsf{M}}_X Y) = \nabla^{\mathsf{S}}_X Y$ for vector fields X and Y that are tangent to S, where $\mathrm{pr}_{\mathsf{TS}}\colon \mathsf{TM}|\mathsf{S} \to \mathsf{TS}$ is the orthogonal projection. To prove this, we show that

$$(X, Y) \mapsto \mathrm{pr}_{\mathsf{TS}}(\nabla^{\mathsf{M}}_X Y),$$

when restricted to S, satisfies the defining conditions (3.8) and (3.9) for the Levi-Civita connection for $\mathbb{G}_{\mathsf{S}}$. Indeed, because $[X, Y]$ is tangent to S whenever X and Y are tangent to S, we determine that, when restricted to S,

$$\mathrm{pr}_{\mathsf{TS}}(\nabla^{\mathsf{M}}_X Y - \nabla^{\mathsf{M}}_Y X) = \mathrm{pr}_{\mathsf{TS}}([X, Y]) = [X, Y]$$

for all vector fields X and Y tangent to S. This shows that $(X, Y) \mapsto \mathrm{pr}_{\mathsf{TS}}(\nabla^{\mathsf{M}}_X Y)$ satisfies (3.9). Also, when we restrict to S, we have

$$\begin{aligned}\mathscr{L}_Z(\mathbb{G}_{\mathsf{S}}(X, Y)) = \mathscr{L}_Z(\mathbb{G}_{\mathsf{M}}(X, Y)) &= \mathbb{G}_{\mathsf{M}}(\nabla^{\mathsf{M}}_Z X, Y) + \mathbb{G}_{\mathsf{M}}(X, \nabla^{\mathsf{S}}_Z Y)\\ &= \mathbb{G}_{\mathsf{S}}(\mathrm{pr}_{\mathsf{TS}}(\nabla^{\mathsf{M}}_Z X), Y) + \mathbb{G}_{\mathsf{S}}(X, \mathrm{pr}_{\mathsf{TS}}(\nabla^{\mathsf{M}}_Z Y))\end{aligned}$$

for all vector fields X, Y, and Z that are tangent to S. This shows that $(X, Y) \mapsto \mathrm{pr}_{\mathsf{TS}}(\nabla^{\mathsf{M}}_X Y)$ satisfies (3.8).

Now we can prove the sublemma. First suppose that S is totally geodesic. Let $v_x \in \mathsf{TS}$ and let $t \mapsto \gamma(t)$ be a geodesic for ∇^{S} satisfying $\gamma'(0) = v_x$. Then γ is also a geodesic for ∇^{M}. Thus

$$\begin{aligned}0 = \nabla^{\mathsf{M}}_{\gamma'(t)}\gamma'(t) &= \nabla^{\mathsf{S}}_{\gamma'(t)}\gamma'(t)\\ &= \mathrm{pr}_{\mathsf{TS}}(\nabla^{\mathsf{M}}_{\gamma'(t)}\gamma'(t))\\ &= \mathrm{pr}_{\mathsf{TS}}(\nabla^{\mathsf{M}}_{\gamma'(t)}\gamma'(t)) + \mathrm{pr}_{\mathsf{NS}}(\nabla^{\mathsf{M}}_{\gamma'(t)}\gamma'(t)),\end{aligned}$$

from which we conclude, evaluating at $t = 0$, that $\Pi_{\mathsf{S}}(v_x, v_x) = 0$. Since Π_{S} is symmetric, $\Pi_{\mathsf{S}} = 0$. Thus

$$\nabla^{\mathsf{S}}_X Y = \mathrm{pr}_{\mathsf{TS}}(\nabla^{\mathsf{M}}_X Y) = \nabla^{\mathsf{M}}_X Y$$

for vector fields X and Y on M tangent to S.

The converse, that S is totally geodesic if $\nabla^{\mathsf{M}}_X Y = \nabla^{\mathsf{S}}_X Y$ for all vector fields X and Y on M tangent to S, is clear. ∇

Given the sublemma, let $x \in \mathsf{M}$ and let $\mathsf{S} = \pi^{-1}(x)$ be the fibre. As we showed in the proof of the sublemma, if U and V are vertical vector fields (in particular, they are tangent to S), then

$$\nabla^{\mathsf{F}}_U V = \mathrm{ver}(\nabla^{\mathsf{F}}_U V) + T_\pi(U, V) = \nabla^{\mathsf{S}}_U V.$$

Matching vertical and horizontal parts on S gives

$$\nabla^{\mathrm{ver}}_U V = \nabla^{\mathsf{S}}_U V, \quad T_\pi(U, V) = 0,$$

as claimed.

(xi) It follows immediately from parts (iv) and (viii) that $\nabla^{\mathsf{F}}_V X^{\mathrm{h}}$ is horizontal. We also have

$$\nabla^{\mathsf{F}}_V X^{\mathrm{h}} = \mathrm{hor}(\nabla^{\mathsf{F}}_V X^{\mathrm{h}}) = \mathrm{hor}(\nabla^{\mathsf{F}}_{X^{\mathrm{h}}} V) + \mathrm{hor}([V, X^{\mathrm{h}}])$$

by (3.9). By part (v), the first term on the right is $A_\pi(X^{\mathrm{h}}, V)$ and, by (3.11), the second term in the right is zero.

(x) By (3.9), we have

$$\mathrm{ver}(\nabla^{\mathsf{F}}_{X^{\mathrm{h}}} V) = \mathrm{ver}(\nabla^{\mathsf{F}}_V X^{\mathrm{h}}) + \mathrm{ver}([X^{\mathrm{h}}, V]).$$

By part (xi), the first term on the right is zero.

(xii) We note here that the fibres of $\pi_{\mathsf{E}}\colon \mathsf{E} \to \mathsf{M}$ are vector spaces and the restriction of $\mathbb{G}_{\mathsf{E}}$ to E_x is just the constant Riemannian metric $\mathbb{G}_{\pi_{\mathsf{E}}}(x)$. Thus covariant derivatives on fibres are just ordinary derivatives. Now, since vertical lifts restricted to fibres are constant, their ordinary derivatives are zero, and this gives the assertion.

(xiii) Here, by part (x), we have

$$\mathrm{ver}(\nabla^{\mathsf{E}}_{X^{\mathrm{h}}} \xi^{\mathrm{v}}) = \mathrm{ver}([X^{\mathrm{h}}, \xi^{\mathrm{v}}]).$$

By (3.11), $[X^{\mathrm{h}}, \xi^{\mathrm{v}}]$ is vertical. By [1, Proposition 4.2.34], we have

$$[\xi^{\mathrm{v}}, X^{\mathrm{h}}] = \frac{1}{2} \frac{\mathrm{d}^2}{\mathrm{d}t^2}\bigg|_{t=0} \Phi^{X^{\mathrm{h}}}_{-t} \circ \Phi^{\xi^{\mathrm{v}}}_{-t} \circ \Phi^{X^{\mathrm{h}}}_{t} \circ \Phi^{\xi^{\mathrm{v}}}_{t}(e).$$

Using the fact that $\Phi^{\xi^{\mathrm{v}}}_t(e) = e + t\xi \circ \pi_{\mathsf{E}}(e)$ and that $\Phi^{X^{\mathrm{h}}}_t(e)$ is the parallel transport $t \mapsto \tau^{\gamma}_t$ along integral curve γ for X through $\pi_{\mathsf{E}}(e)$ cf. [41, Proposition III.1.3], we directly calculate

$$\Phi^{X^{\mathrm{h}}}_{-t} \circ \Phi^{\xi^{\mathrm{v}}}_{-t} \circ \Phi^{X^{\mathrm{h}}}_{t} \circ \Phi^{\xi^{\mathrm{v}}}_{t}(e) = e - t(\tau^{\gamma}_{-t}(\xi \circ \gamma(t)) - \xi \circ \gamma(0)).$$

We recall the relationship

$$\nabla^{\pi_{\mathsf{E}}}_X \xi(x) = \frac{\mathrm{d}}{\mathrm{d}t}\bigg|_{t=0} \tau^{\gamma}_{-t}(\xi \circ \gamma(t)) - \xi \circ \gamma(0)$$

between parallel transport and covariant derivative [41, page 114]. By the Leibniz Rule, we have

$$\frac{\mathrm{d}^2}{\mathrm{d}t^2}\bigg|_{t=0} (t(\tau^{\gamma}_{-t}(\xi \circ \gamma(t)) - \xi \circ \gamma(0))) = 2\frac{\mathrm{d}}{\mathrm{d}t}\bigg|_{t=0} \tau^{\gamma}_{-t}(\xi \circ \gamma(t)) - \xi \circ \gamma(0).$$

Thus we have $[\xi^{\mathrm{v}}, X^{\mathrm{h}}] = -(\nabla^{\pi_{\mathsf{E}}}_X \xi)^{\mathrm{v}}$, as claimed. □

3.2.2 Derivatives of Tensor Contractions

In Sect. 3.1.5 we constructed a tensor contraction/insertion operator. Let us consider the derivative of this operation.

Lemma 3.13 (Covariant Differential of Insertion I) *Let $r \in \{\infty, \omega\}$. Let $\pi_{\mathsf{E}} \colon \mathsf{E} \to \mathsf{M}$ a vector bundle of class C^r, let $\nabla^{\pi_{\mathsf{E}}}$ be a C^r-vector bundle connection in E, let $k, l \in \mathbb{Z}_{>0}$, let $A \in \Gamma^r(\mathrm{T}^k(\mathsf{E}^*))$, and let $S \in \Gamma^r(\mathrm{T}^l(\mathsf{E}^*) \otimes \mathsf{E})$. For $j \in \{1, \dots, k\}$ we have*

$$\nabla^{\pi_{\mathsf{E}}}(\mathrm{Ins}_j(A, S)) = \mathrm{Ins}_j(\nabla^{\pi_{\mathsf{E}}} A, S) + \mathrm{Ins}_j(A, \nabla^{\pi_{\mathsf{E}}} S).$$

Proof We let $\xi_a \in \Gamma^r(\mathsf{E})$, $a \in \{1, \dots, k+l-1\}$, and $X \in \Gamma^r(\mathsf{TM})$. We calculate

$$\begin{aligned}
&\mathscr{L}_X(\mathrm{Ins}_j(A, S)(\xi_1, \dots, \xi_{k+l-1}))\\
&\quad = (\nabla_X^{\pi_{\mathsf{E}}} \mathrm{Ins}_j(A, S))(\xi_1, \dots, \xi_{k+l-1})\\
&\qquad + \sum_{a=1}^{k+l-1} \mathrm{Ins}_j(A, S)(\xi_1, \dots, \nabla_X^{\pi_{\mathsf{E}}} \xi_a, \dots, \xi_{k+l-1})\\
&\quad = (\nabla_X^{\pi_{\mathsf{E}}} \mathrm{Ins}_j(A, S))(\xi_1, \dots, \xi_{k+l-1})\\
&\qquad + \sum_{a=1}^{j-1} A(\xi_1, \dots, \nabla_X^{\pi_{\mathsf{E}}} \xi_a, \dots, \xi_{j-1}, S(\xi_j, \xi_{k+1}, \dots, \xi_{k+l-1}), \xi_{j+1}, \dots, \xi_k)\\
&\qquad + A(\xi_1, \dots, \xi_{j-1}, S(\nabla_X^{\pi_{\mathsf{E}}} \xi_j, \xi_{k+1}, \dots, \xi_{k+l-1}), \xi_{j+1}, \dots, \xi_k)\\
&\qquad + \sum_{a=j+1}^{k} A(\xi_1, \dots, \xi_{j-1}, S(\xi_j, \xi_{k+1}, \dots, \xi_{k+l-1}), \xi_{j+1}, \dots, \nabla_X^{\pi_{\mathsf{E}}} \xi_a, \dots, \xi_k)\\
&\qquad + \sum_{a=k+1}^{k+l-1} A(\xi_1, \dots, \xi_{j-1}, S(\xi_j, \xi_{k+1}, \dots, \nabla_X^{\pi_{\mathsf{E}}} \xi_a, \dots, \xi_{k+l-1}), \xi_{j+1}, \dots, \xi_k).
\end{aligned}$$

We also calculate

$$\begin{aligned}
&\mathscr{L}_X(\mathrm{Ins}_j(A, S)(\xi_1, \dots, \xi_{k+l-1}))\\
&\quad = \mathscr{L}_X(A(\xi_1, \dots, \xi_{j-1}, S(\xi_j, \xi_{k+1}, \dots, \xi_{k+l-1}), \xi_{j+1}, \dots, \xi_k))\\
&\quad = (\nabla_X^{\pi_{\mathsf{E}}} A)(\xi_1, \dots, \xi_{j-1}, S(\xi_j, \xi_{k+1}, \dots, \xi_{k+l-1}), \xi_{j+1}, \dots, \xi_k)\\
&\qquad + \sum_{a=1}^{j-1} A(\xi_1, \dots, \nabla_X^{\pi_{\mathsf{E}}} \xi_a, \dots, \xi_{j-1}, S(\xi_j, \xi_{k+1}, \dots, \xi_{k+l-1}), \xi_{j+1}, \dots, \xi_k)\\
&\qquad + A(\xi_1, \dots, \xi_{j-1}, (\nabla_X^{\pi_{\mathsf{E}}} S)(\xi_j, \xi_{k+1}, \dots, \xi_{k+l-1}), \xi_{j+1}, \dots, \xi_k)
\end{aligned}$$

$$+ A(\xi_1, \dots, \xi_{j-1}, S(\nabla^{\pi_{\mathsf{E}}}_X \xi_j, \xi_{k+1}, \dots, \xi_{k+l-1}), \xi_{j+1}, \dots, \xi_k)$$

$$+ \sum_{a=j+1}^{k} A(\xi_1, \dots, \xi_{j-1}, S(\xi_j, \xi_{k+1}, \dots, \xi_{k+l-1}), \xi_{j+1}, \dots, \nabla^{\pi_{\mathsf{E}}}_X \xi_a, \dots, \xi_k)$$

$$+ \sum_{a=k+1}^{k+l-1} A(\xi_1, \dots, \xi_{j-1}, S(\xi_j, \xi_{k+1}, \dots, \nabla^{\pi_{\mathsf{E}}}_X \xi_a, \dots, \xi_{k+l-1}), \xi_{j+1}, \dots, \xi_k).$$

Comparing the right-hand sides of the preceding calculations gives

$$\begin{aligned}(\nabla^{\pi_{\mathsf{E}}} \mathrm{Ins}_j(A, S))(\xi_1, \dots, \xi_{k+l-1}, X) \\ &= (\nabla^{\pi_{\mathsf{E}}} A)(\xi_1, \dots, \xi_{j-1}, S(\xi_j, \xi_{k+1}, \dots, \xi_{k+l-1}), \xi_{j+1}, \dots, \xi_k, \xi_{k+l-1}, X) \\ &\quad + A(\xi_1, \dots, \xi_{j-1}, (\nabla^{\pi_{\mathsf{E}}} S)(\xi_j, \xi_{k+1}, \dots, \xi_{k+l-1}, X), \xi_{j+1}, \dots, \xi_k) \\ &= \mathrm{Ins}_j(\nabla^{\pi_{\mathsf{E}}} A, S)(\xi_1, \dots, \xi_{k+l-1}, X) + \mathrm{Ins}_j(A, \nabla^{\pi_{\mathsf{E}}} S)(\xi_1, \dots, \xi_{k+l-1}, X),\end{aligned}$$

and this gives the result. □

Using this result, we can easily compute the derivative for tensor insertion with one of the arguments fixed.

Lemma 3.14 (Covariant Differential of Tensor Insertion II) *Let $r \in \{\infty, \omega\}$. Let $\pi_{\mathsf{E}} \colon \mathsf{E} \to \mathsf{M}$ a vector bundle of class C^r, let $\nabla^{\pi_{\mathsf{E}}}$ be a C^r-vector bundle connection in E, let $l \in \mathbb{Z}_{>0}$, and let $S \in \Gamma^r(\mathrm{T}^l(\mathsf{E}^*) \otimes \mathsf{E})$. Then, for $k \in \mathbb{Z}_{>0}$ and $j \in \{1, \dots, k\}$,*

$$(\nabla^{\pi_{\mathsf{E}}} \mathrm{Ins}_{S,j})(A) = \mathrm{Ins}_j(A, \nabla^{\pi_{\mathsf{E}}} S).$$

Proof We have

$$\nabla^{\pi_{\mathsf{E}}}(\mathrm{Ins}_{S,j}(A)) = (\nabla^{\pi_{\mathsf{E}}} \mathrm{Ins}_{S,j})(A) + \mathrm{Ins}_{S,j}(\nabla^{\pi_{\mathsf{E}}} A)$$

and

$$\nabla^{\pi_{\mathsf{E}}}(\mathrm{Ins}_j(A, S)) = \mathrm{Ins}_j(\nabla^{\pi_{\mathsf{E}}} A, S) + \mathrm{Ins}_j(A, \nabla^{\pi_{\mathsf{E}}} S).$$

Comparing the equations, noting that $\mathrm{Ins}_{S,j}(\nabla^{\pi_{\mathsf{E}}} A) = \mathrm{Ins}_j(\nabla^{\pi_{\mathsf{E}}} A, S)$, the result follows. □

Related to tensor contraction is the evaluation of a vector bundle mapping. We shall consider the derivative of this evaluation. In stating the result, we use a bit of tensor notation that we now introduce. Let V be a finite-dimensional $\mathbb{R}$-vector space and let $A \in \mathrm{T}^1_{k+1}(\mathsf{V}^*)$ and $B \in \mathrm{T}^1_l(\mathsf{V})$. We then denote by $A(B) \in \mathrm{T}^{k+l}(\mathsf{V}^*)$ the tensor defined by

$$A(B)(v_1, \dots, v_k, v_{k+1}, \dots, v_{k+l}) = A(v_1, \dots, v_k, B(v_{k+1}, \dots, v_{k+l})) \tag{3.15}$$

Thus $A(B)$ is shorthand for $\mathrm{Ins}_{k+1}(A, B)$. With this notation, we have the following result.

Lemma 3.15 (Leibniz Rule for Tensor Evaluation) *Let $r \in \{\infty, \omega\}$, let $\pi_{\mathsf{E}} \colon \mathsf{E} \to \mathsf{M}$ and $\pi_{\mathsf{F}} \colon \mathsf{F} \to \mathsf{M}$ be C^r-vector bundles, and let $\nabla^{\pi_{\mathsf{E}}}$ and $\nabla^{\pi_{\mathsf{F}}}$ be C^r-vector bundle connections in E and F, respectively. Let ∇^{M} be a C^r-affine connection on M. Let $L \in \Gamma^r(\mathsf{F} \otimes \mathsf{E}^*)$. Then*

$$D^k_{\nabla^{\mathsf{M}},\nabla^{\pi_{\mathsf{F}}}}(L \circ \xi) = \sum_{l=0}^{k} \binom{k}{l} \mathrm{Sym}_k \left(D^l_{\nabla^{\mathsf{M}},\nabla^{\pi_{\mathsf{F}} \otimes \pi_{\mathsf{E}}}}(L)(D^{k-l}_{\nabla^{\mathsf{M}},\nabla^{\pi_{\mathsf{E}}}}(\xi)) \right),$$

for $k \in \mathbb{Z}_{>0}$ and $\xi \in \Gamma^r(\mathsf{E})$.

Proof For $A \in \mathrm{T}^k(\mathsf{V}^*)$ and $\sigma \in \mathfrak{S}_k$, we use the notation

$$\sigma(A)(v_1, \dots, v_k) = A(v_{\sigma(1)}, \dots, v_{\sigma(k)}).$$

First we claim that

$$\nabla^{\mathsf{M},\pi_{\mathsf{F}},k}(L \circ \xi) = \sum_{l=0}^{k} \sum_{\sigma \in \mathfrak{S}_{l,k-l}} \sigma(\nabla^{\mathsf{M},\pi_{\mathsf{F}} \otimes \nabla_{\mathsf{F}},l} L(\nabla^{\mathsf{M},\pi_{\mathsf{E}},k-l} \xi)). \tag{3.16}$$

This clearly holds for $k = 1$. So suppose it is true up to k for $k \geq 1$. We then compute

$$\begin{aligned}
&\nabla^{\pi_{\mathsf{F}}}(\nabla^{\mathsf{M},\pi_{\mathsf{F}},k}(L \circ \xi)(X_1, \dots, X_k))(X_{k+1}) \\
&= \nabla^{\mathsf{M},\pi_{\mathsf{F}},k+1}(L \circ \xi)(X_1, \dots, X_k, X_{k+1}) + \sum_{j=1}^{k} \nabla^{\mathsf{M},\pi_{\mathsf{F}},k}(L \circ \xi)(X_1, \dots, \nabla^{\mathsf{M}}_{X_{k+1}} X_j, \dots, X_k) \\
&= \nabla^{\mathsf{M},\pi_{\mathsf{F}},k+1}(L \circ \xi)(X_1, \dots, X_k, X_{k+1}) + \sum_{j=1}^{k} \sum_{l=0}^{k} \sum_{\sigma \in \mathfrak{S}_{l,k-l}} \nabla^{\mathsf{M},\pi_{\mathsf{F}} \otimes \nabla_{\mathsf{F}},l} L(\nabla^{\mathsf{M},\pi_{\mathsf{E}},k-l} \xi) \\
&\quad (X_{\sigma(1)}, \dots, \nabla^{\mathsf{M}}_{X_{k+1}} X_{\sigma(j)}, \dots, X_{\sigma(k)}),
\end{aligned}$$

using the induction hypothesis. We also compute, still using the induction hypothesis,

$$\begin{aligned}
&\nabla^{\pi_{\mathsf{F}}}(\nabla^{\mathsf{M},\pi_{\mathsf{F}},k}(L \circ \xi)(X_1, \dots, X_k))(X_{k+1}) \\
&\quad = \sum_{l=0}^{k} \sum_{\sigma \in \mathfrak{S}_{l,k-l}} (\nabla^{\mathsf{M},\pi_{\mathsf{F}} \otimes \pi_{\mathsf{E}},l+1} L) \\
&\qquad (X_{\sigma(1)}, \dots, X_{\sigma(l)}, X_{k+1})(\nabla^{\mathsf{M},\pi_{\mathsf{E}},k-l} \xi(X_{\sigma(l+1)}, \dots, X_{\sigma(k)}))
\end{aligned}$$

$$+\sum_{l=0}^{k}\sum_{\sigma\in\mathfrak{S}_{l,k-l}}\sum_{j=1}^{l}(\nabla^{\pi_{\mathsf{F}}\otimes\pi_{\mathsf{E}},l}L(X_{\sigma(l)},\dots,\nabla^{\mathsf{M}}_{X_{k+1}}X_{\sigma(j)},\dots,X_{\sigma(l)}))$$

$$(\nabla^{\mathsf{M},\pi_{\mathsf{E}},k-l}\xi(X_{\sigma(l+1)},\dots,X_{\sigma(k)}))$$

$$+\sum_{l=0}^{k}\sum_{\sigma\in\mathfrak{S}_{l,k-l}}(\nabla^{\mathsf{M},\pi_{\mathsf{F}}\otimes\pi_{\mathsf{E}},l}L(X_{\sigma(1)},\dots,X_{\sigma(l)}))$$

$$(\nabla^{\mathsf{M},\pi_{\mathsf{E}},k-l+1}\xi(X_{\sigma(l+1)},\dots,X_{\sigma(k)},X_{k+1}))$$

$$+\sum_{l=0}^{k}\sum_{\sigma\in\mathfrak{S}_{l,k-l}}\sum_{j=l+1}^{k}(\nabla^{\mathsf{M},\pi_{\mathsf{F}}\otimes\pi_{\mathsf{E}},l}L(X_{\sigma(1)},\dots,X_{\sigma(l)}))$$

$$(\nabla^{\mathsf{M},\pi_{\mathsf{E}},k-l}\xi(X_{\sigma(l+1)},\dots,\nabla^{\mathsf{M}}_{X_{k+1}}X_{\sigma(j)},\dots,X_{\sigma(k)})).$$

Comparing the preceding two equations gives

$$\nabla^{\mathsf{M},\pi_{\mathsf{F}},k+1}(L\circ\xi)(X_1,\dots,X_k,X_{k+1})$$

$$=\sum_{l=0}^{k}\sum_{\sigma\in\mathfrak{S}_{l,k-l}}(\nabla^{\mathsf{M},\pi_{\mathsf{F}}\otimes\pi_{\mathsf{E}},l+1}L)$$

$$(X_{\sigma(1)},\dots,X_{\sigma(l)},X_{k+1})(\nabla^{\mathsf{M},\pi_{\mathsf{E}},k-l}\xi(X_{\sigma(l+1)},\dots,X_{\sigma(k)}))$$

$$+\sum_{l=0}^{k}\sum_{\sigma\in\mathfrak{S}_{l,k-l}}(\nabla^{\mathsf{M},\pi_{\mathsf{F}}\otimes\pi_{\mathsf{E}},l}L(X_{\sigma(1)},\dots,X_{\sigma(l)}))$$

$$(\nabla^{\mathsf{M},\pi_{\mathsf{E}},k-l+1}\xi(X_{\sigma(l+1)},\dots,X_{\sigma(k)},X_{k+1}))$$

$$=\sum_{l=0}^{k+1}\sum_{\sigma\in\mathfrak{S}_{l,k+1-l}}(\nabla^{\mathsf{M},\pi_{\mathsf{F}}\otimes\pi_{\mathsf{E}},l}L(X_{\sigma(1)},\dots,X_{\sigma(l)}))$$

$$(\nabla^{\mathsf{M},\pi_{\mathsf{E}},k+1-l}\xi(X_{\sigma(l+1)},\dots,X_{\sigma(k+1)})),$$

giving (3.16).

For $\sigma\in\mathfrak{S}_k$, write $\sigma=\sigma_1\circ\sigma_2$ for $\sigma_1\in\mathfrak{S}_{k,l}$ and $\sigma_2\in\mathfrak{S}_{k|l}$. Now we compute

$$D^k_{\nabla^{\mathsf{M}},\nabla^{\pi_{\mathsf{F}}}}(L\circ\xi)$$

$$=\frac{1}{k!}\sum_{\sigma\in\mathfrak{S}_k}\sigma(\nabla^{\mathsf{M},\pi_{\mathsf{F}},k}(L\circ\xi))$$

$$=\frac{1}{k!}\sum_{l=0}^{k}\sum_{\sigma\in\mathfrak{S}_k}\sum_{\sigma'\in\mathfrak{S}_{l,k-l}}\sigma'\circ\sigma(\nabla^{\mathsf{M},\pi_{\mathsf{F}}\otimes\pi_{\mathsf{E}},l}L(\nabla^{\mathsf{M},\pi_{\mathsf{E}},k-l}\xi))$$

$$
\begin{aligned}
&= \frac{1}{k!}\sum_{l=0}^{k}\sum_{\sigma'\in\mathfrak{S}_{l,k-l}}\sum_{\sigma_1\in\mathfrak{S}_{l,k-l}}\sum_{\sigma_2\in\mathfrak{S}_{l|k-l}} \sigma'\circ\sigma_1\circ\sigma_2(\nabla^{\mathsf{M},\pi_{\mathsf{F}}\otimes\pi_{\mathsf{E}},l}L(\nabla^{\mathsf{M},\pi_{\mathsf{E}},k-l}\xi))\\
&= \sum_{l=0}^{k}\sum_{\sigma'\in\mathfrak{S}_{l,k-l}}\sum_{\sigma_1\in\mathfrak{S}_{l,k-l}} \frac{l!(k-l)!}{k!}\sigma'\circ\sigma_1(D^{l}_{\nabla^{\mathsf{M}},\nabla^{\pi_{\mathsf{F}}\otimes\pi_{\mathsf{E}}}}L(D^{k-l}_{\nabla^{\mathsf{M}},\nabla^{\pi_{\mathsf{E}}}}(\xi)))\\
&= \sum_{l=0}^{k}\sum_{\sigma'\in\mathfrak{S}_{l,k-l}} \frac{l!(k-l)!}{k!}\sigma'\circ\left(\sum_{\sigma\in\mathfrak{S}_k}\frac{k!}{l!(k-l)!}\mathrm{Sym}_k(D^{l}_{\nabla^{\mathsf{M}},\nabla^{\pi_{\mathsf{F}}\otimes\pi_{\mathsf{E}}}}L(D^{k-l}_{\nabla^{\mathsf{M}},\nabla^{\pi_{\mathsf{E}}}}(\xi)))\right)\\
&= \sum_{l=0}^{k}\frac{k!}{l!(k-l)!}\left(\mathrm{Sym}_k(D^{l}_{\nabla^{\mathsf{M}},\nabla^{\pi_{\mathsf{F}}\otimes\pi_{\mathsf{E}}}}L(D^{k-l}_{\nabla^{\mathsf{M}},\nabla^{\pi_{\mathsf{E}}}}(\xi)))\right),
\end{aligned}
$$

making reference to (1.1) and (1.2) in the penultimate step, and noting that $\mathrm{card}(\mathfrak{S}_{l,k-l}) = \frac{k!}{l!(k-l)!}$. This is the desired result. □

3.2.3 Derivatives of Tensors on the Total Space of a Vector Bundle

In Definition 3.9 we introduced a variety of lifts of tensor fields. Here we give formulae for differentiating these. We shall make ongoing and detailed use of the formulae we develop in this section, and decent notation is an integral part of arriving at useable expressions.

Let $r \in \{\infty, \omega\}$ and let $\pi_{\mathsf{E}} : \mathsf{E} \to \mathsf{M}$ be a C^r-vector bundle. We consider a C^r-affine connection ∇^{M} on M and a C^r-linear connection $\nabla^{\pi_{\mathsf{E}}}$ in E. The connection ∇^{M} induces a covariant derivative for tensor fields $A \in \Gamma^r(\mathrm{T}^k_l(\mathsf{TM}))$ on M, $k, l \in \mathbb{Z}_{\geq 0}$. This covariant derivative we denote by ∇^{M}, dropping any reference to the particular k and l. Similarly, the connection $\nabla^{\pi_{\mathsf{E}}}$ induces a covariant derivative for sections $B \in \Gamma^r(\mathrm{T}^k_l(\mathsf{E}))$ of the tensor bundles associated with E, $k, l \in \mathbb{Z}_{\geq 0}$. This covariant derivative we denote by $\nabla^{\pi_{\mathsf{E}}}$, again dropping reference to the particular k and l. We have already made use of these conventions, e.g., in Lemmata 3.13 and 3.14. We will also consider differentiation of sections of $\mathrm{T}^{k_1}_{l_1}(\mathsf{TM}) \otimes \mathrm{T}^{k_2}_{l_2}(\mathsf{E})$. Here we denote the covariant derivative by $\nabla^{\mathsf{M},\pi_{\mathsf{E}}}$. If we have another C^r-vector bundle $\pi_{\mathsf{F}} : \mathsf{F} \to \mathsf{M}$ with a C^r-affine connection $\nabla^{\pi_{\mathsf{F}}}$, then $\nabla^{\pi_{\mathsf{E}}}$ and $\nabla^{\pi_{\mathsf{F}}}$ induce a covariant derivative in $\mathrm{T}^{k_1}_{l_1}(\mathsf{E}) \otimes \mathrm{T}^{k_2}_{l_2}(\mathsf{F})$, and we denote this covariant derivative by $\nabla^{\pi_{\mathsf{E}}\otimes\pi_{\mathsf{F}}}$.

Another construction we need in this section concerns pull-back bundles. Let $r \in \{\infty, \omega\}$, let M and N be C^r-manifolds, let $\pi_{\mathsf{F}} : \mathsf{F} \to \mathsf{M}$ be a C^r-vector bundle, and let $\Phi \in \mathrm{C}^r(\mathsf{N}; \mathsf{M})$. We then have the pull-back bundle $\Phi^*\pi_{\mathsf{F}} : \Phi^*\mathsf{F} \to \mathsf{N}$, which is a vector bundle over N. Given a section η of F, we have a section $\Phi^*\eta$ of $\Phi^*\mathsf{F}$ defined by $\Phi^*\eta(y) = (y, \eta\circ\Phi(y))$. Given a C^r-vector bundle connection $\nabla^{\pi_{\mathsf{F}}}$ in F, we can define a C^r-connection $\Phi^*\nabla^{\pi_{\mathsf{F}}}$ in $\Phi^*\mathsf{F}$ by requiring that

$$\Phi^*\nabla^{\pi_{\mathsf{F}}}_Z \Phi^*\eta(y) = \Phi^*(\nabla^{\pi_{\mathsf{F}}}_{T_y\Phi(Z)}\eta)$$

for a C^r-section η and for $Z \in \mathsf{T}_y\mathsf{N}$. This is the ***pull-back*** of ∇^{F} by Φ. Given an affine connection ∇^{N} on N, we then have an affine connection on $\mathrm{T}^k(\mathsf{T}^*\mathsf{N}) \otimes \Phi^*\mathsf{F}$ induced by tensor product by ∇^{N} and $\Phi^*\nabla^{\pi_{\mathsf{F}}}$. This connection we denote by $\nabla^{\mathsf{N},\Phi^*\pi_{\mathsf{F}}}$, consistent with our notation above.

If we additionally have an injective vector bundle mapping $\psi\colon \Phi^*\mathsf{F} \to \mathsf{TN}$, then we have

$$\nabla^{\mathsf{N}}_Z(\psi \circ \Phi^*\eta) = \psi \circ (\Phi^*\nabla^{\pi_{\mathsf{F}}}_Z \Phi^*\eta) + B_\psi(\Phi^*\eta, Z)$$

for some tensor $B_\psi \in \Gamma^r(\Phi^*\mathsf{F}^* \otimes \mathrm{T}^1_1(\mathsf{TN}))$. A special case of this is when $\Phi = \pi_{\mathsf{E}}$ for a vector bundle $\pi_{\mathsf{E}}\colon \mathsf{E} \to \mathsf{M}$ and $\mathsf{F} = \mathsf{E}$. In this case, $\pi^*_{\mathsf{E}}\xi = \xi^{\mathrm{v}}$ and $\pi^*_{\mathsf{E}}\mathsf{F} \simeq \mathsf{VE}$ and so we indeed have a natural inclusion of $\pi^*_{\mathsf{E}}\mathsf{F}$ in TE. Moreover, by Lemma 3.12(xiii),

$$\pi^*_{\mathsf{E}}\nabla^{\pi_{\mathsf{E}}}_Z\pi^*_{\mathsf{E}}\xi = (\nabla^{\pi_{\mathsf{E}}}_{T\pi_{\mathsf{E}}(Z)}\xi)^{\mathrm{v}},$$

and so

$$\nabla^{\mathsf{E}}_Z\pi^*_{\mathsf{E}}\xi = \pi^*_{\mathsf{E}}\nabla^{\pi_{\mathsf{E}}}_Z\pi^*_{\mathsf{E}}\xi + A_{\pi_{\mathsf{E}}}(Z, \xi^{\mathrm{v}}). \tag{3.17}$$

With the preceding, we can give formulae for differentiating tensors on vector bundles, rather mirroring what we did in Lemma 3.6 for functions.

Lemma 3.16 (Differentiation of Lifted Tensors on Vector Bundles) *Let $r \in \{\infty, \omega\}$. Let $\pi_{\mathsf{E}}\colon \mathsf{E} \to \mathsf{M}$ and $\pi_{\mathsf{F}}\colon \mathsf{F} \to \mathsf{M}$ be vector bundles of class C^r. Let $\mathbb{G}_{\mathsf{M}}$ be a C^r-Riemannian metric on M, let ∇^{M} be the Levi-Civita connection, let $\mathbb{G}_{\pi_{\mathsf{E}}}$ be a C^r-fibre metric on E, and let $\nabla^{\pi_{\mathsf{E}}}$ be a linear connection of class C^r in E. Let $\nabla^{\pi_{\mathsf{F}}}$ be a C^r-vector bundle connection in F. Let $\mathbb{G}_{\mathsf{E}}$ be the associated C^r-Riemannian metric on E from* (3.5). *Define*

$$B_{\pi_{\mathsf{E}}} = \mathrm{push}_{1,2}\mathrm{Ins}_1(\mathrm{Ins}_2(A_{\pi_{\mathsf{E}}}, \mathrm{hor}), \mathrm{hor}) + \mathrm{Ins}_2(A_{\pi_{\mathsf{E}}}, \mathrm{ver}) + \mathrm{push}_{1,2}\mathrm{Ins}_2(A_{\pi_{\mathsf{E}}}, \mathrm{ver}), \tag{3.18}$$

where $A_{\pi_{\mathsf{E}}}$ is defined as in (3.6).

Then we have the following statements, recalling from (3.2) *the derivation $D_{B_{\pi_{\mathsf{E}}}}$:*

(i) for $k \in \mathbb{Z}_{>0}$ and $A \in \Gamma^r(\mathrm{T}^k(\mathsf{T}^*\mathsf{M}))$, we have

$$\nabla^{\mathsf{E}}(A^{\mathrm{h}}) = (\nabla^{\mathsf{M}}A)^{\mathrm{h}} + D_{B_{\pi_{\mathsf{E}}}}(A^{\mathrm{h}});$$

(ii) for $A \in \Gamma^r(\mathsf{T}^k(\mathsf{T}^*\mathsf{M}) \otimes \mathsf{E})$*, we have*

$$\nabla^{\mathsf{E}}(A^{\mathrm{v}}) = (\nabla^{\mathsf{M},\pi_{\mathsf{E}}} A)^{\mathrm{v}} + D_{B_{\pi_{\mathsf{E}}}}(A^{\mathrm{v}});$$

(iii) for $A \in \Gamma^r(\mathsf{T}^k(\mathsf{T}^*\mathsf{M}) \otimes \mathsf{TM})$*, we have*

$$\nabla^{\mathsf{E}}(A^{\mathrm{h}}) = (\nabla^{\mathsf{M}} A)^{\mathrm{h}} + D_{B_{\pi_{\mathsf{E}}}}(A^{\mathrm{h}});$$

(iv) for $A \in \Gamma^r(\mathsf{T}^k(\mathsf{T}^*\mathsf{M}) \otimes \mathsf{F} \otimes \mathsf{E}^*)$*, we have*

$$\nabla^{\mathsf{E},\pi_{\mathsf{F}}}(A^{\mathrm{v}}) = (\nabla^{\mathsf{M},\pi_{\mathsf{E}}\otimes\pi_{\mathsf{F}}} A)^{\mathrm{v}} + D_{B_{\pi_{\mathsf{E}}}}(A^{\mathrm{v}});$$

(v) for $A \in \Gamma^r(\mathsf{T}^k(\mathsf{T}^*\mathsf{M}) \otimes \mathsf{T}^1_1(\mathsf{E}))$*, we have*

$$\nabla^{\mathsf{E}}(A^{\mathrm{v}}) = (\nabla^{\mathsf{M},\pi_{\mathsf{E}}} A)^{\mathrm{v}} + D_{B_{\pi_{\mathsf{E}}}}(A^{\mathrm{v}});$$

(vi) for $A \in \Gamma^r(\mathsf{T}^k(\mathsf{T}^*\mathsf{M}) \otimes \mathsf{F} \otimes \mathsf{E}^*)$*, we have*

$$\nabla^{\mathsf{E},\pi_{\mathsf{F}}}(A^{\mathrm{e}}) = (\nabla^{\mathsf{M},\pi_{\mathsf{E}}\otimes\pi_{\mathsf{F}}} A)^{\mathrm{e}} + D_{B_{\pi_{\mathsf{E}}}}(A^{\mathrm{e}}) + A^{\mathrm{v}};$$

(vii) for $A \in \Gamma^r(\mathsf{T}^k(\mathsf{T}^*\mathsf{M}) \otimes \mathsf{T}^1_1(\mathsf{E}))$*, we have*

$$\nabla^{\mathsf{E}}(A^{\mathrm{e}}) = (\nabla^{\mathsf{M},\pi_{\mathsf{E}}} A)^{\mathrm{e}} + D_{B_{\pi_{\mathsf{E}}}}(A^{\mathrm{e}}) + A^{\mathrm{v}}.$$

Proof Before we begin the proof proper, let us justify a "without loss of generality" argument that we will make at various points in the proof. The arguments all have to do with assuming that it is sufficient, when working with differential operators on spaces of tensor products, to work with pure tensor products. Let us be a little specific about this. Let $\pi_{\mathsf{E}}\colon \mathsf{E} \to \mathsf{M}$, $\pi_{\mathsf{F}}\colon \mathsf{F} \to \mathsf{M}$, and $\pi_{\mathsf{G}}\colon \mathsf{G} \to \mathsf{M}$ be C^r-vector bundles. Suppose that $\Delta_1, \Delta_2\colon \mathsf{J}^m(\mathsf{E} \otimes \mathsf{F}) \to \mathsf{G}$ are linear differential operators of order m. We wish to give conditions under which $\Delta_1 = \Delta_2$. Of course, this is equivalent to giving conditions under which, for a differential operator $\Delta\colon \mathsf{J}^m(\mathsf{E} \otimes \mathsf{F}) \to \mathsf{G}$, $\Delta = 0$. To do so, we claim that, without loss of generality, we can simply prove that $\Delta(j_m(\xi \otimes \eta)) = 0$ for all $\xi \in \Gamma^r(\mathsf{E})$ and $\eta \in \Gamma^r(\mathsf{E})$. Indeed, suppose that we have proved that $\Delta(j_m(\xi \otimes \eta)) = 0$ for all $\xi \in \Gamma^r(\mathsf{E})$ and $\eta \in \Gamma^r(\mathsf{E})$. Let $x \in \mathsf{M}$ and let $\alpha \in \mathsf{T}^{*m}_x\mathsf{M}$. Let $u \in \mathsf{E}_x$ and $v \in \mathsf{F}_x$. By Lemma 2.1, there exists $f \in \mathrm{C}^r(\mathsf{M})$ such that $j_m f(x) = \alpha$. Then, keeping in mind the identification (1.9),

$$\Delta(\alpha \otimes (u \otimes v)) = \Delta(j_m(f(\xi \otimes \eta))) = \Delta(j_m((f\xi) \otimes \eta)) = 0.$$

Since every element of $\mathsf{J}_x^m\mathsf{E}$ is a finite linear combination of terms of the form $\alpha \otimes (u \otimes v)$ for $\alpha \in \mathsf{T}_x^{*m}\mathsf{M}$, $u \in \mathsf{E}_x$, and $v \in \mathsf{F}_x$, we conclude that $\Delta(j_m A)(x) = 0$ for every $A \in \Gamma^r(\mathsf{E} \otimes \mathsf{F})$.

Now we proceed with the proof.

(i) We have

$$\mathscr{L}_{Z_{k+1}}(A^{\mathrm{h}}(Z_1, \dots, Z_k))$$
$$= (\nabla^{\mathsf{E}}_{Z_{k+1}} A^{\mathrm{h}})(Z_1, \dots, Z_k) + \sum_{j=1}^{k} A^{\mathrm{h}}(Z_1, \dots, \nabla^{\mathsf{E}}_{Z_{k+1}} Z_j, \dots, Z_k).$$

We consider four cases.

1. $Z_j = X_j^{\mathrm{h}}$, $j \in \{1, \dots, k+1\}$: Here we have

$$\mathscr{L}_{X_{k+1}^{\mathrm{h}}}(A^{\mathrm{h}}(X_1^{\mathrm{h}}, \dots, X_k^{\mathrm{h}})) = (\mathscr{L}_{X_{k+1}}(A(X_1, \dots, X_k)))^{\mathrm{h}}$$

(by Lemma 3.6(i)) and

$$A^{\mathrm{h}}(X_1^{\mathrm{h}}, \dots, \nabla^{\mathsf{E}}_{X_{k+1}^{\mathrm{h}}} X_j^{\mathrm{h}}, \dots, X_k^{\mathrm{h}}) = (A(X_1, \dots, \nabla^{\mathsf{M}}_{X_{k+1}} X_j, \dots, X_k))^{\mathrm{h}}$$

(by Lemma 3.12(i)). Thus we conclude that

$$\nabla^{\mathsf{E}} A^{\mathrm{h}}(X_1^{\mathrm{h}}, \dots, X_{k+1}^{\mathrm{h}}) = ((\nabla^{\mathsf{M}} A)(X_1, \dots, X_{k+1}))^{\mathrm{h}}.$$

2. $Z_j = X_j^{\mathrm{h}}$, $j \in \{1, \dots, k\}$, $Z_{k+1} = \xi_{k+1}^{\mathrm{v}}$: Here we calculate

$$\mathscr{L}_{\xi_{k+1}^{\mathrm{v}}}(A^{\mathrm{h}}(X_1^{\mathrm{h}}, \dots, X_k^{\mathrm{h}})) = \mathscr{L}_{\xi_{k+1}^{\mathrm{v}}}(A(X_1, \dots, X_k))^{\mathrm{h}} = 0$$

(using the definition of A^{h} and Lemma 3.6(ii)) and

$$A^{\mathrm{h}}(X_1^{\mathrm{h}}, \dots, \nabla^{\mathsf{E}}_{\xi_{k+1}^{\mathrm{v}}} X_j^{\mathrm{h}}, \dots, X_k^{\mathrm{h}}) = A^{\mathrm{h}}(X_1^{\mathrm{h}}, \dots, A_{\pi_{\mathsf{E}}}(X_j^{\mathrm{h}}, \xi_{k+1}^{\mathrm{v}}), \dots, X_k^{\mathrm{h}})$$

(using Lemma 3.12(xi)). Thus we conclude that

$$\nabla^{\mathsf{E}} A^{\mathrm{h}}(X_1^{\mathrm{h}}, \dots, X_k^{\mathrm{h}}, \xi_{k+1}^{\mathrm{v}}) = -\sum_{j=1}^{k} A^{\mathrm{h}}(X_1^{\mathrm{h}}, \dots, A_{\pi_{\mathsf{E}}}(X_j^{\mathrm{h}}, \xi_{k+1}^{\mathrm{v}}), \dots, X_k^{\mathrm{h}}).$$

3. $Z_j = \xi_j^{\mathrm{v}}$ for some $j \in \{1, \dots, k\}$, $Z_{k+1} = X_{k+1}^{\mathrm{h}}$: We calculate

$$\mathscr{L}_{X_{k+1}^{\mathrm{h}}}(A^{\mathrm{h}}(Z_1, \dots, \xi_j^{\mathrm{v}}, Z_k)) = 0$$

(by definition of A^{h}) and

$$A^{\mathrm{h}}(Z_1, \dots, \nabla^{\mathsf{E}}_{X^{\mathrm{h}}_{k+1}} \xi^{\mathrm{v}}_j, \dots, Z_k) = A^{\mathrm{h}}(Z_1, \dots, A_{\pi_{\mathsf{E}}}(X^{\mathrm{h}}_{k+1}, \xi^{\mathrm{v}}_j), \dots, Z_k)$$

(by Lemma 3.12(v)). Thus

$$\nabla^{\mathsf{E}} A^{\mathrm{h}}(Z_1, \dots, \xi^{\mathrm{v}}_j, \dots, Z_k, X^{\mathrm{h}}_{k+1}) = -A^{\mathrm{h}}(Z_1, \dots, A_{\pi_{\mathsf{E}}}(X^{\mathrm{h}}_{k+1}, \xi^{\mathrm{v}}_j), \dots, Z_k).$$

4. $Z_j = \xi^{\mathrm{v}}_j$ for some $j \in \{1, \dots, k\}$, $Z_{k+1} = \xi^{\mathrm{v}}_{k+1}$: We have

$$\mathscr{L}_{\xi^{\mathrm{v}}_{k+1}}(A^{\mathrm{h}}(Z_1, \dots, \xi^{\mathrm{v}}_j, \dots, Z_k)) = 0$$

(by definition of A^{h}) and

$$A^{\mathrm{h}}(Z_1, \dots, \nabla^{\mathsf{E}}_{\xi^{\mathrm{v}}_{k+1}} \xi^{\mathrm{v}}_j, \dots, Z_k) = 0$$

(by Lemma 3.12(iii)). Thus

$$\nabla^{\mathsf{E}} A^{\mathrm{h}}(Z_1, \dots, \xi^{\mathrm{v}}_j, \dots, Z_k, \xi^{\mathrm{v}}_{k+1}) = 0.$$

Putting this all together, and keeping in mind that $A_{\pi_{\mathsf{E}}}$ is vertical when both arguments are vertical, we have

$$\begin{aligned}\nabla^{\mathsf{E}} A^{\mathrm{h}}(Z_1, \dots, Z_{k+1}) = (\nabla^{\mathsf{M}} A)^{\mathrm{h}}(Z_1, \dots, Z_{k+1})\\ - \sum_{j=1}^{k} A^{\mathrm{h}}(Z_1, \dots, A_{\pi_{\mathsf{E}}}(\mathrm{hor}(Z_j), \mathrm{ver}(Z_{k+1})), \dots, Z_k)\\ - \sum_{j=1}^{k} A^{\mathrm{h}}(Z_1, \dots, A_{\pi_{\mathsf{E}}}(\mathrm{hor}(Z_{k+1}), \mathrm{ver}(Z_j)), \dots, Z_k).\end{aligned}$$

Now we note that

$$\begin{aligned}&B_{\pi_{\mathsf{E}}}(Z_j, Z_{k+1})\\ &\quad = A_{\pi_{\mathsf{E}}}(\mathrm{hor}(Z_{k+1}), \mathrm{hor}(Z_j)) + A_{\pi_{\mathsf{E}}}(Z_j, \mathrm{ver}(Z_{k+1})) + A_{\pi_{\mathsf{E}}}(Z_{k+1}, \mathrm{ver}(Z_j))\\ &\quad = A_{\pi_{\mathsf{E}}}(\mathrm{hor}(Z_j), \mathrm{ver}(Z_{k+1})) + A_{\pi_{\mathsf{E}}}(\mathrm{hor}(Z_{k+1}), \mathrm{ver}(Z_j)) + \text{something vertical},\end{aligned}$$

using Lemma 3.12(ii) and the definition of $A_{\pi_{\mathsf{E}}}$. Thus

$$\nabla^{\mathsf{E}} A^{\mathrm{h}} = (\nabla^{\mathsf{M}} A)^{\mathrm{h}} - \sum_{j=1}^{k} \mathrm{Ins}_j(A^{\mathrm{h}}, B_{\pi_{\mathsf{E}}}),$$

which gives this part of the lemma by Lemma 3.11.

(ii) First we compute, for $Z \in \Gamma^r(\mathsf{TE})$,

$$\begin{aligned}\nabla^{\mathsf{E}}_Z \xi^{\mathrm{v}} &= \nabla^{\mathsf{E}}_{\mathrm{hor}(Z)} \xi^{\mathrm{v}} + \nabla^{\mathsf{E}}_{\mathrm{ver}(Z)} \xi^{\mathrm{v}} = (\nabla^{\mathsf{M},\pi_{\mathsf{E}}}_{T\pi_{\mathsf{E}}(Z)} \xi)^{\mathrm{v}} + A_{\pi_{\mathsf{E}}}(T\pi_{\mathsf{E}}(Z), \xi^{\mathrm{v}}) \\ &= (\nabla^{\mathsf{M},\pi_{\mathsf{E}}}_{T\pi_{\mathsf{E}}(Z)} \xi)^{\mathrm{v}} + A_{\pi_{\mathsf{E}}}(Z, \xi^{\mathrm{v}}),\end{aligned}$$

using Lemma 3.12(iii), (v), and (xiii), and the definition of $A_{\pi_{\mathsf{E}}}$. If we note that

$$\begin{aligned}B_{\pi_{\mathsf{E}}}(\xi^{\mathrm{v}}, Z) &= A_{\pi_{\mathsf{E}}}(\mathrm{hor}(Z), \mathrm{hor}(\xi^{\mathrm{v}})) + A_{\pi_{\mathsf{E}}}(\xi^{\mathrm{v}}, \mathrm{ver}(Z)) + A_{\pi_{\mathsf{E}}}(Z, \mathrm{ver}(\xi^{\mathrm{v}})) \\ &= A_{\pi_{\mathsf{E}}}(Z, \xi^{\mathrm{v}})\end{aligned}$$

(using the definition of $A_{\pi_{\mathsf{E}}}$), we have

$$\nabla^{\mathsf{E}}_Z \xi^{\mathrm{v}} = (\nabla^{\mathsf{M},\pi_{\mathsf{E}}}_{T\pi_{\mathsf{E}}(Z)} \xi)^{\mathrm{v}} + B_{\pi_{\mathsf{E}}}(\xi^{\mathrm{v}}, Z).$$

Now, it suffices to prove this part of the lemma for $A = A_0^{\mathrm{h}} \otimes \xi^{\mathrm{v}}$ for $A_0 \in \Gamma^r(\mathsf{T}^k(\mathsf{T}^*\mathsf{M}))$ and $\xi \in \Gamma^r(\mathsf{E})$. For $Z \in \Gamma^r(\mathsf{TE})$, we have

$$\begin{aligned}\nabla^{\mathsf{E}}_Z (A_0^{\mathrm{h}} \otimes \xi^{\mathrm{v}}) &= (\nabla^{\mathsf{E}}_Z A_0^{\mathrm{h}}) \otimes \xi^{\mathrm{v}} + (A_0^{\mathrm{h}}) \otimes \nabla^{\mathsf{E}}_Z \xi^{\mathrm{v}} \\ &= (\nabla^{\mathsf{M}}_{T\pi_{\mathsf{E}}(Z)} A_0)^{\mathrm{h}} \otimes \xi^{\mathrm{v}} - \sum_{j=1}^{k} \mathrm{Ins}_j(A_0^{\mathrm{h}}, B_{\pi_{\mathsf{E}},Z}) \otimes \xi^{\mathrm{v}} \\ &\quad + A_0^{\mathrm{h}} \otimes (\nabla^{\pi_{\mathsf{E}}}_{T\pi_{\mathsf{E}}(Z)} \xi)^{\mathrm{v}} + A_0^{\mathrm{h}} \otimes B_{\pi_{\mathsf{E}}}(\xi^{\mathrm{v}}, Z).\end{aligned}$$

We have

$$\begin{aligned}(\nabla^{\mathsf{M}}_{T\pi_{\mathsf{E}}(Z)} A_0)^{\mathrm{h}} \otimes \xi^{\mathrm{v}} &+ A_0^{\mathrm{h}} \otimes (\nabla^{\pi_{\mathsf{E}}}_{T\pi_{\mathsf{E}}(Z)} \xi)^{\mathrm{v}} \\ &= ((\nabla^{\mathsf{M}}_{T\pi_{\mathsf{E}}(Z)} A_0) \otimes \xi)^{\mathrm{v}} + (A_0^{\mathrm{h}} \otimes (\nabla^{\pi_{\mathsf{E}}}_{T\pi_{\mathsf{E}}(Z)} \xi))^{\mathrm{v}} \\ &= (\nabla^{\mathsf{M},\pi_{\mathsf{E}}}_{T\pi_{\mathsf{E}}(Z)} (A_0 \otimes \xi))^{\mathrm{v}}.\end{aligned}$$

Thus, by (3.3) and the first part of the lemma, we have

$$A_0^{\mathrm{h}} \otimes B_{\pi_{\mathsf{E}}}(\xi^{\mathrm{v}}, Z) - \sum_{j=1}^{k} \mathrm{Ins}_j(A_0^{\mathrm{h}}, B_{\pi_{\mathsf{E}},Z}) \otimes \xi^{\mathrm{v}} = D_{B_{\pi_{\mathsf{E}},Z}}(A_0^{\mathrm{h}} \otimes \xi^{\mathrm{v}}).$$

Assembling the preceding three computations gives this part of the lemma.

(iii) First note that

$$\begin{aligned}\nabla^{\mathsf{E}}_Z X^{\mathrm{h}} &= \nabla^{\mathsf{E}}_{\mathrm{hor}(Z)} X^{\mathrm{h}} + \nabla^{\mathsf{E}}_{\mathrm{ver}(Z)} X^{\mathrm{h}} \\ &= (\nabla^{\mathsf{M}}_{T\pi_{\mathsf{E}}(Z)} X)^{\mathrm{h}} + A_{\pi_{\mathsf{E}}}(\mathrm{hor}(Z), X^{\mathrm{h}}) + A_{\pi_{\mathsf{E}}}(X^{\mathrm{h}}, \mathrm{ver}(Z)),\end{aligned}$$

using Lemma 3.12(xi). Now we have

$$\begin{aligned}&B_{\pi_{\mathsf{E}}}(X^{\mathrm{h}}, Z) \\ &\quad = A_{\pi_{\mathsf{E}}}(\mathrm{hor}(Z), X^{\mathrm{h}}) + A_{\pi_{\mathsf{E}}}(X^{\mathrm{h}}, \mathrm{ver}(Z)) + A_{\pi_{\mathsf{E}}}(Z, \mathrm{ver}(X^{\mathrm{h}})) \\ &\quad = A_{\pi_{\mathsf{E}}}(\mathrm{hor}(Z), X^{\mathrm{h}}) + A_{\pi_{\mathsf{E}}}(X^{\mathrm{h}}, \mathrm{ver}(Z)).\end{aligned}$$

Thus we have

$$\nabla^{\mathsf{E}}_Z X^{\mathrm{h}} = (\nabla^{\mathsf{M}}_{T\pi_{\mathsf{E}}(Z)} X)^{\mathrm{h}} + B_{\pi_{\mathsf{E}}}(X^{\mathrm{h}}, Z).$$

Now it suffices to prove this part of the lemma for $A = A_0^{\mathrm{h}} \otimes X^{\mathrm{h}}$ for $A_0 \in \Gamma^r(\mathsf{T}^k(\mathsf{T}^*\mathsf{M}))$ and $X \in \Gamma^r(\mathsf{TM})$. In this case we calculate, for $Z \in \Gamma^r(\mathsf{TE})$,

$$\begin{aligned}\nabla^{\mathsf{E}}_Z(A_0^{\mathrm{h}} \otimes X^{\mathrm{h}}) &= (\nabla^{\mathsf{E}}_Z A_0^{\mathrm{h}}) \otimes X^{\mathrm{h}} + A_0^{\mathrm{h}} \otimes \nabla^{\mathsf{E}}_Z X^{\mathrm{h}} \\ &= (\nabla^{\mathsf{M}}_{T\pi_{\mathsf{E}}(Z)} A_0)^{\mathrm{h}} \otimes X^{\mathrm{h}} - \sum_{j=1}^{k} \mathrm{Ins}_j(A_0^{\mathrm{h}}, B_{\pi_{\mathsf{E}},Z}) \otimes X^{\mathrm{h}} \\ &\quad + A_0^{\mathrm{h}} \otimes (\nabla^{\mathsf{M}}_{T\pi_{\mathsf{E}}(Z)} X)^{\mathrm{h}} + A_0^{\mathrm{h}} \otimes B_{\pi_{\mathsf{E}}}(X^{\mathrm{h}}, \mathrm{ver}(Z)).\end{aligned}$$

We have

$$(\nabla^{\mathsf{M}}_{T\pi_{\mathsf{E}}(Z)} A_0)^{\mathrm{h}} \otimes X^{\mathrm{h}} + A_0^{\mathrm{h}} \otimes (\nabla^{\mathsf{M}}_{T\pi_{\mathsf{E}}(Z)} X)^{\mathrm{h}} = (\nabla^{\mathsf{M}}_{T\pi_{\mathsf{E}}(Z)} (A_0 \otimes X))^{\mathrm{h}}.$$

We also have, by (3.3) and the first part of the lemma,

$$A_0^{\mathrm{h}} \otimes B_{\pi_{\mathsf{E}}}(X^{\mathrm{h}}, \mathrm{ver}(Z)) - \sum_{j=1}^{k} \mathrm{Ins}_j(A_0^{\mathrm{h}}, B_{\pi_{\mathsf{E}},Z}) \otimes X^{\mathrm{h}} = D_{(B_{\pi_{\mathsf{E}}})_Z}(A_0^{\mathrm{h}} \otimes X^{\mathrm{h}}).$$

Putting the above computations together gives this part of the lemma.

(iv) First we need to compute $\nabla^{\mathsf{E}}\lambda^{\mathrm{v}}$. We do this by using the formula

$$\mathscr{L}_{Z_1}\langle \lambda^{\mathrm{v}}; Z_2\rangle = \langle \nabla^{\mathsf{E}}_{Z_1}\lambda^{\mathrm{v}}; Z_2\rangle + \langle \lambda^{\mathrm{v}}; \nabla^{\mathsf{E}}_{Z_1} Z_2\rangle$$

in four cases.

1. $Z_1 = X_1^{\mathrm{h}}$ and $Z_2 = X_2^{\mathrm{h}}$: Here we have

$$\mathscr{L}_{X_1^{\mathrm{h}}}\langle\lambda^{\mathrm{v}}; X_2^{\mathrm{h}}\rangle = 0$$

and

$$\langle\lambda^{\mathrm{v}}; \nabla^{\mathsf{E}}_{X_1^{\mathrm{h}}} X_2^{\mathrm{h}}\rangle = \langle\lambda^{\mathrm{v}}; A_{\pi_{\mathsf{E}}}(X_1^{\mathrm{h}}, X_2^{\mathrm{h}})\rangle$$

(by Lemma 3.12(vi)), giving

$$\langle\nabla^{\mathsf{E}}_{X_1^{\mathrm{h}}}\lambda^{\mathrm{v}}; X_2^{\mathrm{h}}\rangle = -\langle\lambda^{\mathrm{v}}; A_{\pi_{\mathsf{E}}}(X_1^{\mathrm{h}}, X_2^{\mathrm{h}})\rangle = \langle\lambda^{\mathrm{v}}; A_{\pi_{\mathsf{E}}}(X_2^{\mathrm{h}}, X_1^{\mathrm{h}})\rangle$$

(by Lemma 3.12(ii)). Thus we have

$$\langle\nabla^{\mathsf{E}}_{X_1^{\mathrm{h}}}\lambda^{\mathrm{v}}; X_2^{\mathrm{h}}\rangle = \langle A^*_{\pi_{\mathsf{E}}}(\lambda^{\mathrm{v}}, X_1^{\mathrm{h}}); X_2^{\mathrm{h}}\rangle.$$

2. $Z_1 = X^{\mathrm{h}}$ and $Z_2 = \xi^{\mathrm{v}}$: We compute

$$\mathscr{L}_{X^{\mathrm{h}}}\langle\lambda^{\mathrm{v}}; \xi^{\mathrm{v}}\rangle = (\mathscr{L}_X\langle\lambda; \xi\rangle)^{\mathrm{h}}$$

(by Lemma 3.6(i)) and

$$\langle\lambda^{\mathrm{v}}; \nabla^{\mathsf{E}}_{X^{\mathrm{h}}}\xi^{\mathrm{v}}\rangle = \langle\lambda^{\mathrm{v}}; (\nabla^{\pi_{\mathsf{E}}}_X \xi)^{\mathrm{v}}\rangle = \langle\lambda; \nabla^{\pi_{\mathsf{E}}}_X \xi\rangle^{\mathrm{h}}$$

(by Lemma 3.12(xiii)). Thus

$$\langle\nabla^{\mathsf{E}}_{X^{\mathrm{h}}}\lambda^{\mathrm{v}}; \xi^{\mathrm{v}}\rangle = (\mathscr{L}_X\langle\lambda; \xi\rangle)^{\mathrm{h}} - \langle\lambda; \nabla^{\pi_{\mathsf{E}}}_X \xi\rangle^{\mathrm{h}}$$

or

$$\langle\nabla^{\mathsf{E}}_{X^{\mathrm{h}}}\lambda^{\mathrm{v}}; \xi^{\mathrm{v}}\rangle = \langle(\nabla^{\pi_{\mathsf{E}}}_X \lambda)^{\mathrm{v}}; \xi^{\mathrm{v}}\rangle.$$

3. $Z_1 = \xi^{\mathrm{v}}$ and $Z_2 = X^{\mathrm{h}}$: In this case we compute

$$\mathscr{L}_{\xi^{\mathrm{v}}}\langle\lambda^{\mathrm{v}}; X^{\mathrm{h}}\rangle = 0$$

and

$$\langle\lambda^{\mathrm{v}}; \nabla^{\mathsf{E}}_{\xi^{\mathrm{v}}} X^{\mathrm{h}}\rangle = 0$$

(by Lemma 3.12(xi)), giving

$$\langle\nabla^{\mathsf{E}}_{\xi^{\mathrm{v}}}\lambda^{\mathrm{v}}; X^{\mathrm{h}}\rangle = 0.$$

4. $Z_1 = \xi_1^{\mathrm{v}}$ and $Z_2 = \xi_2^{\mathrm{v}}$: We have

$$\mathscr{L}_{\xi_1^{\mathrm{v}}}\langle\lambda^{\mathrm{v}}; \xi_2^{\mathrm{v}}\rangle = \mathscr{L}_{\xi_1^{\mathrm{v}}}\langle\lambda; \xi_2\rangle^{\mathrm{h}} = 0$$

(by Lemma 3.6(ii)) and

$$\langle\lambda^{\mathrm{v}}; \nabla^{\mathsf{E}}_{\xi_1^{\mathrm{v}}}\xi_2^{\mathrm{v}}\rangle = 0$$

(using Lemma 3.12(xii)). This gives

$$\langle\nabla^{\mathsf{E}}_{\xi_1^{\mathrm{v}}}\lambda^{\mathrm{v}}; \xi_2^{\mathrm{v}}\rangle = 0.$$

Putting the above together,

$$\nabla^{\mathsf{E}}_Z\lambda^{\mathrm{v}} = (\nabla^{\pi_{\mathsf{E}}}_{T\pi_{\mathsf{E}}(Z)}\lambda)^{\mathrm{v}} + \mathrm{hor}(A^*_{\pi_{\mathsf{E}}}(\lambda^{\mathrm{v}}, \mathrm{hor}(Z))).$$

Now we note that

$$\begin{aligned}
\langle B^*_{\pi_{\mathsf{E}}}(\lambda^{\mathrm{v}}, Z_1); Z_2\rangle &= \langle\lambda^{\mathrm{v}}; B_{\pi_{\mathsf{E}}}(Z_2, Z_1)\rangle \\
&= \langle\lambda^{\mathrm{v}}; A_{\pi_{\mathsf{E}}}(\mathrm{hor}(Z_1), \mathrm{hor}(Z_2))\rangle + \langle\lambda^{\mathrm{v}}; A_{\pi_{\mathsf{E}}}(Z_1, \mathrm{ver}(Z_2))\rangle \\
&\quad + \langle\lambda^{\mathrm{v}}; A_{\pi_{\mathsf{E}}}(Z_2, \mathrm{ver}(Z_1))\rangle \\
&= -\langle\lambda^{\mathrm{v}}; A_{\pi_{\mathsf{E}}}(\mathrm{hor}(Z_2), \mathrm{hor}(Z_1))\rangle \\
&= -\langle A^*_{\pi_{\mathsf{E}}}(\lambda^{\mathrm{v}}, \mathrm{hor}(Z_1)); \mathrm{hor}(Z_2)\rangle,
\end{aligned}$$

using Lemma 3.12(xi). Thus

$$\nabla^{\mathsf{E}}_Z\lambda^{\mathrm{v}} = (\nabla^{\pi_{\mathsf{E}}}_{T\pi_{\mathsf{E}}(Z)}\lambda)^{\mathrm{v}} - B^*_{\pi_{\mathsf{E}}}(\lambda^{\mathrm{v}}, Z).$$

Now, it suffices to prove this part of the lemma for $A = A_0 \otimes \lambda \otimes \eta$ for $A_0 \in \Gamma^r(\mathsf{T}^k(\mathsf{T}^*\mathsf{M}))$, $\lambda \in \Gamma^r(\mathsf{E}^*)$, and $\eta \in \Gamma^r(\mathsf{F})$. Here we calculate, for $Z \in \Gamma^r(\mathsf{TE})$,

$$\begin{aligned}
&\nabla^{\mathsf{E},\pi_{\mathsf{F}}}_Z(A_0^{\mathrm{h}} \otimes \lambda^{\mathrm{v}} \otimes \pi^*_{\mathsf{E}}\eta) \\
&\quad = (\nabla^{\mathsf{E}}_Z A_0^{\mathrm{h}}) \otimes \lambda^{\mathrm{v}} \otimes \pi^*_{\mathsf{E}}\eta + A_0^{\mathrm{h}} \otimes \nabla^{\mathsf{E}}_Z\lambda^{\mathrm{v}} \otimes \eta + A_0^{\mathrm{h}} \otimes \lambda^{\mathrm{v}} + \pi^*_{\mathsf{E}}\nabla^{\pi_{\mathsf{F}}}_Z\pi^*_{\mathsf{E}}\eta \\
&\quad = (\nabla^{\mathsf{M}}_{T\pi_{\mathsf{E}}(Z)}A_0)^{\mathrm{h}} \otimes \lambda^{\mathrm{v}} \otimes \pi^*_{\mathsf{E}}\eta - \sum_{j=1}^k \mathrm{Ins}_j(A_0^{\mathrm{h}}, B_{\pi_{\mathsf{E}},Z}) \otimes \lambda^{\mathrm{v}} \otimes \pi^*_{\mathsf{E}}\eta \\
&\qquad + A_0^{\mathrm{h}} \otimes (\nabla^{\pi_{\mathsf{E}}}_{T\pi_{\mathsf{E}}(Z)}\lambda)^{\mathrm{v}} \otimes \eta - A_0^{\mathrm{h}} \otimes B^*_{\pi_{\mathsf{E}}}(\lambda^{\mathrm{v}}, Z) \\
&\qquad + A_0^{\mathrm{h}} \otimes \lambda^{\mathrm{v}} \otimes \pi^*_{\mathsf{E}}(\nabla^{\pi_{\mathsf{F}}}_{T\pi_{\mathsf{E}}(Z)}\eta).
\end{aligned}$$

We have

$$\begin{aligned}
&(\nabla^{\mathsf{M}}_{T\pi_{\mathsf{E}}(Z)}A_0)^{\mathrm{h}} \otimes \lambda^{\mathrm{v}} + A_0^{\mathrm{h}} \otimes (\nabla^{\pi_{\mathsf{E}}}_{T\pi_{\mathsf{E}}(Z)}\lambda)^{\mathrm{v}} + A_0^{\mathrm{h}} \otimes \lambda^{\mathrm{v}} \otimes \pi_{\mathsf{E}}^*(\nabla^{\pi_{\mathsf{F}}}_{T\pi_{\mathsf{E}}(Z)}\eta)\\
&= (\nabla^{\mathsf{M}}_{T\pi_{\mathsf{E}}(Z)}A_0 \otimes \lambda \otimes \eta)^{\mathrm{v}} + (A_0 \otimes \nabla^{\pi_{\mathsf{E}}}_{T\pi_{\mathsf{E}}(Z)}\lambda \otimes \eta)^{\mathrm{v}} + (A_0 \otimes \lambda^{\mathrm{v}} \otimes \nabla^{\pi_{\mathsf{F}}}_{T\pi_{\mathsf{E}}(Z)}\eta)^{\mathrm{v}}\\
&= (\nabla^{\pi_{\mathsf{E}}\otimes\pi_{\mathsf{F}}}_{T\pi_{\mathsf{E}}(Z)}(A_0 \otimes \lambda \otimes \eta))^{\mathrm{v}}
\end{aligned}$$

and, by (3.4) and the first part of the lemma,

$$-A_0^{\mathrm{h}}\otimes B^*_{\pi_{\mathsf{E}}}(\lambda^{\mathrm{v}}, Z)\otimes\pi_{\mathsf{E}}^*\eta-\sum_{j=1}^{k}\mathrm{Ins}_j(A_0^{\mathrm{h}}, B_{\pi_{\mathsf{E}},Z})\otimes\lambda^{\mathrm{v}}\otimes\pi_{\mathsf{E}}^*\eta = D_{B_{\pi_{\mathsf{E}},Z}}(A_0^{\mathrm{h}}\otimes\lambda^{\mathrm{v}}\otimes\pi_{\mathsf{E}}^*\eta).$$

Assembling the preceding computations gives this part of the lemma.

(v) This is a slight modification of the preceding part of the proof, taking the formula (3.17) into account.

(vi) By Lemma 3.6(iii) and (iv) we have

$$\nabla^{\mathsf{E}}_Z\lambda^{\mathrm{e}} = \mathscr{L}_Z\lambda^{\mathrm{e}} = (\nabla^{\pi_{\mathsf{E}}}_{T\pi_{\mathsf{E}}(Z)}\lambda)^{\mathrm{e}} + \langle\lambda^{\mathrm{v}}; Z\rangle.$$

With the constructions following Definition 3.8 in mind, we work with $A = A_0^{\mathrm{h}} \otimes \lambda^{\mathrm{e}} \otimes \pi_{\mathsf{E}}^*\eta$ for $A_0 \in \Gamma^r(\mathsf{T}^k(\mathsf{T}^*\mathsf{M}))$, $\lambda \in \Gamma^r(\mathsf{E}^*)$, and $\eta \in \Gamma^r(\mathsf{F})$. If we keep in mind that λ^{e} is a function, then we can simply write $A = A_0 \otimes (\lambda^{\mathrm{e}}\pi_{\mathsf{E}}^*\eta)$. We now calculate

$$\begin{aligned}
&\nabla_Z^{\mathsf{E},\pi_{\mathsf{F}}}(A_0^{\mathrm{h}}\otimes\lambda^{\mathrm{e}} \otimes \pi_{\mathsf{E}}^*\eta)\\
&\quad= (\nabla^{\mathsf{E}}_Z A_0^{\mathrm{h}}) \otimes \lambda^{\mathrm{e}} \otimes \pi_{\mathsf{E}}^*\eta + A_0^{\mathrm{h}} \otimes (\nabla^{\mathsf{E}}_Z\lambda^{\mathrm{e}}) \otimes \pi_{\mathsf{E}}^*\eta\\
&\qquad+ A_0^{\mathrm{h}} \otimes \lambda^{\mathrm{e}} \otimes (\pi_{\mathsf{E}}^*\nabla_Z^{\pi_{\mathsf{F}}}\pi_{\mathsf{E}}^*\eta)\\
&\quad= (\nabla^{\mathsf{M}}_{T\pi_{\mathsf{E}}(Z)}A_0)^{\mathrm{h}} \otimes \lambda^{\mathrm{e}} \otimes \pi_{\mathsf{E}}^*\eta - \sum_{j=1}^{k}\mathrm{Ins}_j(A_0^{\mathrm{h}}, B_{\pi_{\mathsf{E}}}) \otimes \lambda^{\mathrm{e}} \otimes \pi_{\mathsf{E}}^*\eta\\
&\qquad+ A_0^{\mathrm{h}} \otimes (\nabla^{\pi_{\mathsf{E}}}_{T\pi_{\mathsf{E}}(Z)}\lambda)^{\mathrm{e}} \otimes \pi_{\mathsf{E}}^*\eta + A_0^{\mathrm{h}} \otimes (\lambda^{\mathrm{v}}(Z)) \otimes \pi_{\mathsf{E}}^*\eta\\
&\qquad+ A_0^{\mathrm{h}} \otimes \lambda^{\mathrm{e}} \otimes \pi_{\mathsf{E}}^*(\nabla^{\pi_{\mathsf{F}}}_{T\pi_{\mathsf{E}}(Z)}\eta) + A_0^{\mathrm{h}} \otimes \lambda^{\mathrm{e}} \otimes B_{\pi_{\mathsf{E}}}(\pi_{\mathsf{E}}^*\eta, Z).
\end{aligned}$$

We have

$$\begin{aligned}
&(\nabla^{\mathsf{M}}_{T\pi_{\mathsf{E}}(Z)}A_0)^{\mathrm{h}} \otimes \lambda^{\mathrm{e}} \otimes \pi_{\mathsf{E}}^*\eta + A_0^{\mathrm{h}} \otimes (\nabla^{\pi_{\mathsf{E}}}_{T\pi_{\mathsf{E}}(Z)}\lambda)^{\mathrm{e}} \otimes \pi_{\mathsf{E}}^*\eta\\
&\quad+ A_0^{\mathrm{h}} \otimes \lambda^{\mathrm{e}} \otimes \pi_{\mathsf{E}}^*(\nabla^{\pi_{\mathsf{F}}}_{T\pi_{\mathsf{E}}(Z)}\eta)
\end{aligned}$$

$$= (\nabla^{\mathsf{M}}_{T\pi_{\mathsf{E}}(Z)} A_0 \otimes \lambda \otimes \eta)^{\mathrm{e}} + (A_0 \otimes \nabla^{\pi_{\mathsf{E}}}_{T\pi_{\mathsf{E}}(Z)} \lambda \otimes \eta)^{\mathrm{e}} + (A_0 \otimes \lambda \otimes \nabla^{\pi_{\mathsf{F}}}_{T\pi_{\mathsf{E}}(Z)} \eta)^{\mathrm{e}}$$
$$= (\nabla^{\mathsf{M},\pi_{\mathsf{E}} \otimes \pi_{\mathsf{F}}}_{T\pi_{\mathsf{E}}(Z)} (A_0 \otimes \lambda \otimes \eta))^{\mathrm{e}}.$$

Next we note that

$$A_0^{\mathrm{h}} \otimes \lambda^{\mathrm{e}} \otimes B_{\pi_{\mathsf{E}}}(\pi_{\mathsf{E}}^* \eta, Z) - \sum_{j=1}^{k} \mathrm{Ins}_j(A_0^{\mathrm{h}}, B_{\pi_{\mathsf{E}}}) \otimes \lambda^{\mathrm{e}} \otimes \pi_{\mathsf{E}}^* \eta = D_{B_{\pi_{\mathsf{E}}}, Z}(A_0 \otimes (\lambda^{\mathrm{e}} \pi_{\mathsf{E}}^* \eta)),$$

keeping in mind that λ^{e} is a function, so the tensor products with λ^{e} are just multiplication. Again making reference to the constructions following Definition 3.8, we have

$$A_0^{\mathrm{h}} \otimes \lambda^{\mathrm{v}} \otimes \pi_{\mathsf{E}}^* \eta = (A_0 \otimes \lambda \otimes \eta)^{\mathrm{v}},$$

and the lemma follows by combining the preceding three formulae.

(vii) This is a slight modification of the preceding part of the proof, taking the formula (3.17) into account. □

The precise complicated definition (3.18) of the tensor B_π is a little immaterial. The point is that it is a tensor associated with the vector bundle E and the induced connection ∇^{E}, and that it is of class C^r.

3.2.4 *Prolongation*

In our geometric setting, differentiation means "prolongation" by taking jets. In this section, we illustrate how our decompositions of Sect. 2.3.1 interact with prolongation. As we shall see, this is one place where our geometric framework makes things a little more complicated, compared to using local coordinates. However, the results are interesting, just for this reason, so we give them in detail and make use of them, e.g., in the proof of Theorem 5.11 below.

We let $r \in \{\infty, \omega\}$ and let $\pi_{\mathsf{E}} \colon \mathsf{E} \to \mathsf{M}$ be a C^r-vector bundle with $\mathsf{J}^m\mathsf{E}$, $m \in \mathbb{Z}_{\geq 0}$, its jet bundles. We suppose that we have a C^r-affine connection ∇^{M} on M and a C^r-vector bundle connection $\nabla^{\pi_{\mathsf{E}}}$ in E. Because we have the decomposition

$$\mathsf{J}^m\mathsf{E} \simeq \bigoplus_{j=0}^{m} (\mathsf{S}^j(\mathsf{T}^*\mathsf{M}) \otimes \mathsf{E}) \tag{3.19}$$

by Lemma 2.15, it follows that the vector bundle $\mathsf{J}^m\mathsf{E}$ has a C^r-connection that we denote by $\nabla^{\pi_{\mathsf{E}},m}$. Explicitly,

$$\nabla_X^{\pi_{\mathsf{E}},m} j_m\xi = (S^m_{\nabla^{\mathsf{M}},\nabla^{\pi_{\mathsf{E}}}})^{-1}(\nabla_X^{\pi_{\mathsf{E}}}\xi, \nabla_X^{\mathsf{M},\pi_{\mathsf{E}}} D^1_{\nabla^{\mathsf{M}},\nabla^{\pi_{\mathsf{E}}}}(\xi), \dots, \nabla_X^{\mathsf{M},\pi_{\mathsf{E}}} D^m_{\nabla^{\mathsf{M}},\nabla^{\pi_{\mathsf{E}}}}(\xi)).$$

Therefore, the construction of Lemma 2.15 can be applied to $\mathsf{J}^m\mathsf{E}$ in place of E, and all that remains to sort out is notation.

To this end, for $k, m \in \mathbb{Z}_{\geq 0}$, let us denote by $\nabla^{\mathsf{M},\pi_{\mathsf{E}},m}$ the connection in $\mathsf{T}^k(\mathsf{T}^*\mathsf{M}) \oplus \mathsf{J}^m\mathsf{E}$ induced, via tensor product, by the connections ∇^{M} and $\nabla^{\pi_{\mathsf{E}},m}$. Then, for $\xi \in \Gamma^r(\mathsf{E})$, denote

$$\nabla^{\mathsf{M},\pi_{\mathsf{E}},m,k} j_m\xi = \underbrace{\nabla^{\mathsf{M},\pi_{\mathsf{E}},m} \cdots (\nabla^{\mathsf{M},\pi_{\mathsf{E}},m}}_{k-1 \text{ times}}(\nabla^{\pi_{\mathsf{E}},m} j_m\xi)) \in \Gamma^r(\mathsf{T}^k(\mathsf{T}^*\mathsf{M}) \otimes \mathsf{J}^m\mathsf{E}).$$

We also denote

$$D^k_{\nabla^{\mathsf{M}},\nabla^{\pi_{\mathsf{E}},m}}(j_m\xi) = \mathrm{Sym}_k \otimes \mathrm{id}_{\mathsf{J}^m\mathsf{E}}(\nabla^{\mathsf{M},\pi_{\mathsf{E}},m,k} j_m\xi) \in \Gamma^r(\mathsf{S}^k(\mathsf{T}^*\mathsf{M}) \otimes \mathsf{J}^m\mathsf{E}).$$

This can be refined further by explicitly decomposing $\mathsf{J}^m\mathsf{E}$, so let us provide the notation for making this refinement. For $m, k \in \mathbb{Z}_{\geq 0}$ and for $A \in \Gamma^r(\mathsf{T}^m(\mathsf{T}^*\mathsf{M}) \otimes \mathsf{E})$, we denote

$$\nabla^{\mathsf{M},\pi_{\mathsf{E}},k} A = \underbrace{\nabla^{\mathsf{M},\pi_{\mathsf{E}}} \cdots \nabla^{\mathsf{M},\pi_{\mathsf{E}}}}_{k \text{ times}} A \in \Gamma^r(\mathsf{T}^{m+k}(\mathsf{T}^*\mathsf{M}) \otimes \mathsf{E})$$

and

$$D^{k,m}_{\nabla^{\mathsf{M}},\nabla^{\pi_{\mathsf{E}}}}(A) = \mathrm{Sym}_k \otimes \mathrm{id}_{\mathsf{T}^m(\mathsf{T}^*\mathsf{M})\otimes\mathsf{E}}(\nabla^{\mathsf{M},\pi_{\mathsf{E}},k} A) \in \Gamma^r(\mathsf{S}^k(\mathsf{T}^*\mathsf{M}) \otimes \mathsf{T}^m(\mathsf{T}^*\mathsf{M}) \otimes \mathsf{E}).$$

Note that, if $A \in \Gamma^r(\mathsf{S}^m(\mathsf{T}^*\mathsf{M}) \otimes \mathsf{E})$, then

$$D^{k,m}_{\nabla^{\mathsf{M}},\nabla^{\pi_{\mathsf{E}}}}(A) \in \Gamma^r(\mathsf{S}^k(\mathsf{T}^*\mathsf{M}) \otimes \mathsf{S}^m(\mathsf{T}^*\mathsf{M}) \otimes \mathsf{E}).$$

An immediate consequence of Lemma 2.15 is then the following result.

Lemma 3.17 (Decompositions of Jet Bundles of Jet Bundles) *The maps*

$$S^k_{\nabla^{\mathsf{M}},\nabla^{\pi_{\mathsf{E}},m}} : \mathsf{J}^k\mathsf{J}^m\mathsf{E} \to \bigoplus_{j=0}^{k}(\mathsf{S}^j(\mathsf{T}^*\mathsf{M}) \otimes \mathsf{J}^m\mathsf{E})$$

defined by

$$S^k_{\nabla^{\mathsf{M}},\nabla^{\pi\mathsf{E}},m}(j_k\Xi(j_m\xi(x)))$$
$$= (\Xi(j_m\xi(x)), D^1_{\nabla^{\mathsf{M}},\nabla^{\pi\mathsf{E}},m}(\Xi)(j_m\xi(x)), \dots, D^k_{\nabla^{\mathsf{M}},\nabla^{\pi\mathsf{E}},m}(\Xi)(j_m\xi(x)))$$

and

$$S^{k,m}_{\nabla^{\mathsf{M}},\nabla^{\pi\mathsf{E}}} : \mathsf{J}^k\left(\bigoplus_{l=0}^m \mathsf{S}^l(\mathsf{T}^*\mathsf{M}) \otimes \mathsf{E}\right) \to \bigoplus_{j=0}^k\left(\mathsf{S}^j(\mathsf{T}^*\mathsf{M}) \otimes \left(\bigoplus_{l=0}^m \mathsf{S}^l(\mathsf{T}^*\mathsf{M}) \otimes \mathsf{E}\right)\right)$$

defined by

$$j_k(A_0, A_1, \dots, A_m)(x) \mapsto ((A_0(x), A_1(x), \dots, A_m(x)),$$
$$(D^{1,0}_{\nabla^{\mathsf{M}},\nabla^{\pi\mathsf{E}}}(A_0)(x), D^{1,1}_{\nabla^{\mathsf{M}},\nabla^{\pi\mathsf{E}}}(A_1)(x), \dots, D^{1,m}_{\nabla^{\mathsf{M}},\nabla^{\pi\mathsf{E}}}(A_m)(x)), \dots,$$
$$(D^{k,0}_{\nabla^{\mathsf{M}},\nabla^{\pi\mathsf{E}}}(A_0)(x), D^{k,1}_{\nabla^{\mathsf{M}},\nabla^{\pi\mathsf{E}}}(A_1)(x), \dots, D^{k,m}_{\nabla^{\mathsf{M}},\nabla^{\pi\mathsf{E}}}(A_m)(x)))$$

are isomorphisms of vector bundles, and, for each $k \in \mathbb{Z}_{>0}$, the diagrams

$$\begin{array}{ccc}
\mathsf{J}^{k+1}\mathsf{J}^m\mathsf{E} & \xrightarrow{S^{k+1}_{\nabla^{\mathsf{M}},\nabla^{\pi\mathsf{E}},m}} & \bigoplus_{j=0}^{k+1}(\mathsf{S}^j(\mathsf{T}^*\mathsf{M}) \otimes \mathsf{J}^m\mathsf{E}) \\
{\scriptstyle(\pi_{\mathsf{E},m})^{k+1}_k}\downarrow & & \downarrow{\scriptstyle \mathrm{pr}^{k+1}_k} \\
\mathsf{J}^k\mathsf{J}^m\mathsf{E} & \xrightarrow[S^k_{\nabla^{\mathsf{M}},\nabla^{\pi\mathsf{E}},m}]{} & \bigoplus_{j=0}^{k}(\mathsf{S}^j(\mathsf{T}^*\mathsf{M}) \otimes \mathsf{J}^m\mathsf{E})
\end{array}$$

and

$$\begin{array}{ccc}
\mathsf{J}^{k+1}(\oplus_{l=0}^m \mathsf{S}^l(\mathsf{T}^*\mathsf{M}) \otimes \mathsf{E}) & \xrightarrow{S^{k+1,m}_{\nabla^{\mathsf{M}},\nabla^{\pi\mathsf{E}}}} & \bigoplus_{j=0}^{k+1}\left(\mathsf{S}^j(\mathsf{T}^*\mathsf{M}) \otimes \left(\bigoplus_{l=0}^m \mathsf{S}^l(\mathsf{T}^*\mathsf{M}) \otimes \mathsf{E}\right)\right) \\
{\scriptstyle(\pi_{\mathsf{E},m})^{k+1}_k}\downarrow & & \downarrow{\scriptstyle \mathrm{pr}^{k+1}_k} \\
\mathsf{J}^{k}(\oplus_{l=0}^m \mathsf{S}^l(\mathsf{T}^*\mathsf{M}) \otimes \mathsf{E}) & \xrightarrow[S^{k,m}_{\nabla^{\mathsf{M}},\nabla^{\pi\mathsf{E}}}]{} & \bigoplus_{j=0}^{k}\left(\mathsf{S}^j(\mathsf{T}^*\mathsf{M}) \otimes \left(\bigoplus_{l=0}^m \mathsf{S}^l(\mathsf{T}^*\mathsf{M}) \otimes \mathsf{E}\right)\right)
\end{array}$$

commute, where pr^{k+1}_k are the obvious projections, stripping off the last component of the direct sum.

Of course, in understanding the meaning of the preceding lemma, one should bear in mind the isomorphism (3.19), which infers that we have vector bundle isomorphisms

$$\mathsf{J}^k\mathsf{J}^m\mathsf{E} \simeq \bigoplus_{j=0}^{k}(\mathsf{S}^j(\mathsf{T}^*\mathsf{M}) \otimes \mathsf{J}^m\mathsf{E}) \simeq \bigoplus_{j=0}^{k}\left(\mathsf{S}^j(\mathsf{T}^*\mathsf{M}) \otimes \left(\bigoplus_{l=0}^{m}\mathsf{S}^l(\mathsf{T}^*\mathsf{M}) \otimes \mathsf{E}\right)\right),$$

and the lemma is just an explicit writing down of these isomorphisms.

Now we recall [62, Definition 6.2.25] the injective vector bundle mapping over id_{M}, for $k, m \in \mathbb{Z}_{\geq 0}$,

$$\begin{aligned} \iota_{\mathsf{E},k,m} \colon \mathsf{J}^{k+m}\mathsf{E} &\to \mathsf{J}^k\mathsf{J}^m\mathsf{E} \\ j_{m+k}\xi(x) &\mapsto j_k j_m \xi(x). \end{aligned} \tag{3.20}$$

Let us understand this mapping using our decompositions of jet bundles. Doing so is a bit more involved than one might imagine due, essentially, to the fact that iterated covariant differentials are not symmetric. In fact, our constructions will illustrate the rôle of curvature of $\nabla^{\pi_{\mathsf{E}}}$ and torsion of ∇^{M} in this lack of symmetry. Indeed, the next lemma gives an explicit form for the lack of symmetry for second derivatives.

Lemma 3.18 (Decomposition of Second Covariant Differential) *Let $r \in \{\infty, \omega\}$, let $\pi_{\mathsf{E}} \colon \mathsf{E} \to \mathsf{M}$ be a C^r-vector bundle, and let ∇^{M} be a C^r-affine connection on M and $\nabla^{\pi_{\mathsf{E}}}$ be a C^r-linear connection in E. The following formula holds for $\xi \in \Gamma^r(\mathsf{E})$:*

$$\nabla^{\mathsf{M},\pi_{\mathsf{E}},2}\xi(X,Y) = D^2_{\nabla^{\mathsf{M}},\nabla^{\pi_{\mathsf{E}}}}(\xi)(X,Y) + \frac{1}{2}R_{\nabla^{\pi_{\mathsf{E}}}}(X,Y)(\xi) - \frac{1}{2}T_{\nabla^{\mathsf{M}}}(X,Y)(\nabla^{\pi_{\mathsf{E}}}\xi),$$
$$X, Y \in \Gamma^r(\mathsf{TM}),$$

where we denote

$$T_{\nabla^{\mathsf{M}}}(X,Y)(\nabla^{\pi_{\mathsf{E}}}\xi) = \nabla^{\pi_{\mathsf{E}}}\xi(T_{\nabla^{\mathsf{M}}}(X,Y)).$$

Proof We have $\nabla^{\pi_{\mathsf{E}}}\xi(Y) = \nabla^{\pi_{\mathsf{E}}}_Y\xi$, whereupon

$$\nabla^{\mathsf{M},\pi_{\mathsf{E}},2}\xi(X,Y) = \nabla^{\pi_{\mathsf{E}}}_X\nabla^{\pi_{\mathsf{E}}}_Y\xi - \nabla^{\pi_{\mathsf{E}}}_{\nabla^{\mathsf{M}}_X Y}\xi.$$

Thus

$$\begin{aligned} &\nabla^{\mathsf{M},\pi_{\mathsf{E}}}\xi(X,Y) \\ &\quad = \frac{1}{2}(\nabla^{\mathsf{M},\pi_{\mathsf{E}}}\xi(X,Y) + \nabla^{\mathsf{M},\pi_{\mathsf{E}}}\xi(Y,X)) + \frac{1}{2}(\nabla^{\mathsf{M},\pi_{\mathsf{E}}}\xi(X,Y) - \nabla^{\mathsf{M},\pi_{\mathsf{E}}}\xi(Y,X)) \end{aligned}$$

$$= D^2_{\nabla^{\mathsf{M}},\nabla^{\pi_{\mathsf{E}}}}(\xi)(X,Y) + \frac{1}{2}(\nabla^{\pi_{\mathsf{E}}}_X \nabla^{\pi_{\mathsf{E}}}_Y \xi - \nabla^{\pi_{\mathsf{E}}}_X \nabla^{\pi_{\mathsf{E}}}_Y \xi - \nabla^{\pi_{\mathsf{E}}}_{[X,Y]}\xi - \nabla^{\pi_{\mathsf{E}}}_{T_{\nabla^{\mathsf{M}}}(X,Y)}\xi)$$

$$= D^2_{\nabla^{\mathsf{M}},\nabla^{\pi_{\mathsf{E}}}}(\xi)(X,Y) + \frac{1}{2}R_{\nabla^{\pi_{\mathsf{E}}}}(X,Y)(\xi) - \frac{1}{2}\nabla^{\pi_{\mathsf{E}}}\xi(T_{\nabla^{\mathsf{M}}}(X,Y)),$$

as desired. □

Aside 3.19 (Relationship to Covariant Exterior Derivative) The formula of the preceding lemma reflects, but is not equal to, a standard construction concerning connections. In this aside, we clarify the relationships. We refer to [42, §III.9.3-4, §III.11.5] for a discussion of the concepts we reference here, but do not fully develop.

One of the many ways to define the notion of a connection in a vector bundle $\pi_{\mathsf{E}}\colon \mathsf{E} \to \mathsf{M}$ is via the connector which is a vector bundle mapping $K_{\nabla^{\pi_{\mathsf{E}}}}: \mathsf{TE} \to \mathsf{E}$ for which the two diagrams (1.6) commute. In this way of thinking about things, we can think of $K_{\nabla^{\pi_{\mathsf{E}}}}$ as defining a VE-valued one-form on E, making use of the vector bundle isomorphism vlft: $\mathsf{E} \oplus \mathsf{E} \to \mathsf{VE}$. The covariant derivative is defined using the connector by the formula (1.7). In this formulation, one can give an alternative definition of the curvature tensor:

$$R_{\nabla^{\pi_{\mathsf{E}}}}(X(\pi_{\mathsf{E}}(e)), Y(\pi_{\mathsf{E}}(e)))(\xi(e)) = -K_{\nabla^{\pi_{\mathsf{E}}}}([X^{\mathrm{h}}, Y^{\mathrm{h}}](e)).$$

In this formulation, we can see that curvature measures the nonintegrability of the horizontal subbundle.

This formulation also admits another differential construction, apart from the covariant derivative. To set this up, we consider E-valued differential forms on M, i.e., sections of $\bigwedge^k(\mathsf{T}^*\mathsf{M}) \otimes \mathsf{E}$. Just as one defines the exterior derivative for differential forms by requiring that it act on functions in the usual way and extends to general differential forms by certain naturality properties [1, Theorem 6.4.1], one can define the ***exterior covariant derivative***, denoted by $\mathrm{d}_{\nabla^{\pi_{\mathsf{E}}}}$, by asking that $\mathrm{d}_{\nabla^{\pi_{\mathsf{E}}}}\xi = \nabla^{\pi_{\mathsf{E}}}\xi$ and that $\mathrm{d}_{\nabla^{\pi_{\mathsf{E}}}}$ have the same naturality conditions as specified by those for the usual exterior derivative. A place where the usual exterior derivative differs from the covariant exterior derivative is that, for the latter, it is not necessarily true that $\mathrm{d}_{\nabla^{\pi_{\mathsf{E}}}} \circ \mathrm{d}_{\nabla^{\pi_{\mathsf{E}}}} = 0$. Indeed, it holds that

$$\mathrm{d}_{\nabla^{\pi_{\mathsf{E}}}} \circ \mathrm{d}_{\nabla^{\pi_{\mathsf{E}}}}\xi(X,Y) = R_{\nabla^{\pi_{\mathsf{E}}}}(X,Y)(\xi).$$

Thus, in this formulation, curvature measures the extent to which it is *not* true that $\mathrm{d}^2_{\nabla^{\pi_{\mathsf{E}}}} = 0$.

With all of this as backdrop, one way to interpret Lemma 3.18 is that curvature plays a rôle in measuring the lack of symmetry of $\nabla^{\mathsf{M},\pi_{\mathsf{E}},2}$. A difference with the constructions of the preceding paragraph is that $\nabla^{\mathsf{M},\pi_{\mathsf{E}}}$ invokes, as well as the linear connection $\nabla^{\pi_{\mathsf{E}}}$, an affine connection ∇^{M} on M. We see, moreover, that this affine connection on M contributes to the lack of symmetry of $\nabla^{\mathsf{M},\pi_{\mathsf{E}},2}$ through its torsion. ○

Note that the first term in the decomposition of $\nabla^{\mathsf{M},\pi_{\mathsf{E}},2}\xi$ from Lemma 3.18 is the symmetric part of this tensor, while the remaining two terms are the skew-symmetric part.

To extend this to higher-order covariant differentials, we require a preliminary construction. For a $\mathbb{R}$-vector space V and for $r, s \in \mathbb{Z}_{\geq 0}$, we have an inclusion,

$$\Delta_{k,m} : \mathrm{S}^{k+m}(\mathsf{V}^*) \to \mathrm{S}^k(\mathsf{V}^*) \otimes \mathrm{S}^m(\mathsf{V}^*). \tag{3.21}$$

Let us give an explicit formula for this inclusion.

Lemma 3.20 (Inclusions for Symmetric Tensors) *For a finite-dimensional $\mathbb{R}$-vector space V and for $r, s \in \mathbb{Z}_{\geq 0}$,*

$$\Delta_{k,m} = (\mathrm{Sym}_k \otimes \mathrm{Sym}_k) \circ \iota_{k,m},$$

where

$$\iota_{k,m} : \mathrm{S}^{k+m}(\mathsf{V}^*) \to \mathrm{T}^{k+m}(\mathsf{V}^*) = \mathrm{T}^k(\mathsf{V}^*) \otimes \mathrm{T}^m(\mathsf{V}^*)$$

is the inclusion. Explicitly,

$$\Delta_{k,m}(\alpha^1 \odot \cdots \odot \alpha^{k+m}) = \sum_{\sigma \in \mathfrak{S}_{k,m}} (\alpha^{\sigma(1)} \odot \cdots \odot \alpha^{\sigma(k)}) \otimes (\alpha^{\sigma(k+1)} \odot \cdots \odot \alpha^{\sigma(k+m)})$$

for $\alpha^1, \dots, \alpha^{k+m} \in \mathsf{V}^*$.

Proof We note that $\mathrm{Sym}_{k+m} \circ \Delta_{k,m} = \mathrm{id}_{\mathrm{S}^{k+m}(\mathsf{V}^*)}$, simply since $\Delta_{k,m}$ is the inclusion and Sym_{k+m} is projection onto $\mathrm{S}^{k+m}(\mathsf{V}^*) \subseteq \mathrm{T}^{k+m}(\mathsf{V}^*)$. Thus, for the first part of the lemma, it will suffice to show that

$$\mathrm{Sym}_{k+m} \circ (\mathrm{Sym}_k \otimes \mathrm{Sym}_m) \circ \iota_{k,m} = \mathrm{id}_{\mathrm{S}^{k+m}(\mathsf{V}^*)}.$$

For $A \in \mathrm{S}^{k+m}(\mathsf{V}^*)$ we have

$$\begin{aligned}
&\mathrm{Sym}_k \otimes \mathrm{Sym}_m(A)(v_1, \dots, v_{k+m}) \\
&\qquad = \frac{1}{k!m!} \sum_{\sigma_1 \in \mathfrak{S}_k} \sum_{\sigma_2 \in \mathfrak{S}_k} A(v_{\sigma_1(1)}, \dots, v_{\sigma_1(k)}, v_{k+\sigma_2(1)}, \dots, v_{k+\sigma_2(m)}) \\
&\qquad = A(v_1, \dots, v_k, v_{k+1}, \dots, v_{k+m}),
\end{aligned}$$

since A is symmetric, and so symmetric on the first k and last m entries. Since $\mathrm{Sym}_{k+m}(A) = A$, our claim follows, and so does the first part of the lemma.

For the asserted explicit formula, it is evident that the mapping

$$\alpha^1 \odot \cdots \odot \alpha^{k+m} \mapsto \sum_{\sigma \in \mathfrak{S}_{k,m}} (\alpha^{\sigma(1)} \odot \cdots \odot \alpha^{\sigma(k)}) \otimes (\alpha^{\sigma(k+1)} \odot \cdots \odot \alpha^{\sigma(k+m)})$$

takes values in $\mathsf{S}^k(\mathsf{V}^*) \otimes \mathsf{S}^m(\mathsf{V}^*)$, and so it suffices to show that

$$\mathrm{Sym}_{k+m}\left(\sum_{\sigma \in \mathfrak{S}_{k,m}} (\alpha^{\sigma(1)} \odot \cdots \odot \alpha^{\sigma(k)}) \otimes (\alpha^{\sigma(k+1)} \odot \cdots \odot \alpha^{\sigma(k+m)})\right) = \alpha^1 \odot \cdots \odot \alpha^{k+m}. \tag{3.22}$$

To this end, we calculate

$$\begin{aligned}
&\sum_{\sigma \in \mathfrak{S}_{k,m}} (\alpha^{\sigma(1)} \odot \cdots \odot \alpha^{\sigma(k)}) \otimes (\alpha^{\sigma(k+1)} \odot \cdots \odot \alpha^{\sigma(k+m)}) \\
&= \frac{1}{k!m!} \sum_{\sigma_1 \in \mathfrak{S}_{k,m}} \sum_{\sigma_2 \in \mathfrak{S}_{k|m}} \sigma_1 \circ \sigma_2((\alpha^1 \odot \cdots \odot \alpha^k) \otimes (\alpha^{k+1} \odot \cdots \odot \alpha^{k+m})) \\
&= \frac{(k+m)!}{k!m!} \mathrm{Sym}_{k+m}((\alpha^1 \odot \cdots \odot \alpha^k) \otimes (\alpha^{k+1} \odot \cdots \odot \alpha^{k+m})) \\
&= \alpha^1 \odot \cdots \odot \alpha^{k+m},
\end{aligned}$$

using (1.2). Applying Sym_{k+m} to the leftmost and rightmost components of this string of equalities gives (3.22). □

Using the preceding constructions, we next see how the decomposition of Lemma 3.18 for iterated covariant differentials of order 2 are reflected in higher-order iterated derivatives. To state the result, we use the abbreviation

$$\theta_{k,m} = \frac{1}{2}(\Delta_{k-1,1} \otimes \Delta_{1,m-1}) \circ (\mathrm{Sym}_k \otimes \mathrm{Sym}_m)$$

for $k, m \in \mathbb{Z}_{>0}$.

Lemma 3.21 (Decomposition of Iterated Derivative) *Let $r \in \{\infty, \omega\}$, let $\pi_{\mathsf{E}} \colon \mathsf{E} \to \mathsf{M}$ be a C^r-vector bundle, and let ∇^{M} be a C^r-affine connection on M and $\nabla^{\pi_{\mathsf{E}}}$ be a C^r-linear connection in E. The following formulae hold for $\xi \in \Gamma^r(\mathsf{E})$ and for $k, m \in \mathbb{Z}_{>0}$:*

(i) $D^{k,1}_{\nabla^{\mathsf{M}},\nabla^{\pi_{\mathsf{E}}}}(D^1_{\nabla^{\mathsf{M}},\nabla^{\pi_{\mathsf{E}}}}(\xi)) = D^{k+1}_{\nabla^{\mathsf{M}},\nabla^{\pi_{\mathsf{E}}}}(\xi) +$
$\theta_{k,1} \otimes \mathrm{id}_{\mathsf{E}}\left(D^{k-1}_{\nabla^{\mathsf{M}},\nabla^{\pi_{\mathsf{E}}}} R_{\nabla^{\pi_{\mathsf{E}}}}(\xi) - D^{k-1}_{\nabla^{\mathsf{M}},\nabla^{\pi_{\mathsf{E}}}} T_{\nabla^{\mathsf{M}}}(\nabla^{\mathsf{M},\pi_{\mathsf{E}}}\xi)\right);$

(ii) $D^{k,m}_{\nabla^{\mathsf{M}},\nabla^{\pi_{\mathsf{E}}}}(D^m_{\nabla^{\mathsf{M}},\nabla^{\pi_{\mathsf{E}}}}(\xi)) = D^{k+m}_{\nabla^{\mathsf{M}},\nabla^{\pi_{\mathsf{E}}}}(\xi) + \theta_{k,m} \otimes \mathrm{id}_{\mathsf{E}}\Big(D^{k-1}_{\nabla^{\mathsf{M}},\nabla^{\pi_{\mathsf{E}}}} R_{\nabla^{\pi_{\mathsf{E}}}}(D^{m-1}_{\nabla^{\mathsf{M}},\nabla^{\pi_{\mathsf{E}}}}(\xi))$
$- D^{k-1}_{\nabla^{\mathsf{M}},\nabla^{\pi_{\mathsf{E}}}} T_{\nabla^{\mathsf{M}}}\left(D^m_{\nabla^{\mathsf{M}},\nabla^{\pi_{\mathsf{E}}}}(\xi)\right)$
$+\theta_{m-1,1} \otimes \mathrm{id}_{\mathsf{E}}\left(D^{m-2}_{\nabla^{\mathsf{M}},\nabla^{\pi_{\mathsf{E}}}} R_{\nabla^{\pi_{\mathsf{E}}}}(\xi) - D^{m-2}_{\nabla^{\mathsf{M}},\nabla^{\pi_{\mathsf{E}}}} T_{\nabla^{\mathsf{M}}}(\nabla^{\mathsf{M},\pi_{\mathsf{E}}}\xi)\right)\Big)\Big)$

Moreover, in both cases the second term on the right takes values in

$$\mathsf{S}^{k-1}(\mathsf{T}^*\mathsf{M}) \otimes \textstyle\bigwedge^2(\mathsf{T}^*\mathsf{M}) \otimes \mathsf{S}^{m-1}(\mathsf{T}^*\mathsf{M}) \otimes \mathsf{E};$$

in particular, its symmetric part is zero.

Proof Write

$$\begin{aligned}
\nabla^{\mathsf{M},\pi_{\mathsf{E}},k}(\nabla^{\mathsf{M},\pi_{\mathsf{E}},m}\xi) &= \nabla^{\mathsf{M},\pi_{\mathsf{E}},k+m}\xi \\
&= \nabla^{\mathsf{M},\pi_{\mathsf{E}},k-1}\nabla^{\mathsf{M},\pi_{\mathsf{E}},2}\nabla^{\mathsf{M},\pi_{\mathsf{E}},m-1}\xi \\
&= \nabla^{\mathsf{M},\pi_{\mathsf{E}},k-1} D^2_{\nabla^{\mathsf{M}},\nabla^{\pi_{\mathsf{E}}}} \nabla^{\mathsf{M},\pi_{\mathsf{E}},m-1}\xi \\
&\quad + \frac{1}{2}\nabla^{\mathsf{M},\pi_{\mathsf{E}},k-1} R_{\nabla^{\pi_{\mathsf{E}}}}(\nabla^{\mathsf{M},\pi_{\mathsf{E}},m-1}\xi) \\
&\quad - \frac{1}{2}\nabla^{\mathsf{M},k-1} T_{\nabla^{\mathsf{M}}}(\nabla^{\mathsf{M},\pi_{\mathsf{E}},m}\xi).
\end{aligned}$$

(The precise meaning of the final terms will become apparent below when we examine these carefully below.)

We first consider the tensor

$$\mathrm{Sym}_k \otimes \mathrm{Sym}_m \otimes \mathrm{id}_{\mathsf{E}}(\nabla^{\mathsf{M},\pi_{\mathsf{E}},k-1} D^2_{\nabla^{\mathsf{M}},\nabla^{\pi_{\mathsf{E}}}} \nabla^{\mathsf{M},\pi_{\mathsf{E}},m-1}\xi). \tag{3.23}$$

We note that this tensor is symmetric in its first k entries, its last m entries, and in the "middle" two entries $(k, k+1)$. Since permutations of the forms

$$\begin{pmatrix} 1 & \cdots & k & k+1 & \cdots & k+m \\ \sigma_1(1) & \cdots & \sigma_1(k) & k+1 & \cdots & k+m \end{pmatrix}, \quad \begin{pmatrix} 1 & \cdots & k & k+1 & \cdots & k+m \\ 1 & \cdots & k & k+\sigma_2(1) & \cdots & k+\sigma_2(m) \end{pmatrix},$$
$$\begin{pmatrix} 1 & \cdots & k & k+1 & \cdots & k+m \\ 1 & \cdots & k+1 & k & \cdots & k+m \end{pmatrix}, \qquad \sigma_1 \in \mathfrak{S}_k,\ \sigma_2 \in \mathfrak{S}_m,$$

generate $\mathfrak{S}_{k+m}$, we conclude that the tensor (3.23) is symmetric. Moreover, we compute, for $v \in \mathsf{TM}$,

$$\begin{aligned}
&\mathrm{Sym}_k \otimes \mathrm{Sym}_m \otimes \mathrm{id}_{\mathsf{E}}(\nabla^{\mathsf{M},\pi_{\mathsf{E}},k-1} D^2_{\nabla^{\mathsf{M}},\nabla^{\pi_{\mathsf{E}}}} \nabla^{\mathsf{M},\pi_{\mathsf{E}},m-1}\xi)(v, \dots, v) \\
&\quad = \frac{1}{k!m!}\sum_{\sigma_1 \in \mathfrak{S}_k}\sum_{\sigma_2 \in \mathfrak{S}_m} \nabla^{\mathsf{M},\pi_{\mathsf{E}},k-1} D^2_{\nabla^{\mathsf{M}},\nabla^{\pi_{\mathsf{E}}}} \nabla^{\mathsf{M},\pi_{\mathsf{E}},m-1}\xi(v, \dots, v, v, \dots, v)
\end{aligned}$$

$$= \frac{1}{2k!m!} \sum_{\sigma_1 \in \mathfrak{S}_k} \sum_{\sigma_2 \in \mathfrak{S}_m} \nabla^{\mathsf{M},\pi_{\mathsf{E}},k-1}(\nabla^{\mathsf{M},\pi_{\mathsf{E}},2} + \mathrm{push}_{k,k+1} \circ \nabla^{\mathsf{M},\pi_{\mathsf{E}},2})$$
$$\nabla^{\mathsf{M},\pi_{\mathsf{E}},m-1}\xi(v, \dots, v, v, \dots, v)$$
$$= \frac{1}{k!m!} \sum_{\sigma_1 \in \mathfrak{S}_k} \sum_{\sigma_2 \in \mathfrak{S}_m} \nabla^{\mathsf{M},\pi_{\mathsf{E}},k+m}\xi(v, \dots, v)$$
$$= \nabla^{\mathsf{M},\pi_{\mathsf{E}},k+m}(v, \dots, v) = \mathrm{Sym}_{k+m}(\nabla^{\mathsf{M},\pi_{\mathsf{E}},k+m}\xi)(v, \dots, v).$$

Since the tensors (3.23) and $D^{k+m}_{\nabla^{\mathsf{M}},\nabla^{\pi_{\mathsf{E}}}}(\xi)$ are both symmetric, this is sufficient to conclude that

$$\mathrm{Sym}_k \otimes \mathrm{Sym}_m \otimes \mathrm{id}_{\mathsf{E}}(\nabla^{\mathsf{M},\pi_{\mathsf{E}},k-1} D^2_{\nabla^{\mathsf{M}},\nabla^{\pi_{\mathsf{E}}}} \nabla^{\mathsf{M},\pi_{\mathsf{E}},m-1}\xi) = D^{k+m}_{\nabla^{\mathsf{M}},\nabla^{\pi_{\mathsf{E}}}}(\xi)$$

cf. [10, Proposition IV.5.4.3].

Next we consider the tensor

$$\mathrm{Sym}_k \otimes \mathrm{Sym}_m \otimes \mathrm{id}_{\mathsf{E}}(\nabla^{\mathsf{M},\pi_{\mathsf{E}},k-1} R_{\nabla^{\pi_{\mathsf{E}}}}(\nabla^{\mathsf{M},\pi_{\mathsf{E}},m-1}\xi) - \nabla^{\mathsf{M},k-1} T_{\nabla^{\mathsf{M}}}(\nabla^{\mathsf{M},\pi_{\mathsf{E}},m}\xi)). \tag{3.24}$$

Here we note that

$$\begin{aligned}
&\mathrm{Sym}_k \otimes \mathrm{Sym}_m \otimes \mathrm{id}_{\mathsf{E}} \\
&\quad \left(\nabla^{\mathsf{M},\pi_{\mathsf{E}},k-1} R_{\nabla^{\pi_{\mathsf{E}}}}(\nabla^{\mathsf{M},\pi_{\mathsf{E}},m-1}\xi) - \nabla^{\mathsf{M},k-1} T_{\nabla^{\mathsf{M}}}(\nabla^{\mathsf{M},\pi_{\mathsf{E}},m}\xi)\right)(v_1, \dots, v_{k+m}) \\
&= \frac{1}{k!m!} \sum_{\sigma_1 \in \mathfrak{S}_{k-1,1}} \sum_{\sigma'_1 \in \mathfrak{S}_{k-1|1}} \sum_{\sigma_2 \in \mathfrak{S}_{1,m-1}} \sum_{\sigma'_2 \in \mathfrak{S}_{1|m-1}} \\
&\quad \left(\nabla^{\mathsf{M},\pi_{\mathsf{E}},k-1} R_{\nabla^{\pi_{\mathsf{E}}}}(v_{\sigma_1 \circ \sigma'_1(k)}, v_{k+\sigma_2 \circ \sigma'_2(1)})(v_{\sigma_1 \circ \sigma'_1(1)}, \dots, v_{\sigma_1 \circ \sigma'_1(k-1)})\right. \\
&\quad \left.(\nabla^{\mathsf{M},\pi_{\mathsf{E}},m-1}\xi(v_{k+\sigma_2 \circ \sigma'_2(2)}, \dots, v_{k+\sigma_2 \circ \sigma'_2(m)}))\right) \\
&\quad - \frac{1}{k!m!} \sum_{\sigma_1 \in \mathfrak{S}_{k-1,1}} \sum_{\sigma'_1 \in \mathfrak{S}_{k-1|1}} \sum_{\sigma_2 \in \mathfrak{S}_{1,m-1}} \sum_{\sigma'_2 \in \mathfrak{S}_{1|m-1}} \\
&\quad \left(\nabla^{\mathsf{M},k-1} T_{\nabla^{\mathsf{M}}}(v_{\sigma_1 \circ \sigma'_1(k)}, v_{k+\sigma_2 \circ \sigma'_2(1)})(v_{\sigma_1 \circ \sigma'_1(1)}, \dots, v_{\sigma_1 \circ \sigma'_1(k-1)})\right. \\
&\quad \left.(\nabla^{\mathsf{M},\pi_{\mathsf{E}},m-1}\nabla^{\pi_{\mathsf{E}}}\xi)(v_{k+\sigma_2 \circ \sigma'_2(2)}, \dots, v_{k+\sigma_2 \circ \sigma'_2(m)})\right)
\end{aligned}$$

$$= \frac{1}{km} \sum_{j=1}^{k} \sum_{l=1}^{m} \Big(D^{k-1}_{\nabla^{\mathsf{M}},\nabla^{\pi\mathsf{E}}} R_{\nabla^{\pi\mathsf{E}}}(v_j, v_{k+l})(v_1, \dots, \widehat{v}_j, \dots, v_{k-1})$$

$$(D^{m-1}_{\nabla^{\mathsf{M}},\nabla^{\pi\mathsf{E}}}(\xi))(v_{k+1}, \dots, \widehat{v}_{k+l}, \dots, v_m)$$

$$+ D^{k-1}_{\nabla^{\mathsf{M}},\nabla^{\pi\mathsf{E}}} T_{\nabla^{\mathsf{M}}}(v_j, v_{k+l})(v_1, \dots, \widehat{v}_j, \dots, v_{k-1})$$

$$(D^{m-1}_{\nabla^{\mathsf{M}},\nabla^{\pi\mathsf{E}}}(\nabla^{\pi\mathsf{E}}\xi))(v_{k+1}, \dots, \widehat{v}_{k+l}, \dots, v_m)\Big),$$

where the "hat" over an element of a list means that element is omitted. Combining the preceding two computations, and applying Lemma 3.20 to the second of these gives

$$D^{k,m}_{\nabla^{\mathsf{M}},\nabla^{\pi\mathsf{E}}}(D^{m}_{\nabla^{\mathsf{M}},\nabla^{\pi\mathsf{E}}}(\xi)) = D^{k+m}_{\nabla^{\mathsf{M}},\nabla^{\pi\mathsf{E}}}(\xi) + \theta_{k,m} \otimes \mathrm{id}_{\mathsf{E}} \Big(D^{k-1}_{\nabla^{\mathsf{M}},\nabla^{\pi\mathsf{E}}} R_{\nabla^{\pi\mathsf{E}}}(D^{m-1}_{\nabla^{\mathsf{M}},\nabla^{\pi\mathsf{E}}}(\xi))$$
$$- D^{k-1}_{\nabla^{\mathsf{M}},\nabla^{\pi\mathsf{E}}} T_{\nabla^{\mathsf{M}}}(D^{m-1}_{\nabla^{\mathsf{M}},\nabla^{\pi\mathsf{E}}}(\nabla^{\pi\mathsf{E}}\xi))\Big). \tag{3.25}$$

Now, for $m \geq 2$, we consider the expression $D^{m-1}_{\nabla^{\mathsf{M}},\nabla^{\pi\mathsf{E}}}(\nabla^{\pi\mathsf{E}}\xi)$ that arises in the last term in the preceding equation. For this, the computations from the preceding two paragraphs simplify to

$$D^{m-1,1}_{\nabla^{\mathsf{M}},\nabla^{\pi\mathsf{E}}}(D^{1}_{\nabla^{\mathsf{M}},\nabla^{\pi\mathsf{E}}}(\xi))$$
$$= D^{m}_{\nabla^{\mathsf{M}},\nabla^{\pi\mathsf{E}}}(\xi) + \theta_{m-1,1} \otimes \mathrm{id}_{\mathsf{E}} \Big(D^{m-2}_{\nabla^{\mathsf{M}},\nabla^{\pi\mathsf{E}}} R_{\nabla^{\pi\mathsf{E}}}(\xi) - D^{m-2}_{\nabla^{\mathsf{M}},\nabla^{\pi\mathsf{E}}} T_{\nabla^{\mathsf{M}}}(\nabla^{\mathsf{M},\pi\mathsf{E}}\xi)\Big).$$

This, incidentally, gives the formula in the first part of the lemma. Additionally, a substitution of this into (3.25) gives the second part of the lemma. □

We can see from the preceding lemma that, if one works in local coordinates with the canonical flat connections in the resulting trivial bundles, then the lemma simply tells us that "the kth-derivative of the mth-derivative is the $(k+m)$th-derivative." This agrees with calculus, fortunately.

The preceding lemma simplifies tremendously when the affine connection ∇^{M} is torsion-free, and results in an elegant formula simply involving curvature. However, we shall retain torsion since we do not, at this point in our presentation, understand the rôle of differing connections in our characterisations of the seminorms for the real analytic topology. Based on the lemma, we define a vector bundle mapping

$$\hat{\iota}_{\mathsf{E},k,m} \colon \bigoplus_{r=0}^{k+m} \mathsf{S}^r(\mathsf{T}^*\mathsf{M}) \otimes \mathsf{E} \to \bigoplus_{j=0}^{k} \mathsf{S}^j(\mathsf{T}^*\mathsf{M}) \otimes \left(\bigoplus_{l=0}^{m} \mathsf{S}^l(\mathsf{T}^*\mathsf{M}) \otimes \mathsf{E} \right)$$

by

$$
\begin{aligned}
\hat{\iota}_{\mathsf{E},k,m}(A_0, A_1, \dots, A_{k+m}) = ([A_0, A_1, \dots, A_m],&\\
[A_1, A_2 + \theta_{1,1} \otimes \mathrm{id}_{\mathsf{E}}(R_{\nabla^{\pi_{\mathsf{E}}}}(A_0) - T_{\nabla^{\mathsf{M}}}(A_1)), \dots,&\\
A_{m+1} + \theta_{1,m} \otimes \mathrm{id}_{\mathsf{E}}(R_{\nabla^{\pi_{\mathsf{E}}}}(A_{m-1}) -&\\
T_{\nabla^{\mathsf{M}}}(A_m + \theta_{m-1,1} \otimes \mathrm{id}_{\mathsf{E}}(D^{m-2}_{\nabla^{\mathsf{M}},\nabla^{\pi_{\mathsf{E}}}} R_{\nabla^{\pi_{\mathsf{E}}}}(A_0) - D^{m-2}_{\nabla^{\mathsf{M}},\nabla^{\pi_{\mathsf{E}}}} T_{\nabla^{\mathsf{M}}}(A_1))))], \dots,&\\
[A_{k+1}, A_{k+2} + \theta_{k,1} \otimes \mathrm{id}_{\mathsf{E}}(D^{k-1}_{\nabla^{\mathsf{M}},\nabla^{\pi_{\mathsf{E}}}} R_{\nabla^{\pi_{\mathsf{E}}}}(A_0) - D^{k-1}_{\nabla^{\mathsf{M}}} T_{\nabla^{\mathsf{M}}}(A_1)), \dots,&\\
A_{k+m} + \theta_{k,m} \otimes \mathrm{id}_{\mathsf{E}}(D^{k-1}_{\nabla^{\mathsf{M}},\nabla^{\pi_{\mathsf{E}}}} R_{\nabla^{\pi_{\mathsf{E}}}}(A_{m-1}) -&\\
D^{k-1}_{\nabla^{\mathsf{M}},\nabla^{\pi_{\mathsf{E}}}} T_{\nabla^{\mathsf{M}}}(A_m + \theta_{m-1,1} \otimes \mathrm{id}_{\mathsf{E}}(D^{m-2}_{\nabla^{\mathsf{M}},\nabla^{\pi_{\mathsf{E}}}} R_{\nabla^{\pi_{\mathsf{E}}}}(A_0) - D^{m-2}_{\nabla^{\mathsf{M}}} T_{\nabla^{\mathsf{M}}}(A_1)))))]).&
\end{aligned}
$$

(Here we break a rule we have about parentheses, using square brackets for some of the groupings for the sake of clarity.) To better understand the meaning of this complicated (though elementary) formula, let us observe that it has been designed so that

$$
\begin{aligned}
\hat{\iota}_{\mathsf{E},k,m}(\xi(x), D^1_{\nabla^{\mathsf{M}},\nabla^{\pi_{\mathsf{E}}}}(\xi)(x), \dots, D^{k+m}_{\nabla^{\mathsf{M}},\nabla^{\pi_{\mathsf{E}}}}(\xi)(x))&\\
= ([\xi(x), D^1_{\nabla^{\mathsf{M}},\nabla^{\pi_{\mathsf{E}}}}(\xi)(x), \dots, D^m_{\nabla^{\mathsf{M}},\nabla^{\pi_{\mathsf{E}}}}(\xi)(x)],&\\
[D^{1,0}_{\nabla^{\mathsf{M}},\nabla^{\pi_{\mathsf{E}}}}(\xi)(x), D^{1,1}_{\nabla^{\mathsf{M}},\nabla^{\pi_{\mathsf{E}}}}(D^1_{\nabla^{\mathsf{M}},\nabla^{\pi_{\mathsf{E}}}}(\xi))(x), \dots, D^{1,m}_{\nabla^{\mathsf{M}},\nabla^{\pi_{\mathsf{E}}}}(D^m_{\nabla^{\mathsf{M}},\nabla^{\pi_{\mathsf{E}}}}(\xi))(x)], \dots,&\\
[D^{k,0}_{\nabla^{\mathsf{M}},\nabla^{\pi_{\mathsf{E}}}}(\xi)(x), D^{k,1}_{\nabla^{\mathsf{M}},\nabla^{\pi_{\mathsf{E}}}}(D^1_{\nabla^{\mathsf{M}},\nabla^{\pi_{\mathsf{E}}}}(\xi))(x), \dots, D^{k,m}_{\nabla^{\mathsf{M}},\nabla^{\pi_{\mathsf{E}}}}(D^m_{\nabla^{\mathsf{M}},\nabla^{\pi_{\mathsf{E}}}}(\xi))(x)]),&
\end{aligned}
\tag{3.26}
$$

for $\xi \in \Gamma^r(\mathsf{E})$.

We now have the following result.

Lemma 3.22 (Decomposition of Prolongation of Jet Bundles) *Let $r \in \{\infty, \omega\}$, let $\pi_{\mathsf{E}} \colon \mathsf{E} \to \mathsf{M}$ be a C^r-vector bundle, let ∇^{M} be a C^r-affine connection on M, and let $\nabla^{\pi_{\mathsf{E}}}$ be a C^r-vector bundle connection on E. Then, for $k, m \in \mathbb{Z}_{\geq 0}$, the diagram*

$$
\begin{array}{ccc}
\mathsf{J}^{k+m}\mathsf{E} & \xrightarrow{\iota_{\mathsf{E},k,m}} & \mathsf{J}^k\mathsf{J}^m\mathsf{E} \\
\Big\downarrow S^{k+m}_{\nabla^{\mathsf{M}},\nabla^{\pi_{\mathsf{E}}}} & & \Big\downarrow S^{k,m}_{\nabla^{\mathsf{M}},\nabla^{\pi_{\mathsf{E}}}} \\
\bigoplus_{r=0}^{k+m} \mathsf{S}^r(\mathsf{T}^*\mathsf{M}) \otimes \mathsf{E} & \xrightarrow[\hat{\iota}_{\mathsf{E},k,m}]{} & \bigoplus_{j=0}^{k} \mathsf{S}^j(\mathsf{T}^*\mathsf{M}) \otimes \left(\bigoplus_{l=0}^{m} \mathsf{S}^l(\mathsf{T}^*\mathsf{M}) \otimes \mathsf{E}\right)
\end{array}
$$

commutes.

Proof This follows from Lemmata 3.17 and 3.21, taking note of the connection between these lemmata via (3.26). □

3.3 Isomorphisms Defined by Lifts and Pull-Backs

Let $r \in \{\infty, \omega\}$ and let $\pi_{\mathsf{E}}: \mathsf{E} \to \mathsf{M}$ be a C^r-vector bundle. In this section we carefully study isomorphisms that arise from lifts of objects on M to objects on E, of the sorts introduced in Sects. 3.1.1, 3.1.2, and 3.1.3. In particular, we shall see that jets of geometric objects can be decomposed (as in Sect. 2.3.1) before or after lifting. We wish here to relate these two sorts of decompositions for all of the lifts we consider in the book. This makes use of our constructions of Sect. 3.2 to give explicit decompositions for jets of certain sections of certain jet bundles on the total space of a vector bundle. Indeed, it is the results in the current section that provide the motivation for the rather intricate constructions of Sect. 3.2. For these constructions, we additionally suppose that we have a C^r-Riemannian metric $\mathbb{G}_{\mathsf{M}}$ on M and a C^r-fibre metric $\mathbb{G}_{\pi_{\mathsf{E}}}$ on E. We suppose that ∇^{M} is the Levi-Civita connection for $\mathbb{G}_{\mathsf{M}}$ and that we have a C^r-linear connection $\nabla^{\pi_{\mathsf{E}}}$ in E. This data gives rise to a Riemannian metric $\mathbb{G}_{\mathsf{E}}$ on E with its Levi-Civita connection ∇^{E}. We break the discussion into nine cases, the first seven of which correspond to the seven parts of Lemma 3.16. The eighth section provides a construction for pull-backs of functions and the ninth provides constructions involving two different affine connections and two different vector bundle connections. The constructions, statements, and proofs are somewhat repetitive, so we do not provide explicit proofs that are essentially identical to previous proofs. While the results are similar, they are not the same, so we elect to go through all of the cases. There is probably a "meta" result here, but it would take a small journey in itself to setup the framework for this. For our purposes, we stick to a treatment that is concrete at the cost of being dull.

3.3.1 Isomorphisms for Horizontal Lifts of Functions

Here we consider the horizontal lift mapping

$$\mathrm{C}^r(\mathsf{M}) \ni f \mapsto \pi_{\mathsf{E}}^* f \in \mathrm{C}^r(\mathsf{E}).$$

We wish to relate the decomposition associated with the jets of f to those associated with the jets of $\pi_{\mathsf{E}}^* f$. Associated with this, let us denote by $\mathsf{P}^{*m}\mathsf{E}$ the subbundle of $\mathbb{R}_{\mathsf{E}} \oplus \mathsf{T}^{*m}\mathsf{E}$ defined by

$$\mathsf{P}^{*m}_e\mathsf{E} = \{j_m(\pi_{\mathsf{E}}^* f)(e) \mid f \in \mathrm{C}^m(\mathsf{M})\}.$$

Following Lemma 2.15, our constructions have to do with iterated covariant differentials. The basis of all of our formulae will be a formula for iterated covariant differentials of horizontal lifts of functions on M. Thus we let $f \in \mathrm{C}^\infty(\mathsf{M})$ and consider

$$\nabla^{\mathsf{E},m}\pi_{\mathsf{E}}^* f \triangleq \underbrace{\nabla^{\mathsf{E}}\cdots\nabla^{\mathsf{E}}}_{m \text{ times}}\pi_{\mathsf{E}}^* f, \qquad m \in \mathbb{Z}_{>0}.$$

We state the first two lemmata that we will use. We recall from Lemma 3.16 the definition of $B_{\pi_{\mathsf{E}}}$.

Lemma 3.23 (Iterated Covariant Differentials of Horizontal Lifts of Functions I) *Let $r \in \{\infty, \omega\}$ and let $\pi_{\mathsf{E}}\colon \mathsf{E} \to \mathsf{M}$ be a C^r-vector bundle, with the data prescribed in Sect. 3.2.1 to define the Riemannian metric $\mathbb{G}_{\mathsf{E}}$ on E. For $m \in \mathbb{Z}_{\geq 0}$, there exist C^r-vector bundle mappings*

$$(A_s^m, \mathrm{id}_{\mathsf{E}}) \in \mathrm{VB}^r(\mathrm{T}^s(\pi_{\mathsf{E}}^*\mathsf{T}^*\mathsf{M}); \mathrm{T}^m(\mathsf{T}^*\mathsf{E})), \qquad s \in \{0, 1, \dots, m\},$$

such that

$$\nabla^{\mathsf{E},m}\pi_{\mathsf{E}}^* f = \sum_{s=0}^{m} A_s^m(\pi_{\mathsf{E}}^*\nabla^{\mathsf{M},s} f)$$

for all $f \in \mathrm{C}^m(\mathsf{M})$. Moreover, the vector bundle mappings $A_0^m, A_1^m, \dots, A_m^m$ satisfy the recursion relations prescribed by

$$A_0^0(\beta_0) = \beta_0, \quad A_1^1(\beta_1) = \beta_1, \quad A_0^1 = 0,$$

and

$$\begin{aligned} A_{m+1}^{m+1}(\beta_{m+1}) &= \beta_{m+1}, \\ A_s^{m+1}(\beta_s) &= (\nabla^{\mathsf{E}} A_s^m)(\beta_s) + A_{s-1}^m \otimes \mathrm{id}_{\mathsf{T}^*\mathsf{E}}(\beta_s) \\ &\quad - \sum_{j=1}^{s} A_s^m \otimes \mathrm{id}_{\mathsf{T}^*\mathsf{E}}(\mathrm{Ins}_j(\beta_s, B_{\pi_{\mathsf{E}}})), \\ &\quad s \in \{1, \dots, m\}, \\ A_0^{m+1}(\beta_0) &= (\nabla^{\mathsf{E}} A_0^m)(\beta_0), \end{aligned}$$

where $\beta_s \in \mathrm{T}^s(\pi_{\mathsf{E}}^*\mathsf{T}^*\mathsf{M})$, $s \in \{0, 1, \dots, m\}$.

Proof The assertion clearly holds for the initial conditions of the recursion, simply because

$$\pi^* f = \pi^* f, \quad \mathrm{d}(\pi^* f) = \pi^* \mathrm{d} f + 0 f.$$

So suppose it true for $m \in \mathbb{Z}_{>0}$. Thus

$$\nabla^{\mathsf{E},m} \pi_{\mathsf{E}}^* f = \sum_{s=0}^{m} A_s^m (\pi_{\mathsf{E}}^* \nabla^{\mathsf{M},s} f),$$

where the vector bundle mappings A_s^a, $a \in \{0, 1, \dots, m\}$, $s \in \{0, 1, \dots, a\}$, satisfy the recursion relations from the statement of the lemma. Then

$$\begin{aligned}
\nabla^{\mathsf{E},m+1} \pi_{\mathsf{E}}^* f &= \sum_{s=0}^{m} (\nabla^{\mathsf{E}} A_s^m)(\pi_{\mathsf{E}}^* \nabla^{\mathsf{M},s} f) + \sum_{s=0}^{m} A_s^m \otimes \mathrm{id}_{\mathsf{T}^*\mathsf{E}} (\nabla^{\mathsf{E}} \pi_{\mathsf{E}}^* \nabla^{\mathsf{M},s} f) \\
&= \sum_{s=0}^{m} (\nabla^{\mathsf{E}} A_s^m)(\pi_{\mathsf{E}}^* \nabla^{\mathsf{M},s} f) + \sum_{s=0}^{m} A_s^m \otimes \mathrm{id}_{\mathsf{T}^*\mathsf{E}} (\pi_{\mathsf{E}}^* \nabla^{\mathsf{M},s+1} f) \\
&\quad - \sum_{s=0}^{m} \sum_{j=1}^{s} A_s^m \otimes \mathrm{id}_{\mathsf{T}^*\mathsf{E}} (\mathrm{Ins}_j (\pi_{\mathsf{E}}^* \nabla^{\mathsf{M},s} f, B_{\pi_{\mathsf{E}}})) \\
&= \pi_{\mathsf{E}}^* \nabla^{\mathsf{M},m+1} f \\
&\quad + \sum_{s=1}^{m} \Bigg((\nabla^{\mathsf{E}} A_s^m)(\pi_{\mathsf{E}}^* \nabla^{\mathsf{M},s} f) + A_{s-1}^m \otimes \mathrm{id}_{\mathsf{T}^*\mathsf{E}} (\pi_{\mathsf{E}}^* \nabla^{\mathsf{M},s} f) \\
&\quad - \sum_{j=1}^{s} A_s^m \otimes \mathrm{id}_{\mathsf{T}^*\mathsf{E}} (\mathrm{Ins}_j (\pi_{\mathsf{E}}^* \nabla^{\mathsf{M},s} f, B_{\pi_{\mathsf{E}}})) \Bigg) + (\nabla^{\mathsf{E}} A_0^m)(\pi_{\mathsf{E}}^* f)
\end{aligned}$$

by Lemma 3.16(i). From this, the lemma follows. □

We shall also need to "invert" the relationship of the preceding lemma.

Lemma 3.24 (Iterated Covariant Differentials of Horizontal Lifts of Functions II) *Let $r \in \{\infty, \omega\}$ and let $\pi_{\mathsf{E}} \colon \mathsf{E} \to \mathsf{M}$ be a C^r-vector bundle, with the data prescribed in Sect. 3.2.1 to define the Riemannian metric $\mathbb{G}_{\mathsf{E}}$ on E. For $m \in \mathbb{Z}_{\geq 0}$, there exist C^r-vector bundle mappings*

$$(B_s^m, \mathrm{id}_{\mathsf{E}}) \in \mathrm{VB}^r(\mathrm{T}^s(\mathsf{T}^*\mathsf{E}); \mathrm{T}^m(\pi_{\mathsf{E}}^* \mathsf{T}^*\mathsf{M})), \qquad s \in \{0, 1, \dots, m\},$$

such that

$$\pi_{\mathsf{E}}^* \nabla^{\mathsf{M},m} f = \sum_{s=0}^{m} B_s^m (\nabla^{\mathsf{E},s} \pi_{\mathsf{E}}^* f)$$

for all $f \in \mathrm{C}^m(\mathsf{M})$*. Moreover, the vector bundle mappings* $B_0^m, B_1^m, \dots, B_m^m$ *satisfy the recursion relations prescribed by*

$$B_0^0(\alpha_0) = \alpha_0, \quad B_1^1(\alpha_1) = \alpha_1, \quad B_0^1 = 0,$$

and

$$B_{m+1}^{m+1}(\alpha_{m+1}) = \alpha_{m+1},$$

$$B_s^{m+1}(\alpha_s) = (\nabla^{\mathsf{E}} B_s^m)(\alpha_s) + B_{s-1}^m \otimes \mathrm{id}_{\mathsf{T}^*\mathsf{E}}(\alpha_s) + \sum_{j=1}^{m} \mathrm{Ins}_j(B_s^m(\alpha_s), B_{\pi_{\mathsf{E}}}),$$

$$s \in \{1, \dots, m\},$$

$$B_0^{m+1}(\alpha_0) = (\nabla^{\mathsf{E}} B_0^m)(\alpha_0) + \sum_{j=1}^{m} \mathrm{Ins}_j(B_0^m(\alpha_0), B_{\pi_{\mathsf{E}}}),$$

where $\alpha_s \in \mathrm{T}^s(\mathsf{T}^*\mathsf{E})$*,* $s \in \{0, 1, \dots, m\}$*.*

Proof The assertion clearly holds for the initial conditions for the recursion since

$$\pi^* f = \pi^* f, \quad \pi^*(\mathrm{d}f) = \mathrm{d}(\pi^* f) + 0 f.$$

So suppose it true for $m \in \mathbb{Z}_{>0}$. Thus

$$\pi_{\mathsf{E}}^* \nabla^{\mathsf{M},m} f = \sum_{s=0}^{m} B_s^m (\nabla^{\mathsf{E},s} \pi_{\mathsf{E}}^* f), \tag{3.27}$$

where the vector bundle mappings B_s^a, $a \in \{0, 1, \dots, m\}$, $s \in \{0, 1, \dots, a\}$, satisfy the recursion relations from the statement of the lemma. Then, by Lemma 3.16(i), we can work on the left-hand side of (3.27) to give

$$\begin{aligned}
\nabla^{\mathsf{E}} \pi_{\mathsf{E}}^* \nabla^{\mathsf{M},m} f &= \pi_{\mathsf{E}}^* \nabla^{\mathsf{M},m+1} f - \sum_{j=1}^{m} \mathrm{Ins}_j(\pi_{\mathsf{E}}^* \nabla^{\mathsf{M},m} f, B_{\pi_{\mathsf{E}}}) \\
&= \pi_{\mathsf{E}}^* \nabla^{\mathsf{M},m+1} f - \sum_{s=0}^{m} \sum_{j=1}^{m} \mathrm{Ins}_j(B_s^m(\nabla^{\mathsf{E},s} \pi_{\mathsf{E}}^* f), B_{\pi_{\mathsf{E}}}).
\end{aligned}$$

Working on the right-hand side of (3.27) gives

$$\nabla^{\mathsf{E}}\pi_{\mathsf{E}}^{*}\nabla^{\mathsf{M},m}f = \sum_{s=0}^{m}\nabla^{\mathsf{E}}B_{s}^{m}(\nabla^{\mathsf{E},s}\pi_{\mathsf{E}}^{*}f) + \sum_{s=0}^{m}B_{s}^{m}\otimes \mathrm{id}_{\mathsf{T}^{*}\mathsf{E}}(\nabla^{\mathsf{E},s+1}\pi_{\mathsf{E}}^{*}f).$$

Combining the preceding two equations gives

$$\begin{aligned}
&\pi_{\mathsf{E}}^{*}\nabla^{\mathsf{M},m+1}f\\
&= \sum_{s=0}^{m}\nabla^{\mathsf{E}}B_{s}^{m}(\nabla^{\mathsf{E},s}\pi_{\mathsf{E}}^{*}f) + \sum_{s=0}^{m}B_{s}^{m}\otimes \mathrm{id}_{\mathsf{T}^{*}\mathsf{E}}(\nabla^{\mathsf{E},s+1}\pi_{\mathsf{E}}^{*}f)\\
&\quad + \sum_{s=0}^{m}\sum_{j=1}^{m}\mathrm{Ins}_{j}(B_{s}^{m}(\nabla^{\mathsf{E},s}\pi_{\mathsf{E}}^{*}f), B_{\pi_{\mathsf{E}}})\\
&= \nabla^{\mathsf{E},m+1}\pi_{\mathsf{E}}^{*}f + \sum_{s=1}^{m}\Bigg(\nabla^{\mathsf{E}}B_{s}^{m}(\nabla^{\mathsf{E},s}\pi_{\mathsf{E}}^{*}f) + B_{s-1}^{m}\otimes \mathrm{id}_{\mathsf{T}^{*}\mathsf{E}}(\nabla^{\mathsf{E},s}\pi_{\mathsf{E}}^{*}f)\\
&\quad + \sum_{j=1}^{m}\mathrm{Ins}_{j}(B_{s}^{m}(\nabla^{\mathsf{E},s}\pi_{\mathsf{E}}^{*}f), B_{\pi_{\mathsf{E}}})\Bigg) + \nabla^{\mathsf{E}}B_{0}^{m}(\pi_{\mathsf{E}}^{*}f) + \sum_{j=1}^{m}\mathrm{Ins}_{j}(B_{0}^{m}(\pi_{\mathsf{E}}^{*}f), B_{\pi_{\mathsf{E}}}),
\end{aligned}$$

and the lemma follows from this. □

Next we turn to symmetrised versions of the preceding lemmata. We show that the preceding two lemmata induce corresponding mappings between symmetric tensors.

Lemma 3.25 (Iterated Symmetrised Covariant Differentials of Horizontal Lifts of Functions I) *Let $r \in \{\infty, \omega\}$ and let $\pi_{\mathsf{E}}\colon \mathsf{E} \to \mathsf{M}$ be a C^{r}-vector bundle, with the data prescribed in Sect. 3.2.1 to define the Riemannian metric $\mathbb{G}_{\mathsf{E}}$ on E. For $m \in \mathbb{Z}_{\geq 0}$, there exist C^{r}-vector bundle mappings*

$$(\widehat{A}_{s}^{m}, \mathrm{id}_{\mathsf{E}}) \in \mathrm{VB}^{r}(\mathrm{S}^{s}(\pi_{\mathsf{E}}^{*}\mathsf{T}^{*}\mathsf{M}); \mathrm{S}^{m}(\mathsf{T}^{*}\mathsf{E})), \qquad s \in \{0, 1, \dots, m\},$$

such that

$$\mathrm{Sym}_{m}\circ\nabla^{\mathsf{E},m}\pi_{\mathsf{E}}^{*}f = \sum_{s=0}^{m}\widehat{A}_{s}^{m}(\mathrm{Sym}_{s}\circ\pi_{\mathsf{E}}^{*}\nabla^{\mathsf{M},s}f)$$

for all $f \in \mathrm{C}^{m}(\mathsf{M})$.

Proof We define $A^m : \mathsf{T}^{\leq m}(\pi_{\mathsf{E}}^*\mathsf{T}^*\mathsf{M}) \to \mathsf{T}^{\leq m}(\mathsf{T}^*\mathsf{E})$ by

$$\begin{aligned} &A^m(\pi_{\mathsf{E}}^* f, \pi_{\mathsf{E}}^* \nabla^{\mathsf{M}} f, \dots, \pi_{\mathsf{E}}^* \nabla^{\mathsf{M},m} f) \\ &\quad = \left(A_0^0(\pi_{\mathsf{E}}^* f), \sum_{s=0}^{1} A_s^1(\pi_{\mathsf{E}}^* \nabla^{\mathsf{M},s} f), \dots, \sum_{s=0}^{m} A_s^m(\pi_{\mathsf{E}}^* \nabla^{\mathsf{M},s} f) \right). \end{aligned}$$

Let us organise the mappings we require into the following diagram:

$$\begin{array}{ccccc} \mathsf{T}^{\leq m}(\pi_{\mathsf{E}}^*\mathsf{T}^*\mathsf{M}) & \xrightarrow{\mathrm{Sym}_{\leq m}} & \mathsf{S}^{\leq m}(\pi_{\mathsf{E}}^*\mathsf{T}^*\mathsf{M}) & \xrightarrow{S^m_{\nabla\mathsf{M}}} & \pi_{\mathsf{E}}^*(\mathbb{R}_{\mathsf{M}} \oplus \mathsf{T}^{*m}\mathsf{M}) \\ {\scriptstyle A^m}\big\downarrow & & \big\downarrow{\scriptstyle \widehat{A}^m} & & \big\downarrow{\scriptstyle \mathrm{id}_{\mathbb{R}} \oplus j_m \pi_{\mathsf{E}}} \\ \mathsf{T}^{\leq m}(\mathsf{T}^*\mathsf{E}) & \xrightarrow{\mathrm{Sym}_{\leq m}} & \mathsf{S}^{\leq m}(\mathsf{T}^*\mathsf{E}) & \xrightarrow{S^m_{\nabla\mathsf{E}}} & \mathbb{R}_{\mathsf{E}} \oplus \mathsf{T}^{*m}\mathsf{E} \end{array} \tag{3.28}$$

Here $\widehat{A}^m$ is defined so that the right square commutes. We shall show that the left square also commutes. Indeed,

$$\begin{aligned} &\widehat{A}^m \circ \mathrm{Sym}_{\leq m}(\pi_{\mathsf{E}}^* f, \pi_{\mathsf{E}}^* \nabla^{\mathsf{M}} f, \dots, \pi_{\mathsf{E}}^* \nabla^{\mathsf{M},m} f) \\ &\quad = (S^m_{\nabla\mathsf{E}})^{-1} \circ (\mathrm{id}_{\mathbb{R}} \oplus j_m \pi_{\mathsf{E}}) \circ S^m_{\nabla\mathsf{M}} \circ \mathrm{Sym}_{\leq m}(\pi_{\mathsf{E}}^* f, \pi_{\mathsf{E}}^* \nabla^{\mathsf{M}} f, \dots, \pi_{\mathsf{E}}^* \nabla^{\mathsf{M},m} f) \\ &\quad = \mathrm{Sym}_{\leq m}(\pi_{\mathsf{E}}^* f, \nabla^{\mathsf{E}} \pi_{\mathsf{E}}^* f, \dots, \nabla^{\mathsf{E},m} \pi_{\mathsf{E}}^* f) \\ &\quad = \mathrm{Sym}_{\leq m} \circ A^m(\pi_{\mathsf{E}}^* f, \pi_{\mathsf{E}}^* \nabla^{\mathsf{M}} f, \dots, \pi_{\mathsf{E}}^* \nabla^{\mathsf{M},m} f). \end{aligned}$$

Thus the diagram (3.28) commutes. Now we have

$$\begin{aligned} &\widehat{A}^m \circ \mathrm{Sym}_{\leq m}(\pi_{\mathsf{E}}^* f, \pi_{\mathsf{E}}^* \nabla^{\mathsf{M}} f, \dots, \pi_{\mathsf{E}}^* \nabla^{\mathsf{M},m} f) = \\ &\left(\mathrm{Sym}_1 \circ A_0^0(\pi_{\mathsf{E}}^* f), \sum_{s=0}^{1} \mathrm{Sym}_2 \circ A_s^1(\pi_{\mathsf{E}}^* \nabla^{\mathsf{M},s} f), \dots, \sum_{s=0}^{m} \mathrm{Sym}_m \circ A_s^m(\pi_{\mathsf{E}}^* \nabla^{\mathsf{M},s} f) \right). \end{aligned}$$

Thus, if we define

$$\widehat{A}_s^m(\mathrm{Sym}_s \circ \pi_{\mathsf{E}}^* \nabla^{\mathsf{M},s} f) = \mathrm{Sym}_m \circ A_s^m(\pi_{\mathsf{E}}^* \nabla^{\mathsf{M},s} f), \tag{3.29}$$

then we have

$$\mathrm{Sym}_m \circ \nabla^{\mathsf{E},m} \pi_{\mathsf{E}}^* f = \sum_{s=0}^{m} \widehat{A}_s^m(\mathrm{Sym}_s \circ \pi_{\mathsf{E}}^* \nabla^{\mathsf{M},s} f),$$

as desired. □

Next we consider the "inverse" of the preceding lemma.

Lemma 3.26 (Iterated Symmetrised Covariant Differentials of Horizontal Lifts of Functions II) *Let $r \in \{\infty, \omega\}$ and let $\pi_{\mathsf{E}} \colon \mathsf{E} \to \mathsf{M}$ be a C^r-vector bundle, with the data prescribed in Sect. 3.2.1 to define the Riemannian metric $\mathbb{G}_{\mathsf{E}}$ on E. For $m \in \mathbb{Z}_{\geq 0}$, there exist C^r-vector bundle mappings*

$$(\widehat{B}^m_s, \mathrm{id}_{\mathsf{E}}) \in \mathrm{VB}^r(\mathrm{S}^s(\mathsf{T}^*\mathsf{E}); \mathrm{S}^m(\pi_{\mathsf{E}}^*\mathsf{T}^*\mathsf{M})), \qquad s \in \{0, 1, \dots, m\},$$

such that

$$\mathrm{Sym}_m \circ \pi_{\mathsf{E}}^* \nabla^{\mathsf{M},m} f = \sum_{s=0}^{m} \widehat{B}^m_s (\mathrm{Sym}_s \circ \nabla^{\mathsf{E},s} \pi_{\mathsf{E}}^* f)$$

for all $f \in \mathrm{C}^m(\mathsf{M})$.

Proof We define $B^m \colon \mathrm{T}^{\leq m}(\mathsf{T}^*\mathsf{E}) \to \mathrm{T}^{\leq m}(\pi_{\mathsf{E}}^*\mathsf{T}^*\mathsf{M})$ by requiring that

$$\begin{aligned} &B^m(\pi_{\mathsf{E}}^* f, \dots, \nabla^{\mathsf{E},m}\pi_{\mathsf{E}}^* f) \\ &= \left(B^0_0(\pi_{\mathsf{E}}^* f), \sum_{s=0}^{1} B^1_s(\nabla^{\mathsf{E},s}\pi_{\mathsf{E}}^* f), \dots, \sum_{s=0}^{m} B^m_s(\nabla^{\mathsf{E},m}\pi_{\mathsf{E}}^* f) \right), \end{aligned} \tag{3.30}$$

as in Lemma 3.24. Note that the mapping

$$\mathrm{id}_{\mathbb{R}} \oplus j_m\pi_{\mathsf{E}} \colon \pi_{\mathsf{E}}^*(\mathbb{R}_{\mathsf{M}} \oplus \mathsf{T}^{*m}\mathsf{M}) \to \mathsf{P}^{*m}\mathsf{E}$$

is well-defined and a vector bundle isomorphism. Let us organise the mappings we require into the following diagram:

$$\begin{array}{ccccc} \mathrm{T}^{\leq m}(\mathsf{T}^*\mathsf{E}) & \xrightarrow{\mathrm{Sym}_{\leq m}} & \mathrm{S}^{\leq m}(\mathsf{T}^*\mathsf{E}) & \xrightarrow{S^m_{\nabla\mathsf{E}}} & \mathsf{P}^{*m}\mathsf{E} \\ \Big\downarrow B^m & & \Big\downarrow \widehat{B}^m & & \Big\uparrow \mathrm{id}_{\mathbb{R}} \oplus j_m\pi_{\mathsf{E}} \\ \mathrm{T}^{\leq m}(\pi_{\mathsf{E}}^*\mathsf{T}^*\mathsf{M}) & \xrightarrow{\mathrm{Sym}_{\leq m}} & \mathrm{S}^{\leq m}(\pi_{\mathsf{E}}^*\mathsf{T}^*\mathsf{M}) & \xrightarrow{S^m_{\nabla\mathsf{M}}} & \pi_{\mathsf{E}}^*(\mathbb{R}_{\mathsf{M}} \oplus \mathsf{T}^{*m}\mathsf{M}) \end{array} \tag{3.31}$$

Here $\widehat{B}^m$ is defined so that the right square commutes. We shall show that the left square also commutes. Indeed,

$$\begin{aligned} &\widehat{B}^m \circ \mathrm{Sym}_{\leq m}(\pi_{\mathsf{E}}^* f, \nabla^{\mathsf{E}}\pi_{\mathsf{E}}^* f, \dots, \nabla^{\mathsf{E},m}\pi_{\mathsf{E}}^* f) \\ &= (S^m_{\nabla\mathsf{M}})^{-1} \circ (\mathrm{id}_{\mathbb{R}} \oplus j_m\pi_{\mathsf{E}})^{-1} \circ S^m_{\nabla\mathsf{E}} \circ \mathrm{Sym}_{\leq m}(\pi_{\mathsf{E}}^* f, \nabla^{\mathsf{E}}\pi_{\mathsf{E}}^* f, \dots, \nabla^{\mathsf{E},m}\pi_{\mathsf{E}}^* f) \\ &= \mathrm{Sym}_{\leq m}(\pi_{\mathsf{E}}^* f, \pi_{\mathsf{E}}^*\nabla^{\mathsf{M}} f, \dots, \pi_{\mathsf{E}}^*\nabla^{\mathsf{M},m} f) \\ &= \mathrm{Sym}_{\leq m} \circ B^m(\pi_{\mathsf{E}}^* f, \nabla^{\mathsf{E}}\pi_{\mathsf{E}}^* f, \dots, \nabla^{\mathsf{E},m}\pi_{\mathsf{E}}^* f). \end{aligned}$$

Thus the diagram (3.31) commutes. Thus, if we define $\widehat{B}^m_s$ so as to satisfy

$$\widehat{B}^m_s(\mathrm{Sym}_s \circ \nabla^{\mathsf{E},s}\pi_{\mathsf{E}}^* f) = \mathrm{Sym}_m \circ B^m_s(\nabla^{\mathsf{E},s}\pi_{\mathsf{E}}^* f),$$

then we have

$$\mathrm{Sym}_m \circ \pi_{\mathsf{E}}^* \nabla^{\mathsf{M},m} f = \sum_{s=0}^{m} \widehat{B}^m_s(\mathrm{Sym}_s \circ \nabla^{\mathsf{E},s}\pi_{\mathsf{E}}^* f),$$

as desired. □

Remark 3.27 (Nonuniqueness of Inverses) We are being a little sloppy in the preceding lemma, and will be similarly sloppy in subsequent related results. The sloppiness that arises is that the mapping B^m in the lemma is not uniquely defined by the condition (3.30). Indeed, B^m is only uniquely defined on image(A^m). This can be resolved by giving notation to the vector bundle image(A^m) and then defining B^m uniquely on this vector bundle. Alternatively, a vector bundle mapping on image(A^m) can be arbitrarily extended to $\mathsf{T}^{\leq m}(\mathsf{T}^*\mathsf{E})$ and one can work with this since the conditions defining it only depend on its values on image(A^m). Thus the sloppiness arises from an unwillingness to introduce even more notation than we already use. This is cleaned up in Lemma 3.28 below. ∘

The following lemma provides two decompositions of $\mathsf{P}^{*m}\mathsf{E}$, one "downstairs" and one "upstairs," and the relationship between them. The assertion simply results from an examination of the preceding four lemmata.

Lemma 3.28 (Decomposition of Jets of Horizontal Lifts of Functions) *Let $r \in \{\infty, \omega\}$ and let $\pi_{\mathsf{E}} \colon \mathsf{E} \to \mathsf{M}$ be a C^r-vector bundle, with the data prescribed in Sect. 3.2.1 to define the Riemannian metric $\mathbb{G}_{\mathsf{E}}$ on E. Then there exist C^r-vector bundle mappings*

$$A^m_{\nabla^{\mathsf{E}}} \in \mathrm{VB}^r(\mathsf{P}^{*m}\mathsf{E}; \mathrm{S}^{\leq m}(\pi_{\mathsf{E}}^*\mathsf{T}^*\mathsf{M})), \quad B^m_{\nabla^{\mathsf{E}}} \in \mathrm{VB}^r(\mathsf{P}^{*m}\mathsf{E}; \mathrm{S}^{\leq m}(\mathsf{T}^*\mathsf{E})),$$

defined by

$$A^m_{\nabla^{\mathsf{E}}}(j_m(\pi_{\mathsf{E}}^* f)(e)) = \mathrm{Sym}_{\leq m}(\pi_{\mathsf{E}}^* f(e), \pi_{\mathsf{E}}^*\nabla^{\mathsf{M}} f(e), \dots, \pi_{\mathsf{E}}^*\nabla^{\mathsf{M},m} f(e)),$$
$$B^m_{\nabla^{\mathsf{E}}}(j_m(\pi_{\mathsf{E}}^* f)(e)) = \mathrm{Sym}_{\leq m}(\pi_{\mathsf{E}}^* f(e), \nabla^{\mathsf{E}}\pi_{\mathsf{E}}^* f(e), \dots, \nabla^{\mathsf{E},m}\pi_{\mathsf{E}}^* f(e)).$$

Moreover, $A^m_{\nabla \mathsf{E}}$ is an isomorphism, $B^m_{\nabla \mathsf{E}}$ is injective, and

$$B^m_{\nabla \mathsf{E}} \circ (A^m_{\nabla \mathsf{E}})^{-1} \circ (\mathrm{Sym}_{\leq m}(\pi_{\mathsf{E}}^* f(e), \pi_{\mathsf{E}}^* \nabla^{\mathsf{M}} f(e), \dots, \pi_{\mathsf{E}}^* \nabla^{\mathsf{M},m} f(e))$$
$$= \left(A_0^0(\pi_{\mathsf{E}}^* f(e)), \sum_{s=0}^{1} \widehat{A}_s^1(\mathrm{Sym}_s \circ \pi_{\mathsf{E}}^* \nabla^{\mathsf{M},s} f(e)), \dots, \right.$$
$$\left. \sum_{s=0}^{m} \widehat{A}_s^m(\mathrm{Sym}_s \circ \pi_{\mathsf{E}}^* \nabla^{\mathsf{M},s} f(e)) \right)$$

and

$$A^m_{\nabla \mathsf{E}} \circ (B^m_{\nabla \mathsf{E}})^{-1} \circ \mathrm{Sym}_{\leq m}(\pi_{\mathsf{E}}^* f(e), \nabla^{\mathsf{E}} \pi_{\mathsf{E}}^* f(e), \dots, \nabla^{\mathsf{E},m} \pi_{\mathsf{E}}^* f(e))$$
$$= \left(B_0^0(\pi_{\mathsf{E}}^* f(e)), \sum_{s=0}^{1} \widehat{B}_s^1(\mathrm{Sym}_s \circ \nabla^{\mathsf{E},s} \pi_{\mathsf{E}}^* f(e)), \dots, \right.$$
$$\left. \sum_{s=0}^{m} \widehat{B}_s^m(\mathrm{Sym}_s \circ \nabla^{\mathsf{E},s} \pi_{\mathsf{E}}^* f(e)) \right),$$

where the vector bundle mappings $\widehat{A}_s^m$ and $\widehat{B}_s^m$, $s \in \{0, 1, \dots, m\}$, are as in Lemmata 3.25 and 3.26.

3.3.2 Isomorphisms for Vertical Lifts of Sections

Next we consider vertical lifts of sections, i.e., the mapping

$$\Gamma^r(\mathsf{E}) \ni \xi \mapsto \xi^{\mathrm{v}} \in \Gamma^r(\mathsf{TE}).$$

We wish to relate the decomposition of the jets of ξ with those of ξ^{v}. Associated with this, we denote

$$\mathsf{V}_e^{*m}\mathsf{E} = \{j_m \xi^{\mathrm{v}}(e) \mid \xi \in \Gamma^m(\mathsf{E})\}.$$

By (1.9), we have

$$\mathsf{V}_e^{*m}\mathsf{E} \simeq \mathsf{P}_e^{*m}\mathsf{E} \otimes \mathsf{V}_e\mathsf{E}.$$

As with the constructions of the preceding section, we wish to use Lemma 2.15 to provide a decomposition of $\mathsf{V}^{*m}\mathsf{E}$, and to do so we need to understand the covariant derivatives

$$\nabla^{\mathsf{E},m}\xi^{\mathrm{v}} \triangleq \underbrace{\nabla^{\mathsf{E}}\cdots\nabla^{\mathsf{E}}}_{m\ \text{times}}\xi^{\mathrm{v}}, \qquad m \in \mathbb{Z}_{\geq 0}.$$

In our development, we shall use the notation used in the preceding section in a slightly different, but similar, context. This seems reasonable since we have to do more or less the same thing six times, and using six different pieces of notation will be excessively burdensome.

The first result we give is the following.

Lemma 3.29 (Iterated Covariant Differentials of Vertical Lifts of Sections I) *Let $r \in \{\infty, \omega\}$ and let $\pi_{\mathsf{E}}\colon \mathsf{E} \to \mathsf{M}$ be a C^r-vector bundle, with the data prescribed in Sect. 3.2.1 to define the Riemannian metric $\mathbb{G}_{\mathsf{E}}$ on E. For $m \in \mathbb{Z}_{\geq 0}$, there exist C^r-vector bundle mappings*

$$(A^m_s, \mathrm{id}_{\mathsf{E}}) \in \mathrm{VB}^r(\mathrm{T}^s(\pi_{\mathsf{E}}^*\mathrm{T}^*\mathsf{M}) \otimes \mathsf{VE}; \mathrm{T}^m(\mathrm{T}^*\mathsf{E}) \otimes \mathsf{VE}), \qquad s \in \{0, 1, \ldots, m\},$$

such that

$$\nabla^{\mathsf{E},m}\xi^{\mathrm{v}} = \sum_{s=0}^{m} A^m_s((\nabla^{\mathsf{M},\pi_{\mathsf{E}},s}\xi)^{\mathrm{v}})$$

for all $\xi \in \Gamma^m(\mathsf{E})$. Moreover, the vector bundle mappings $A^m_0, A^m_1, \ldots, A^m_m$ satisfy the recursion relations prescribed by $A^0_0(\beta_0) = \beta_0$ and

$$\begin{aligned}
A^{m+1}_{m+1}(\beta_{m+1}) &= \beta_{m+1},\\
A^{m+1}_s(\beta_s) &= (\nabla^{\mathsf{E}} A^m_s)(\beta_s) + A^m_{s-1} \otimes \mathrm{id}_{\mathrm{T}^*\mathsf{E}}(\beta_s) - \sum_{j=1}^{s} A^m_s \otimes \mathrm{id}_{\mathrm{T}^*\mathsf{E}}(\mathrm{Ins}_j(\beta_s, B_{\pi_{\mathsf{E}}}))\\
&\quad + A^m_s \otimes \mathrm{id}_{\mathrm{T}^*\mathsf{E}}(\mathrm{Ins}_{s+1}(\beta_s, B^*_{\pi_{\mathsf{E}}})), \quad s \in \{1, \ldots, m\},\\
A^{m+1}_0(\beta_0) &= (\nabla^{\mathsf{E}} A^m_0)(\beta_0) + A^m_0 \otimes \mathrm{id}_{\mathrm{T}^*\mathsf{E}}(\mathrm{Ins}_1(\beta_0, B^*_{\pi_{\mathsf{E}}})),
\end{aligned}$$

where $\beta_s \in \mathrm{T}^s(\pi_{\mathsf{E}}^*\mathrm{T}^*\mathsf{M}) \otimes \mathsf{VE}$, $s \in \{0, 1, \ldots, m+1\}$.

Proof The assertion clearly holds for $m = 0$, so suppose it true for $m \in \mathbb{Z}_{>0}$. Thus

$$\nabla^{\mathsf{E},m}\xi^{\mathrm{v}} = \sum_{s=0}^{m} A^m_s((\nabla^{\mathsf{M},\pi_{\mathsf{E}},s}\xi)^{\mathrm{v}}),$$

where the vector bundle mappings A^a_s, $a \in \{0, 1, \dots, m\}$, $s \in \{0, 1, \dots, a\}$, satisfy the recursion relations from the statement of the lemma. Then

$$\begin{aligned}
\nabla^{\mathsf{E},m+1}\xi^{\mathrm{v}} &= \sum_{s=0}^{m}(\nabla^{\mathsf{E}}A^m_s)((\nabla^{\mathsf{M},\pi_{\mathsf{E}},s}\xi)^{\mathrm{v}}) + \sum_{s=0}^{m} A^m_s \otimes \mathrm{id}_{\mathsf{T}^*\mathsf{E}}(\nabla^{\mathsf{E}}(\nabla^{\mathsf{M},\pi_{\mathsf{E}},s}\xi)^{\mathrm{v}}) \\
&= \sum_{s=0}^{m}(\nabla^{\mathsf{E}}A^m_s)((\nabla^{\mathsf{M},\pi_{\mathsf{E}},s}\xi)^{\mathrm{v}}) + \sum_{s=0}^{m} A^m_s \otimes \mathrm{id}_{\mathsf{T}^*\mathsf{E}}((\nabla^{\mathsf{M},\pi_{\mathsf{E}},s+1}\xi)^{\mathrm{v}}) \\
&\quad - \sum_{s=1}^{m}\sum_{j=1}^{s} A^m_s \otimes \mathrm{id}_{\mathsf{T}^*\mathsf{E}}(\mathrm{Ins}_j((\nabla^{\mathsf{M},\pi_{\mathsf{E}},s}\xi)^{\mathrm{v}}, B_{\pi_{\mathsf{E}}})) \\
&\quad + \sum_{s=1}^{m} A^m_s \otimes \mathrm{id}_{\mathsf{T}^*\mathsf{E}}(\mathrm{Ins}_{s+1}((\nabla^{\mathsf{M},\pi_{\mathsf{E}},s}\xi)^{\mathrm{v}}, B^*_{\pi_{\mathsf{E}}})) \\
&\quad + A^m_0 \otimes \mathrm{id}_{\mathsf{T}^*\mathsf{E}}(\mathrm{Ins}_1(\xi^{\mathrm{v}}, B^*_{\pi_{\mathsf{E}}})) \\
&= (\nabla^{\mathsf{M},\pi_{\mathsf{E}},m+1}\xi)^{\mathrm{v}} \\
&\quad + \sum_{s=1}^{m}\Bigg((\nabla^{\mathsf{E}}A^m_s)((\nabla^{\mathsf{M},\pi_{\mathsf{E}},s}\xi)^{\mathrm{v}}) + A^m_{s-1} \otimes \mathrm{id}_{\mathsf{T}^*\mathsf{E}}((\nabla^{\mathsf{M},\pi_{\mathsf{E}},s}\xi)^{\mathrm{v}}) \\
&\quad - \sum_{j=1}^{s} A^m_s \otimes \mathrm{id}_{\mathsf{T}^*\mathsf{E}}(\mathrm{Ins}_j((\nabla^{\mathsf{M},\pi_{\mathsf{E}},s}\xi)^{\mathrm{v}}, B_{\pi_{\mathsf{E}}})) \\
&\quad + A^m_s \otimes \mathrm{id}_{\mathsf{T}^*\mathsf{E}}(\mathrm{Ins}_{s+1}((\nabla^{\mathsf{M},\pi_{\mathsf{E}},s}\xi)^{\mathrm{v}}, B^*_{\pi_{\mathsf{E}}}))\Bigg) \\
&\quad + (\nabla^{\mathsf{E}}A^m_0)(\xi^{\mathrm{v}}) + A^m_0 \otimes \mathrm{id}_{\mathsf{T}^*\mathsf{E}}(\mathrm{Ins}_1(\xi^{\mathrm{v}}, B^*_{\pi_{\mathsf{E}}}))
\end{aligned}$$

by Lemma 3.16(ii). From this, the lemma follows. □

Now we "invert" the constructions from the preceding lemma.

Lemma 3.30 (Iterated Covariant Differentials of Vertical Lifts of Sections II) *Let $r \in \{\infty, \omega\}$ and let $\pi_{\mathsf{E}}\colon \mathsf{E} \to \mathsf{M}$ be a C^r-vector bundle, with the data prescribed in Sect. 3.2.1 to define the Riemannian metric $\mathbb{G}_{\mathsf{E}}$ on E. For $m \in \mathbb{Z}_{\geq 0}$, there exist C^r-vector bundle mappings*

$$(B^m_s, \mathrm{id}_{\mathsf{E}}) \in \mathrm{VB}^r(\mathrm{T}^m(\mathrm{T}^*\mathsf{E}) \otimes \mathsf{VE}; \mathrm{T}^m(\pi^*_{\mathsf{E}}\mathrm{T}^*\mathsf{M}) \otimes \mathsf{VE}), \qquad s \in \{0, 1, \dots, m\},$$

such that

$$(\nabla^{\mathsf{M},\pi_{\mathsf{E}},m}\xi)^{\mathrm{v}} = \sum_{s=0}^{m} B^m_s(\nabla^{\mathsf{E},s}\xi^{\mathrm{v}})$$

for all $\xi \in \Gamma^m(\mathsf{E})$. *Moreover, the vector bundle mappings* $B_0^m, B_1^m, \dots, B_m^m$ *satisfy the recursion relations prescribed by* $B_0^0(\alpha_0) = \alpha_0$ *and*

$$\begin{aligned}
B_{m+1}^{m+1}(\alpha_{m+1}) &= \alpha_{m+1},\\
B_s^{m+1}(\alpha_s) &= (\nabla^{\mathsf{E}} B_s^m)(\alpha_s) + B_{s-1}^m \otimes \mathrm{id}_{\mathsf{T}^*\mathsf{E}}(\alpha_s) + \sum_{j=1}^m \mathrm{Ins}_j(B_s^m(\alpha_s), B_{\pi_{\mathsf{E}}})\\
&\quad - \mathrm{Ins}_{m+1}(B_s^m(\alpha_s), B_{\pi_{\mathsf{E}}}^*), \quad s \in \{1, \dots, m\},\\
B_0^{m+1}(\alpha_0) &= (\nabla^{\mathsf{E}} B_0^m)(\alpha_0) + \sum_{j=1}^m \mathrm{Ins}_j(B_0^m(\alpha_0), B_{\pi_{\mathsf{E}}}) - \mathrm{Ins}_{m+1}(B_0^m(\alpha_0), B_{\pi_{\mathsf{E}}}^*),
\end{aligned}$$

where $\alpha_s \in \mathrm{T}^s(\mathsf{T}^*\mathsf{E}) \otimes \mathsf{VE}$, $s \in \{0, 1, \dots, m+1\}$.

Proof The assertion clearly holds for $m = 0$, so suppose it true for $m \in \mathbb{Z}_{>0}$. Thus

$$(\nabla^{\mathsf{M},\pi_{\mathsf{E}},m}\xi)^{\mathrm{v}} = \sum_{s=0}^m B_s^m(\nabla^{\mathsf{E},s}\xi^{\mathrm{v}}), \tag{3.32}$$

where the vector bundle mappings B_s^a, $a \in \{0, 1, \dots, m\}$, $s \in \{0, 1, \dots, a\}$, satisfy the recursion relations from the statement of the lemma. Then, by Lemma 3.16(ii), we can work on the left-hand side of (3.32) to give

$$\begin{aligned}
\nabla^{\mathsf{E}}(\nabla^{\mathsf{M},\pi_{\mathsf{E}},m}\xi)^{\mathrm{v}} &= (\nabla^{\mathsf{M},\pi_{\mathsf{E}},m+1}\xi)^{\mathrm{v}} - \sum_{j=1}^m \mathrm{Ins}_j((\nabla^{\mathsf{M},\pi_{\mathsf{E}},m}\xi)^{\mathrm{v}}, B_{\pi_{\mathsf{E}}})\\
&\quad + \mathrm{Ins}_{m+1}((\nabla^{\mathsf{M},\pi_{\mathsf{E}},m}\xi)^{\mathrm{v}}, B_{\pi_{\mathsf{E}}}^*)\\
&= (\nabla^{\mathsf{M},\pi_{\mathsf{E}},m+1}\xi)^{\mathrm{v}} - \sum_{s=0}^m \sum_{j=1}^m \mathrm{Ins}_j(B_s^m(\nabla^{\mathsf{E},s}\xi^{\mathrm{v}}), B_{\pi_{\mathsf{E}}})\\
&\quad + \sum_{s=0}^m \mathrm{Ins}_{m+1}(B_s^m(\nabla^{\mathsf{E},s}\xi^{\mathrm{v}}), B_{\pi_{\mathsf{E}}}^*).
\end{aligned}$$

Working on the right-hand side of (3.32) gives

$$\nabla^{\mathsf{E}}(\nabla^{\mathsf{M},\pi_{\mathsf{E}},m}\xi)^{\mathrm{v}} = \sum_{s=0}^m \nabla^{\mathsf{E}} B_s^m(\nabla^{\mathsf{E},s}\xi^{\mathrm{v}}) + \sum_{s=0}^m B_s^m \otimes \mathrm{id}_{\mathsf{T}^*\mathsf{E}}(\nabla^{\mathsf{E},s+1}\xi^{\mathrm{v}}).$$

Combining the preceding two equations gives

$$\begin{aligned}
\nabla^{\mathsf{M},\pi_{\mathsf{E}},m+1}\xi^{\mathrm{v}} &= \sum_{s=0}^{m} \nabla^{\mathsf{E}} B_s^m(\nabla^{\mathsf{E},s}\xi^{\mathrm{v}}) + \sum_{s=0}^{m} B_s^m \otimes \mathrm{id}_{\mathsf{T}^*\mathsf{E}}(\nabla^{\mathsf{E},s+1}\xi^{\mathrm{v}}) \\
&\quad + \sum_{s=0}^{m}\sum_{j=1}^{m} \mathrm{Ins}_j(B_s^m(\nabla^{\mathsf{E},s}\xi^{\mathrm{v}}), B_{\pi_{\mathsf{E}}}) - \mathrm{Ins}_{m+1}((\nabla^{\mathsf{M},\pi_{\mathsf{E}},m}\xi)^{\mathrm{v}}, B_{\pi_{\mathsf{E}}}^*) \\
&= \nabla^{\mathsf{E},m+1}\xi^{\mathrm{v}} + \sum_{s=1}^{m} \Bigg(\nabla^{\mathsf{E}} B_s^m(\nabla^{\mathsf{E},s}\xi^{\mathrm{v}}) + B_{s-1}^m \otimes \mathrm{id}_{\mathsf{T}^*\mathsf{E}}(\nabla^{\mathsf{E},s}\xi^{\mathrm{v}}) \\
&\quad + \sum_{j=1}^{m} \mathrm{Ins}_j(B_s^m(\nabla^{\mathsf{E},s}\xi^{\mathrm{v}}), B_{\pi_{\mathsf{E}}}) - \mathrm{Ins}_{m+1}(B_s^m(\nabla^{\mathsf{E},s}\xi^{\mathrm{v}}), B_{\pi_{\mathsf{E}}}^*) \Bigg) \\
&\quad + \nabla^{\mathsf{E}} B_0^m(\xi^{\mathrm{v}}) + \sum_{j=1}^{m} \mathrm{Ins}_j(B_0^m(\xi^{\mathrm{v}}), B_{\pi_{\mathsf{E}}}) - \mathrm{Ins}_{m+1}(B_0^m(\xi^{\mathrm{v}}), B_{\pi_{\mathsf{E}}}^*),
\end{aligned}$$

and the lemma follows from this. □

Next we turn to symmetrised versions of the preceding lemmata. We show that the preceding two lemmata induce corresponding mappings between symmetric tensors.

Lemma 3.31 (Iterated Symmetrised Covariant Differentials of Vertical Lifts of Sections I) *Let $r \in \{\infty, \omega\}$ and let $\pi_{\mathsf{E}}\colon \mathsf{E} \to \mathsf{M}$ be a C^r-vector bundle, with the data prescribed in Sect. 3.2.1 to define the Riemannian metric $\mathbb{G}_{\mathsf{E}}$ on E. For $m \in \mathbb{Z}_{\geq 0}$, there exist C^r-vector bundle mappings*

$$(\widehat{A}_s^m, \mathrm{id}_{\mathsf{E}}) \in \mathrm{VB}^r(\mathrm{S}^s(\pi_{\mathsf{E}}^*\mathsf{T}^*\mathsf{M}) \otimes \mathsf{VE}; \mathrm{S}^m(\mathsf{T}^*\mathsf{E}) \otimes \mathsf{VE}), \qquad s \in \{0, 1, \dots, m\},$$

such that

$$(\mathrm{Sym}_m \otimes \mathrm{id}_{\mathsf{VE}}) \circ \nabla^{\mathsf{E},m}\xi^{\mathrm{v}} = \sum_{s=0}^{m} \widehat{A}_s^m((\mathrm{Sym}_s \otimes \mathrm{id}_{\mathsf{VE}}) \circ (\nabla^{\mathsf{M},\pi_{\mathsf{E}},s}\xi)^{\mathrm{v}})$$

for all $\xi \in \Gamma^m(\mathsf{E})$.

Proof The proof follows very similarly to that of Lemma 3.25, but taking the tensor product of everything with VE. We shall present the complete construction here, but will not repeat it for similar proofs that follow.

We define $A^m : \mathsf{T}^{\leq m}(\pi_{\mathsf{E}}^*\mathsf{T}^*\mathsf{M}) \otimes \mathsf{VE} \to \mathsf{T}^{\leq m}(\mathsf{T}^*\mathsf{E}) \otimes \mathsf{VE}$ by

$$A^m(\xi^{\mathrm{v}}, (\nabla^{\pi_{\mathsf{E}}}\xi)^{\mathrm{v}}, \dots, (\nabla^{\mathsf{M},\pi_{\mathsf{E}},m}\xi)^{\mathrm{v}}) \\ = \left(A_0^0(\xi^{\mathrm{v}}), \sum_{s=0}^{1} A_s^1((\nabla^{\mathsf{M},\pi_{\mathsf{E}},s}\xi)^{\mathrm{v}}), \dots, \sum_{s=0}^{m} A_s^m((\nabla^{\mathsf{M},\pi_{\mathsf{E}},s}\xi)^{\mathrm{v}})\right)$$

Let us organise the mappings we require into the following diagram:

$$\begin{array}{ccccc} \mathsf{T}^{\leq m}(\pi_{\mathsf{E}}^*\mathsf{T}^*\mathsf{M}) \otimes \mathsf{VE} & \xrightarrow{\mathrm{Sym}_{\leq m} \otimes \mathrm{id}_{\mathsf{VE}}} & \mathsf{S}^{\leq m}(\pi_{\mathsf{E}}^*\mathsf{T}^*\mathsf{M}) \otimes \mathsf{VE} & \xrightarrow{S^m_{\nabla^{\mathsf{M}},\nabla^{\pi_{\mathsf{E}}}} \otimes \mathrm{id}_{\mathsf{VE}}} & \pi_{\mathsf{E}}^*(\mathbb{R}_{\mathsf{M}} \oplus \mathsf{T}^{*m}\mathsf{M}) \otimes \mathsf{VE} \\ \Big\downarrow{\scriptstyle A^m} & & \Big\downarrow{\scriptstyle \widehat{A}^m} & & \Big\downarrow{\scriptstyle (\mathrm{id}_{\mathbb{R}} \oplus j_m\pi_{\mathsf{E}}) \otimes \mathrm{id}_{\mathsf{VE}}} \\ \mathsf{T}^{\leq m}(\mathsf{T}^*\mathsf{E}) \otimes \mathsf{VE} & \xrightarrow{\mathrm{Sym}_{\leq m} \otimes \mathrm{id}_{\mathsf{VE}}} & \mathsf{S}^{\leq m}(\mathsf{T}^*\mathsf{E}) \otimes \mathsf{VE} & \xrightarrow{S^m_{\nabla^{\mathsf{E}}} \otimes \mathrm{id}_{\mathsf{VE}}} & (\mathbb{R}_{\mathsf{M}} \oplus \mathsf{T}^{*m}\mathsf{E}) \otimes \mathsf{VE} \end{array} \tag{3.33}$$

Here $\widehat{A}^m$ is defined so that the right square commutes. We shall show that the left square also commutes. Indeed,

$$\begin{aligned} \widehat{A}^m \circ \mathrm{Sym}_{\leq m} \otimes \mathrm{id}_{\mathsf{VE}}&(\xi^{\mathrm{v}}, (\nabla^{\pi_{\mathsf{E}}}\xi)^{\mathrm{v}}, \dots, (\nabla^{\mathsf{M},\pi_{\mathsf{E}},m}\xi)^{\mathrm{v}}) \\ &= (S^m_{\nabla^{\mathsf{E}}} \otimes \mathrm{id}_{\mathsf{VE}})^{-1} \circ ((\mathrm{id}_{\mathbb{R}} \otimes j_m\pi_{\mathsf{E}}) \otimes \mathrm{id}_{\mathsf{VE}}) \circ (S^m_{\nabla^{\mathsf{M}},\nabla^{\pi_{\mathsf{E}}}} \otimes \mathrm{id}_{\mathsf{VE}}) \\ &\quad \circ (\mathrm{Sym}_{\leq m} \otimes \mathrm{id}_{\mathsf{VE}})(\xi^{\mathrm{v}}, (\nabla^{\pi_{\mathsf{E}}}\xi)^{\mathrm{v}}, \dots, (\nabla^{\mathsf{M},\pi_{\mathsf{E}},m}\xi)^{\mathrm{v}}) \\ &= \mathrm{Sym}_{\leq m} \otimes \mathrm{id}_{\mathsf{VE}}(\xi^{\mathrm{v}}, \nabla^{\mathsf{E}}\xi^{\mathrm{v}}, \dots, \nabla^{\mathsf{E},m}\xi^{\mathrm{v}}) \\ &= (\mathrm{Sym}_{\leq m} \otimes \mathrm{id}_{\mathsf{VE}}) \circ A^m(\xi^{\mathrm{v}}, (\nabla^{\pi_{\mathsf{E}}}\xi)^{\mathrm{v}}, \dots, (\nabla^{\mathsf{M},\pi_{\mathsf{E}},m}\xi)^{\mathrm{v}}). \end{aligned}$$

Thus the diagram (3.33) commutes. Thus, if we define

$$\widehat{A}_s^m((\mathrm{Sym}_s \otimes \mathrm{id}_{\mathsf{VE}}) \circ (\nabla^{\mathsf{M},\pi_{\mathsf{E}},s}\xi)^{\mathrm{v}}) = (\mathrm{Sym}_m \otimes \mathrm{id}_{\mathsf{VE}}) \circ A_s^m((\nabla^{\mathsf{M},\pi_{\mathsf{E}},s}\xi)^{\mathrm{v}}),$$

then we have

$$(\mathrm{Sym}_m \otimes \mathrm{id}_{\mathsf{VE}}) \circ \nabla^{\mathsf{E},m}\xi^{\mathrm{v}} = \sum_{s=0}^{m} \widehat{A}_s^m((\mathrm{Sym}_s \otimes \mathrm{id}_{\mathsf{VE}}) \circ (\nabla^{\mathsf{M},\pi_{\mathsf{E}},s}\xi)^{\mathrm{v}}),$$

as desired. □

The preceding lemma gives rise to an "inverse," which we state in the following lemma.

Lemma 3.32 (Iterated Symmetrised Covariant Differentials of Vertical Lifts of Sections II) *Let $r \in \{\infty, \omega\}$ and let $\pi_{\mathsf{E}} : \mathsf{E} \to \mathsf{M}$ be a C^r-vector bundle, with the data prescribed in Sect. 3.2.1 to define the Riemannian metric $\mathbb{G}_{\mathsf{E}}$ on E. For $m \in \mathbb{Z}_{\geq 0}$, there exist C^r-vector bundle mappings*

$$(\widehat{B}^m_s, \mathrm{id}_{\mathsf{E}}) \in \mathrm{VB}^r(\mathsf{S}^s(\mathsf{T}^*\mathsf{E}) \otimes \mathsf{VE}; \mathsf{S}^m(\pi_{\mathsf{E}}^*\mathsf{T}^*\mathsf{M}) \otimes \mathsf{VE}), \qquad s \in \{0, 1, \dots, m\},$$

such that

$$(\mathrm{Sym}_m \otimes \mathrm{id}_{\mathsf{VE}}) \circ (\nabla^{\mathsf{M},\pi_{\mathsf{E}},m}\xi)^{\mathrm{v}} = \sum_{s=0}^{m} \widehat{B}^m_s((\mathrm{Sym}_s \otimes \mathrm{id}_{\mathsf{VE}}) \circ \nabla^{\mathsf{E},s}\xi^{\mathrm{v}})$$

for all $\xi \in \Gamma^m(\mathsf{E})$.

Proof This follows along the lines of Lemma 3.26 in the same manner as Lemma 3.31 follows from Lemma 3.25, by taking tensor products with VE. □

We can put together the previous four lemmata into the following decomposition result, which is to be regarded as the main result of this section.

Lemma 3.33 (Decomposition of Jets of Vertical Lifts of Sections) *Let* $r \in \{\infty, \omega\}$ *and let* $\pi_{\mathsf{E}} : \mathsf{E} \to \mathsf{M}$ *be a* C^r*-vector bundle, with the data prescribed in Sect. 3.2.1 to define the Riemannian metric* $\mathbb{G}_{\mathsf{E}}$ *on* E*. Then there exist* C^r*-vector bundle mappings*

$$A^m_{\nabla\mathsf{E}} \in \mathrm{VB}^r(\mathsf{P}^{*m}\mathsf{E} \otimes \mathsf{VE}; \mathsf{S}^{\leq m}(\pi_{\mathsf{E}}^*\mathsf{T}^*\mathsf{M}) \otimes \mathsf{VE}),$$
$$B^m_{\nabla\mathsf{E}} \in \mathrm{VB}^r(\mathsf{P}^{*m}\mathsf{E} \otimes \mathsf{VE}; \mathsf{S}^{\leq m}(\mathsf{T}^*\mathsf{E}) \otimes \mathsf{VE}),$$

defined by

$$A^m_{\nabla\mathsf{E}}(j_m(\xi^{\mathrm{v}})(e)) = \mathrm{Sym}_{\leq m} \otimes \mathrm{id}_{\mathsf{VE}}(\xi^{\mathrm{v}}(e), (\nabla^{\pi_{\mathsf{E}}}\xi)^{\mathrm{v}}(e), \dots, (\nabla^{\mathsf{M},\pi_{\mathsf{E}},m}\xi)^{\mathrm{v}}(e)),$$
$$B^m_{\nabla\mathsf{E}}(j_m(\xi^{\mathrm{v}})(e)) = \mathrm{Sym}_{\leq m} \otimes \mathrm{id}_{\mathsf{VE}}(\xi^{\mathrm{v}}(e), \nabla^{\mathsf{E}}\xi^{\mathrm{v}}(e), \dots, \nabla^{\mathsf{E},m}\xi^{\mathrm{v}}(e)).$$

Moreover, $A^m_{\nabla\mathsf{E}}$ *is an isomorphism,* $B^m_{\nabla\mathsf{E}}$ *is injective, and*

$$B^m_{\nabla\mathsf{E}} \circ (A^m_{\nabla\mathsf{E}})^{-1} \circ (\mathrm{Sym}_{\leq m} \otimes \mathrm{id}_{\mathsf{VE}})(\xi^{\mathrm{v}}(e), (\nabla^{\pi_{\mathsf{E}}}\xi)^{\mathrm{v}}(e), \dots, (\nabla^{\mathsf{M},\pi_{\mathsf{E}},m}\xi)^{\mathrm{v}}(e))$$
$$= \Bigg(\xi^{\mathrm{v}}(e), \sum_{s=0}^{1} \widehat{A}^1_s((\mathrm{Sym}_s \otimes \mathrm{id}_{\mathsf{VE}}) \circ (\nabla^{\mathsf{M},\pi_{\mathsf{E}},s}\xi)^{\mathrm{v}}(e)), \dots,$$
$$\sum_{s=0}^{m} \widehat{A}^m_s((\mathrm{Sym}_s \otimes \mathrm{id}_{\mathsf{VE}}) \circ (\nabla^{\mathsf{M},\pi_{\mathsf{E}},s}\xi)^{\mathrm{v}}(e))\Bigg)$$

and

$$A^m_{\nabla\mathsf{E}} \circ (B^m_{\nabla\mathsf{E}})^{-1} \circ (\mathrm{Sym}_{\leq m} \otimes \mathrm{id}_{\mathsf{VE}})(\xi^{\mathrm{v}}(e), \nabla^{\mathsf{E}}\xi^{\mathrm{v}}(e), \dots, \nabla^{\mathsf{E},m}\xi^{\mathrm{v}}(e))$$
$$= \Bigg(\xi^{\mathrm{v}}(e), \sum_{s=0}^{1} \widehat{B}^1_s((\mathrm{Sym}_s \otimes \mathrm{id}_{\mathsf{VE}}) \circ \nabla^{\mathsf{E},s}\xi^{\mathrm{v}}(e)), \dots,$$
$$\sum_{s=0}^{m} \widehat{B}^m_s((\mathrm{Sym}_s \otimes \mathrm{id}_{\mathsf{VE}}) \circ \nabla^{\mathsf{E},s}\xi^{\mathrm{v}}(e))\Bigg),$$

where the vector bundle mappings $\widehat{A}^m_s$ and $\widehat{B}^m_s$, $s \in \{0, 1, \dots, m\}$, are as in Lemmata 3.31 and 3.32.

3.3.3 *Isomorphisms for Horizontal Lifts of Vector Fields*

Next we consider horizontal lifts of vector fields via the mapping

$$\Gamma^r(\mathsf{TM}) \ni X \mapsto X^{\mathrm{h}} \in \Gamma^r(\mathsf{TE}).$$

We wish to relate the decomposition of the jets of X with the jets of X^{h}. Associated with this, we denote

$$\mathsf{H}^{*m}_e\mathsf{E} = \{j_m X^{\mathrm{h}}(e) \mid X \in \Gamma^m(\mathsf{TM})\}.$$

By (1.9), we have

$$\mathsf{H}^{*m}_e\mathsf{E} \simeq \mathsf{P}^{*m}_e\mathsf{E} \otimes \mathsf{H}_e\mathsf{E}.$$

As with the constructions of the preceding sections, we wish to use Lemma 2.15 to provide a decomposition of $\mathsf{H}^{*m}\mathsf{E}$, and to do so we need to understand the covariant derivatives

$$\nabla^{\mathsf{E},m}X^{\mathrm{h}} \triangleq \underbrace{\nabla^{\mathsf{E}} \cdots \nabla^{\mathsf{E}}}_{m \text{ times}} X^{\mathrm{h}}, \qquad m \in \mathbb{Z}_{\geq 0}.$$

In this section we omit proofs, since proofs follow along entirely similar lines to those of the preceding section.

The first result we give is the following.

Lemma 3.34 (Iterated Covariant Differentials of Horizontal Lifts of Vector Fields I) *Let $r \in \{\infty, \omega\}$ and let $\pi_{\mathsf{E}}\colon \mathsf{E} \to \mathsf{M}$ be a C^r-vector bundle, with the data prescribed in Sect. 3.2.1 to define the Riemannian metric $\mathbb{G}_{\mathsf{E}}$ on E. For $m \in \mathbb{Z}_{\geq 0}$,* *there exist* C^r*-vector bundle mappings*

$$(A_s^m, \mathrm{id}_{\mathsf{E}}) \in \mathrm{VB}^r(\mathrm{T}^s(\pi_{\mathsf{E}}^*\mathsf{T}^*\mathsf{M}) \otimes \mathsf{HE}; \mathrm{T}^m(\mathsf{T}^*\mathsf{E}) \otimes \mathsf{HE}), \qquad s \in \{0, 1, \dots, m\},$$

such that

$$\nabla^{\mathsf{E},m} X^{\mathrm{h}} = \sum_{s=0}^{m} A_s^m((\nabla^{\mathsf{M},s} X)^{\mathrm{h}})$$

for all $X \in \Gamma^m(\mathsf{TM})$*. Moreover, the vector bundle mappings* $A_0^m, A_1^m, \dots, A_m^m$ *satisfy the recursion relations prescribed by* $A_0^0(\beta_0) = \beta_0$ *and*

$$\begin{aligned}
A_{m+1}^{m+1}(\beta_{m+1}) &= \beta_{m+1}, \\
A_s^{m+1}(\beta_s) &= (\nabla^{\mathsf{E}} A_s^m)(\beta_s) + A_{s-1}^m \otimes \mathrm{id}_{\mathsf{T}^*\mathsf{E}}(\beta_s) \\
&\quad - \sum_{j=1}^{s} A_s^m \otimes \mathrm{id}_{\mathsf{T}^*\mathsf{E}}(\mathrm{Ins}_j(\beta_s, B_{\pi_{\mathsf{E}}})) \\
&\quad + A_s^m \otimes \mathrm{id}_{\mathsf{T}^*\mathsf{E}}(\mathrm{Ins}_{s+1}(\beta_s, B_{\pi_{\mathsf{E}}}^*)), \quad s \in \{1, \dots, m\}, \\
A_0^{m+1}(\beta_0) &= (\nabla^{\mathsf{E}} A_0^m)(\beta_0) + A_0^m \otimes \mathrm{id}_{\mathsf{T}^*\mathsf{E}}(\mathrm{Ins}_1(\beta_0, B_{\pi_{\mathsf{E}}}^*)),
\end{aligned}$$

where $\beta_s \in \mathrm{T}^s(\pi_{\mathsf{E}}^*\mathsf{T}^*\mathsf{M}) \otimes \mathsf{HE}$*,* $s \in \{0, 1, \dots, m+1\}$*.*

Proof This follows in the same manner as Lemma 3.29, making use of Lemma 3.16(iii). □

The following lemma "inverts" the relations from the preceding one.

Lemma 3.35 (Iterated Covariant Differentials of Horizontal Lifts of Vector Fields II) *Let* $r \in \{\infty, \omega\}$ *and let* $\pi_{\mathsf{E}} \colon \mathsf{E} \to \mathsf{M}$ *be a* C^r*-vector bundle, with the data prescribed in Sect. 3.2.1 to define the Riemannian metric* $\mathbb{G}_{\mathsf{E}}$ *on* E*. For* $m \in \mathbb{Z}_{\geq 0}$*, there exist* C^r*-vector bundle mappings*

$$(B_s^m, \mathrm{id}_{\mathsf{E}}) \in \mathrm{VB}^r(\mathrm{T}^s(\mathsf{T}^*\mathsf{E}) \otimes \mathsf{HE}; \mathrm{T}^m(\pi_{\mathsf{E}}^*\mathsf{T}^*\mathsf{M}) \otimes \mathsf{HE}), \qquad s \in \{0, 1, \dots, m\},$$

such that

$$(\nabla^{\mathsf{M},m} X)^{\mathrm{h}} = \sum_{s=0}^{m} B_s^m(\nabla^{\mathsf{E},s} X^{\mathrm{h}})$$

for all $X \in \Gamma^m(\mathsf{TM})$*. Moreover, the vector bundle mappings* $B_0^m, B_1^m, \dots, B_m^m$ *satisfy the recursion relations prescribed by* $B_0^0(\alpha_0) = \alpha_0$ *and*

$$B_{m+1}^{m+1}(\alpha_{m+1}) = \alpha_{m+1},$$

$$B_s^{m+1}(\alpha_s) = (\nabla^{\mathsf{E}} B_s^m)(\alpha_s) + B_{s-1}^m \otimes \mathrm{id}_{\mathsf{T}^*\mathsf{E}}(\alpha_s) + \sum_{j=1}^m \mathrm{Ins}_j(B_s^m(\alpha_s), B_{\pi_{\mathsf{E}}})$$

$$- \mathrm{Ins}_{m+1}(B_s^m(\alpha_s), B_{\pi_{\mathsf{E}}}^*), \quad s \in \{1, \dots, m\},$$

$$B_0^{m+1}(\alpha_0) = (\nabla^{\mathsf{E}} B_0^m)(\alpha_0) + \sum_{j=1}^m \mathrm{Ins}_j(B_0^m(\alpha_0), B_{\pi_{\mathsf{E}}}) - \mathrm{Ins}_{m+1}(B_0^m(\alpha_0), B_{\pi_{\mathsf{E}}}^*),$$

where $\alpha_s \in \mathrm{T}^s(\mathsf{T}^*\mathsf{E}) \otimes \mathsf{HE}$*,* $s \in \{0, 1, \dots, m+1\}$*.*

Proof This follows in the same manner as Lemma 3.30, making use of Lemma 3.16(iii). □

Now we can give the symmetrised versions of the preceding lemmata.

Lemma 3.36 (Iterated Symmetrised Covariant Differentials of Horizontal Lifts of Vector Fields I) *Let* $r \in \{\infty, \omega\}$ *and let* $\pi_{\mathsf{E}} \colon \mathsf{E} \to \mathsf{M}$ *be a* C^r*-vector bundle, with the data prescribed in Sect. 3.2.1 to define the Riemannian metric* $\mathbb{G}_{\mathsf{E}}$ *on* E*. For* $m \in \mathbb{Z}_{\geq 0}$*, there exist* C^r*-vector bundle mappings*

$$(\widehat{A}_s^m, \mathrm{id}_{\mathsf{E}}) \in \mathrm{VB}^r(\mathrm{S}^s(\pi_{\mathsf{E}}^*\mathsf{T}^*\mathsf{M}) \otimes \mathsf{HE}; \mathrm{S}^m(\mathsf{T}^*\mathsf{E}) \otimes \mathsf{HE}), \qquad s \in \{0, 1, \dots, m\},$$

such that

$$(\mathrm{Sym}_m \otimes \mathrm{id}_{\mathsf{HE}}) \circ \nabla^{\mathsf{E},m} X^{\mathrm{h}} = \sum_{s=0}^m \widehat{A}_s^m((\mathrm{Sym}_s \otimes \mathrm{id}_{\mathsf{HE}}) \circ (\nabla^{\mathsf{M},s} X)^{\mathrm{h}})$$

for all $X \in \Gamma^m(\mathsf{TM})$*.*

Proof This follows along the lines of Lemma 3.25 in the same manner as Lemma 3.31 follows from Lemma 3.25, by taking tensor products with HE. □

Lemma 3.37 (Iterated Symmetrised Covariant Differentials of Horizontal Lifts of Vector Fields II) *Let* $r \in \{\infty, \omega\}$ *and let* $\pi_{\mathsf{E}} \colon \mathsf{E} \to \mathsf{M}$ *be a* C^r*-vector bundle, with the data prescribed in Sect. 3.2.1 to define the Riemannian metric* $\mathbb{G}_{\mathsf{E}}$ *on* E*. For* $m \in \mathbb{Z}_{\geq 0}$*, there exist* C^r*-vector bundle mappings*

$$(\widehat{B}_s^m, \mathrm{id}_{\mathsf{E}}) \in \mathrm{VB}^r(\mathrm{S}^s(\mathsf{T}^*\mathsf{E}) \otimes \mathsf{HE}; \mathrm{S}^m(\pi_{\mathsf{E}}^*\mathsf{T}^*\mathsf{M}) \otimes \mathsf{HE}), \qquad s \in \{0, 1, \dots, m\},$$

such that

$$(\mathrm{Sym}_m \otimes \mathrm{id}_{\mathsf{HE}}) \circ (\nabla^{\mathsf{M},m} X)^{\mathrm{h}} = \sum_{s=0}^{m} \widehat{B}_s^m((\mathrm{Sym}_s \otimes \mathrm{id}_{\mathsf{HE}}) \circ \nabla^{\mathsf{E},s} X^{\mathrm{h}})$$

for all $X \in \Gamma^m(\mathsf{TM})$.

Proof This follows along the lines of Lemma 3.26 in the same manner as Lemma 3.31 follows from Lemma 3.25, by taking tensor products with HE. □

We can put together the previous four lemmata into the following decomposition result, which is to be regarded as the main result of this section.

Lemma 3.38 (Decomposition of Jets of Horizontal Lifts of Vector Fields) *Let* $r \in \{\infty, \omega\}$ *and let* $\pi_{\mathsf{E}} \colon \mathsf{E} \to \mathsf{M}$ *be a* C^r*-vector bundle, with the data prescribed in Sect. 3.2.1 to define the Riemannian metric* $\mathbb{G}_{\mathsf{E}}$ *on* E*. Then there exist* C^r*-vector bundle mappings*

$$A^m_{\nabla\mathsf{E}} \in \mathrm{VB}^r(\mathsf{P}^{*m}\mathsf{E} \otimes \mathsf{HE}; \mathsf{S}^{\leq m}(\pi_{\mathsf{E}}^*\mathsf{T}^*\mathsf{M}) \otimes \mathsf{HE}),$$

$$B^m_{\nabla\mathsf{E}} \in \mathrm{VB}^r(\mathsf{P}^{*m}\mathsf{E} \otimes \mathsf{HE}; \mathsf{S}^{\leq m}(\mathsf{T}^*\mathsf{E}) \otimes \mathsf{HE}),$$

defined by

$$A^m_{\nabla\mathsf{E}}(j_m(X^{\mathrm{h}})(e)) = \mathrm{Sym}_{\leq m} \otimes \mathrm{id}_{\mathsf{HE}}(X^{\mathrm{h}}(e), (\nabla^{\mathsf{M}} X)^{\mathrm{h}}(e), \dots, (\nabla^{\mathsf{M},m} X)^{\mathrm{h}}(e)),$$

$$B^m_{\nabla\mathsf{E}}(j_m(X^{\mathrm{h}})(e)) = \mathrm{Sym}_{\leq m} \otimes \mathrm{id}_{\mathsf{HE}}(X^{\mathrm{h}}(e), \nabla^{\mathsf{E}} X^{\mathrm{h}}(e), \dots, \nabla^{\mathsf{E},m} X^{\mathrm{h}}(e)).$$

Moreover, $A^m_{\nabla\mathsf{E}}$ *is an isomorphism,* $B^m_{\nabla\mathsf{E}}$ *is injective, and*

$$B^m_{\nabla\mathsf{E}} \circ (A^m_{\nabla\mathsf{E}})^{-1} \circ (\mathrm{Sym}_{\leq m} \otimes \mathrm{id}_{\mathsf{HE}})(X^{\mathrm{h}}(e), (\nabla^{\mathsf{M}} X)^{\mathrm{h}}(e), \dots, (\nabla^{\mathsf{M},m} X)^{\mathrm{h}}(e))$$
$$= \Bigg(X^{\mathrm{h}}(e), \sum_{s=0}^{1} \widehat{A}_s^1((\mathrm{Sym}_s \otimes \mathrm{id}_{\mathsf{HE}}) \circ (\nabla^{\mathsf{M},s} X)(e)), \dots,$$
$$\sum_{s=0}^{m} \widehat{A}_s^m((\mathrm{Sym}_s \otimes \mathrm{id}_{\mathsf{HE}}) \circ (\nabla^{\mathsf{M},s} X)^{\mathrm{h}}(e))\Bigg)$$

and

$$A^m_{\nabla^{\mathsf{E}}} \circ (B^m_{\nabla^{\mathsf{E}}})^{-1} \circ (\mathrm{Sym}_{\le m} \otimes \mathrm{id}_{\mathsf{HE}})(X^{\mathrm{h}}(e), \nabla^{\mathsf{E}} X^{\mathrm{h}}(e), \dots, \nabla^{\mathsf{E},m} X^{\mathrm{h}}(e))$$
$$= \left(X^{\mathrm{h}}(e), \sum_{s=0}^{1} \widehat{B}^1_s((\mathrm{Sym}_s \otimes \mathrm{id}_{\mathsf{HE}}) \circ \nabla^{\mathsf{E},s} X^{\mathrm{h}}(e)), \dots, \right.$$
$$\left. \sum_{s=0}^{m} \widehat{B}^m_s((\mathrm{Sym}_s \otimes \mathrm{id}_{\mathsf{HE}}) \circ \nabla^{\mathsf{E},s} X^{\mathrm{h}}(e)) \right),$$

where the vector bundle mappings $\widehat{A}^m_s$ and $\widehat{B}^m_s$, $s \in \{0, 1, \dots, m\}$, are as in Lemmata 3.36 and 3.37.

3.3.4 Isomorphisms for Vertical Lifts of Dual Sections

Next we consider vertical lifts of sections of the dual bundle, i.e., the mapping defined by

$$\Gamma^r(\mathsf{E}^*) \ni \lambda \mapsto \lambda^{\mathrm{v}} \in \Gamma^r(\mathsf{T}^*\mathsf{E}).$$

Our objective is to relate the decomposition of the jets of λ with the decomposition of the jets of λ^{v}. To do this, we denote

$$\mathsf{F}^{*m}_e\mathsf{E} = \{j_m\lambda^{\mathrm{v}}(e) \mid \lambda \in \Gamma^m(\mathsf{E}^*)\}.$$

By (1.9), we have

$$\mathsf{F}^{*m}_e\mathsf{E} \simeq \mathsf{P}^{*m}_e\mathsf{E} \otimes \mathsf{V}^*_e\mathsf{E}.$$

As with the constructions of the preceding sections, we wish to use Lemma 2.15 to provide a decomposition of $\mathsf{F}^{*m}\mathsf{E}$, and to do so we need to understand the covariant derivatives

$$\nabla^{\mathsf{E},m}\lambda^{\mathrm{v}} \triangleq \underbrace{\nabla^{\mathsf{E}} \cdots \nabla^{\mathsf{E}}}_{m \text{ times}} \lambda^{\mathrm{v}}, \qquad m \in \mathbb{Z}_{\ge 0}.$$

In this section we omit proofs, since proofs follow along entirely similar lines to those of preceding sections.

The first result we give is the following.

Lemma 3.39 (Iterated Covariant Differentials of Vertical Lifts of Dual Sections I) *Let $r \in \{\infty, \omega\}$ and let $\pi_{\mathsf{E}} \colon \mathsf{E} \to \mathsf{M}$ be a C^r-vector bundle, with the data* *prescribed in Sect. 3.2.1 to define the Riemannian metric* $\mathbb{G}_{\mathsf{E}}$ *on* E*. For* $m \in \mathbb{Z}_{\geq 0}$*, there exist* C^r*-vector bundle mappings*

$$(A_s^m, \mathrm{id}_{\mathsf{E}}) \in \mathrm{VB}^r(\mathrm{T}^s(\pi_{\mathsf{E}}^*\mathsf{T}^*\mathsf{M}) \otimes \mathsf{V}^*\mathsf{E}; \mathrm{T}^m(\mathsf{T}^*\mathsf{E}) \otimes \mathsf{V}^*\mathsf{E}), \qquad s \in \{0, 1, \dots, m\},$$

such that

$$\nabla^{\mathsf{E},m}\lambda^{\mathrm{v}} = \sum_{s=0}^{m} A_s^m((\nabla^{\mathsf{M},\pi_{\mathsf{E}},s}\lambda)^{\mathrm{v}})$$

for all $\lambda \in \Gamma^m(\mathsf{E}^*)$*. Moreover, the vector bundle mappings* $A_0^m, A_1^m, \dots, A_m^m$ *satisfy the recursion relations prescribed by* $A_0^0(\beta_0) = \beta_0$ *and*

$$\begin{aligned}
A_{m+1}^{m+1}(\beta_{m+1}) &= \beta_{m+1},\\
A_s^{m+1}(\beta_s) &= (\nabla^{\mathsf{E}} A_s^m)(\beta_s) + A_{s-1}^m \otimes \mathrm{id}_{\mathsf{T}^*\mathsf{E}}(\beta_s)\\
&\quad - \sum_{j=1}^{s} A_s^m \otimes \mathrm{id}_{\mathsf{T}^*\mathsf{E}}(\mathrm{Ins}_j(\beta_s, B_{\pi_{\mathsf{E}}})),\\
&\quad s \in \{1, \dots, m\},\\
A_0^{m+1}(\beta_0) &= (\nabla^{\mathsf{E}} A_0^m)(\beta_0) - A_0^m \otimes \mathrm{id}_{\mathsf{T}^*\mathsf{E}}(\mathrm{Ins}_1(\beta_0, B_{\pi_{\mathsf{E}}})),
\end{aligned}$$

where $\beta_s \in \mathrm{T}^s(\pi_{\mathsf{E}}^*\mathsf{T}^*\mathsf{M}) \otimes \mathsf{V}^*\mathsf{E}$*,* $s \in \{0, 1, \dots, m+1\}$*.*

Proof This follows in the same manner as Lemma 3.29, making use of Lemma 3.16(iv). □

The "inverse" of the preceding lemma is as follows.

Lemma 3.40 (Iterated Covariant Differentials of Vertical Lifts of Dual Sections II) *Let* $r \in \{\infty, \omega\}$ *and let* $\pi_{\mathsf{E}} \colon \mathsf{E} \to \mathsf{M}$ *be a* C^r*-vector bundle, with the data prescribed in Sect. 3.2.1 to define the Riemannian metric* $\mathbb{G}_{\mathsf{E}}$ *on* E*. For* $m \in \mathbb{Z}_{\geq 0}$*, there exist* C^r*-vector bundle mappings*

$$(B_s^m, \mathrm{id}_{\mathsf{E}}) \in \mathrm{VB}^r(\mathrm{T}^s(\mathsf{T}^*\mathsf{E}) \otimes \mathsf{V}^*\mathsf{E}; \mathrm{T}^m(\pi_{\mathsf{E}}^*\mathsf{T}^*\mathsf{M}) \otimes \mathsf{V}^*\mathsf{E}), \qquad s \in \{0, 1, \dots, m\},$$

such that

$$(\nabla^{\mathsf{M},\pi_{\mathsf{E}},m}\lambda)^{\mathrm{v}} = \sum_{s=0}^{m} B_s^m(\nabla^{\mathsf{E},s}\lambda^{\mathrm{v}})$$

for all $\lambda \in \Gamma^m(\mathsf{E}^*)$*. Moreover, the vector bundle mappings* $B_0^m, B_1^m, \dots, B_m^m$ *satisfy the recursion relations prescribed by* $B_0^0(\alpha_0) = \alpha_0$ *and*

$$B_{m+1}^{m+1}(\alpha_{m+1}) = \alpha_{m+1},$$

$$B_s^{m+1}(\alpha_s) = (\nabla^{\mathsf{E}} B_s^m)(\alpha_s) + B_{s-1}^m \otimes \mathrm{id}_{\mathsf{T}^*\mathsf{E}}(\alpha_s) + \sum_{j=1}^{m} \mathrm{Ins}_j(B_s^m(\alpha_s), B_{\pi_{\mathsf{E}}}),$$

$$s \in \{1, \dots, m\},$$

$$B_0^{m+1}(\alpha_0) = (\nabla^{\mathsf{E}} B_0^m)(\alpha_0) + \sum_{j=1}^{m+1} \mathrm{Ins}_j(B_0^m(\alpha_0), B_{\pi_{\mathsf{E}}}),$$

where $\alpha_s \in \mathrm{T}^s(\mathsf{T}^*\mathsf{E}) \otimes \mathsf{V}^*\mathsf{E}$*,* $s \in \{0, 1, \dots, m+1\}$*.*

Proof This follows in the same manner as Lemma 3.30, making use of Lemma 3.16(iv). □

Next we turn to symmetrised versions of the preceding lemmata. We show that the preceding two lemmata induce corresponding mappings between symmetric tensors.

Lemma 3.41 (Iterated Symmetrised Covariant Differentials of Vertical Lifts of Dual Sections I) *Let* $r \in \{\infty, \omega\}$ *and let* $\pi_{\mathsf{E}} \colon \mathsf{E} \to \mathsf{M}$ *be a* C^r*-vector bundle, with the data prescribed in Sect. 3.2.1 to define the Riemannian metric* $\mathbb{G}_{\mathsf{E}}$ *on* E*. For* $m \in \mathbb{Z}_{\geq 0}$*, there exist* C^r*-vector bundle mappings*

$$(\widehat{A}_s^m, \mathrm{id}_{\mathsf{E}}) \in \mathrm{VB}^r(\mathrm{S}^s(\pi_{\mathsf{E}}^*\mathsf{T}^*\mathsf{M}) \otimes \mathsf{V}^*\mathsf{E}; \mathrm{S}^m(\mathsf{T}^*\mathsf{E}) \otimes \mathsf{V}^*\mathsf{E}), \qquad s \in \{0, 1, \dots, m\},$$

such that

$$(\mathrm{Sym}_m \otimes \mathrm{id}_{\mathsf{V}^*\mathsf{E}}) \circ \nabla^{\mathsf{E},m} \lambda^{\mathrm{v}} = \sum_{s=0}^{m} \widehat{A}_s^m((\mathrm{Sym}_s \otimes \mathrm{id}_{\mathsf{V}^*\mathsf{E}}) \circ (\nabla^{\mathsf{M},\pi_{\mathsf{E}},s}\lambda)^{\mathrm{v}})$$

for all $\lambda \in \Gamma^m(\mathsf{E}^*)$*.*

Proof This follows along the lines of Lemma 3.25 in the same manner as Lemma 3.31 follows from Lemma 3.25, by taking tensor products with $\mathsf{V}^*\mathsf{E}$. □

The preceding lemma gives rise to an "inverse," which we state in the following lemma.

Lemma 3.42 (Iterated Symmetrised Covariant Differentials of Vertical Lifts of Dual Sections II) *Let* $r \in \{\infty, \omega\}$ *and let* $\pi_{\mathsf{E}} \colon \mathsf{E} \to \mathsf{M}$ *be a* C^r*-vector bundle, with the data prescribed in Sect. 3.2.1 to define the Riemannian metric* $\mathbb{G}_{\mathsf{E}}$ *on* E*.*

For $m \in \mathbb{Z}_{\geq 0}$*, there exist* C^r*-vector bundle mappings*

$$(\widehat{B}^m_s, \mathrm{id}_{\mathsf{E}}) \in \mathrm{VB}^r(\mathsf{S}^s(\mathsf{T}^*\mathsf{E}) \otimes \mathsf{V}^*\mathsf{E}; \mathsf{S}^m(\pi_{\mathsf{E}}^*\mathsf{T}^*\mathsf{M}) \otimes \mathsf{V}^*\mathsf{E}), \qquad s \in \{0, 1, \dots, m\},$$

such that

$$(\mathrm{Sym}_m \otimes \mathrm{id}_{\mathsf{V}^*\mathsf{E}}) \circ (\nabla^{\mathsf{M},\pi_{\mathsf{E}},m}\lambda)^{\mathrm{v}} = \sum_{s=0}^{m} \widehat{B}^m_s((\mathrm{Sym}_s \otimes \mathrm{id}_{\mathsf{V}^*\mathsf{E}}) \circ \nabla^{\mathsf{E},s}\lambda^{\mathrm{v}})$$

for all $\lambda \in \Gamma^m(\mathsf{E}^*)$.

Proof This follows along the lines of Lemma 3.26 in the same manner as Lemma 3.31 follows from Lemma 3.25, by taking tensor products with $\mathsf{V}^*\mathsf{E}$. □

We can put together the previous four lemmata into the following decomposition result, which is to be regarded as the main result of this section.

Lemma 3.43 (Decomposition of Jets of Vertical Lifts of Dual Sections) *Let* $r \in \{\infty, \omega\}$ *and let* $\pi_{\mathsf{E}} \colon \mathsf{E} \to \mathsf{M}$ *be a* C^r*-vector bundle, with the data prescribed in Sect. 3.2.1 to define the Riemannian metric* $\mathbb{G}_{\mathsf{E}}$ *on* E*. Then there exist* C^r*-vector bundle mappings*

$$A^m_{\nabla\mathsf{E}} \in \mathrm{VB}^r(\mathsf{P}^{*m}\mathsf{E} \otimes \mathsf{V}^*\mathsf{E}; \mathsf{S}^{\leq m}(\pi_{\mathsf{E}}^*\mathsf{T}^*\mathsf{M}) \otimes \mathsf{V}^*\mathsf{E}),$$
$$B^m_{\nabla\mathsf{E}} \in \mathrm{VB}^r(\mathsf{P}^{*m}\mathsf{E} \otimes \mathsf{V}^*\mathsf{E}; \mathsf{S}^{\leq m}(\mathsf{T}^*\mathsf{E}) \otimes \mathsf{V}^*\mathsf{E}),$$

defined by

$$A^m_{\nabla\mathsf{E}}(j_m(\lambda^{\mathrm{v}})(e)) = \mathrm{Sym}_{\leq m} \otimes \mathrm{id}_{\mathsf{V}^*\mathsf{E}}(\lambda^{\mathrm{v}}(e), (\nabla^{\pi_{\mathsf{E}}}\lambda)^{\mathrm{v}}(e), \dots, (\nabla^{\mathsf{M},\pi_{\mathsf{E}},m}\lambda)^{\mathrm{v}}(e)),$$
$$B^m_{\nabla\mathsf{E}}(j_m(\lambda^{\mathrm{v}})(e)) = \mathrm{Sym}_{\leq m} \otimes \mathrm{id}_{\mathsf{V}^*\mathsf{E}}(\lambda^{\mathrm{v}}(e), \nabla^{\mathsf{E}}\lambda^{\mathrm{v}}(e), \dots, \nabla^{\mathsf{E},m}\lambda^{\mathrm{v}}(e)).$$

Moreover, $A^m_{\nabla\mathsf{E}}$ *is an isomorphism,* $B^m_{\nabla\mathsf{E}}$ *is injective, and*

$$B^m_{\nabla\mathsf{E}} \circ (A^m_{\nabla\mathsf{E}})^{-1} \circ (\mathrm{Sym}_{\leq m} \otimes \mathrm{id}_{\mathsf{V}^*\mathsf{E}})(\lambda^{\mathrm{v}}(e), (\nabla^{\pi_{\mathsf{E}}}\lambda)^{\mathrm{v}}(e), \dots, (\nabla^{\mathsf{M},\pi_{\mathsf{E}},m}\lambda)^{\mathrm{v}}(e)) =$$
$$\Bigg(\lambda^{\mathrm{v}}(e), \sum_{s=0}^{1} \widehat{A}^1_s((\mathrm{Sym}_s \otimes \mathrm{id}_{\mathsf{V}^*\mathsf{E}}) \circ (\nabla^{\mathsf{M},\pi_{\mathsf{E}},s}\lambda)^{\mathrm{v}}(e)), \dots,$$
$$\sum_{s=0}^{m} \widehat{A}^m_s((\mathrm{Sym}_s \otimes \mathrm{id}_{\mathsf{V}^*\mathsf{E}}) \circ (\nabla^{\mathsf{M},\pi_{\mathsf{E}},s}\lambda)^{\mathrm{v}}(e))\Bigg)$$

and

$$A^m_{\nabla^{\mathsf{E}}} \circ (B^m_{\nabla^{\mathsf{E}}})^{-1} \circ (\mathrm{Sym}_{\leq m} \otimes \mathrm{id}_{\mathsf{V}^*\mathsf{E}})(\lambda^{\mathrm{v}}(e), \nabla^{\mathsf{E}}\lambda^{\mathrm{v}}(e), \dots, \nabla^{\mathsf{E},m}\lambda^{\mathrm{v}}(e))$$
$$= \left(\lambda^{\mathrm{v}}(e), \sum_{s=0}^{1} \widehat{B}^1_s((\mathrm{Sym}_s \otimes \mathrm{id}_{\mathsf{V}^*\mathsf{E}}) \circ \nabla^{\mathsf{E},s}\lambda^{\mathrm{v}}(e)), \dots, \sum_{s=0}^{m} \widehat{B}^m_s((\mathrm{Sym}_s \otimes \mathrm{id}_{\mathsf{V}^*\mathsf{E}}) \circ \nabla^{\mathsf{E},s}\lambda^{\mathrm{v}}(e))\right),$$

where the vector bundle mappings $\widehat{A}^m_s$ *and* $\widehat{B}^m_s$, $s \in \{0, 1, \dots, m\}$, *are as in Lemmata 3.41 and 3.42.*

3.3.5 Isomorphisms for Vertical Lifts of Endomorphisms

Next we consider vertical lifts of endomorphisms defined by the mapping

$$\Gamma^r(\mathrm{T}^1_1(\mathsf{E})) \ni L \mapsto L^{\mathrm{v}} \in \Gamma^r(\mathrm{T}^1_1(\mathsf{TE})).$$

We wish to relate the decomposition of the jets of L with those of L^{v}. Associated with this, we denote

$$\mathsf{L}^{*m}_e\mathsf{E} = \{j_m L^{\mathrm{v}}(e) \mid L \in \Gamma^m(\mathrm{T}^1_1(\mathsf{E}))\}.$$

By (1.9), we have

$$\mathsf{L}^{*m}_e\mathsf{E} \simeq \mathsf{P}^{*m}_e\mathsf{E} \otimes \mathrm{T}^1_1(\mathsf{V}_e\mathsf{E}).$$

As with the constructions of the preceding sections, we wish to use Lemma 2.15 to provide a decomposition of $\mathsf{L}^{*m}\mathsf{E}$, and to do so we need to understand the covariant derivatives

$$\nabla^{\mathsf{E},m}L^{\mathrm{v}} \triangleq \underbrace{\nabla^{\mathsf{E}} \cdots \nabla^{\mathsf{E}}}_{m \text{ times}} L^{\mathrm{v}}, \qquad m \in \mathbb{Z}_{\geq 0}.$$

In this section we omit proofs, since proofs follow along entirely similar lines to those of preceding sections.

The first result we give is the following.

Lemma 3.44 (Iterated Covariant Differentials of Vertical Lifts of Endomorphisms I) *Let* $r \in \{\infty, \omega\}$ *and let* $\pi_{\mathsf{E}}\colon \mathsf{E} \to \mathsf{M}$ *be a* C^r*-vector bundle, with the data prescribed in Sect. 3.2.1 to define the Riemannian metric* $\mathbb{G}_{\mathsf{E}}$ *on* E. *For* $m \in \mathbb{Z}_{\geq 0}$, *there exist* C^r*-vector bundle mappings*

$$(A_s^m, \mathrm{id}_{\mathsf{E}}) \in \mathrm{VB}^r(\mathrm{T}^s(\pi_{\mathsf{E}}^*\mathsf{T}^*\mathsf{M}) \otimes \mathrm{T}_1^1(\mathsf{VE}); \mathrm{T}^m(\mathsf{T}^*\mathsf{E}) \otimes \mathrm{T}_1^1(\mathsf{VE})), \quad s \in \{0, 1, \dots, m\},$$

such that

$$\nabla^{\mathsf{E},m} L^{\mathrm{v}} = \sum_{s=0}^{m} A_s^m((\nabla^{\mathsf{M},\pi_{\mathsf{E}},s} L)^{\mathrm{v}})$$

for all $L \in \Gamma^m(\mathrm{T}_1^1(\mathsf{E}))$*. Moreover, the vector bundle mappings* $A_0^m, A_1^m, \dots, A_m^m$ *satisfy the recursion relations prescribed by* $A_0^0(\beta_0) = \beta_0$ *and*

$$\begin{aligned}
&A_{m+1}^{m+1}(\beta_{m+1}) = \beta_{m+1},\\
&\quad A_s^{m+1}(\beta_s) = (\nabla^{\mathsf{E}} A_s^m)(\beta_s) + A_{s-1}^m \otimes \mathrm{id}_{\mathsf{T}^*\mathsf{E}}(\beta_s) - \sum_{j=1}^{s} A_s^m \otimes \mathrm{id}_{\mathsf{T}^*\mathsf{E}}(\mathrm{Ins}_j(\beta_s, B_{\pi_{\mathsf{E}}}))\\
&\qquad\qquad + A_s^m \otimes \mathrm{id}_{\mathsf{T}^*\mathsf{E}}(\mathrm{Ins}_{s+1}(\beta_s, B_{\pi_{\mathsf{E}}}^*)), \quad s \in \{1, \dots, m\},\\
&\quad A_0^{m+1}(\beta_0) = (\nabla^{\mathsf{E}} A_0^m)(\beta_0) - A_0^m \otimes \mathrm{id}_{\mathsf{T}^*\mathsf{E}}(\mathrm{Ins}_1(\beta_0, B_{\pi_{\mathsf{E}}}))\\
&\qquad\qquad + A_0^m \otimes \mathrm{id}_{\mathsf{T}^*\mathsf{E}}(\mathrm{Ins}_2(\beta_0, B_{\pi_{\mathsf{E}}}^*)),
\end{aligned}$$

where $\beta_s \in \mathrm{T}^s(\pi_{\mathsf{E}}^*\mathsf{T}^*\mathsf{M}) \otimes \mathrm{T}_1^1(\mathsf{VE})$*,* $s \in \{0, 1, \dots, m+1\}$*.*

Proof This follows in the same manner as Lemma 3.29, making use of Lemma 3.16(v). □

The “inverse” of the preceding lemma is as follows.

Lemma 3.45 (Iterated Covariant Differentials of Vertical Lifts of Endomorphisms II) *Let* $r \in \{\infty, \omega\}$ *and let* $\pi_{\mathsf{E}} \colon \mathsf{E} \to \mathsf{M}$ *be a* C^r*-vector bundle, with the data prescribed in Sect. 3.2.1 to define the Riemannian metric* $\mathbb{G}_{\mathsf{E}}$ *on* E*. For* $m \in \mathbb{Z}_{\geq 0}$*, there exist* C^r*-vector bundle mappings*

$$(B_s^m, \mathrm{id}_{\mathsf{E}}) \in \mathrm{VB}^r(\mathrm{T}^s(\mathsf{T}^*\mathsf{E}) \otimes \mathrm{T}_1^1(\mathsf{VE}); \mathrm{T}^m(\pi_{\mathsf{E}}^*\mathsf{T}^*\mathsf{M}) \otimes \mathrm{T}_1^1(\mathsf{VE})), \quad s \in \{0, 1, \dots, m\},$$

such that

$$(\nabla^{\mathsf{M},\pi_{\mathsf{E}},m} L)^{\mathrm{v}} = \sum_{s=0}^{m} B_s^m(\nabla^{\mathsf{E},s} L^{\mathrm{v}})$$

for all $L \in \Gamma^m(\mathrm{T}^1_1(\mathsf{E}))$. *Moreover, the vector bundle mappings* $B^m_0, B^m_1, \dots, B^m_m$ *satisfy the recursion relations prescribed by* $B^0_0(\alpha_0) = \alpha_0$ *and*

$$B^{m+1}_{m+1}(\alpha_{m+1}) = \alpha_{m+1},$$

$$B^{m+1}_s(\alpha_s) = (\nabla^{\mathsf{E}} B^m_s)(\alpha_s) + B^m_{s-1} \otimes \mathrm{id}_{\mathsf{T}^*\mathsf{E}}(\alpha_s) + \sum_{j=1}^{m} \mathrm{Ins}_j(B^m_s(\alpha_s), B_{\pi_{\mathsf{E}}})$$
$$- \mathrm{Ins}_{m+1}(B^m_s(\alpha_s), B^*_{\pi_{\mathsf{E}}}), \quad s \in \{1, \dots, m\},$$

$$B^{m+1}_0(\alpha_0) = (\nabla^{\mathsf{E}} B^m_0)(\alpha_0) + \sum_{j=1}^{m} \mathrm{Ins}_j(B^m_0(\alpha_0), B_{\pi_{\mathsf{E}}}) - \mathrm{Ins}_{m+1}(B^m_0(\alpha_0), B_{\pi_{\mathsf{E}}}),$$

where $\alpha_s \in \mathrm{T}^s(\mathsf{T}^*\mathsf{E}) \otimes \mathrm{T}^1_1(\mathsf{VE})$, $s \in \{0, 1, \dots, m+1\}$.

Proof This follows in the same manner as Lemma 3.30, making use of Lemma 3.16(v). □

Next we turn to symmetrised versions of the preceding lemmata. We show that the preceding two lemmata induce corresponding mappings between symmetric tensors.

Lemma 3.46 (Iterated Symmetrised Covariant Differentials of Vertical Lifts of Endomorphisms I) *Let* $r \in \{\infty, \omega\}$ *and let* $\pi_{\mathsf{E}} \colon \mathsf{E} \to \mathsf{M}$ *be a* C^r*-vector bundle, with the data prescribed in Sect. 3.2.1 to define the Riemannian metric* $\mathbb{G}_{\mathsf{E}}$ *on* E. *For* $m \in \mathbb{Z}_{\geq 0}$, *there exist* C^r*-vector bundle mappings*

$$(\widehat{A}^m_s, \mathrm{id}_{\mathsf{E}}) \in \mathrm{VB}^r(\mathrm{S}^s(\pi^*_{\mathsf{E}}\mathsf{T}^*\mathsf{M}) \otimes \mathrm{T}^1_1(\mathsf{VE}); \mathrm{S}^m(\mathsf{T}^*\mathsf{E}) \otimes \mathrm{T}^1_1(\mathsf{VE})), \quad s \in \{0, 1, \dots, m\},$$

such that

$$(\mathrm{Sym}_m \otimes \mathrm{id}_{\mathrm{T}^1_1(\mathsf{VE})}) \circ \nabla^{\mathsf{E},m} L^{\mathrm{v}} = \sum_{s=0}^{m} \widehat{A}^m_s((\mathrm{Sym}_s \otimes \mathrm{id}_{\mathrm{T}^1_1(\mathsf{VE})}) \circ (\nabla^{\mathsf{M},\pi_{\mathsf{E}},s} L)^{\mathrm{v}})$$

for all $L \in \Gamma^m(\mathrm{T}^1_1(\mathsf{E}))$.

Proof This follows along the lines of Lemma 3.25 in the same manner as Lemma 3.31 follows from Lemma 3.25, by taking tensor products with $\mathrm{T}^1_1(\mathsf{VE})$. □

The preceding lemma gives rise to an "inverse," which we state in the following lemma.

Lemma 3.47 (Iterated Symmetrised Covariant Differentials of Vertical Lifts of Endomorphisms II) *Let* $r \in \{\infty, \omega\}$ *and let* $\pi_{\mathsf{E}} \colon \mathsf{E} \to \mathsf{M}$ *be a* C^r*-vector bundle, with the data prescribed in Sect. 3.2.1 to define the Riemannian metric* $\mathbb{G}_{\mathsf{E}}$ *on* E. *For*

$m \in \mathbb{Z}_{\geq 0}$, there exist C^r-vector bundle mappings

$$(\widehat{B}_s^m, \mathrm{id}_{\mathsf{E}}) \in \mathrm{VB}^r(\mathrm{S}^s(\mathrm{T}^*\mathsf{E}) \otimes \mathrm{T}_1^1(\mathsf{VE}); \mathrm{S}^m(\pi_{\mathsf{E}}^*\mathrm{T}^*\mathsf{M}) \otimes \mathrm{T}_1^1(\mathsf{VE})), \quad s \in \{0, 1, \dots, m\},$$

such that

$$(\mathrm{Sym}_m \otimes \mathrm{id}_{\mathrm{T}_1^1(\mathsf{VE})}) \circ (\nabla^{\mathsf{M},\pi_{\mathsf{E}},m} L)^{\mathrm{v}} = \sum_{s=0}^{m} \widehat{B}_s^m((\mathrm{Sym}_s \otimes \mathrm{id}_{\mathrm{T}_1^1(\mathsf{VE})}) \circ \nabla^{\mathsf{E},s} L^{\mathrm{v}})$$

for all $L \in \Gamma^m(\mathrm{T}_1^1(\mathsf{E}))$.

Proof This follows along the lines of Lemma 3.26 in the same manner as Lemma 3.31 follows from Lemma 3.25, by taking tensor products with $\mathrm{T}_1^1(\mathsf{VE})$. □

We can put together the previous four lemmata into the following decomposition result, which is to be regarded as the main result of this section.

Lemma 3.48 (Decomposition of Jets of Vertical Lifts of Endomorphisms) *Let $r \in \{\infty, \omega\}$ and let $\pi_{\mathsf{E}} \colon \mathsf{E} \to \mathsf{M}$ be a C^r-vector bundle, with the data prescribed in Sect. 3.2.1 to define the Riemannian metric $\mathbb{G}_{\mathsf{E}}$ on E. Then there exist C^r-vector bundle mappings*

$$A_{\nabla\mathsf{E}}^m \in \mathrm{VB}^r(\mathrm{P}^{*m}\mathsf{E} \otimes \mathrm{T}_1^1(\mathsf{VE}); \mathrm{S}^{\leq m}(\pi_{\mathsf{E}}^*\mathrm{T}^*\mathsf{M}) \otimes \mathrm{T}_1^1(\mathsf{VE})),$$
$$B_{\nabla\mathsf{E}}^m \in \mathrm{VB}^r(\mathrm{P}^{*m}\mathsf{E} \otimes \mathrm{T}_1^1(\mathsf{VE}); \mathrm{S}^{\leq m}(\mathrm{T}^*\mathsf{E}) \otimes \mathrm{T}_1^1(\mathsf{VE})),$$

defined by

$$A_{\nabla\mathsf{E}}^m(j_m(L^{\mathrm{v}})(e)) = \mathrm{Sym}_{\leq m} \otimes \mathrm{id}_{\mathrm{T}_1^1(\mathsf{VE})}(L^{\mathrm{v}}(e), (\nabla^{\pi_{\mathsf{E}}} L)^{\mathrm{v}}(e), \dots, (\nabla^{\mathsf{M},\pi_{\mathsf{E}},m} L)^{\mathrm{v}}(e)),$$
$$B_{\nabla\mathsf{E}}^m(j_m(L^{\mathrm{v}})(e)) = \mathrm{Sym}_{\leq m} \otimes \mathrm{id}_{\mathrm{T}_1^1(\mathsf{VE})}(L^{\mathrm{v}}(e), \nabla^{\mathsf{E}} L^{\mathrm{v}}(e), \dots, \nabla^{\mathsf{E},m} L^{\mathrm{v}}(e)).$$

Moreover, $A_{\nabla\mathsf{E}}^m$ is an isomorphism, $B_{\nabla\mathsf{E}}^m$ is injective, and

$$\begin{aligned} B_{\nabla\mathsf{E}}^m \circ (A_{\nabla\mathsf{E}}^m)^{-1} \circ (\mathrm{Sym}_{\leq m} \otimes \mathrm{id}_{\mathrm{T}_1^1(\mathsf{VE})})(L^{\mathrm{v}}(e), (\nabla^{\pi_{\mathsf{E}}} L)^{\mathrm{v}}(e), \dots, (\nabla^{\mathsf{M},\pi_{\mathsf{E}},m} L)^{\mathrm{v}}(e)) \\ = \Bigg(L^{\mathrm{v}}(e), \sum_{s=0}^{1} \widehat{A}_s^1((\mathrm{Sym}_s \otimes \mathrm{id}_{\mathrm{T}_1^1(\mathsf{VE})}) \circ (\nabla^{\mathsf{M},\pi_{\mathsf{E}},s} L)^{\mathrm{v}}(e)), \dots, \\ \sum_{s=0}^{m} \widehat{A}_s^m((\mathrm{Sym}_s \otimes \mathrm{id}_{\mathrm{T}_1^1(\mathsf{VE})}) \circ (\nabla^{\mathsf{M},\pi_{\mathsf{E}},s} L)^{\mathrm{v}}(e)) \Bigg) \end{aligned}$$

and

$$A^m_{\nabla^{\mathsf{E}}} \circ (B^m_{\nabla^{\mathsf{E}}})^{-1} \circ (\mathrm{Sym}_{\leq m} \otimes \mathrm{id}_{\mathsf{T}^1_1(\mathsf{VE})})(L^{\mathrm{v}}(e), \nabla^{\mathsf{E}} L^{\mathrm{v}}(e), \dots, \nabla^{\mathsf{E},m} L^{\mathrm{v}}(e))$$
$$= \Bigg(L^{\mathrm{v}}(e), \sum_{s=0}^{1} \widehat{B}^1_s((\mathrm{Sym}_s \otimes \mathrm{id}_{\mathsf{T}^1_1(\mathsf{VE})}) \circ \nabla^{\mathsf{E},s} L^{\mathrm{v}}(e)), \dots,$$
$$\sum_{s=0}^{m} \widehat{B}^m_s((\mathrm{Sym}_s \otimes \mathrm{id}_{\mathsf{T}^1_1(\mathsf{VE})}) \circ \nabla^{\mathsf{E},s} L^{\mathrm{v}}(e)) \Bigg),$$

where the vector bundle mappings $\widehat{A}^m_s$ *and* $\widehat{B}^m_s$, $s \in \{0, 1, \dots, m\}$, *are as in Lemmata 3.46 and 3.47.*

3.3.6 Isomorphisms for Vertical Evaluations of Dual Sections

Next we consider vertical evaluations of endomorphisms given by the mapping

$$\Gamma^r(\mathsf{E}^*) \ni \lambda \mapsto \lambda^{\mathrm{e}} \in \mathsf{C}^r(\mathsf{E}).$$

To study the relationship between the decomposition of the jets of λ with those of the jets of λ^{e}, we denote

$$\mathsf{D}^{*m}_e \mathsf{E} = \{j_m \lambda^{\mathrm{e}}(e) \mid \lambda \in \Gamma^m(\mathsf{E}^*)\}.$$

By (1.9), we have

$$\mathsf{D}^{*m}_e \mathsf{E} \subseteq \mathsf{P}^{*m}_e \mathsf{E}.$$

As we shall see, one can be a little more explicit about the nature of $\mathsf{D}^{*m}_e \mathsf{E}$, and see that

$$\mathsf{D}^{*m}_e \mathsf{E} \simeq (\mathsf{P}^{*m}_e \mathsf{E} \otimes \mathsf{V}^* \mathsf{E}) \oplus (\mathsf{P}^{*\mathsf{E}}_{m-1} \otimes \mathsf{V}^* \mathsf{E}).$$

However, this sort of isomorphism is too cumbersome to make explicit, and so we will just keep the notation $\mathsf{D}^{*m}_e \mathsf{E}$. As with the constructions of the preceding sections, we wish to use Lemma 2.15 to provide a decomposition of $\mathsf{D}^{*m} \mathsf{E}$, and to do so we need to understand the covariant derivatives

$$\nabla^{\mathsf{E},m} \lambda^{\mathrm{e}} \triangleq \underbrace{\nabla^{\mathsf{E}} \cdots \nabla^{\mathsf{E}}}_{m \text{ times}} \lambda^{\mathrm{e}}, \qquad m \in \mathbb{Z}_{\geq 0}.$$

The results in this section have a slightly different character than in the preceding sections. We will not give the complete proofs, but will note that they are similar to the complete proofs given in the next section.

Our first result is the following.

Lemma 3.49 (Iterated Covariant Differentials of Vertical Evaluations of Dual Sections I) *Let $r \in \{\infty, \omega\}$ and let $\pi_{\mathsf{E}}\colon \mathsf{E} \to \mathsf{M}$ be a C^r-vector bundle, with the data prescribed in Sect. 3.2.1 to define the Riemannian metric $\mathbb{G}_{\mathsf{E}}$ on E. For $m \in \mathbb{Z}_{\geq 0}$, there exist C^r-vector bundle mappings*

$$(A^m_s, \mathrm{id}_{\mathsf{E}}) \in \mathrm{VB}^r(\mathrm{T}^s(\pi^*_{\mathsf{E}}\mathsf{T}^*\mathsf{M}); \mathrm{T}^m(\mathsf{T}^*\mathsf{E})), \quad s \in \{0, 1, \dots, m\},$$

and

$$(C^m_s, \mathrm{id}_{\mathsf{E}}) \in \mathrm{VB}^r(\mathrm{T}^s(\pi^*_{\mathsf{E}}\mathsf{T}^*\mathsf{M}) \otimes \mathsf{V}^*\mathsf{E}; \mathrm{T}^{m-1}(\mathsf{T}^*\mathsf{E}) \otimes \mathsf{V}^*\mathsf{E}), \quad s \in \{0, 1, \dots, m-1\},$$

such that

$$\nabla^{\mathsf{E},m}\lambda^{\mathrm{e}} = \sum_{s=0}^{m} A^m_s((\nabla^{\mathsf{M},\pi_{\mathsf{E}},s}\lambda)^{\mathrm{e}}) + \sum_{s=0}^{m-1} C^m_s((\nabla^{\mathsf{M},\pi_{\mathsf{E}},s}\lambda)^{\mathrm{v}})$$

for all $\lambda \in \Gamma^m(\mathsf{E}^*)$ (Here we regard $\mathsf{V}^*\mathsf{E}$ as a subbundle of $\mathsf{T}^*\mathsf{E}$). Moreover, the vector bundle mappings $A^m_0, A^m_1, \dots, A^m_m$ and $C^m_0, C^m_1, \dots, C^m_{m-1}$ satisfy the recursion relations prescribed by

$$A^0_0(\beta_0) = \beta_0, \quad A^1_1(\beta_1) = \beta_1, \quad A^1_0(\beta_0) = \mathrm{Ins}_1(\beta_0, B_{\pi_{\mathsf{E}}}), \quad C^1_0(\gamma_0) = \gamma_0,$$

and, for $m \geq 2$,

$$\begin{aligned}
A^{m+1}_{m+1}(\beta_{m+1}) &= \beta_{m+1}\\
A^{m+1}_m(\beta_m) &= A^m_{m-1} \otimes \mathrm{id}_{\mathsf{T}^*\mathsf{E}}(\beta_m) - \sum_{j=1}^{m} \mathrm{Ins}_j(\beta_m, B_{\pi_{\mathsf{E}}})\\
A^{m+1}_s(\beta_s) &= (\nabla^{\mathsf{E}} A^m_s)(\beta_m) + A^m_{s-1} \otimes \mathrm{id}_{\mathsf{T}^*\mathsf{E}}(\beta_s)\\
&\quad - \sum_{j=1}^{s} A^m_s \otimes \mathrm{id}_{\mathsf{T}^*\mathsf{E}}(\mathrm{Ins}_j(\beta_s, B_{\pi_{\mathsf{E}}})),\\
&\qquad s \in \{1, \dots, m-1\},\\
A^{m+1}_0(\beta_0) &= (\nabla^{\mathsf{E}} A^m_0)(\beta_0)
\end{aligned}$$

and

$$\begin{aligned}
C_m^{m+1}(\gamma_m) &= C_{m-1}^m \otimes \mathrm{id}_{\mathsf{T}^*\mathsf{E}}(\gamma_m) + \gamma_m \\
C_s^{m+1}(\gamma_s) &= A_s^m \otimes \mathrm{id}_{\mathsf{T}^*\mathsf{E}}(\gamma_s) + (\nabla^{\mathsf{E}} C_s^m)(\gamma_s) + C_{s-1}^m \otimes \mathrm{id}_{\mathsf{T}^*\mathsf{E}}(\gamma_s) \\
&\quad - \sum_{j=1}^{s+1} C_s^m \otimes \mathrm{id}_{\mathsf{T}^*\mathsf{E}}(\mathrm{Ins}_j(\gamma_s, B_{\pi_{\mathsf{E}}})), \quad s \in \{1, \dots, m-1\}, \\
C_0^{m+1}(\gamma_0) &= A_0^m \otimes \mathrm{id}_{\mathsf{T}^*\mathsf{E}}(\gamma_0) + (\nabla^{\mathsf{E}} C_0^m)(\gamma_0) - C_0^m \otimes \mathrm{id}_{\mathsf{T}^*\mathsf{E}}(\mathrm{Ins}_1(\gamma_0, B_{\pi_{\mathsf{E}}})),
\end{aligned}$$

where $\beta_s \in \mathrm{T}^s(\pi_{\mathsf{E}}^*\mathsf{T}^*\mathsf{M})$, $s \in \{0, 1, \dots, m+1\}$, *and* $\gamma_s \in \mathrm{T}^s(\pi_{\mathsf{E}}^*\mathsf{T}^*\mathsf{M}) \otimes \mathsf{V}^*\mathsf{E}$, $s \in \{0, 1, \dots, m-1\}$.

Proof This follows in the same manner as Lemma 3.54 below, making use of Lemma 3.16(vi). □

Now we "invert" the constructions from the preceding lemma.

Lemma 3.50 (Iterated Covariant Differentials of Vertical Evaluations of Dual Sections II) *Let $r \in \{\infty, \omega\}$ and let $\pi_{\mathsf{E}} \colon \mathsf{E} \to \mathsf{M}$ be a C^r-vector bundle, with the data prescribed in Sect. 3.2.1 to define the Riemannian metric $\mathbb{G}_{\mathsf{E}}$ on E. For $m \in \mathbb{Z}_{\geq 0}$, there exist C^r-vector bundle mappings*

$$(B_s^m, \mathrm{id}_{\mathsf{E}}) \in \mathrm{VB}^r(\mathrm{T}^s(\mathsf{T}^*\mathsf{E}); \mathrm{T}^m(\pi_{\mathsf{E}}^*\mathsf{T}^*\mathsf{M})), \qquad s \in \{0, 1, \dots, m\},$$

and

$$(D_s^m, \mathrm{id}_{\mathsf{E}}) \in \mathrm{VB}^r(\mathrm{T}^s(\mathsf{T}^*\mathsf{E}) \otimes \mathsf{V}^*\mathsf{E}; \mathrm{T}^{m-1}(\pi_{\mathsf{E}}^*\mathsf{T}^*\mathsf{M}) \otimes \mathsf{V}^*\mathsf{E}), \quad s \in \{0, 1, \dots, m-1\},$$

such that

$$(\nabla^{\mathsf{M},\pi_{\mathsf{E}},m}\lambda)^{\mathrm{e}} = \sum_{s=0}^{m} B_s^m(\nabla^{\mathsf{E},s}\lambda^{\mathrm{e}}) + \sum_{s=0}^{m-1} D_s^m(\nabla^{\mathsf{E},s}\lambda^{\mathrm{v}})$$

for all $\lambda \in \Gamma^m(\mathsf{E}^*)$. Moreover, the vector bundle mappings $B_0^m, B_1^m, \dots, B_m^m$ and $D_0^m, D_1^m, \dots, D_{m-1}^m$ satisfy the recursion relations prescribed by $B_0^0(\alpha_0) = \alpha_0$, $D_0^1(\gamma_0) = \gamma_0$,

$$\begin{aligned}
B_{m+1}^{m+1}(\alpha_{m+1}) &= \alpha_{m+1} \\
B_m^{m+1}(\alpha_m) &= B_{m-1}^m \otimes \mathrm{id}_{\mathsf{T}^*\mathsf{E}}(\alpha_m) + \sum_{j=1}^{m} \mathrm{Ins}_j(\alpha_m, B_{\pi_{\mathsf{E}}}) - \mathrm{Ins}_{m+1}(\alpha_m, B_{\pi_{\mathsf{E}}}^*)
\end{aligned}$$

$$B_s^{m+1} = (\nabla^{\mathsf{E}} B_s^m)(\alpha_s) + B_{s-1}^m \otimes \mathrm{id}_{\mathsf{T}^*\mathsf{E}}(\alpha_s) + \sum_{j=1}^m \mathrm{Ins}_j(B_s^m(\alpha_s), B_{\pi_{\mathsf{E}}})$$
$$- \mathrm{Ins}_{m+1}(B_s^m(\alpha_s), B_{\pi_{\mathsf{E}}}^*), \quad s \in \{1, \dots, m-1\},$$

$$B_0^{m+1}(\alpha_0) = (\nabla^{\mathsf{E}} B_0^m)(\alpha_0) + \sum_{j=1}^m \mathrm{Ins}_j(B_0^m(\alpha_0), B_{\pi_{\mathsf{E}}}) - \mathrm{Ins}_{m+1}(B_0^m(\alpha_0), B_{\pi_{\mathsf{E}}}^*)$$

and

$$D_m^{m+1}(\gamma_m) = D_{m-1}^m \otimes \mathrm{id}_{\mathsf{T}^*\mathsf{E}}(\gamma_m) - \gamma_m$$
$$D_s^m(\gamma_s) = (\nabla^{\mathsf{E}} D_s^m)(\gamma_s) + D_{s-1}^m \otimes \mathrm{id}_{\mathsf{T}^*\mathsf{E}}(\gamma_s) - \overline{B}_s^m(\gamma_s), \quad s \in \{1, \dots, m-1\},$$
$$D_0^{m+1} = (\nabla^{\mathsf{E}} D_0^m)(\gamma_0) - \overline{B}_0^m(\gamma_0)$$

for $\alpha_s \in \mathrm{T}^s(\mathsf{T}^*\mathsf{E})$, $s \in \{0, 1, \dots, m\}$, *and* $\gamma_s \in \mathrm{T}^s(\mathsf{T}^*\mathsf{E}) \otimes \mathsf{V}^*\mathsf{E}$, $s \in \{0, 1, \dots, m-1\}$, *and where*

$$(\overline{B}_s^m, \mathrm{id}_{\mathsf{E}}) \in \mathrm{VB}^r(\mathrm{T}^s(\mathsf{T}^*\mathsf{E}) \otimes \mathsf{V}^*\mathsf{E}; \mathrm{T}^m(\pi_{\mathsf{E}}^*\mathsf{T}^*\mathsf{M}) \otimes \mathsf{V}^*\mathsf{E}), \qquad s \in \{0, 1, \dots, m\},$$

are the vector bundle mappings from Lemma 3.40.

Proof This follows in the same manner as Lemma 3.55 below, making use of Lemma 3.16(vi). □

Next we turn to symmetrised versions of the preceding lemmata. We show that the preceding two lemmata induce corresponding mappings between symmetric tensors.

Lemma 3.51 (Iterated Symmetrised Covariant Differentials of Vertical Evaluations of Dual Sections I) *Let* $r \in \{\infty, \omega\}$ *and let* $\pi_{\mathsf{E}}\colon \mathsf{E} \to \mathsf{M}$ *be a* C^r*-vector bundle, with the data prescribed in Sect. 3.2.1 to define the Riemannian metric* $\mathbb{G}_{\mathsf{E}}$ *on* E. *For* $m \in \mathbb{Z}_{\geq 0}$, *there exist* C^r*-vector bundle mappings*

$$(\widehat{A}_s^m, \mathrm{id}_{\mathsf{E}}) \in \mathrm{VB}^r(\mathrm{S}^s(\pi_{\mathsf{E}}^*\mathsf{T}^*\mathsf{M}); \mathrm{S}^m(\mathsf{T}^*\mathsf{E})), \qquad s \in \{0, 1, \dots, m\},$$

and

$$(\widehat{C}_s^m, \mathrm{id}_{\mathsf{E}}) \in \mathrm{VB}^r(\mathrm{S}^s(\pi_{\mathsf{E}}^*\mathsf{T}^*\mathsf{M}) \otimes \mathsf{V}^*\mathsf{E}; \mathrm{S}^m(\mathsf{T}^*\mathsf{E})), \qquad s \in \{0, 1, \dots, m-1\},$$

such that

$$\mathrm{Sym}_m \circ \nabla^{\mathsf{E},m} \lambda^{\mathrm{e}} = \sum_{s=0}^{m} \widehat{A}_s^m (\mathrm{Sym}_s \circ (\nabla^{\mathsf{M},\pi_{\mathsf{E}},s} \lambda)^{\mathrm{e}})$$
$$+ \sum_{s=0}^{m-1} \widehat{C}_s^m ((\mathrm{Sym}_s \otimes \mathrm{id}_{\mathsf{V}^*\mathsf{E}}) \circ (\nabla^{\mathsf{M},\pi_{\mathsf{E}},s} \lambda)^{\mathrm{v}})$$

for all $\lambda \in \Gamma^m(\mathsf{E}^*)$.

Proof The proof here follows along the lines of Lemma 3.56 below. □

The preceding lemma gives rise to an "inverse," which we state in the following lemma.

Lemma 3.52 (Iterated Symmetrised Covariant Differentials of Vertical Evaluations of Dual Sections II) *Let* $r \in \{\infty, \omega\}$ *and let* $\pi_{\mathsf{E}} \colon \mathsf{E} \to \mathsf{M}$ *be a* C^r*-vector bundle, with the data prescribed in Sect. 3.2.1 to define the Riemannian metric* $\mathbb{G}_{\mathsf{E}}$ *on* E*. For* $m \in \mathbb{Z}_{\geq 0}$*, there exist* C^r*-vector bundle mappings*

$$(\widehat{B}_s^m, \mathrm{id}_{\mathsf{E}}) \in \mathrm{VB}^r(\mathrm{S}^s(\mathsf{T}^*\mathsf{E}); \mathrm{S}^m(\pi_{\mathsf{E}}^*\mathsf{T}^*\mathsf{M})), \qquad s \in \{0, 1, \dots, m\},$$

and

$$(\widehat{D}_s^m, \mathrm{id}_{\mathsf{E}}) \in \mathrm{VB}^r(\mathrm{S}^s(\mathsf{T}^*\mathsf{E}) \otimes \mathsf{V}^*\mathsf{E}; \mathrm{S}^m(\pi_{\mathsf{E}}^*\mathsf{T}^*\mathsf{M})), \qquad s \in \{0, 1, \dots, m-1\},$$

such that

$$\mathrm{Sym}_m \circ (\nabla^{\mathsf{M},\pi_{\mathsf{E}},m} \lambda)^{\mathrm{e}} = \sum_{s=0}^{m} \widehat{B}_s^m (\mathrm{Sym}_s \circ \nabla^{\mathsf{E},s} \lambda^{\mathrm{e}})$$
$$+ \sum_{s=0}^{m-1} \widehat{D}_s^m ((\mathrm{Sym}_s \otimes \mathrm{id}_{\mathsf{V}^*\mathsf{E}}) \circ \nabla^{\mathsf{E},s} \lambda^{\mathrm{v}})$$

for all $\lambda \in \Gamma^m(\mathsf{E}^*)$.

Proof The proof here follows along the lines of Lemma 3.56 below. □

We can put together the previous four lemmata, along with Lemma 3.48, into the following decomposition result, which is to be regarded as the main result of this section.

Lemma 3.53 (Decomposition of Jets of Vertical Evaluations of Dual Sections) *Let* $r \in \{\infty, \omega\}$ *and let* $\pi_{\mathsf{E}} \colon \mathsf{E} \to \mathsf{M}$ *be a* C^r*-vector bundle, with the data prescribed in Sect. 3.2.1 to define the Riemannian metric* $\mathbb{G}_{\mathsf{E}}$ *on* E*. Then there exist* C^r*-vector* *bundle mappings*

$$A^m_{\nabla^{\mathsf{E}}} \in \mathrm{VB}^r(\mathsf{D}^{*m}\mathsf{E}; \mathsf{S}^{\leq m}(\pi^*_{\mathsf{E}}\mathsf{T}^*\mathsf{M})), \quad B^m_{\nabla^{\mathsf{E}}} \in \mathrm{VB}^r(\mathsf{D}^{*m}\mathsf{E}; \mathsf{S}^{\leq m}(\mathsf{T}^*\mathsf{E}))$$

defined by

$$A^m_{\nabla^{\mathsf{E}}}(j_m(\lambda^{\mathrm{e}})(e)) = \mathrm{Sym}_{\leq m}(\lambda^{\mathrm{e}}(e), (\nabla^{\pi_{\mathsf{E}}}\lambda)^{\mathrm{e}}(e), \dots, (\nabla^{\mathsf{M},\pi_{\mathsf{E}},m}\lambda)^{\mathrm{e}}(e)),$$

$$B^m_{\nabla^{\mathsf{E}}}(j_m(\lambda^{\mathrm{e}})(e)) = \mathrm{Sym}_{\leq m}(\lambda^{\mathrm{e}}(e), \nabla^{\mathsf{E}}\lambda^{\mathrm{e}}(e), \dots, \nabla^{\mathsf{E},m}\lambda^{\mathrm{e}}(e)).$$

Moreover, $A^m_{\nabla^{\mathsf{E}}}$ and $B^m_{\nabla^{\mathsf{E}}}$ are injective, and

$$\begin{aligned}
&B^m_{\nabla^{\mathsf{E}}} \circ (A^m_{\nabla^{\mathsf{E}}})^{-1} \circ \mathrm{Sym}_{\leq m}(\lambda^{\mathrm{e}}(e), (\nabla^{\pi_{\mathsf{E}}}\lambda)^{\mathrm{e}}(e), \dots, (\nabla^{\mathsf{M},\pi_{\mathsf{E}},m}\lambda)^{\mathrm{e}}(e)) \\
&= \left(\lambda^{\mathrm{e}}(e), \sum_{s=0}^{1} \widehat{A}^1_s(\mathrm{Sym}_s \circ (\nabla^{\mathsf{M},\pi_{\mathsf{E}},s}\lambda)^{\mathrm{e}}(e)), \dots, \sum_{s=0}^{m} \widehat{A}^m_s(\mathrm{Sym}_s \circ (\nabla^{\mathsf{M},\pi_{\mathsf{E}},s}\lambda)^{\mathrm{e}}(e))\right) \\
&\quad + \Bigg(0, \lambda^{\mathrm{v}}(e), \sum_{s=0}^{1} \widehat{C}^2_s((\mathrm{Sym}_s \otimes \mathrm{id}_{\mathsf{V}^*\mathsf{E}}) \circ (\nabla^{\mathsf{M},\pi_{\mathsf{E}},s}\lambda)^{\mathrm{v}}(e)), \dots, \\
&\qquad\qquad \sum_{s=0}^{m-1} \widehat{C}^m_s((\mathrm{Sym}_s \otimes \mathrm{id}_{\mathsf{V}^*\mathsf{E}}) \circ (\nabla^{\mathsf{M},\pi_{\mathsf{E}},s}\lambda)^{\mathrm{v}}(e))\Bigg)
\end{aligned}$$

and

$$\begin{aligned}
&A^m_{\nabla^{\mathsf{E}}} \circ (B^m_{\nabla^{\mathsf{E}}})^{-1} \circ \mathrm{Sym}_{\leq m}(\lambda^{\mathrm{e}}(e), \nabla^{\mathsf{E}}\lambda^{\mathrm{e}}(e), \dots, \nabla^{\mathsf{E},m}\lambda^{\mathrm{e}}(e)) \\
&= \left(\lambda^{\mathrm{e}}(e), \sum_{s=0}^{1} \widehat{B}^1_s(\mathrm{Sym}_s \circ \nabla^{\mathsf{E},s}\lambda^{\mathrm{e}}(e)), \dots, \sum_{s=0}^{m} \widehat{B}^m_s(\mathrm{Sym}_s \circ \nabla^{\mathsf{E},s}\lambda^{\mathrm{e}}(e))\right) \\
&\quad + \Bigg(0, \lambda^{\mathrm{v}}(e), \sum_{s=0}^{1} \widehat{D}^2_s((\mathrm{Sym}_s \otimes \mathrm{id}_{\mathsf{V}^*\mathsf{E}}) \circ \nabla^{\mathsf{E},s}\lambda^{\mathrm{v}}(e)), \dots, \\
&\qquad\qquad \sum_{s=0}^{m-1} \widehat{D}^m_s((\mathrm{Sym}_s \otimes \mathrm{id}_{\mathsf{V}^*\mathsf{E}}) \circ \nabla^{\mathsf{E},s}\lambda^{\mathrm{v}}(e))\Bigg),
\end{aligned}$$

where the vector bundle mappings $\widehat{A}^m_s$ and $\widehat{B}^m_s$, $s \in \{0, 1, \dots, m\}$, and $\widehat{C}^m_s$ and $\widehat{D}^m_s$, $s \in \{0, 1, \dots, m-1\}$, are as in Lemmata 3.51 and 3.52.

3.3.7 Isomorphisms for Vertical Evaluations of Endomorphisms

Next we consider vertical evaluations of endomorphisms, i.e., the mapping given by

$$\Gamma^r(\mathrm{T}^1_1(\mathsf{E})) \ni L \mapsto L^{\mathrm{e}} \in \Gamma^r(\mathsf{TE}).$$

To study the relationship between the decomposition of jets of L with those of L^{e}, we denote

$$\mathsf{C}^{*m}_e\mathsf{E} = \{j_m L^{\mathrm{e}}(e) \mid L \in \Gamma^m(\mathrm{T}^1_1(\mathsf{E}))\}.$$

By (1.9), we have

$$\mathsf{C}^{*m}_e\mathsf{E} \subseteq \mathsf{P}^{*m}_e\mathsf{E} \otimes \mathsf{V}_e\mathsf{E}.$$

As we shall see, one can be a little more explicit about the nature of $\mathsf{C}^{*m}_e\mathsf{E}$, and see that

$$\mathsf{C}^{*m}_e\mathsf{E} \simeq (\mathsf{P}^{*m}_e\mathsf{E} \otimes \mathsf{V}^*\mathsf{E} \otimes \mathsf{V}_e\mathsf{E}) \oplus (\mathsf{P}^{*\mathsf{E}}_{m-1} \otimes \mathsf{V}^*\mathsf{E} \otimes \mathsf{V}_e\mathsf{E}).$$

However, as in the previous section, this sort of isomorphism is too cumbersome to make explicit. As with the constructions of the preceding sections, we wish to use Lemma 2.15 to provide a decomposition of $\mathsf{C}^{*m}\mathsf{E}$, and to do so we need to understand the covariant derivatives

$$\nabla^{\mathsf{E},m} L^{\mathrm{e}} \triangleq \underbrace{\nabla^{\mathsf{E}} \cdots \nabla^{\mathsf{E}}}_{m \text{ times}} L^{\mathrm{e}}, \qquad m \in \mathbb{Z}_{\geq 0}.$$

The results in this section have a slightly different character than in the preceding sections, so we provide complete proofs.

The first result we give is the following.

Lemma 3.54 (Iterated Covariant Differentials of Vertical Evaluations of Endomorphisms I) *Let $r \in \{\infty, \omega\}$ and let $\pi_{\mathsf{E}} \colon \mathsf{E} \to \mathsf{M}$ be a C^r-vector bundle, with the data prescribed in Sect. 3.2.1 to define the Riemannian metric $\mathbb{G}_{\mathsf{E}}$ on E. For $m \in \mathbb{Z}_{\geq 0}$, there exist C^r-vector bundle mappings*

$$(A^m_s, \mathrm{id}_{\mathsf{E}}) \in \mathrm{VB}^r(\mathrm{T}^s(\pi^*_{\mathsf{E}}\mathsf{T}^*\mathsf{M}) \otimes \mathsf{VE}; \mathrm{T}^m(\mathsf{T}^*\mathsf{E}) \otimes \mathsf{VE}), \qquad s \in \{0, 1, \dots, m\},$$

and

$$(C^m_s, \mathrm{id}_{\mathsf{E}}) \in \mathrm{VB}^r(\mathrm{T}^s(\pi^*_{\mathsf{E}}\mathsf{T}^*\mathsf{M}) \otimes \mathrm{T}^1_1(\mathsf{VE}); \mathrm{T}^{m-1}(\mathsf{T}^*\mathsf{E}) \otimes \mathrm{T}^1_1(\mathsf{VE})),$$
$$s \in \{0, 1, \dots, m-1\},$$

such that

$$\nabla^{\mathsf{E},m} L^{\mathrm{e}} = \sum_{s=0}^{m} A_s^m((\nabla^{\mathsf{M},\pi_{\mathsf{E}},s} L)^{\mathrm{e}}) + \sum_{s=0}^{m-1} C_s^m((\nabla^{\mathsf{M},\pi_{\mathsf{E}},s} L)^{\mathrm{v}})$$

for all $L \in \Gamma^m(\mathrm{T}_1^1(\mathsf{E}))$. *(Here we regard* $\mathrm{T}_1^1(\mathsf{VE})$ *as a subbundle of* $\mathsf{T}^*\mathsf{E} \otimes \mathsf{VE}$ *by the mapping*

$$\mathrm{T}_1^1(\mathsf{VE}) \ni A \mapsto A \circ \mathrm{ver} \in \mathsf{T}^*\mathsf{E} \otimes \mathsf{VE}).$$

Moreover, the vector bundle mappings $A_0^m, A_1^m, \dots, A_m^m$ *and* $C_0^m, C_1^m, \dots, C_{m-1}^m$ *satisfy the recursion relations prescribed by*

$$A_0^0(\beta_0) = \beta_0, \quad A_1^1(\beta_1) = \beta_1, \quad A_0^1(\beta_0) = \mathrm{Ins}_1(\beta_0, B_{\pi_{\mathsf{E}}}), \quad C_0^1(\gamma_0) = \gamma_0,$$

and, for $m \geq 2$,

$$\begin{aligned}
A_{m+1}^{m+1}(\beta_{m+1}) &= \beta_{m+1} \\
A_m^{m+1}(\beta_m) &= A_{m-1}^m \otimes \mathrm{id}_{\mathsf{T}^*\mathsf{E}}(\beta_m) - \sum_{j=1}^{m} \mathrm{Ins}_j(\beta_m, B_{\pi_{\mathsf{E}}}) + \mathrm{Ins}_{m+1}(\beta_m, B_{\pi_{\mathsf{E}}}^*) \\
A_s^{m+1}(\beta_s) &= (\nabla^{\mathsf{E}} A_s^m)(\beta_m) + A_{s-1}^m \otimes \mathrm{id}_{\mathsf{T}^*\mathsf{E}}(\beta_s) \\
&\quad - \sum_{j=1}^{s} A_s^m \otimes \mathrm{id}_{\mathsf{T}^*\mathsf{E}}(\mathrm{Ins}_j(\beta_s, B_{\pi_{\mathsf{E}}})) \\
&\quad + A_s^m \otimes \mathrm{id}_{\mathsf{T}^*\mathsf{E}}(\mathrm{Ins}_{s+1}(\beta_s, B_{\pi_{\mathsf{E}}}^*)), \quad s \in \{1, \dots, m-1\}, \\
A_0^{m+1}(\beta_0) &= (\nabla^{\mathsf{E}} A_0^m)(\beta_0) - A_0^m \otimes \mathrm{id}_{\mathsf{T}^*\mathsf{E}}(\mathrm{Ins}_1(\beta_0, B_{\pi_{\mathsf{E}}}^*))
\end{aligned}$$

and

$$\begin{aligned}
C_m^{m+1}(\gamma_m) &= C_{m-1}^m \otimes \mathrm{id}_{\mathsf{T}^*\mathsf{E}}(\gamma_m) + \gamma_m \\
C_s^{m+1}(\gamma_s) &= A_s^m \otimes \mathrm{id}_{\mathsf{T}^*\mathsf{E}}(\gamma_s) + (\nabla^{\mathsf{E}} C_s^m)(\gamma_s) + C_{s-1}^m \otimes \mathrm{id}_{\mathsf{T}^*\mathsf{E}}(\gamma_s) \\
&\quad - \sum_{j=1}^{s+1} C_s^m \otimes \mathrm{id}_{\mathsf{T}^*\mathsf{E}}(\mathrm{Ins}_j(\gamma_s, B_{\pi_{\mathsf{E}}})) + C_s^m \otimes \mathrm{id}_{\mathsf{T}^*\mathsf{E}}(\mathrm{Ins}_{s+1}(\gamma_s, B_{\pi_{\mathsf{E}}}^*)), \\
&\quad s \in \{1, \dots, m-1\}, \\
C_0^{m+1}(\gamma_0) &= A_0^m \otimes \mathrm{id}_{\mathsf{T}^*\mathsf{E}}(\gamma_0) + (\nabla^{\mathsf{E}} C_0^m)(\gamma_0) - C_0^m \otimes \mathrm{id}_{\mathsf{T}^*\mathsf{E}}(\mathrm{Ins}_1(\gamma_0, B_{\pi_{\mathsf{E}}})) \\
&\quad + C_0^m \otimes \mathrm{id}_{\mathsf{T}^*\mathsf{E}}(\mathrm{Ins}_2(\gamma_0, B_{\pi_{\mathsf{E}}}^*)),
\end{aligned}$$

where $\beta_s \in \mathrm{T}^s(\pi_{\mathsf{E}}^*\mathsf{T}^*\mathsf{M}) \otimes \mathsf{VE}$, $s \in \{0, 1, \dots, m\}$, *and* $\gamma_s \in \mathrm{T}^s(\pi_{\mathsf{E}}^*\mathsf{T}^*\mathsf{M}) \otimes \mathrm{T}^1_1(\mathsf{VE})$, $s \in \{0, 1, \dots, m-1\}$.

Proof The assertion is clearly true for $m = 0$ and, for $m = 1$, we have

$$\nabla^{\mathsf{E}} L^{\mathrm{e}} = (\nabla^{\pi_{\mathsf{E}}} L)^{\mathrm{e}} + \mathrm{Ins}_1(L, B_{\pi_{\mathsf{E}}}) + L^{\mathrm{v}}$$

by Lemma 3.16(vii), which gives the result for $m = 1$. Thus suppose the result true for $m \geq 2$ so that

$$\nabla^{\mathsf{E},m} L^{\mathrm{e}} = \sum_{s=0}^{m} A_s^m((\nabla^{\mathsf{M},\pi_{\mathsf{E}},s} L)^{\mathrm{e}}) + \sum_{s=0}^{m-1} C_s^m((\nabla^{\mathsf{M},\pi_{\mathsf{E}},s} L)^{\mathrm{v}})$$

for vector bundle mappings A_s^m and C_s^m satisfying the stated recursion relations. We then compute

$$\begin{aligned}
\nabla^{\mathsf{E},m+1} L^{\mathrm{e}} = &\sum_{s=0}^{m} (\nabla^{\mathsf{E}} A_s^m)((\nabla^{\mathsf{M},\pi_{\mathsf{E}},s} L)^{\mathrm{e}}) + \sum_{s=0}^{m} A_s^m \otimes \mathrm{id}_{\mathsf{T}^*\mathsf{E}}(\nabla^{\mathsf{E}}(\nabla^{\mathsf{M},\pi_{\mathsf{E}},s} L)^{\mathrm{e}}) \\
&+ \sum_{s=0}^{m-1} (\nabla^{\mathsf{E}} C_s^m)((\nabla^{\mathsf{M},\pi_{\mathsf{E}},s} L)^{\mathrm{v}}) + \sum_{s=0}^{m-1} C_s^m \otimes \mathrm{id}_{\mathsf{T}^*\mathsf{E}}(\nabla^{\mathsf{E}}(\nabla^{\mathsf{M},\pi_{\mathsf{E}},s} L)^{\mathrm{v}}) \\
= &\sum_{s=0}^{m} (\nabla^{\mathsf{E}} A_s^m)((\nabla^{\mathsf{M},\pi_{\mathsf{E}},s} L)^{\mathrm{e}}) + \sum_{s=0}^{m} A_s^m \otimes \mathrm{id}_{\mathsf{T}^*\mathsf{E}}((\nabla^{\mathsf{M},\pi_{\mathsf{E}},s+1} L)^{\mathrm{e}}) \\
&- \sum_{s=1}^{m} \sum_{j=1}^{s} A_s^m \otimes \mathrm{id}_{\mathsf{T}^*\mathsf{E}}(\mathrm{Ins}_j((\nabla^{\mathsf{M},\pi_{\mathsf{E}},s} L)^{\mathrm{e}}, B_{\pi_{\mathsf{E}}})) \\
&+ \sum_{s=0}^{m} A_s^m \otimes \mathrm{id}_{\mathsf{T}^*\mathsf{E}}(\mathrm{Ins}_{s+1}((\nabla^{\mathsf{M},\pi_{\mathsf{E}},s} L)^{\mathrm{e}}, B^*_{\pi_{\mathsf{E}}})) \\
&+ \sum_{s=0}^{m} A_s^m \otimes \mathrm{id}_{\mathsf{T}^*\mathsf{E}}((\nabla^{\mathsf{M},\pi_{\mathsf{E}},s} L)^{\mathrm{v}}) + \sum_{s=0}^{m-1} (\nabla^{\mathsf{E}} C_s^m)((\nabla^{\mathsf{M},\pi_{\mathsf{E}},s} L)^{\mathrm{v}}) \\
&+ \sum_{s=0}^{m-1} C_s^m \otimes \mathrm{id}_{\mathsf{T}^*\mathsf{E}}((\nabla^{\mathsf{M},\pi_{\mathsf{E}},s+1} L)^{\mathrm{v}}) \\
&- \sum_{s=1}^{m-1} \sum_{j=1}^{s} C_s^m \otimes \mathrm{id}_{\mathsf{T}^*\mathsf{E}}(\mathrm{Ins}_j((\nabla^{\mathsf{M},\pi_{\mathsf{E}},s} L)^{\mathrm{v}}, B_{\pi_{\mathsf{E}}})) \\
&+ \sum_{s=0}^{m-1} C_s^m \otimes \mathrm{id}_{\mathsf{T}^*\mathsf{E}}(\mathrm{Ins}_{s+1}((\nabla^{\mathsf{M},\pi_{\mathsf{E}},s} L)^{\mathrm{v}}, B^*_{\pi_{\mathsf{E}}}))
\end{aligned}$$

$$
\begin{aligned}
&= (\nabla^{\mathsf{M},\pi_{\mathsf{E}},m+1}L)^{\mathrm{e}} + \Bigg(A^m_{m-1} \otimes \mathrm{id}_{\mathsf{T}^*\mathsf{E}}((\nabla^{\mathsf{M},\pi_{\mathsf{E}},m}L)^{\mathrm{e}}) \\
&\quad - \sum_{j=1}^{m} \mathrm{Ins}_j((\nabla^{\mathsf{M},\pi_{\mathsf{E}},m}L)^{\mathrm{e}}, B_{\pi_{\mathsf{E}}}) + \mathrm{Ins}_{m+1}((\nabla^{\mathsf{M},\pi_{\mathsf{E}},m}L)^{\mathrm{e}}, B^*_{\pi_{\mathsf{E}}}) \\
&\quad + (\nabla^{\mathsf{M},\pi_{\mathsf{E}},m}L)^{\mathrm{v}} + C^m_{m-1} \otimes \mathrm{id}_{\mathsf{T}^*\mathsf{E}}((\nabla^{\mathsf{M},\pi_{\mathsf{E}},m}L)^{\mathrm{v}}) \Bigg) \\
&\quad + \Bigg(\sum_{s=1}^{m-1} (\nabla^{\mathsf{E}} A^m_s)((\nabla^{\mathsf{M},\pi_{\mathsf{E}},s}L)^{\mathrm{e}}) + \sum_{s=1}^{m-1} A^m_{s-1} \otimes \mathrm{id}_{\mathsf{T}^*\mathsf{E}}((\nabla^{\mathsf{M},\pi_{\mathsf{E}},s}L)^{\mathrm{e}}) \\
&\quad - \sum_{s=1}^{m-1}\sum_{j=1}^{s} A^m_s \otimes \mathrm{id}_{\mathsf{T}^*\mathsf{E}}(\mathrm{Ins}_j((\nabla^{\mathsf{M},\pi_{\mathsf{E}},s}L)^{\mathrm{e}}), B_{\pi_{\mathsf{E}}}) \\
&\quad + \sum_{s=1}^{m-1} A^m_s \otimes \mathrm{id}_{\mathsf{T}^*\mathsf{E}}(\mathrm{Ins}_{s+1}((\nabla^{\mathsf{M},\pi_{\mathsf{E}},s}L)^{\mathrm{e}}), B^*_{\pi_{\mathsf{E}}}) \\
&\quad + \sum_{s=1}^{m-1} A^m_s \otimes \mathrm{id}_{\mathsf{T}^*\mathsf{E}}((\nabla^{\mathsf{M},\pi_{\mathsf{E}},s}L)^{\mathrm{v}}) + \sum_{s=1}^{m-1} (\nabla^{\mathsf{E}} C^m_s)((\nabla^{\mathsf{M},\pi_{\mathsf{E}},s}L)^{\mathrm{v}}) \\
&\quad + \sum_{s=1}^{m-1} C^m_{s-1} \otimes \mathrm{id}_{\mathsf{T}^*\mathsf{E}}((\nabla^{\mathsf{M},\pi_{\mathsf{E}},s}L)^{\mathrm{v}}) \\
&\quad - \sum_{s=0}^{m-1}\sum_{j=1}^{s} C^m_s \otimes \mathrm{id}_{\mathsf{T}^*\mathsf{E}}(\mathrm{Ins}_j((\nabla^{\mathsf{M},\pi_{\mathsf{E}},s}L)^{\mathrm{v}}, B_{\pi_{\mathsf{E}}})) \\
&\quad + \sum_{s=1}^{m-1} C^m_s \otimes \mathrm{id}_{\mathsf{T}^*\mathsf{E}}(\mathrm{Ins}_{s+1}((\nabla^{\mathsf{M},\pi_{\mathsf{E}},s}L)^{\mathrm{v}}, B^*_{\pi_{\mathsf{E}}})) \Bigg) \\
&\quad + (\nabla^{\mathsf{E}} A^m_0)(L^{\mathrm{e}}) + A^m_0 \otimes \mathrm{id}_{\mathsf{T}^*\mathsf{E}}(\mathrm{Ins}_1(L^{\mathrm{e}}, B^*_{\pi_{\mathsf{E}}})) + A^m_0 \otimes \mathrm{id}_{\mathsf{T}^*\mathsf{E}}(L^{\mathrm{v}}) \\
&\quad + (\nabla^{\mathsf{E}} C^m_0)(L^{\mathrm{v}}) - C^m_0 \otimes \mathrm{id}_{\mathsf{T}^*\mathsf{E}}(\mathrm{Ins}_1(L^{\mathrm{v}}, B_{\pi_{\mathsf{E}}})) \\
&\quad + C^m_0 \otimes \mathrm{id}_{\mathsf{T}^*\mathsf{E}}(\mathrm{Ins}_2(L^{\mathrm{v}}, B^*_{\pi_{\mathsf{E}}})).
\end{aligned}
$$

From these calculations, the lemma follows. □

Now we "invert" the constructions from the preceding lemma.

Lemma 3.55 (Iterated Covariant Differentials of Vertical Evaluations of Endomorphisms II) *Let $r \in \{\infty, \omega\}$ and let $\pi_{\mathsf{E}} \colon \mathsf{E} \to \mathsf{M}$ be a C^r-vector bundle, with the data prescribed in Sect. 3.2.1 to define the Riemannian metric $\mathbb{G}_{\mathsf{E}}$ on E. For*

$m \in \mathbb{Z}_{\geq 0}$, there exist C^r-vector bundle mappings

$$(B_s^m, \mathrm{id}_{\mathsf{E}}) \in \mathrm{VB}^r(\mathrm{T}^s(\mathsf{T}^*\mathsf{E}) \otimes \mathsf{VE}; \mathrm{T}^m(\pi_{\mathsf{E}}^*\mathsf{T}^*\mathsf{M}) \otimes \mathsf{VE}), \qquad s \in \{0, 1, \dots, m\},$$

and

$$(D_s^m, \mathrm{id}_{\mathsf{E}}) \in \mathrm{VB}^r(\mathrm{T}^s(\mathsf{T}^*\mathsf{E}) \otimes \mathrm{T}_1^1(\mathsf{VE}); \mathrm{T}^{m-1}(\pi_{\mathsf{E}}^*\mathsf{T}^*\mathsf{M}) \otimes \mathrm{T}_1^1(\mathsf{VE})), \qquad s \in \{0, 1, \dots, m-1\},$$

such that

$$(\nabla^{\mathsf{M},\pi_{\mathsf{E}},m} L)^{\mathrm{e}} = \sum_{s=0}^{m} B_s^m(\nabla^{\mathsf{E},s} L^{\mathrm{e}}) + \sum_{s=0}^{m-1} D_s^m(\nabla^{\mathsf{E},s} L^{\mathrm{v}})$$

for all $L \in \Gamma^m(\mathrm{T}_1^1(\mathsf{E}))$. Moreover, the vector bundle mappings $B_0^m, B_1^m, \dots, B_m^m$ and $D_0^m, D_1^m, \dots, D_{m-1}^m$ satisfy the recursion relations prescribed by $B_0^0(\alpha_0) = \alpha_0$, $D_0^1(\gamma_0) = \gamma_0$,

$$\begin{aligned}
B_{m+1}^{m+1}(\alpha_{m+1}) &= \alpha_{m+1} \\
B_m^{m+1}(\alpha_m) &= B_{m-1}^m \otimes \mathrm{id}_{\mathsf{T}^*\mathsf{E}}(\alpha_m) + \sum_{j=1}^{m} \mathrm{Ins}_j(\alpha_m, B_{\pi_{\mathsf{E}}}) - \mathrm{Ins}_{m+1}(\alpha, B_{\pi_{\mathsf{E}}}^*) \\
B_s^{m+1} &= (\nabla^{\mathsf{E}} B_s^m)(\alpha_s) + B_{s-1}^m \otimes \mathrm{id}_{\mathsf{T}^*\mathsf{E}}(\alpha_s) + \sum_{j=1}^{m} \mathrm{Ins}_j(B_s^m(\alpha_s), B_{\pi_{\mathsf{E}}}) \\
&\quad - \mathrm{Ins}_{m+1}(B_s^m(\alpha_s), B_{\pi_{\mathsf{E}}}^*), \quad s \in \{1, \dots, m-1\}, \\
B_0^{m+1}(\alpha_0) &= (\nabla^{\mathsf{E}} B_0^m)(\alpha_0) + \sum_{j=1}^{m} \mathrm{Ins}_j(B_0^m(\alpha_0), B_{\pi_{\mathsf{E}}}) - \mathrm{Ins}_{m+1}(B_0^m(\alpha_0), B_{\pi_{\mathsf{E}}}^*)
\end{aligned}$$

and

$$\begin{aligned}
D_m^{m+1}(\gamma_m) &= D_{m-1}^m \otimes \mathrm{id}_{\mathsf{T}^*\mathsf{E}}(\gamma_m) - \gamma_m \\
D_s^m(\gamma_s) &= (\nabla^{\mathsf{E}} D_s^m)(\gamma_s) + D_{s-1}^m \otimes \mathrm{id}_{\mathsf{T}^*\mathsf{E}}(\gamma_s) - \overline{B}_s^m(\gamma_s), \quad s \in \{1, \dots, m-1\}, \\
D_0^{m+1} &= (\nabla^{\mathsf{E}} D_0^m)(\gamma_0) - \overline{B}_0^m(\gamma_0)
\end{aligned}$$

for $\alpha_s \in \mathrm{T}^s(\mathsf{T}^*\mathsf{E} \otimes \mathsf{VE})$, $s \in \{0, 1, \dots, m+1\}$, *and* $\gamma_s \in \mathrm{T}^s(\mathsf{T}^*\mathsf{E}) \otimes \mathrm{T}^1_1(\mathsf{VE})$, $s \in \{0, 1, \dots, m\}$, *and where*

$$(\overline{B}^m_s, \mathrm{id}_{\mathsf{E}}) \in \mathrm{VB}^r(\mathrm{T}^s(\mathsf{T}^*\mathsf{E}) \otimes \mathrm{T}^1_1(\mathsf{VE}); \mathrm{T}^m(\pi_{\mathsf{E}}^*\mathsf{T}^*\mathsf{M}) \otimes \mathrm{T}^1_1(\mathsf{VE})), \quad s \in \{0, 1, \dots, m\},$$

are the vector bundle mappings from Lemma 3.45.

Proof The assertion is clearly true for $m = 0$, so suppose it true for $m \in \mathbb{Z}_{>0}$. Thus

$$(\nabla^{\mathsf{M},\pi_{\mathsf{E}},m} L)^{\mathrm{e}} = \sum_{s=0}^{m} B^m_s(\nabla^{\mathsf{E},s} L^{\mathrm{e}}) + \sum_{s=0}^{m-1} D^m_s((\nabla^{\mathsf{E},s} L)^{\mathrm{v}}). \tag{3.34}$$

Working on the left-hand side of this equation, using Lemma 3.16(vii), we have

$$\begin{aligned}
\nabla^{\mathsf{E}}(\nabla^{\mathsf{M},\pi_{\mathsf{E}},m} L)^{\mathrm{e}} &= (\nabla^{\mathsf{M},\pi_{\mathsf{E}},m+1} L)^{\mathrm{e}} - \sum_{j=1}^{m} \mathrm{Ins}_j((\nabla^{\mathsf{M},\pi_{\mathsf{E}},m} L)^{\mathrm{e}}, B_{\pi_{\mathsf{E}}}) \\
&\quad + \mathrm{Ins}_{m+1}((\nabla^{\mathsf{M},\pi_{\mathsf{E}},m} L)^{\mathrm{e}}, B^*_{\pi_{\mathsf{E}}}) + (\nabla^{\mathsf{M},\pi_{\mathsf{E}},m} L)^{\mathrm{v}} \\
&= (\nabla^{\mathsf{M},\pi_{\mathsf{E}},m+1} L)^{\mathrm{e}} - \sum_{s=0}^{m}\sum_{j=1}^{m} \mathrm{Ins}_j(B^m_s(\nabla^{\mathsf{E},s} L^{\mathrm{e}}), B_{\pi_{\mathsf{E}}}) \\
&\quad + \sum_{s=0}^{m} \mathrm{Ins}_{m+1}(B^m_s(\nabla^{\mathsf{E},s} L^{\mathrm{e}}), B^*_{\pi_{\mathsf{E}}}) + \sum_{s=0}^{m} \overline{B}^m_s(\nabla^{\mathsf{E},s} L^{\mathrm{v}}).
\end{aligned}$$

Working on the right-hand side of (3.34),

$$\begin{aligned}
\nabla^{\mathsf{E}}(\nabla^{\mathsf{M},\pi_{\mathsf{E}},m} L)^{\mathrm{e}} &= \sum_{s=0}^{m} (\nabla^{\mathsf{E}} B^m_s)(\nabla^{\mathsf{E},s} L^{\mathrm{e}}) + \sum_{s=0}^{m} B^m_s \otimes \mathrm{id}_{\mathsf{T}^*\mathsf{E}}(\nabla^{\mathsf{E},s+1} L^{\mathrm{e}}) \\
&\quad + \sum_{s=0}^{m-1} (\nabla^{\mathsf{E}} D^m_s)(\nabla^{\mathsf{E},s} L^{\mathrm{v}}) + \sum_{s=0}^{m-1} D^m_s \otimes \mathrm{id}_{\mathsf{T}^*\mathsf{E}}(\nabla^{\mathsf{E},s+1} L^{\mathrm{v}}).
\end{aligned}$$

Combining the preceding two computations,

$$\begin{aligned}
&(\nabla^{\mathsf{M},\pi_{\mathsf{E}},m+1} L)^{\mathrm{e}} \\
&\quad = \sum_{s=0}^{m} (\nabla^{\mathsf{E}} B^m_s)(\nabla^{\mathsf{E},s} L^{\mathrm{e}}) + \sum_{s=0}^{m} B^m_s \otimes \mathrm{id}_{\mathsf{T}^*\mathsf{E}}(\nabla^{\mathsf{E},s+1} L^{\mathrm{e}}) \\
&\qquad + \sum_{s=0}^{m-1} (\nabla^{\mathsf{E}} D^m_s)(\nabla^{\mathsf{E},s} L^{\mathrm{v}}) + \sum_{s=0}^{m-1} D^m_s \otimes \mathrm{id}_{\mathsf{T}^*\mathsf{E}}(\nabla^{\mathsf{E},s+1} L^{\mathrm{v}})
\end{aligned}$$

$$
\begin{aligned}
&+ \sum_{s=0}^{m} \sum_{j=1}^{m} \mathrm{Ins}_j(B_s^m(\nabla^{\mathsf{E},s} L^{\mathrm{e}}), B_{\pi_{\mathsf{E}}}) - \sum_{s=1}^{m} \mathrm{Ins}_{m+1}(B_s^m(\nabla^{\mathsf{E},s} L^{\mathrm{e}}), B_{\pi_{\mathsf{E}}}^*) \\
&- \sum_{s=0}^{m} \overline{B}_s^m(\nabla^{\mathsf{E},s} L^{\mathrm{v}}) \\
&= \nabla^{\mathsf{E},m+1} L^{\mathrm{e}} + \Bigg(B_{m-1}^m \otimes \mathrm{id}_{\mathsf{T}^*\mathsf{E}}(\nabla^{\mathsf{E},m} L^{\mathrm{e}}) + D_{m-1}^m \otimes \mathrm{id}_{\mathsf{T}^*\mathsf{E}}(\nabla^{\mathsf{E},m} L^{\mathrm{v}}) \\
&+ \sum_{j=1}^{m} \mathrm{Ins}_j(\nabla^{\mathsf{E},m} L^{\mathrm{e}}, B_{\pi_{\mathsf{E}}}) - \mathrm{Ins}_{m+1}(\nabla^{\mathsf{E},m} L^{\mathrm{e}}, B_{\pi_{\mathsf{E}}}^*) - (\nabla^{\mathsf{E},m} L^{\mathrm{v}}) \Bigg) \\
&+ \Bigg(\sum_{s=1}^{m-1} (\nabla^{\mathsf{E}} B_s^m)(\nabla^{\mathsf{E},s} L^{\mathrm{e}}) + \sum_{s=1}^{m-1} B_{s-1}^m \otimes \mathrm{id}_{\mathsf{T}^*\mathsf{E}}(\nabla^{\mathsf{E},s} L^{\mathrm{e}}) \\
&+ \sum_{s=1}^{m-1} (\nabla^{\mathsf{E}} D_s^m)(\nabla^{\mathsf{E},s} L^{\mathrm{v}}) + \sum_{s=1}^{m-1} D_{s-1}^m \otimes \mathrm{id}_{\mathsf{T}^*\mathsf{E}}(\nabla^{\mathsf{E},s} L^{\mathrm{v}}) \\
&+ \sum_{s=1}^{m-1} \sum_{j=1}^{m} \mathrm{Ins}_j(B_s^m(\nabla^{\mathsf{E},s} L^{\mathrm{e}}), B_{\pi_{\mathsf{E}}}) \\
&- \sum_{s=1}^{m-1} \mathrm{Ins}_{m+1}(B_s^m(\nabla^{\mathsf{E},s} L^{\mathrm{e}}), B_{\pi_{\mathsf{E}}}^*) - \sum_{s=1}^{m-1} \overline{B}_s^m(\nabla^{\mathsf{E},s} L^{\mathrm{v}}) \Bigg) \\
&+ \Bigg((\nabla^{\mathsf{E}} B_0^m)(L^{\mathrm{e}}) + (\nabla^{\mathsf{E}} D_0^m)(L^{\mathrm{v}}) + \sum_{j=1}^{m} \mathrm{Ins}_j(B_0^m(L^{\mathrm{e}}), B_{\pi_{\mathsf{E}}}) \\
&- \mathrm{Ins}_{m+1}(B_0^m(L^{\mathrm{e}}), B_{\pi_{\mathsf{E}}}^*) - \overline{B}_0^m(L^{\mathrm{v}}) \Bigg).
\end{aligned}
$$

The lemma follows from these computations. □

Next we turn to symmetrised versions of the preceding lemmata. We show that the preceding two lemmata induce corresponding mappings between symmetric tensors.

Lemma 3.56 (Iterated Symmetrised Covariant Differentials of Vertical Evaluations of Endomorphisms I) *Let $r \in \{\infty, \omega\}$ and let $\pi_{\mathsf{E}} \colon \mathsf{E} \to \mathsf{M}$ be a C^r-vector bundle, with the data prescribed in Sect. 3.2.1 to define the Riemannian metric $\mathbb{G}_{\mathsf{E}}$ on E. For $m \in \mathbb{Z}_{\geq 0}$, there exist C^r-vector bundle mappings*

$$
(\widehat{A}_s^m, \mathrm{id}_{\mathsf{E}}) \in \mathrm{VB}^r(\mathrm{S}^s(\pi_{\mathsf{E}}^*\mathsf{T}^*\mathsf{M}) \otimes \mathsf{VE}; \mathrm{S}^m(\mathsf{T}^*\mathsf{E}) \otimes \mathsf{VE}), \qquad s \in \{0, 1, \dots, m\},
$$

and

$$(\widehat{C}^m_s, \mathrm{id}_{\mathsf{E}}) \in \mathrm{VB}^r(\mathrm{S}^s(\pi_{\mathsf{E}}^*\mathsf{T}^*\mathsf{M}) \otimes \mathrm{T}^1_1(\mathsf{VE}); \mathrm{S}^m(\mathsf{T}^*\mathsf{E}) \otimes \mathsf{VE}), \quad s \in \{0, 1, \dots, m-1\},$$

such that

$$(\mathrm{Sym}_m \otimes \mathrm{id}_{\mathsf{VE}}) \circ \nabla^{\mathsf{E},m} L^{\mathrm{e}} = \sum_{s=0}^{m} \widehat{A}^m_s((\mathrm{Sym}_s \otimes \mathrm{id}_{\mathsf{VE}}) \circ (\nabla^{\mathsf{M},\pi_{\mathsf{E}},s} L)^{\mathrm{e}} + \sum_{s=0}^{m-1} \widehat{C}^m_s((\mathrm{Sym}_s \otimes \mathrm{id}_{\mathrm{T}^1_1(\mathsf{VE})}) \circ (\nabla^{\mathsf{M},\pi_{\mathsf{E}},s} L)^{\mathrm{v}})$$

for all $L \in \Gamma^m(\mathrm{T}^1_1(\mathsf{E}))$.

Proof Following along the lines of the proof of Lemma 3.31, we define $\widehat{A}^m_s$ by requiring that

$$\widehat{A}^m_s((\mathrm{Sym}_s \otimes \mathrm{id}_{\mathsf{VE}}) \circ (\nabla^{\mathsf{M},\pi_{\mathsf{E}},s} L)^{\mathrm{v}}) = (\mathrm{Sym}_m \otimes \mathrm{id}_{\mathsf{VE}}) \circ A^m_s((\nabla^{\mathsf{M},\pi_{\mathsf{E}},s} L)^{\mathrm{v}}),$$

and $\widehat{C}^m_s$ by requiring that

$$\widehat{C}^m_s((\mathrm{Sym}_s \otimes \mathrm{id}_{\mathrm{T}^1_1(\mathsf{VE})}) \circ (\nabla^{\mathsf{M},\pi_{\mathsf{E}},s} L)^{\mathrm{e}}) = (\mathrm{Sym}_m \otimes \mathrm{id}_{\mathsf{VE}}) \circ C^m_s((\nabla^{\mathsf{M},\pi_{\mathsf{E}},s} L)^{\mathrm{e}}).$$

That this definition of $\widehat{A}^m_s$ makes sense follows exactly as in the proof of Lemma 3.31. Let us see how the same arguments also apply to the definition of $\widehat{C}^m_s$.

For $m \in \mathbb{Z}_{>0}$, we define $C^m : \mathrm{T}^{\leq m-1}(\pi_{\mathsf{E}}^*\mathsf{T}^*\mathsf{M}) \otimes \mathrm{T}^1_1(\mathsf{VE}) \to \mathrm{T}^{\leq m}(\mathsf{T}^*\mathsf{E}) \otimes \mathsf{VE}$ by

$$C^m(L^{\mathrm{v}}, (\nabla^{\pi_{\mathsf{E}}} L)^{\mathrm{v}}, \dots, (\nabla^{\mathsf{M},\pi_{\mathsf{E}},m-1} L)^{\mathrm{v}}) = \left(C^1_0(L^{\mathrm{v}}), \sum_{s=0}^{1} C^2_s((\nabla^{\mathsf{M},\pi_{\mathsf{E}},s} L)^{\mathrm{v}}), \dots, \sum_{s=0}^{m-1} C^m_s((\nabla^{\mathsf{M},\pi_{\mathsf{E}},s} L)^{\mathrm{v}})\right),$$

making the identification of $\mathrm{T}^1_1(\mathsf{VE})$ with a subspace of $\mathsf{T}^*\mathsf{E} \otimes \mathsf{VE}$ as in the statement of Lemma 3.54. Note that we have a natural mapping

$$\mathsf{T}^*\mathsf{E} \otimes \mathsf{T}^{*m-1}\mathsf{E} \to \mathsf{T}^{*m}\mathsf{E}$$

as in Sect. 3.2.4. This then induces a mapping

$$P_m : (\mathbb{R}_{\mathsf{E}} \oplus \mathsf{T}^{*m-1}\mathsf{E}) \otimes \mathrm{T}^1_1(\mathsf{VE}) \to (\mathbb{R}_{\mathsf{E}} \otimes \mathsf{T}^{*m}\mathsf{E}) \otimes \mathsf{VE}.$$

Now define

$$\widehat{P}_m\colon \pi_{\mathsf{E}}^*(\mathbb{R}_{\mathsf{M}} \oplus \mathsf{T}^{*m-1}\mathsf{M}) \otimes \mathsf{T}^1_1(\mathsf{VE}) \to (\mathbb{R}_{\mathsf{E}} \oplus \mathsf{T}^{*m}\mathsf{E}) \otimes \mathsf{VE}$$

by

$$\widehat{P}_m = P_m \circ ((\mathrm{id}_{\mathbb{R}} \oplus j_{m-1}\pi_{\mathsf{E}}) \otimes \mathrm{id}_{\mathsf{T}^1_1(\mathsf{VE})}),$$

noting that

$$\mathrm{id}_{\mathbb{R}} \oplus j_{m-1}\pi_{\mathsf{E}}\colon \pi_{\mathsf{E}}^*(\mathbb{R}_{\mathsf{M}} \oplus \mathsf{T}^{*m-1}\mathsf{M}) \to \mathbb{R}_{\mathsf{E}} \oplus \mathsf{T}^{*m-1}\mathsf{E}$$

is injective. Also define

$$Q_m\colon \mathsf{S}^{\leq m-1}(\mathsf{T}^*\mathsf{E}) \otimes \mathsf{T}^1_1(\mathsf{VE}) \to \mathsf{S}^m(\mathsf{T}^*\mathsf{E}) \otimes \mathsf{VE}$$

by

$$\begin{aligned} Q_m(A_0 \otimes \alpha_0 \otimes u_0, \dots, A_{m-1} \otimes \alpha_{m-1} \otimes u_{m-1}) \\ = (\mathrm{Sym}_1(A_0 \otimes \alpha_0) \otimes u_0, \dots, \mathrm{Sym}_m(A_{m-1} \otimes \alpha_{m-1}) \otimes u_{m-1}). \end{aligned}$$

Note that the diagram

$$\begin{CD} \mathsf{S}^{\leq m-1}(\mathsf{T}^*\mathsf{E}) \otimes \mathsf{T}^1_1(\mathsf{VE}) @>{S^{m-1}_{\nabla^{\mathsf{E}}} \otimes \mathrm{id}_{\mathsf{T}^1_1(\mathsf{VE})}}>> (\mathbb{R}_{\mathsf{E}} \oplus \mathsf{T}^{*m-1}\mathsf{E}) \otimes \mathsf{T}^1_1(\mathsf{VE}) \\ @V{Q_m}VV @VV{P_m}V \\ \mathsf{S}^{\leq m}(\mathsf{T}^*\mathsf{E}) \otimes \mathsf{VE} @>>{S^m_{\nabla^{\mathsf{E}}} \otimes \mathrm{id}_{\mathsf{VE}}}> (\mathbb{R}_{\mathsf{E}} \oplus \mathsf{T}^{*m}\mathsf{E}) \otimes \mathsf{VE} \end{CD}$$

commutes. We also define

$$\widehat{Q}_m = Q_m \circ (\pi^*_{m-1} \otimes \mathrm{id}_{\mathsf{T}^1_1(\mathsf{VE})}),$$

where

$$\pi^*_{m-1}\colon \mathsf{S}^{\leq m-1}(\pi_{\mathsf{E}}^*\mathsf{T}^*\mathsf{M}) \to \mathsf{S}^{\leq m-1}(\mathsf{T}^*\mathsf{E})$$

is the inclusion. Note that the diagram

$$
\begin{array}{ccc}
\mathsf{S}^{\leq m-1}(\pi_{\mathsf{E}}^{*}\mathsf{T}^{*}\mathsf{M}) & \xrightarrow{\pi_{m-1}^{*}} & \mathsf{S}^{\leq m-1}(\mathsf{T}^{*}\mathsf{E}) \\
{\scriptstyle S^{m-1}_{\nabla^{\mathsf{M}},\nabla^{\pi_{\mathsf{E}}}}}\big\downarrow & & \big\downarrow{\scriptstyle S^{m-1}_{\nabla^{\mathsf{E}}}} \\
\pi_{\mathsf{E}}^{*}(\mathbb{R}_{\mathsf{M}} \oplus \mathsf{T}^{*m-1}\mathsf{M}) & \xrightarrow[\mathrm{id}_{\mathbb{R}} \oplus j_{m-1}\pi_{\mathsf{E}}]{} & \mathbb{R}_{\mathsf{E}} \oplus \mathsf{T}^{*m-1}\mathsf{E}
\end{array}
$$

commutes.

Let us organise the mappings we require into the following diagram:

$$
\begin{array}{ccccc}
\mathsf{T}^{\leq m-1}(\pi_{\mathsf{E}}^{*}\mathsf{T}^{*}\mathsf{M}) \otimes \mathsf{T}^{1}_{1}(\mathsf{VE}) & \xrightarrow{\mathrm{Sym}_{\leq m-1} \otimes \mathrm{id}_{\mathsf{T}^{1}_{1}(\mathsf{VE})}} & \mathsf{S}^{\leq m-1}(\pi_{\mathsf{E}}^{*}\mathsf{T}^{*}\mathsf{M}) \otimes \mathsf{T}^{1}_{1}(\mathsf{VE}) & \xrightarrow{S^{m-1}_{\nabla^{\mathsf{M}},\nabla^{\pi_{\mathsf{E}}}} \otimes \mathrm{id}_{\mathsf{T}^{1}_{1}(\mathsf{VE})}} & \pi_{\mathsf{E}}^{*}(\mathbb{R}_{\mathsf{M}} \oplus \mathsf{T}^{*m-1}\mathsf{M}) \otimes \mathsf{T}^{1}_{1}(\mathsf{VE}) \\
{\scriptstyle C^{m}}\big\downarrow & & \big\downarrow{\scriptstyle \widehat{C}^{m}} & & \big\downarrow{\scriptstyle \widehat{P}_{m}} \\
\mathsf{T}^{\leq m}(\mathsf{T}^{*}\mathsf{E}) \otimes \mathsf{VE} & \xrightarrow{\mathrm{Sym}_{\leq m} \otimes \mathrm{id}_{\mathsf{VE}}} & \mathsf{S}^{\leq m}(\mathsf{T}^{*}\mathsf{E}) \otimes \mathsf{VE} & \xrightarrow{S^{m}_{\nabla^{\mathsf{E}}} \otimes \mathrm{id}_{\mathsf{VE}}} & (\mathbb{R}_{\mathsf{E}} \oplus \mathsf{T}^{*m}\mathsf{E}) \otimes \mathsf{VE}
\end{array}
\tag{3.35}
$$

Here $\widehat{C}^{m}$ is defined so that the right square commutes, which is possible since the horizontal arrows in the right square are isomorphisms. We shall show that the left square also commutes. Indeed,

$$
\begin{aligned}
\widehat{C}^{m} &\circ (\mathrm{Sym}_{\leq m-1} \otimes \mathrm{id}_{\mathsf{T}^{1}_{1}(\mathsf{VE})})(L^{\mathrm{v}}, (\nabla^{\pi_{\mathsf{E}}} L)^{\mathrm{v}}, \dots, (\nabla^{\mathsf{M},\pi_{\mathsf{E}},m} L)^{\mathrm{v}}) \\
&= (S^{m}_{\nabla^{\mathsf{E}}} \otimes \mathrm{id}_{\mathsf{VE}})^{-1} \circ \widehat{P}_{m} \circ (S^{m-1}_{\nabla^{\mathsf{M}},\nabla^{\pi_{\mathsf{E}}}} \otimes \mathrm{id}_{\mathsf{T}^{1}_{1}(\mathsf{VE})}) \\
&\qquad \circ (\mathrm{Sym}_{\leq m-1} \otimes \mathrm{id}_{\mathsf{T}^{1}_{1}(\mathsf{VE})})(L^{\mathrm{v}}, (\nabla^{\pi_{\mathsf{E}}} L)^{\mathrm{v}}, \dots, (\nabla^{\mathsf{M},\pi_{\mathsf{E}},m} L^{\mathrm{v}}) \\
&= (\mathrm{Sym}_{\leq m-1} \otimes \mathrm{id}_{\mathsf{VE}})(L^{\mathrm{v}}, \nabla^{\mathsf{E}} L^{\mathrm{v}}, \dots, \nabla^{\mathsf{E},m} L^{\mathrm{v}}) \\
&= (\mathrm{Sym}_{\leq m} \otimes \mathrm{id}_{\mathsf{VE}}) \circ C^{m}(L^{\mathrm{v}}, (\nabla^{\pi_{\mathsf{E}}} L)^{\mathrm{v}}, \dots, (\nabla^{\mathsf{M},\pi_{\mathsf{E}},m} L)^{\mathrm{v}}).
\end{aligned}
$$

Thus the diagram (3.35) commutes. Thus, if we define

$$
\widehat{C}^{m}_{s}((\mathrm{Sym}_{s} \otimes \mathrm{id}_{\mathsf{T}^{1}_{1}(\mathsf{VE})}) \circ (\nabla^{\mathsf{M},\pi_{\mathsf{E}},s} L)^{\mathrm{v}}) = (\mathrm{Sym}_{m} \otimes \mathrm{id}_{\mathsf{VE}}) \circ C^{m}_{s}((\nabla^{\mathsf{M},\pi_{\mathsf{E}},s} L)^{\mathrm{v}}),
$$

then we have

$$
(\mathrm{Sym}_{m} \otimes \mathrm{id}_{\mathsf{VE}}) \circ \nabla^{\mathsf{E},m} L^{\mathrm{e}} = \sum_{s=0}^{m} \widehat{C}^{m}_{s}((\mathrm{Sym}_{s} \otimes \mathrm{id}_{\mathsf{T}^{1}_{1}(\mathsf{VE})}) \circ (\nabla^{\mathsf{M},\pi_{\mathsf{E}},s} L)^{\mathrm{v}}),
$$

as desired. □

The preceding lemma gives rise to an "inverse," which we state in the following lemma.

Lemma 3.57 (Iterated Symmetrised Covariant Differentials of Vertical Evaluations of Endomorphisms II) *Let $r \in \{\infty, \omega\}$ and let $\pi_{\mathsf{E}}\colon \mathsf{E} \to \mathsf{M}$ be a C^{r}-vector bundle, with the data prescribed in Sect. 3.2.1 to define the Riemannian metric $\mathbb{G}_{\mathsf{E}}$*

on E. *For* $m \in \mathbb{Z}_{\geq 0}$, *there exist* C^r*-vector bundle mappings*

$$(\widehat{B}^m_s, \mathrm{id}_{\mathsf{E}}) \in \mathrm{VB}^r(\mathrm{S}^s(\mathsf{T}^*\mathsf{E}) \otimes \mathsf{VE}; \mathrm{S}^m(\pi_{\mathsf{E}}^*\mathsf{T}^*\mathsf{M}) \otimes \mathsf{VE}), \quad s \in \{0, 1, \dots, m\},$$

and

$$(\widehat{D}^m_s, \mathrm{id}_{\mathsf{E}}) \in \mathrm{VB}^r(\mathrm{S}^s(\mathsf{T}^*\mathsf{E}) \otimes \mathrm{T}^1_1(\mathsf{VE}); \mathrm{S}^m(\pi_{\mathsf{E}}^*\mathsf{T}^*\mathsf{M}) \otimes \mathsf{VE}), \quad s \in \{0, 1, \dots, m-1\},$$

such that

$$\begin{aligned}&(\mathrm{Sym}_m \otimes \mathrm{id}_{\mathsf{VE}}) \circ (\nabla^{\mathsf{M},\pi_{\mathsf{E}},m} L)^{\mathrm{e}} \\ &\quad = \sum_{s=0}^{m} \widehat{B}^m_s((\mathrm{Sym}_s \otimes \mathrm{id}_{\mathsf{VE}}) \circ \nabla^{\mathsf{E},s} L^{\mathrm{e}}) + \sum_{s=0}^{m-1} \widehat{D}^m_s((\mathrm{Sym}_s \otimes \mathrm{id}_{\mathrm{T}^1_1(\mathsf{VE})}) \circ \nabla^{\mathsf{E},s} L^{\mathrm{v}})\end{aligned}$$

for all $L \in \Gamma^m(\mathrm{T}^1_1(\mathsf{E}))$.

Proof Following along the lines of the proof of Lemma 3.31, we define $\widehat{B}^m_s$ by requiring that

$$\widehat{B}^m_s((\mathrm{Sym}_s \otimes \mathrm{id}_{\mathsf{VE}}) \circ \nabla^{\mathsf{E},s} L^{\mathrm{e}} = (\mathrm{Sym}_m \otimes \mathrm{id}_{\mathsf{VE}}) \circ B^m_s(\nabla^{\mathsf{E},s} L^{\mathrm{e}}),$$

and $\widehat{C}^m_s$ by requiring that

$$\widehat{C}^m_s((\mathrm{Sym}_s \otimes \mathrm{id}_{\mathrm{T}^1_1(\mathsf{VE})}) \circ \nabla^{\mathsf{E},s} L^{\mathrm{v}}) = (\mathrm{Sym}_m \otimes \mathrm{id}_{\mathsf{VE}}) \circ C^m_s(\nabla^{\mathsf{E},s} L^{\mathrm{v}}).$$

That these definitions make sense follows along the same lines as the proof of Lemma 3.56. □

We can put together the previous four lemmata into the following decomposition result, which is to be regarded as the main result of this section.

Lemma 3.58 (Decomposition of Jets of Vertical Evaluations of Endomorphisms) *Let* $r \in \{\infty, \omega\}$ *and let* $\pi_{\mathsf{E}} \colon \mathsf{E} \to \mathsf{M}$ *be a* C^r*-vector bundle, with the data prescribed in Sect. 3.2.1 to define the Riemannian metric* $\mathbb{G}_{\mathsf{E}}$ *on* E. *Then there exist* C^r*-vector bundle mappings*

$$A^m_{\nabla^{\mathsf{E}}} \in \mathrm{VB}^r(\mathrm{C}^{*m}\mathsf{E}; \mathrm{S}^{\leq m}(\pi_{\mathsf{E}}^*\mathsf{T}^*\mathsf{M}) \otimes \mathsf{VE}), \quad B^m_{\nabla^{\mathsf{E}}} \in \mathrm{VB}^r(\mathrm{C}^{*m}\mathsf{E}; \mathrm{S}^{\leq m}(\mathsf{T}^*\mathsf{E}) \otimes \mathsf{VE})$$

defined by

$$\begin{aligned}A^m_{\nabla^{\mathsf{E}}}(j_m(L^{\mathrm{e}})(e)) &= \mathrm{Sym}_{\leq m} \otimes \mathrm{id}_{\mathsf{VE}}(L^{\mathrm{e}}(e), (\nabla^{\pi_{\mathsf{E}}} L)^{\mathrm{e}}(e), \dots, (\nabla^{\mathsf{M},\pi_{\mathsf{E}},m} L)^{\mathrm{e}}(e)),\\ B^m_{\nabla^{\mathsf{E}}}(j_m(L^{\mathrm{e}})(e)) &= \mathrm{Sym}_{\leq m} \otimes \mathrm{id}_{\mathsf{VE}}(L^{\mathrm{e}}(e), \nabla^{\mathsf{E}} L^{\mathrm{e}}(e), \dots, \nabla^{\mathsf{E},m} L^{\mathrm{e}}(e)).\end{aligned}$$

Moreover, $A^m_{\nabla\mathsf{E}}$ and $B^m_{\nabla\mathsf{E}}$ are injective, and

$$\begin{aligned}
B^m_{\nabla\mathsf{E}} \circ (A^m_{\nabla\mathsf{E}})^{-1} \circ (\mathrm{Sym}_{\le m} \otimes \mathrm{id}_{\mathsf{VE}})(L^{\mathrm{e}}(e), (\nabla^{\pi_{\mathsf{E}}} L)^{\mathrm{e}}(e), \dots, (\nabla^{\mathsf{M},\pi_{\mathsf{E}},m} L)^{\mathrm{e}}(e)) \\
= \Bigg(L^{\mathrm{e}}(e), \sum_{s=0}^{1} \widehat{A}^1_s((\mathrm{Sym}_s \otimes \mathrm{id}_{\mathsf{VE}}) \circ (\nabla^{\mathsf{M},\pi_{\mathsf{E}},s} L)^{\mathrm{e}}(e)), \dots, \\
\sum_{s=0}^{m} \widehat{A}^m_s((\mathrm{Sym}_s \otimes \mathrm{id}_{\mathsf{VE}}) \circ (\nabla^{\mathsf{M},\pi_{\mathsf{E}},s} L)^{\mathrm{e}}(e)) \Bigg) \\
+ \Bigg(0, L^{\mathrm{v}}(e), \sum_{s=0}^{1} \widehat{C}^2_s((\mathrm{Sym}_s \otimes \mathrm{id}_{\mathsf{T}^1_1(\mathsf{VE})}) \circ (\nabla^{\mathsf{M},\pi_{\mathsf{E}},s} L)^{\mathrm{v}}(e)), \dots, \\
\sum_{s=0}^{m-1} \widehat{C}^m_s((\mathrm{Sym}_s \otimes \mathrm{id}_{\mathsf{T}^1_1(\mathsf{VE})}) \circ (\nabla^{\mathsf{M},\pi_{\mathsf{E}},s} L)^{\mathrm{v}}(e)) \Bigg)
\end{aligned}$$

and

$$\begin{aligned}
A^m_{\nabla\mathsf{E}} \circ (B^m_{\nabla\mathsf{E}})^{-1} \circ (\mathrm{Sym}_{\le m} \otimes \mathrm{id}_{\mathsf{VE}})(L^{\mathrm{e}}(e), \nabla^{\mathsf{E}} L^{\mathrm{e}}(e), \dots, \nabla^{\mathsf{E},m} L^{\mathrm{e}}(e)) \\
= \Bigg(L^{\mathrm{e}}(e), \sum_{s=0}^{1} \widehat{B}^1_s((\mathrm{Sym}_s \otimes \mathrm{id}_{\mathsf{VE}}) \circ \nabla^{\mathsf{E},s} L^{\mathrm{e}}(e)), \dots, \\
\sum_{s=0}^{m} \widehat{B}^m_s((\mathrm{Sym}_s \otimes \mathrm{id}_{\mathsf{VE}}) \circ \nabla^{\mathsf{E},s} L^{\mathrm{e}}(e)) \Bigg) \\
+ \Bigg(0, L^{\mathrm{v}}(e), \sum_{s=0}^{1} \widehat{D}^2_s((\mathrm{Sym}_s \otimes \mathrm{id}_{\mathsf{T}^1_1(\mathsf{VE})}) \circ \nabla^{\mathsf{E},s} L^{\mathrm{v}}(e)), \dots, \\
\sum_{s=0}^{m-1} \widehat{D}^m_s((\mathrm{Sym}_s \otimes \mathrm{id}_{\mathsf{T}^1_1(\mathsf{VE})}) \circ \nabla^{\mathsf{E},s} L^{\mathrm{v}}(e)) \Bigg),
\end{aligned}$$

where the vector bundle mappings $\widehat{A}^m_s$ and $\widehat{B}^m_s$, $s \in \{0, 1, \dots, m\}$, and $\widehat{C}^m_s$ and $\widehat{D}^m_s$, $s \in \{0, 1, \dots, m-1\}$, are as in Lemmata 3.56 and 3.57.

3.3.8 Isomorphisms for Pull-Backs of Functions

Next we generalise the presentation of Sect. 3.3.1 from the pull-back of a vector bundle projection to the pull-back by a general mapping. The development here is a little different from the preceding sections, so we first have a little bit of setting up to do. For C^r-manifolds M and N, and for $\Phi \in \mathsf{C}^r(\mathsf{M}; \mathsf{N})$, we consider the mapping

$$\mathsf{C}^r(\mathsf{N}) \ni f \mapsto \Phi^* f \in \mathsf{C}^r(\mathsf{M}).$$

We wish to compare the decomposition of jets of f with those of $\Phi^* f$, and to do so we consider the subbundle $\mathsf{T}^{*m}_{\Phi}\mathsf{M}$ of $\mathsf{T}^{*m}\mathsf{M}$ defined by

$$\mathsf{T}^{*m}_{\Phi,x}\mathsf{M} = \{j_m(\Phi^* f)(x) \mid f \in \mathsf{C}^m(\mathsf{N})\}.$$

Following Lemma 2.15, we shall give a formula for iterated covariant differentials of pull-backs of functions on N. To do this, we let ∇^{M} and ∇^{N} be affine connections on M and N, respectively. We note that we have the linear connection $\Phi^*\nabla^{\mathsf{N}}$ in the vector bundle $\Phi^*\mathsf{TN}$ over M. Explicitly,

$$(\Phi^*\nabla^{\mathsf{N}}_X \Phi^* Y)(x) = (x, \nabla^{\mathsf{N}}_{T_x\Phi(X(x))} Y).$$

Following our usual mild notational abuse, we shall also denote by $\Phi^*\nabla^{\mathsf{N}}$ the connection in the dual bundle $(\Phi^*\mathsf{TN})^* \simeq \Phi^*\mathsf{T}^*\mathsf{N}$. We have a natural mapping

$$\begin{aligned} \widehat{\Phi}\colon \mathsf{TM} &\to \Phi^*\mathsf{TN} \\ v_x &\mapsto (x, T_x\Phi(v_x)). \end{aligned}$$

This mapping induces a mappings on sections which we denote by the same symbol; thus we have the mapping

$$\widehat{\Phi}\colon \Gamma^\infty(\mathsf{TM}) \to \Gamma^\infty(\Phi^*\mathsf{TN}).$$

The following lemma gives an important tensor for our analysis.

Lemma 3.59 (Tensor for Pull-Back Connection) *Let $r \in \{\infty, \omega\}$. Let M and N be C^r-manifolds and let ∇^{M} and ∇^{N} be C^r-affine connections on M and N, respectively. Let $\Phi \in \mathsf{C}^r(\mathsf{M}; \mathsf{N})$. Then there exists $A_\Phi \in \Gamma^r(\mathsf{T}^2(\mathsf{T}^*\mathsf{M}) \otimes \Phi^*\mathsf{TN})$ such that, for $x \in \mathsf{M}$,*

$$\widehat{\Phi}(\nabla^{\mathsf{M}}_X Y)(x) - \Phi^*\nabla^{\mathsf{N}}_X \widehat{\Phi}(Y)(x) = A_\Phi(X(x), Y(x))$$

for $X, Y \in \Gamma^\infty(\mathsf{TM})$.

Proof Let $K^{\mathsf{M}}\colon \mathsf{TTM} \to \mathsf{TM}$ and $K^{\mathsf{N}}\colon \mathsf{TTN} \to \mathsf{TN}$ be the connectors for ∇^{M} and ∇^{N} so that

$$\nabla^{\mathsf{M}}_X Y = K^{\mathsf{M}} \circ TY \circ X, \qquad X, Y \in \Gamma^\infty(\mathsf{TM}),$$

and

$$\nabla^{\mathsf{N}}_U V = K^{\mathsf{N}} \circ TV \circ U, \qquad U, V \in \Gamma^\infty(\mathsf{TN}).$$

We, moreover, have

$$\widehat{\Phi}(\nabla^{\mathsf{M}}_X Y) = T\Phi \circ K^{\mathsf{M}} \circ TY \circ X, \qquad X, Y \in \Gamma^\infty(\mathsf{TM}),$$

and

$$\Phi^* \nabla^{\mathsf{N}}_X \widehat{\Phi}(Y) = K^{\mathsf{N}} \circ T(T\Phi \circ Y) \circ X, \qquad X, Y \in \Gamma^\infty(\mathsf{TM})$$

[51, §10.12]. In preparation to use these formulae, we have the following results.

Sublemma 1 *If* $\pi_{\mathsf{E}} \colon \mathsf{E} \to \mathsf{M}$ *is a smooth vector bundle, if* $\xi \in \Gamma^\infty(\mathsf{E})$*, and if* $f \in \mathrm{C}^\infty(\mathsf{M})$*, then*

$$T_x(f\xi)(v_x) = f(x) T_x\xi(v_x) + \langle df(x); v_x\rangle \xi^{\mathrm{v}}(x).$$

Proof Let $\nabla^{\pi_{\mathsf{E}}}$ be a linear connection in the vector bundle E which gives the decomposition $\mathsf{TE} = \mathsf{HE} \oplus \mathsf{VE}$. Let hor and ver be the horizontal and vertical projections. Let $v_x \in \mathsf{T}_x\mathsf{M}$ and let $\gamma \colon I \to \mathsf{M}$ be a smooth curve for which $\gamma'(0) = v_x$. Denote $\Xi(t) = (f \circ \gamma(t))(\xi \circ \gamma(t))$ the corresponding curve in E. Then

$$\begin{aligned} \operatorname{hor}(\Xi'(t)) &= \operatorname{hlft}(f \circ \gamma(t)\xi \circ \gamma(t), \gamma'(t)), \\ \operatorname{ver}(\Xi'(t)) &= \operatorname{vlft}(f \circ \gamma(t)\xi \circ \gamma(t), \nabla^{\pi_{\mathsf{E}}}_{\gamma'(t)} \Xi(t)). \end{aligned}$$

We now have

$$\nabla^{\pi_{\mathsf{E}}}_{\gamma'(t)} \Xi(t) = f \circ \gamma(t) \nabla^{\pi_{\mathsf{E}}}_{\gamma'(t)} \xi \circ \gamma(t) + \langle \mathrm{d} f \circ \gamma(t); \gamma'(t)\rangle \xi \circ \gamma(t).$$

Thus

$$\begin{aligned} &T_x(f\xi)(v_x) \\ &\quad = \frac{\mathrm{d}}{\mathrm{d}t}\bigg|_{t=0} f \circ \gamma(t))\xi \circ \gamma(t) \\ &\quad = f(x)\operatorname{hlft}(f(x)\xi(x), v_x) + \operatorname{vlft}(f(x)\xi(x), f(x)\nabla^{\pi_{\mathsf{E}}}_{v_x}\xi + \langle \mathrm{d} f(x); v_x\rangle \xi(x)) \\ &\quad = f(x)\Xi'(0) + \langle \mathrm{d} f(x); v_x\rangle \xi(x) = f(x) T_x\xi(v_x) + \langle \mathrm{d} f(x); v_x\rangle \xi(x)), \end{aligned}$$

as claimed. ▽

Sublemma 2 *If* M *and* N *are smooth manifolds, if* $\Phi \in \mathrm{C}^\infty(\mathsf{M};\mathsf{N})$*, and if* $X \in \Gamma^\infty(\mathsf{TM})$*, then*

$$TT\Phi \circ X^{\mathrm{v}}(v_x) = \mathrm{vlft}(T_x\Phi(v_x), T_x\Phi(X(x))).$$

Proof We have

$$\begin{aligned} TT\Phi \circ X^{\mathrm{v}}(v_x) &= \left.\frac{\mathrm{d}}{\mathrm{d}t}\right|_{t=0} T_x\Phi(v_x + tX(x)) \\ &= \left.\frac{\mathrm{d}}{\mathrm{d}t}\right|_{t=0} (T_x\Phi(v_x) + tT_x\Phi(X(x))) \\ &= \mathrm{vlft}(T_x\Phi(v_x), T_x\Phi(X(x))), \end{aligned}$$

as claimed. ∇

We now directly compute, using Sublemma 1,

$$\begin{aligned} \widehat{\Phi}(\nabla^{\mathsf{M}}_X fY)(x) &= T\Phi \circ K^{\mathsf{M}} \circ T(fY) \circ X(x) \\ &= f(x)T\Phi \circ K^{\mathsf{M}} \circ TY \circ X(x) + \langle \mathrm{d}f(x); X(x)\rangle T\Phi \circ K^{\mathsf{M}} \circ Y^{\mathrm{v}} \circ X(x) \\ &= f(x)\widehat{\Phi}(\nabla^{\mathsf{M}}_X Y)(x) + \langle \mathrm{d}f(x); X(x)\rangle T\Phi \circ X(x), \end{aligned}$$

noting that K^{M} is a left-inverse for vertical lift. We also directly compute, using both of the sublemmata above,

$$\begin{aligned} \Phi^*\nabla^{\mathsf{N}}_X \widehat{fY}(x) &= K^{\mathsf{N}} \circ TT\Phi \circ T(fY) \circ X(x) \\ &= f(x)K^{\mathsf{N}} \circ T(T\Phi \circ Y) \circ X(x) \\ &\quad + \langle \mathrm{d}f(x); X(x)\rangle K^{\mathsf{N}} \circ TT\Phi \circ Y^{\mathrm{v}} \circ X(x) \\ &= f(x)\Phi^*\nabla^{\mathsf{N}}_X \widehat{Y}(x) \\ &\quad + \langle \mathrm{d}f(x); X(x)\rangle K^{\mathsf{N}}(\mathrm{vlft}(T_x\Phi(X(x)), T_x\Phi(X(x)))) \\ &= f(x)\Phi^*\nabla^{\mathsf{N}}_X \widehat{Y}(x) + \langle \mathrm{d}f(x); X(x)\rangle T\Phi \circ Y \circ X(x), \end{aligned}$$

again noting that K^{N} is the left-inverse for the vertical lift. Combining the preceding two computations gives the tensoriality of

$$(X, Y) \mapsto \widehat{\Phi}(\nabla^{\mathsf{M}}_X Y)(x) - \Phi^*\nabla^{\mathsf{N}}_X \widehat{\Phi}(Y)(x),$$

and so gives $A_\Phi \in \Gamma^r(\mathrm{T}^2(\mathsf{TM}) \otimes \Phi^*\mathsf{TN})$ satisfying the assertion of the lemma. □

Note that, if $A \in \Gamma^\infty(\mathsf{T}^k(\mathsf{T}^*\mathsf{N}))$, then Φ^*A denotes the pull-back of A to $\Gamma^\infty(\mathsf{T}^k(\mathsf{T}^*\mathsf{M}))$ and also the section of the tensor bundle $\mathsf{T}^k(\Phi^*\mathsf{T}^*\mathsf{N})$. Let $x \in \mathsf{T}_x\mathsf{M}$, let $v_1, \dots, v_k \in \mathsf{T}_x\mathsf{M}$, and denote $u_j = \mathsf{T}_x\Phi(v_j)$, $j \in \{1, \dots, k\}$. Note that

$$\begin{aligned}\Phi^*A((x, u_1), \dots, (x, u_k)) = A(u_1, \dots, u_k) &= A(T_x\Phi(v_1), \dots, T_x\Phi(v_k)) \\ &= \Phi^*A(v_1, \dots, v_k),\end{aligned} \tag{3.36}$$

where we are using the two interpretations of the symbol Φ^*A.

With the above as background, we can now understand the iterated covariant derivatives

$$\nabla^{\mathsf{M},m}\Phi^*f = \underbrace{\nabla^{\mathsf{M}}\cdots\nabla^{\mathsf{M}}}_{m \text{ times}}\Phi^*f, \qquad m \in \mathbb{Z}_{>0},$$

and

$$\nabla^{\mathsf{N},m}f = \underbrace{\nabla^{\mathsf{N}}\cdots\nabla^{\mathsf{N}}}_{m \text{ times}}f, \qquad m \in \mathbb{Z}_{>0},$$

for $f \in \mathrm{C}^\infty(\mathsf{N})$. The following lemma gives the first part of this development, playing the rôle of Lemma 3.16 in this case.

Lemma 3.60 (Differentiation of Pull-Backs of Covariant Tensors) *Let $r \in \{\infty, \omega\}$. Let M and N be C^r-manifolds with C^r-affine connections ∇^{M} and ∇^{N}, respectively. Define $B_\Phi = \mathrm{push}_{1,2}A_\Phi$ with A_Φ as in Lemma 3.59. Then, for $k \in \mathbb{Z}_{>0}$ and $A \in \Gamma^r(\mathsf{T}^k(\mathsf{T}^*\mathsf{N}))$,*

$$\nabla^{\mathsf{M}}\Phi^*A = \Phi^*\nabla^{\mathsf{N}}A + D_{B_\Phi}(\Phi^*A).$$

Proof Let $x \in \mathsf{M}$. Let $X_1, \dots, X_k \in \Gamma^\infty(\mathsf{TM})$. For $X_{k+1} \in \Gamma^\infty(\mathsf{TM})$, we have

$$\begin{aligned}&\mathscr{L}_{X_{k+1}}(\Phi^*A(X_1, \dots, X_k)) \\ &\quad = (\nabla^{\mathsf{M}}_{X_{k+1}}\Phi^*A)(X_1, \dots, X_k) + \sum_{j=1}^k \Phi^*A(X_1, \dots, \nabla^{\mathsf{M}}_{X_{k+1}}X_j, \dots, X_k)\end{aligned}$$

and

$$\begin{aligned}\mathscr{L}_{X_{k+1}}(\Phi^*A(\widehat{\Phi}(X_1), \dots, \widehat{\Phi}(X_k))) &= (\Phi^*\nabla^{\mathsf{N}}_{X_{k+1}}\Phi^*A)(\widehat{\Phi}(X_1), \dots, \widehat{\Phi}(X_k)) \\ &\quad + \sum_{j=1}^k \Phi^*A(\widehat{\Phi}(X_1), \dots, \Phi^*\nabla^{\mathsf{N}}_{X_{k+1}}\widehat{\Phi}(X_j), \dots, \widehat{\Phi}(X_k)),\end{aligned}$$

using the two interpretations of $\Phi^* A$. By (3.36) we have, in the above expressions,

$$\Phi^* A(X_1, \dots, X_k) = \Phi^* A(\widehat{\Phi}(X_1), \dots, \widehat{\Phi}(X_k)).$$

By (3.36) again, we have

$$\Phi^* A(X_1, \dots, \nabla^{\mathsf{M}}_{X_{k+1}} X_j, \dots, X_k) = \Phi^* A(\widehat{\Phi}(X_1), \dots, \widehat{\Phi}(\nabla^{\mathsf{M}}_{X_{k+1}} X_j), \dots, \widehat{\Phi}(X_k)).$$

Also note that

$$\begin{aligned}
&(\Phi^* \nabla^{\mathsf{N}}_{X_{k+1}} \Phi^* A)(\widehat{\Phi}(X_1), \dots, \widehat{\Phi}(X_k))(x) \\
&\qquad = (\Phi^* \nabla^{\mathsf{N}}_{X_{k+1}} \Phi^* A)(T_x\Phi(X_1(x)), \dots, T_x\Phi(X_k(x))) \\
&\qquad = \nabla^{\mathsf{N}}_{T_x\Phi(X_{k+1}(x))} A(T_x\Phi(X_1(x)), \dots, T_x\Phi(X_{k+1}(x))) \\
&\qquad = \nabla^{\mathsf{N}} A(T_x\Phi(X_1(x)), \dots, T_x\Phi(X_{k+1}(x))) \\
&\qquad = \Phi^* \nabla^{\mathsf{N}} A(X_1, \dots, X_{k+1})(x).
\end{aligned}$$

Combining the above gives

$$\begin{aligned}
&\nabla^{\mathsf{M}} \Phi^* A(X_1, \dots, X_{k+1}) \\
&\quad = \Phi^* \nabla^{\mathsf{N}} A(X_1, \dots, X_{k+1}) + \sum_{j=1}^{k} \Phi^* A(\widehat{\Phi}(X_1), \dots, \Phi^* \nabla^{\mathsf{N}}_{X_{k+1}} \widehat{\Phi}(X_j) \\
&\qquad - \widehat{\Phi}(\nabla^{\mathsf{M}}_{X_{k+1}} X_j), \dots, \widehat{\Phi}(X_k)) \\
&\quad = \Phi^* \nabla^{\mathsf{N}} A(X_1, \dots, X_{k+1}) \\
&\qquad - \sum_{j=1}^{k} \Phi^* A(\widehat{\Phi}(X_1), \dots, A_\Phi(X_{k+1}, X_j), \dots, \widehat{\Phi}(X_k)).
\end{aligned}$$

Thus

$$\nabla^{\mathsf{M}} \Phi^* A = \Phi^* \nabla^{\mathsf{N}} A - \sum_{j=1}^{k} \operatorname{Ins}_j(\Phi^* A, B_\Phi),$$

giving the result by Lemma 3.11. □

We now have the following lemma, the first of two regarding iterated covariant differentials.

Lemma 3.61 (Iterated Covariant Differentials of Pull-Backs of Functions I) *Let $r \in \{\infty, \omega\}$ and let M and N be C^r-manifolds with C^r-affine connections ∇^{M} and ∇^{N}, respectively. For $m \in \mathbb{Z}_{\geq 0}$, there exist C^r-vector bundle mappings*

$$(A_s^m, \mathrm{id}_{\mathsf{M}}) \in \mathrm{VB}^r(\mathrm{T}^s(\Phi^*\mathsf{T}^*\mathsf{N}); \mathrm{T}^m(\mathsf{T}^*\mathsf{M})), \qquad s \in \{0, 1, \dots, m\},$$

such that

$$\nabla^{\mathsf{M},m}\Phi^* f = \sum_{s=0}^{m} A_s^m(\Phi^*\nabla^{\mathsf{N},s} f)$$

for all $f \in \mathrm{C}^m(\mathsf{N})$. Moreover, the vector bundle mappings $A_0^m, A_1^m, \dots, A_m^m$ satisfy the recursion relations prescribed by

$$A_0^0(\beta_0) = \beta_0, \quad A_1^1(\beta_1) = \beta_1, \quad A_0^1 = 0,$$

and

$$\begin{aligned}
A_{m+1}^{m+1}(\beta_{m+1}) &= \beta_{m+1}, \\
A_s^{m+1}(\beta_s) &= (\nabla^{\mathsf{M}} A_s^m)(\beta_s) + A_{s-1}^m \otimes \mathrm{id}_{\mathsf{T}^*\mathsf{M}}(\beta_s) \\
&\quad - \sum_{j=1}^{s} A_s^m \otimes \mathrm{id}_{\mathsf{T}^*\mathsf{M}}(\mathrm{Ins}_j(\beta_s, B_\Phi)), \qquad s \in \{1, \dots, m\}, \\
A_0^{m+1}(\beta_0) &= (\nabla^{\mathsf{M}} A_0^m)(\beta_0),
\end{aligned}$$

where $\beta_s \in \mathrm{T}^s(\Phi^*\mathsf{T}^*\mathsf{N})$, $s \in \{0, 1, \dots, m\}$.

Proof The assertion clearly holds for the initial conditions of the recursion, simply because

$$\Phi^* f = \Phi^* f, \quad \mathrm{d}(\Phi^* f) = \Phi^*\mathrm{d}f + 0f.$$

So suppose that it holds for $m \in \mathbb{Z}_{>0}$. Thus

$$\nabla^{\mathsf{M},m}\Phi^* f = \sum_{s=0}^{m} A_s^m(\Phi^*\nabla^{\mathsf{N},s} f),$$

where the vector bundle mappings A^a_s, $a \in \{0, 1, \dots, m\}$, $s \in \{0, 1, \dots, a\}$, satisfy the stated recursion relations. Then

$$
\begin{aligned}
\nabla^{\mathsf{M},m+1}\Phi^* f &= \sum_{s=0}^{m}(\nabla^{\mathsf{M}} A^m_s)(\Phi^*\nabla^{\mathsf{N},s} f) + \sum_{s=0}^{m} A^m_s \otimes \mathrm{id}_{\mathsf{T}^*\mathsf{M}}(\nabla^{\mathsf{M}}\Phi^*\nabla^{\mathsf{N},s} f)\\
&= \sum_{s=0}^{m}(\nabla^{\mathsf{M}} A^m_s)(\Phi^*\nabla^{\mathsf{N},s} f) + \sum_{s=0}^{m} A^m_s \otimes \mathrm{id}_{\mathsf{T}^*\mathsf{M}}(\Phi^*\nabla^{\mathsf{N},s+1} f)\\
&\quad - \sum_{s=0}^{m}\sum_{j=1}^{s} A^m_s \otimes \mathrm{id}_{\mathsf{T}^*\mathsf{M}} \operatorname{Ins}_j(\Phi^*\nabla^{\mathsf{N},s} f, B_\Phi)\\
&= \Phi^*\nabla^{\mathsf{N},m+1} f\\
&\quad + \sum_{s=1}^{m}\Bigg((\nabla^{\mathsf{M}} A^m_s)(\Phi^*\nabla^{\mathsf{N},s} f) + A^m_{s-1} \otimes \mathrm{id}_{\mathsf{T}^*\mathsf{M}}(\Phi^*\nabla^{\mathsf{N},s} f)\\
&\quad - \sum_{j=1}^{s} A^m_s \otimes \mathrm{id}_{\mathsf{T}^*\mathsf{M}}(\operatorname{Ins}_j(\Phi^*\nabla^{\mathsf{N},s} f, B_\Phi))\Bigg) + (\nabla^{\mathsf{M}} A^m_0)(\Phi^* f)
\end{aligned}
$$

by Lemma 3.60. From this the lemma follows. □

We shall also need to "invert" the relationship of the preceding lemma.

Lemma 3.62 (Iterated Covariant Differentials of Pull-Backs of Functions II) *Let $r \in \{\infty, \omega\}$ and let M and N be C^r-manifolds with C^r-affine connections ∇^{M} and ∇^{N}, respectively. For $m \in \mathbb{Z}_{\geq 0}$, there exist C^r-vector bundle mappings*

$$(B^m_s, \mathrm{id}_{\mathsf{M}}) \in \mathrm{VB}^r(\mathrm{T}^s(\mathsf{T}^*\mathsf{M}); \mathrm{T}^m(\Phi^*\mathsf{T}^*\mathsf{N})), \qquad s \in \{0, 1, \dots, m\},$$

such that

$$\Phi^*\nabla^{\mathsf{N},m} f = \sum_{s=0}^{m} B^m_s(\nabla^{\mathsf{M},s}\Phi^* f)$$

for all $f \in \mathrm{C}^m(\mathsf{N})$. Moreover, the vector bundle mappings $B^m_0, B^m_1, \dots, B^m_m$ satisfy the recursion relations prescribed by

$$B^0_0(\alpha_0) = \alpha_0, \quad B^1_1(\alpha_1) = \alpha_1, \quad B^1_0 = 0,$$

and

$$B^{m+1}_{m+1}(\alpha_{m+1}) = \alpha_{m+1},$$

$$B^{m+1}_s(\alpha_s) = (\nabla^{\mathsf{M}} B^m_s)(\alpha_s) + B^m_{s-1} \otimes \mathrm{id}_{\mathsf{T}^*\mathsf{M}}(\alpha_s) + \sum_{j=1}^m \mathrm{Ins}_j(B^m_s(\alpha_s), B_\Phi),$$

$$s \in \{1, \dots, m\},$$

$$B^{m+1}_0(\alpha_0) = (\nabla^{\mathsf{M}} B^m_0)(\alpha_0) + \sum_{j=1}^m \mathrm{Ins}_j(B^m_0(\alpha_0), B_\Phi),$$

where $\alpha_s \in \mathrm{T}^s(\mathsf{T}^*\mathsf{M})$, $s \in \{0, 1, \dots, m\}$.

Proof The assertion clearly holds for the initial conditions for the recursion because

$$\Phi^* f = \Phi^* f, \quad \Phi^*(\mathrm{d}f) = \mathrm{d}(\Phi^* f) + 0f.$$

So suppose it true for $m \in \mathbb{Z}_{>0}$. Thus

$$\Phi^* \nabla^{\mathsf{N},m} f = \sum_{s=0}^m B^m_s(\nabla^{\mathsf{M},s} \Phi^* f), \tag{3.37}$$

where the vector bundle mappings B^a_s, $a \in \{0, 1, \dots, m\}$, $s \in \{0, 1, \dots, a\}$, satisfy the recursion relations from the statement of the lemma. Then, by Lemma 3.60, we can work on the left-hand side of (3.37) to give

$$\begin{aligned}\nabla^{\mathsf{M}} \Phi^* \nabla^{\mathsf{N},m} f &= \Phi^* \nabla^{\mathsf{N},m+1} f - \sum_{j=1}^m \mathrm{Ins}_j(\Phi^* \nabla^{\mathsf{N},m} f, B_\Phi) \\ &= \Phi^* \nabla^{\mathsf{N},m+1} f - \sum_{s=0}^m \sum_{j=1}^m \mathrm{Ins}_j(B^m_s(\nabla^{\mathsf{M},s} \Phi^* f), B_\Phi).\end{aligned}$$

Working on the right-hand side of (3.27) gives

$$\nabla^{\mathsf{M}} \Phi^* \nabla^{\mathsf{N},m} f = \sum_{s=0}^m \nabla^{\mathsf{M}} B^m_s(\nabla^{\mathsf{M},s} \Phi^* f) + \sum_{s=0}^m B^m_s \otimes \mathrm{id}_{\mathsf{T}^*\mathsf{M}}(\nabla^{\mathsf{M},s+1} \Phi^* f).$$

Combining the preceding two equations gives

$$\begin{aligned}\Phi^*\nabla^{\mathsf{N},m+1}f &= \sum_{s=0}^{m}\nabla^{\mathsf{M}}B_s^m(\nabla^{\mathsf{M},s}\Phi^*f)+\sum_{s=0}^{m}B_s^m\otimes\mathrm{id}_{\mathsf{T}^*\mathsf{M}}(\nabla^{\mathsf{M},s+1}\Phi^*f)\\ &\quad+\sum_{s=0}^{m}\sum_{j=1}^{m}\mathrm{Ins}_j(B_s^m(\nabla^{\mathsf{M},s}\Phi^*f),B_\Phi)\\ &= \nabla^{\mathsf{M},m+1}\Phi^*f\\ &\quad+\sum_{s=1}^{m}\Bigg(\nabla^{\mathsf{M}}B_s^m(\nabla^{\mathsf{M},s}\Phi^*f)+B_{s-1}^m\otimes\mathrm{id}_{\mathsf{T}^*\mathsf{M}}(\nabla^{\mathsf{M},s}\Phi^*f)\\ &\quad+\sum_{j=1}^{m}\mathrm{Ins}_j(B_s^m(\nabla^{\mathsf{M},s}\Phi^*f),B_\Phi)\Bigg)+\nabla^{\mathsf{M}}B_0^m(\Phi^*f)\\ &\quad+\sum_{j=1}^{m}\mathrm{Ins}_j(B_0^m(\Phi^*f),B_\Phi),\end{aligned}$$

and the lemma follows from this. □

With this data, we have the following result.

Lemma 3.63 (Iterated Symmetrised Covariant Differentials of Pull-Backs of Functions I) *Let $r\in\{\infty,\omega\}$ and let M and N be C^r-manifolds with C^r-affine connections ∇^{M} and ∇^{N}, respectively. For $m\in\mathbb{Z}_{\geq 0}$, there exist C^r-vector bundle mappings*

$$(\widehat{A}_s^m,\mathrm{id}_{\mathsf{M}})\in\mathrm{VB}^r(\mathsf{S}^s(\Phi^*\mathsf{T}^*\mathsf{N});\mathsf{S}^m(\mathsf{T}^*\mathsf{M})),\qquad s\in\{0,1,\dots,m\},$$

such that

$$\mathrm{Sym}_m\circ\nabla^{\mathsf{M},m}\Phi^*f=\sum_{s=0}^{m}\widehat{A}_s^m(\mathrm{Sym}_s\circ\Phi^*\nabla^{\mathsf{N},s}f)$$

for all $f\in\mathrm{C}^m(\mathsf{N})$.

Proof This follows from Lemma 3.61 in the same way as Lemma 3.25 follows from Lemma 3.23. □

Next we consider the "inverse" of the preceding lemma.

Lemma 3.64 (Iterated Symmetrised Covariant Differentials of Horizontal Lifts of Functions II) *Let $r\in\{\infty,\omega\}$ and let M and N be C^r-manifolds with C^r-affine connections ∇^{M} and ∇^{N}, respectively. For $m\in\mathbb{Z}_{\geq 0}$, there exist C^r-vector bundle* *mappings*

$$(\widehat{B}^m_s, \mathrm{id}_{\mathsf{M}}) \in \mathrm{VB}^r(\mathsf{S}^s(\mathsf{T}^*\mathsf{M}); \mathsf{S}^m(\Phi^*\mathsf{T}^*\mathsf{N})), \qquad s \in \{0, 1, \dots, m\},$$

such that

$$\mathrm{Sym}_m \circ \Phi^* \nabla^{\mathsf{N},m} f = \sum_{s=0}^{m} \widehat{B}^m_s (\mathrm{Sym}_s \circ \nabla^{\mathsf{M},s} \Phi^* f)$$

for all $f \in \mathrm{C}^m(\mathsf{N})$.

Proof This follows from Lemma 3.62 in the same way as Lemma 3.26 follows from Lemma 3.24. □

The following lemma provides two decompositions of $\mathsf{T}^{*m}_{\Phi}\mathsf{M}$, one "in the domain" and one "in the codomain," and the relationship between them. The assertion simply results from an examination of the preceding four lemmata.

Lemma 3.65 (Decomposition of Jets of Pull-Backs of Functions) *Let* $r \in \{\infty, \omega\}$ *and let* M *and* N *be* C^r*-manifolds with* C^r*-affine connections* ∇^{M} *and* ∇^{N}*, respectively. Then there exist* C^r*-vector bundle mappings*

$$A^m_{\nabla^{\mathsf{M}},\nabla^{\mathsf{N}}} \in \mathrm{VB}^r(\mathsf{T}^{*m}_{\Phi}\mathsf{M}; \mathsf{S}^{\leq m}(\Phi^*\mathsf{T}^*\mathsf{N})), \quad B^m_{\nabla^{\mathsf{M}},\nabla^{\mathsf{N}}} \in \mathrm{VB}^r(\mathsf{T}^{*m}_{\Phi}\mathsf{M}; \mathsf{S}^{\leq m}(\mathsf{T}^*\mathsf{M})),$$

defined by

$$A^m_{\nabla^{\mathsf{M}},\nabla^{\mathsf{N}}}(j_m(\Phi^* f)(x)) = \mathrm{Sym}_{\leq m}(\Phi^* f(x), \Phi^*\nabla^{\mathsf{N}} f(x), \dots, \Phi^*\nabla^{\mathsf{N},m} f(x)),$$

$$B^m_{\nabla^{\mathsf{M}},\nabla^{\mathsf{N}}}(j_m(\Phi^* f)(x)) = \mathrm{Sym}_{\leq m}(\Phi^* f(x), \nabla^{\mathsf{M}}\Phi^* f(x), \dots, \nabla^{\mathsf{M},m}\Phi^* f(x)).$$

Moreover, $A^m_{\nabla^{\mathsf{M}},\nabla^{\mathsf{N}}}$ *is an isomorphism,* $B^m_{\nabla^{\mathsf{M}},\nabla^{\mathsf{N}}}$ *is injective, and*

$$\begin{aligned} B^m_{\nabla^{\mathsf{M}},\nabla^{\mathsf{N}}} \circ (A^m_{\nabla^{\mathsf{M}},\nabla^{\mathsf{N}}})^{-1} &\circ (\mathrm{Sym}_{\leq m}(\Phi^* f(e), \Phi^*\nabla^{\mathsf{N}} f(x), \dots, \Phi^*\nabla^{\mathsf{N},m} f(x)) \\ &= \Bigg(A^0_0(\Phi^* f(x)), \sum_{s=0}^{1} \widehat{A}^1_s(\mathrm{Sym}_s \circ \Phi^*\nabla^{\mathsf{N},s} f(x)), \dots, \\ &\qquad \sum_{s=0}^{m} \widehat{A}^m_s(\mathrm{Sym}_s \circ \Phi^*\nabla^{\mathsf{N},s} f(x)) \Bigg) \end{aligned}$$

and

$$A^m_{\nabla^{\mathsf{M}},\nabla^{\mathsf{N}}} \circ (B^m_{\nabla^{\mathsf{M}},\nabla^{\mathsf{N}}})^{-1} \circ \mathrm{Sym}_{\le m}(\Phi^* f(x), \nabla^{\mathsf{M}}\Phi^* f(x), \dots, \nabla^{\mathsf{M},m}\Phi^* f(x))$$
$$= \left(B^0_0(\Phi^* f(x)), \sum_{s=0}^{1} \widehat{B}^1_s(\mathrm{Sym}_s \circ \nabla^{\mathsf{M},s}\Phi^* f(x)), \dots, \right.$$
$$\left. \sum_{s=0}^{m} \widehat{B}^m_s(\mathrm{Sym}_s \circ \nabla^{\mathsf{M},s}\Phi^* f(x)) \right),$$

where the vector bundle mappings $\widehat{A}^m_s$ and $\widehat{B}^m_s$, $s \in \{0, 1, \dots, m\}$, are as in Lemmata 3.63 and 3.64.

3.3.9 Comparison of Iterated Covariant Derivatives for Different Connections

In Sect. 4.3 we will prove that the seminorms for the C^ω-topology are well-defined, in that they do not depend on choices of metrics and connections. In doing this, it is useful to have at hand formulae that relate arbitrary-order covariant differentials with respect to different connections. This is what we do in this section.

We let $r \in \{\infty, \omega\}$ and let $\pi_{\mathsf{E}}: \mathsf{E} \to \mathsf{M}$ be a C^r-vector bundle. We consider C^r-affine connections ∇^{M} and $\overline{\nabla}^{\mathsf{M}}$ on M, and C^r-linear connections $\nabla^{\pi_{\mathsf{E}}}$ and $\overline{\nabla}^{\pi_{\mathsf{E}}}$ in E. It then holds that

$$\overline{\nabla}^{\mathsf{M}}_X Y = \nabla^{\mathsf{M}}_X Y + S_{\mathsf{M}}(Y, X), \quad \overline{\nabla}^{\pi_{\mathsf{E}}}_X \xi = \nabla^{\pi_{\mathsf{E}}}_X \xi + S_{\pi_{\mathsf{E}}}(\xi, X)$$

for $S_{\mathsf{M}} \in \Gamma^r(\mathrm{T}^1_2(\mathsf{TM}))$ an $S_{\pi_{\mathsf{E}}} \in \Gamma^r(\mathsf{E}^* \otimes \mathsf{T}^*\mathsf{M} \otimes \mathsf{E})$.

First we relate covariant derivatives of higher-order tensors.

Lemma 3.66 (Covariant Derivatives of Higher-Order Tensors with Respect to Different Connections) *Let $r \in \{\infty, \omega\}$ and let $\pi_{\mathsf{E}}: \mathsf{E} \to \mathsf{M}$ be a C^r-vector bundle. Consider C^r-affine connections ∇^{M} and $\overline{\nabla}^{\mathsf{M}}$ on M, and C^r-linear connections $\nabla^{\pi_{\mathsf{E}}}$ and $\overline{\nabla}^{\pi_{\mathsf{E}}}$ in E. If $k \in \mathbb{Z}_{>0}$ and if $B \in \Gamma^1(\mathrm{T}^k(\mathsf{T}^*\mathsf{M}) \otimes \mathsf{E})$, then*

$$\overline{\nabla}^{\mathsf{M},\pi_{\mathsf{E}}} B = \nabla^{\mathsf{M},\pi_{\mathsf{E}}} B - \sum_{j=1}^{k} \mathrm{Ins}_j(B, S_{\mathsf{M}}) - \mathrm{Ins}_{k+1}(B, S_{\pi_{\mathsf{E}}}).$$

Proof We have

$$\begin{aligned}
\mathscr{L}_{X_{k+1}}(B(X_1,\dots,X_k,\alpha)) &= (\overline{\nabla}^{\mathsf{M},\pi_{\mathsf{E}}}_{X_{k+1}}B)(X_1,\dots,X_k,\alpha)\\
&\quad+\sum_{j=1}^{k}B(X_1,\dots,\overline{\nabla}^{\mathsf{M}}_{X_{k+1}}X_j,\dots,X_k,\alpha)+B(X_1,\dots,X_k,\overline{\nabla}^{\pi_{\mathsf{E}}}_{X_{k+1}}\alpha)\\
&= (\overline{\nabla}^{\mathsf{M},\pi_{\mathsf{E}}}_{X_{k+1}}B)(X_1,\dots,X_k,\alpha)+\sum_{j=1}^{k}B(X_1,\dots,\nabla^{\mathsf{M}}_{X_{k+1}}X_j,\dots,X_k,\alpha)\\
&\quad+\sum_{j=1}^{k}B(X_1,\dots,S_{\mathsf{M}}(X_j,X_{k+1}),\dots,X_k,\alpha)+B(X_1,\dots,X_k,\nabla^{\pi_{\mathsf{E}}}_{X_{k+1}}\alpha)\\
&\quad+B(X_1,\dots,X_k,S_{\pi_{\mathsf{E}}}(\alpha,X_{k+1})).
\end{aligned}$$

This gives

$$\overline{\nabla}^{\mathsf{M},\pi_{\mathsf{E}}}B=\nabla^{\mathsf{M},\pi_{\mathsf{E}}}B-\sum_{j=1}^{k}\mathrm{Ins}_j(B,S_{\mathsf{M}})-\mathrm{Ins}_{k+1}(B,S_{\pi_{\mathsf{E}}}),$$

as claimed. □

With this lemma, we can provide the following characterisation of iterated covariant differentials of sections of E with respect to different connections.

Lemma 3.67 (Iterated Covariant Differentials of Sections with Respect to Different Connections I) *Let $r\in\{\infty,\omega\}$ and let $\pi_{\mathsf{E}}\colon\mathsf{E}\to\mathsf{M}$ be a C^r-vector bundle. Consider C^r-affine connections ∇^{M} and $\overline{\nabla}^{\mathsf{M}}$ on M, and C^r-linear connections $\nabla^{\pi_{\mathsf{E}}}$ and $\overline{\nabla}^{\pi_{\mathsf{E}}}$ in E. For $m\in\mathbb{Z}_{\geq 0}$, there exist C^r-vector bundle mappings*

$$(A^m_s,\mathrm{id}_{\mathsf{E}})\in\mathrm{VB}^r(\mathrm{T}^s(\mathrm{T}^*\mathsf{M}\otimes\mathsf{E});\mathrm{T}^m(\mathrm{T}^*\mathsf{M})\otimes\mathsf{E}),\qquad s\in\{0,1,\dots,m\},$$

such that

$$\overline{\nabla}^{\mathsf{M},\pi_{\mathsf{E}},m}\xi=\sum_{s=0}^{m}A^m_s(\nabla^{\mathsf{M},\pi_{\mathsf{E}},s}\xi)$$

for all $\xi\in\Gamma^m(\mathsf{E})$. Moreover, the vector bundle mappings $A^m_0,A^m_1,\dots,A^m_m$ satisfy the recursion relations prescribed by $A^0_0(\beta_0)=\beta_0$ and

$$\begin{aligned}
&A^{m+1}_{m+1}(\beta_{m+1})=\beta_{m+1},\\
&\quad A^{m+1}_s(\beta_s)=(\overline{\nabla}^{\mathsf{M},\pi_{\mathsf{E}}}A^m_s)(\beta_s)+A^m_{s-1}\otimes\mathrm{id}_{\mathrm{T}^*\mathsf{M}}(\beta_s)
\end{aligned}$$

$$- \sum_{j=1}^{s} A_s^m \otimes \mathrm{id}_{\mathsf{T}^*\mathsf{M}}(\mathrm{Ins}_j(\beta_s, S_{\mathsf{M}})) - A_s^m \otimes \mathrm{id}_{\mathsf{T}^*\mathsf{M}}(\mathrm{Ins}_{s+1}(\beta_s, S_{\pi_{\mathsf{E}}})),$$

$$s \in \{1, \dots, m\},$$

$$A_0^{m+1}(\beta_0) = (\overline{\nabla}^{\mathsf{M},\pi_{\mathsf{E}}} A_0^m)(\beta_0) - A_0^m \otimes \mathrm{id}_{\mathsf{T}^*\mathsf{M}}(\mathrm{Ins}_1(\beta_0, S_{\pi_{\mathsf{E}}})),$$

where $\beta_s \in \mathrm{T}^s(\mathsf{T}^*\mathsf{M}) \otimes \mathsf{E}$, $s \in \{0, 1, \dots, m\}$.

Proof The assertion clearly holds for $m = 0$, so suppose it true for $m \in \mathbb{Z}_{>0}$. Thus

$$\overline{\nabla}^{\mathsf{M},\pi_{\mathsf{E}},m}\xi = \sum_{s=0}^{m} A_s^m(\nabla^{\mathsf{M},\pi_{\mathsf{E}},s}\xi),$$

where the vector bundle mappings A_s^a, $a \in \{0, 1, \dots, m\}$, $s \in \{0, 1, \dots, a\}$, satisfy the recursion relations from the statement of the lemma. Then

$$\begin{aligned}
\overline{\nabla}^{\mathsf{M},\pi_{\mathsf{E}},m+1}\xi &= \sum_{s=0}^{m}(\overline{\nabla}^{\mathsf{M},\pi_{\mathsf{E}}} A_s^m)(\nabla^{\mathsf{M},\pi_{\mathsf{E}},s}\xi) + \sum_{s=0}^{m} A_s^m \otimes \mathrm{id}_{\mathsf{T}^*\mathsf{M}}(\overline{\nabla}^{\mathsf{M},\pi_{\mathsf{E}}}\nabla^{\mathsf{M},\pi_{\mathsf{E}},s}\xi) \\
&= \sum_{s=0}^{m}(\overline{\nabla}^{\mathsf{M},\pi_{\mathsf{E}}} A_s^m)(\nabla^{\mathsf{M},\pi_{\mathsf{E}},s}\xi) + \sum_{s=0}^{m} A_s^m \otimes \mathrm{id}_{\mathsf{T}^*\mathsf{M}}(\nabla^{\mathsf{M},\pi_{\mathsf{E}},s+1}\xi) \\
&\quad - \sum_{s=1}^{m}\sum_{j=1}^{s} A_s^m \otimes \mathrm{id}_{\mathsf{T}^*\mathsf{M}}(\mathrm{Ins}_j(\nabla^{\mathsf{M},\pi_{\mathsf{E}},s}\xi, S_{\mathsf{M}})) \\
&\quad - \sum_{s=1}^{m} A_s^m \otimes \mathrm{id}_{\mathsf{T}^*\mathsf{M}}(\mathrm{Ins}_{s+1}(\nabla^{\mathsf{M},\pi_{\mathsf{E}},s}\xi, S_{\pi_{\mathsf{E}}})) - A_0^m \otimes \mathrm{id}_{\mathsf{T}^*\mathsf{M}}(\mathrm{Ins}_1(\xi, S_{\pi_{\mathsf{E}}})) \\
&= \nabla^{\mathsf{M},\pi_{\mathsf{E}},m+1}\xi + \sum_{s=1}^{m}\Bigg((\overline{\nabla}^{\mathsf{M},\pi_{\mathsf{E}}} A_s^m)(\nabla^{\mathsf{M},\pi_{\mathsf{E}},s}\xi) + A_{s-1}^m \otimes \mathrm{id}_{\mathsf{T}^*\mathsf{M}}(\nabla^{\mathsf{M},\pi_{\mathsf{E}},s}\xi) \\
&\quad - \sum_{j=1}^{s} A_s^m \otimes \mathrm{id}_{\mathsf{T}^*\mathsf{M}}(\mathrm{Ins}_j(\nabla^{\mathsf{M},\pi_{\mathsf{E}},s}\xi, S_{\mathsf{M}})) \\
&\quad - A_s^m \otimes \mathrm{id}_{\mathsf{T}^*\mathsf{M}}(\mathrm{Ins}_{s+1}(\nabla^{\mathsf{M},\pi_{\mathsf{E}},s}\xi, S_{\pi_{\mathsf{E}}}))\Bigg) - (\overline{\nabla}^{\mathsf{M},\pi_{\mathsf{E}}} A_0^m)(\xi) \\
&\quad - A_0^m \otimes \mathrm{id}_{\mathsf{T}^*\mathsf{M}}(\mathrm{Ins}_1(\xi, S_{\pi_{\mathsf{E}}}))
\end{aligned}$$

by Lemma 3.66. From this, the lemma follows. □

The lemma has an "inverse" which we state next.

Lemma 3.68 (Iterated Covariant Differentials of Sections with Respect to Different Connections II) *Let $r \in \{\infty, \omega\}$ and let $\pi_{\mathsf{E}} \colon \mathsf{E} \to \mathsf{M}$ be a C^r-vector bundle. Consider C^r-affine connections ∇^{M} and $\overline{\nabla}^{\mathsf{M}}$ on M, and C^r-linear connections $\nabla^{\pi_{\mathsf{E}}}$ and $\overline{\nabla}^{\pi_{\mathsf{E}}}$ in E. For $m \in \mathbb{Z}_{\geq 0}$, there exist C^r-vector bundle mappings*

$$(B_s^m, \mathrm{id}_{\mathsf{E}}) \in \mathrm{VB}^r(\mathrm{T}^s(\mathrm{T}^*\mathsf{M} \otimes \mathsf{E}); \mathrm{T}^m(\mathrm{T}^*\mathsf{M}) \otimes \mathsf{E}), \qquad s \in \{0, 1, \dots, m\},$$

such that

$$\nabla^{\mathsf{M},\pi_{\mathsf{E}},m}\xi = \sum_{s=0}^{m} B_s^m(\overline{\nabla}^{\mathsf{M},\pi_{\mathsf{E}},s}\xi)$$

for all $\xi \in \Gamma^m(\mathsf{E})$. Moreover, the vector bundle mappings $B_0^m, B_1^m, \dots, B_m^m$ satisfy the recursion relations prescribed by $B_0^0(\alpha_0) = \beta_0$ and

$$B_{m+1}^{m+1}(\alpha_{m+1}) = \alpha_{m+1},$$

$$B_s^{m+1}(\alpha_s) = (\overline{\nabla}^{\mathsf{M},\pi_{\mathsf{E}}} B_s^m)(\alpha_s) + B_{s-1}^m \otimes \mathrm{id}_{\mathrm{T}^*\mathsf{M}}(\alpha_s) + \sum_{j=1}^{m} \mathrm{Ins}_j(B_s^m(\alpha_s), S_{\mathsf{M}})$$
$$+ \mathrm{Ins}_{m+1}(B_s^m(\alpha_s\xi), S_{\pi_{\mathsf{E}}}), \quad s \in \{1, \dots, m\},$$

$$B_0^{m+1}(\alpha_0) = (\overline{\nabla}^{\mathsf{M},\pi_{\mathsf{E}}} B_0^m)(\alpha_0) + \sum_{j=1}^{m} \mathrm{Ins}_j(B_0^m(\alpha_0), S_{\mathsf{M}}) + \mathrm{Ins}_{m+1}(B_0^m(\alpha_0), S_{\pi_{\mathsf{E}}}),$$

where $\alpha_s \in \mathrm{T}^s(\mathrm{T}^*\mathsf{M}) \otimes \mathsf{E}$, $s \in \{0, 1, \dots, m\}$.

Proof The lemma is clearly true for $m = 0$, so suppose it true for $m \in \mathbb{Z}_{>0}$. Thus

$$\nabla^{\mathsf{M},\pi_{\mathsf{E}},m}\xi = \sum_{s=0}^{m} B_s^m(\overline{\nabla}^{\mathsf{M},\pi_{\mathsf{E}},s}\xi), \tag{3.38}$$

where the vector bundle mappings B_s^a, $a \in \{0, 1, \dots, m\}$, $s \in \{0, 1, \dots, a\}$, satisfy the recursion relations given in the lemma. Then, working with the left-hand side of this relation,

$$\overline{\nabla}^{\mathsf{M},\pi_{\mathsf{E}}}\nabla^{\mathsf{M},\pi_{\mathsf{E}},m}\xi = \nabla^{\mathsf{M},\pi_{\mathsf{E}},m+1}\xi$$
$$- \sum_{j=1}^{m} \mathrm{Ins}_j(\nabla^{\mathsf{M},\pi_{\mathsf{E}},m}\xi, S_{\mathsf{M}}) - \mathrm{Ins}_{m+1}(\nabla^{\mathsf{M},\pi_{\mathsf{E}},m}\xi, S_{\pi_{\mathsf{E}}})$$

$$= \nabla^{\mathsf{M},\pi_{\mathsf{E}},m+1}\xi - \sum_{s=0}^{m}\sum_{j=1}^{m} \mathrm{Ins}_j(B_s^m(\overline{\nabla}^{\mathsf{M},\pi_{\mathsf{E}},s}\xi), S_{\mathsf{M}})$$
$$- \sum_{s=0}^{m} \mathrm{Ins}_{m+1}(B_s^m(\overline{\nabla}^{\mathsf{M},\pi_{\mathsf{E}},s}\xi), S_{\pi_{\mathsf{E}}}),$$

by Lemma 3.66. Now, working with the right-hand side of (3.38),

$$\overline{\nabla}^{\mathsf{M},\pi_{\mathsf{E}}}\nabla^{\mathsf{M},\pi_{\mathsf{E}},m}\xi = \sum_{s=0}^{m}(\overline{\nabla}^{\mathsf{M},\pi_{\mathsf{E}}}B_s^m)(\overline{\nabla}^{\mathsf{M},\pi_{\mathsf{E}},m}\xi) + \sum_{s=0}^{m} B_s^m \otimes \mathrm{id}_{\mathsf{T}^*\mathsf{M}}(\overline{\nabla}^{\mathsf{M},\pi_{\mathsf{E}},m+1}\xi).$$

Combining the preceding two computations,

$$\begin{aligned}\nabla^{\mathsf{M},\pi_{\mathsf{E}},m+1}\xi &= \overline{\nabla}^{\mathsf{M},\pi_{\mathsf{E}},m+1}\xi \\ &+ \sum_{s=1}^{m}\Bigg((\overline{\nabla}^{\mathsf{M},\pi_{\mathsf{E}}}B_s^m)(\overline{\nabla}^{\mathsf{M},\pi_{\mathsf{E}},s}\xi) + B_{s-1}^m \otimes \mathrm{id}_{\mathsf{T}^*\mathsf{M}}(\overline{\nabla}^{\mathsf{M},\pi_{\mathsf{E}},s}\xi) \\ &+ \sum_{j=1}^{m}\mathrm{Ins}_j(B_s^m(\overline{\nabla}^{\mathsf{M},\pi_{\mathsf{E}},s}\xi), S_{\mathsf{M}}) + \mathrm{Ins}_{m+1}(B_s^m(\overline{\nabla}^{\mathsf{M},\pi_{\mathsf{E}},s}\xi), S_{\pi_{\mathsf{E}}})\Bigg) \\ &+ (\overline{\nabla}^{\mathsf{M},\pi_{\mathsf{E}}}B_0^m)(\xi) + \sum_{j=1}^{m}\mathrm{Ins}_j(B_0^m(\xi), S_{\mathsf{M}}) + \mathrm{Ins}_{m+1}(B_0^m(\xi), S_{\pi_{\mathsf{E}}}),\end{aligned}$$

and from this the lemma follows. □

Now we give symmetrised versions of the preceding lemmata, since it is these that are required for computations with jets.

Lemma 3.69 (Iterated Symmetrised Covariant Differentials of Sections with Respect to Different Connections I) *Let $r \in \{\infty, \omega\}$ and let $\pi_{\mathsf{E}} \colon \mathsf{E} \to \mathsf{M}$ be a C^r-vector bundle. Consider C^r-affine connections ∇^{M} and $\overline{\nabla}^{\mathsf{M}}$ on M, and C^r-linear connections $\nabla^{\pi_{\mathsf{E}}}$ and $\overline{\nabla}^{\pi_{\mathsf{E}}}$ in E. For $m \in \mathbb{Z}_{\geq 0}$, there exist C^r-vector bundle mappings*

$$(\widehat{A}_s^m, \mathrm{id}_{\mathsf{E}}) \in \mathrm{VB}^r(\mathrm{T}^s(\mathsf{T}^*\mathsf{M} \otimes \mathsf{E}); \mathrm{T}^m(\mathsf{T}^*\mathsf{M}) \otimes \mathsf{E}), \qquad s \in \{0, 1, \dots, m\},$$

such that

$$(\mathrm{Sym}_m \otimes \mathrm{id}_{\mathsf{E}}) \circ \overline{\nabla}^{\mathsf{M},\pi_{\mathsf{E}},m}\xi = \sum_{s=0}^{m}\widehat{A}_s^m((\mathrm{Sym}_s \otimes \mathrm{id}_{\mathsf{E}}) \circ \nabla^{\mathsf{M},\pi_{\mathsf{E}},s}\xi)$$

for all $\xi \in \Gamma^m(\mathsf{E})$.

Proof We define $A^m\colon \mathsf{T}^{\leq m}(\mathsf{T}^*\mathsf{M}) \otimes \mathsf{E} \to \mathsf{T}^{\leq m}(\mathsf{T}^*\mathsf{M}) \otimes \mathsf{E}$ by

$$\begin{aligned} &A^m(\xi, \nabla^{\pi_{\mathsf{E}}}\xi, \dots, \nabla^{\mathsf{M},\pi_{\mathsf{E}},m}\xi) \\ &= \left(A_0^0(\xi), \sum_{s=0}^{1} A_s^1(\nabla^{\mathsf{M},\pi_{\mathsf{E}},s}\xi), \dots, \sum_{s=0}^{m} A_s^m(\nabla^{\mathsf{M},\pi_{\mathsf{E}},s}\xi)\right). \end{aligned}$$

Let us organise the mappings we require into the following diagram:

$$\begin{array}{ccccc} \mathsf{T}^{\leq m}(\mathsf{T}^*\mathsf{M}) \otimes \mathsf{E} & \xrightarrow{\mathrm{Sym}_{\leq m} \otimes \mathrm{id}_{\mathsf{E}}} & \mathsf{S}^{\leq m}(\mathsf{T}^*\mathsf{M}) \otimes \mathsf{E} & \xrightarrow{S^m_{\overline{\nabla}^{\mathsf{M}},\overline{\nabla}^{\pi_{\mathsf{E}}}}} & \mathsf{J}^m\mathsf{E} \\ \downarrow{\scriptstyle A^m} & & \downarrow{\scriptstyle \widehat{A}^m} & & \| \\ \mathsf{T}^{\leq m}(\mathsf{T}^*\mathsf{M}) \otimes \mathsf{E} & \xrightarrow{\mathrm{Sym}_{\leq m} \otimes \mathrm{id}_{\mathsf{E}}} & \mathsf{S}^{\leq m}(\mathsf{T}^*\mathsf{M}) \otimes \mathsf{E} & \xrightarrow{S^m_{\nabla^{\mathsf{M}},\nabla^{\pi_{\mathsf{E}}}}} & \mathsf{J}^m\mathsf{E} \end{array} \tag{3.39}$$

Here $\widehat{A}^m$ is defined so that the right square commutes. We shall show that the left square also commutes. Indeed,

$$\begin{aligned} &\widehat{A}^m \circ \mathrm{Sym}_{\leq m} \otimes \mathrm{id}_{\mathsf{E}}(\xi, \nabla^{\pi_{\mathsf{E}}}\xi, \dots, \nabla^{\mathsf{M},\pi_{\mathsf{E}},m}\xi) \\ &\qquad = (S^m_{\overline{\nabla}^{\mathsf{M}},\overline{\nabla}^{\pi_{\mathsf{E}}}})^{-1} \circ S^m_{\nabla^{\mathsf{M}},\nabla^{\pi_{\mathsf{E}}}} \circ (\mathrm{Sym}_{\leq m} \otimes \mathrm{id}_{\mathsf{E}})(\xi, \overline{\nabla}^{\pi_{\mathsf{E}}}\xi, \dots, \nabla^{\mathsf{M},\pi_{\mathsf{E}},m}\xi) \\ &\qquad = \mathrm{Sym}_{\leq m} \otimes \mathrm{id}_{\mathsf{VE}}(\xi, \overline{\nabla}^{\pi_{\mathsf{E}}}\xi, \dots, \overline{\nabla}^{\mathsf{M},\pi_{\mathsf{E}},m}\xi) \\ &\qquad = (\mathrm{Sym}_{\leq m} \otimes \mathrm{id}_{\mathsf{E}}) \circ A^m(\xi, \nabla^{\pi_{\mathsf{E}}}\xi, \dots, \nabla^{\mathsf{M},\pi_{\mathsf{E}},m}\xi). \end{aligned}$$

Thus the diagram (3.39) commutes. Thus, if we define

$$\widehat{A}_s^m((\mathrm{Sym}_s \otimes \mathrm{id}_{\mathsf{E}}) \circ \nabla^{\mathsf{M},\pi_{\mathsf{E}},s}\xi) = (\mathrm{Sym}_m \otimes \mathrm{id}_{\mathsf{E}}) \circ A_s^m(\nabla^{\mathsf{M},\pi_{\mathsf{E}},s}\xi), \tag{3.40}$$

then we have

$$(\mathrm{Sym}_m \otimes \mathrm{id}_{\mathsf{E}}) \circ \overline{\nabla}^{\mathsf{M},\pi_{\mathsf{E}},m}\xi = \sum_{s=0}^{m} \widehat{A}_s^m((\mathrm{Sym}_s \otimes \mathrm{id}_{\mathsf{E}}) \circ \nabla^{\mathsf{M},\pi_{\mathsf{E}},s}\xi),$$

as desired. □

The previous lemma has an “inverse” which we state next.

Lemma 3.70 (Iterated Symmetrised Covariant Differentials of Sections with Respect to Different Connections II) *Let $r \in \{\infty, \omega\}$ and let $\pi_{\mathsf{E}}\colon \mathsf{E} \to \mathsf{M}$ be a C^r-vector bundle. Consider C^r-affine connections ∇^{M} and $\overline{\nabla}^{\mathsf{M}}$ on M, and C^r-linear connections $\nabla^{\pi_{\mathsf{E}}}$ and $\overline{\nabla}^{\pi_{\mathsf{E}}}$ in E. For $m \in \mathbb{Z}_{\geq 0}$, there exist C^r-vector bundle* *mappings*

$$(\widehat{B}_s^m, \mathrm{id}_{\mathsf{E}}) \in \mathrm{VB}^r(\mathsf{T}^s(\mathsf{T}^*\mathsf{M} \otimes \mathsf{E}); \mathsf{T}^m(\mathsf{T}^*\mathsf{M}) \otimes \mathsf{E}), \qquad s \in \{0, 1, \dots, m\},$$

such that

$$(\mathrm{Sym}_m \otimes \mathrm{id}_{\mathsf{E}}) \circ \nabla^{\mathsf{M},\pi_{\mathsf{E}},m}\xi = \sum_{s=0}^{m} \widehat{B}_s^m((\mathrm{Sym}_s \otimes \mathrm{id}_{\mathsf{E}}) \circ \overline{\nabla}^{\mathsf{M},\pi_{\mathsf{E}},s}\xi)$$

for all $\xi \in \Gamma^m(\mathsf{E})$.

Proof The proof here is identical with the proof of Lemma 3.69, making the obvious notational transpositions. □

The preceding four lemmata combine to give the following result.

Lemma 3.71 (Decompositions of Jets of Sections with Respect to Different Connections) *Let* $r \in \{\infty, \omega\}$ *and let* $\pi_{\mathsf{E}} \colon \mathsf{E} \to \mathsf{M}$ *be a* C^r*-vector bundle. Consider* C^r*-affine connections* ∇^{M} *and* $\overline{\nabla}^{\mathsf{M}}$ *on* M*, and* C^r*-linear connections* $\nabla^{\pi_{\mathsf{E}}}$ *and* $\overline{\nabla}^{\pi_{\mathsf{E}}}$ *in* E*. For* $m \in \mathbb{Z}_{\geq 0}$*, there exist* C^r*-vector bundle mappings*

$$A^m \in \mathrm{VB}^r(\mathsf{J}^m\mathsf{E}; \mathsf{S}^{\leq m}(\mathsf{T}^*\mathsf{M}) \otimes \mathsf{E}), \quad B^m \in \mathrm{VB}^r(\mathsf{J}^m\mathsf{E}; \mathsf{S}^{\leq m}(\mathsf{T}^*\mathsf{M}) \otimes \mathsf{E}),$$

defined by

$$A^m(j_m\xi(x)) = \mathrm{Sym}_{\leq m} \otimes \mathrm{id}_{\mathsf{E}}(\xi(x), \nabla^{\pi_{\mathsf{E}}}\xi(x), \dots, \nabla^{\mathsf{M},\pi_{\mathsf{E}},m}\xi(x)),$$
$$B^m(j_m\xi(x)) = \mathrm{Sym}_{\leq m} \otimes \mathrm{id}_{\mathsf{VE}}(\xi(x), \overline{\nabla}^{\pi_{\mathsf{E}}}\xi(x), \dots, \overline{\nabla}^{\mathsf{M},\pi_{\mathsf{E}},m}\xi(x)).$$

Moreover, A^m *and* B^m *are isomorphisms, and*

$$\begin{aligned} B^m \circ (A^m)^{-1} &\circ (\mathrm{Sym}_{\leq m} \otimes \mathrm{id}_{\mathsf{E}})(\xi(x), \nabla^{\pi_{\mathsf{E}}}\xi(x), \dots, \nabla^{\mathsf{M},\pi_{\mathsf{E}},m}\xi(x)) \\ &= \Bigg(\xi(x), \sum_{s=0}^{1} \widehat{A}_s^1((\mathrm{Sym}_s \otimes \mathrm{id}_{\mathsf{E}}) \circ \nabla^{\mathsf{M},\pi_{\mathsf{E}},s}\xi(x)), \dots, \\ &\qquad \sum_{s=0}^{m} \widehat{A}_s^m((\mathrm{Sym}_s \otimes \mathrm{id}_{\mathsf{E}}) \circ \nabla^{\mathsf{M},\pi_{\mathsf{E}},s}\xi(x))\Bigg) \end{aligned}$$

and

$$A^m \circ (B^m)^{-1} \circ (\mathrm{Sym}_{\leq m} \otimes \mathrm{id}_{\mathsf{E}})(\xi(x), \overline{\nabla}^{\pi_{\mathsf{E}}}\xi(x), \dots, \overline{\nabla}^{\mathsf{M},\pi_{\mathsf{E}},m}\xi(x))$$
$$= \left(\xi(x), \sum_{s=0}^{1} \widehat{B}_s^1((\mathrm{Sym}_s \otimes \mathrm{id}_{\mathsf{E}}) \circ \overline{\nabla}^{\mathsf{M},\pi_{\mathsf{E}},s}\xi(e)), \dots, \right.$$
$$\left. \sum_{s=0}^{m} \widehat{B}_s^m((\mathrm{Sym}_s \otimes \mathrm{id}_{\mathsf{E}}) \circ \nabla^{\mathsf{M},\pi_{\mathsf{E}},s}\xi(x))\right),$$

where the vector bundle mappings $\widehat{A}_s^m$ and $\widehat{B}_s^m$, $s \in \{0, 1, \dots, m\}$, are as in Lemmata 3.69 and 3.70.

Chapter 4
Analysis: Norm Estimates for Derivatives

In this chapter, we carry out the principal analytical constructions that are needed to prove the continuity results of Chap. 5. The main objective is to obtain useful bounds for the fibre norms for the tensors obtained in Sect. 3.3 that relate derivatives on the total space of a vector bundle to derivatives on the base space. This bound is given in Lemma 4.17. When we say "usefulness" of the bounds, the property of usefulness is that it permits one to prove the continuity results of Chap. 5. Another measure of "usefulness" of the norm estimates we obtain is that they permit us to show that the topologies we have defined are independent of the data—namely the metrics and connections—used to define them. This important analysis is made in Sect. 4.3.

Our geometric constructions in Chap. 3 were made in both the smooth and real analytic setting. In this chapter, a number of results require real analyticity and/or give conclusions that are really only interesting in the real analytic setting. In most such cases, however, there are analogous (and significantly easier) results in the smooth setting. We will occasionally allude to these analogous results.

4.1 Fibre Norms for Some Useful Jet Bundles

In Sect. 3.3 we saw how to make decompositions for jets of sections of vector bundles and jets of various lifts to the total space of a vector bundle $\pi_{\mathsf{E}} : \mathsf{E} \to \mathsf{M}$, using the Levi-Civita affine connection induced by a natural Riemannian metric on E. In this section we consider fibre norms for these jet bundles. The fibre norm for the space of jets of sections of a vector bundle is deduced in a natural way from a Riemannian metric on M and a fibre metric in $\pi_{\mathsf{E}} : \mathsf{E} \to \mathsf{M}$, as we saw in Sect. 2.3.2. For fibre norms of lifted objects, the story is more complicated. Since the objects are lifted from M, there are two natural fibre norms in each case, one coming from the Riemannian metric on E (for the jets of the lifted objects), and the other coming from the Riemannian metric on M and the fibre metric on the vector bundle (for the

A. D. Lewis, *Geometric Analysis on Real Analytic Manifolds*, Lecture Notes in Mathematics 2333, https://doi.org/10.1007/978-3-031-37913-0_4

jets of the unlifted objects). It is important to understand how these fibre metrics are related, and that is the purpose of this section.

The setup is the following. We let $r \in \{\infty, \omega\}$ and let $\pi_{\mathsf{E}}\colon \mathsf{E} \to \mathsf{M}$ be a C^r-vector bundle. We consider a Riemannian metric $\mathbb{G}_{\mathsf{M}}$ on M, a fibre metric $\mathbb{G}_{\pi_{\mathsf{E}}}$ on E, the Levi-Civita connection ∇^{M} on M, and a vector bundle connection $\nabla^{\pi_{\mathsf{E}}}$ in E, all being of class C^r. This gives the Riemannian metric $\mathbb{G}_{\mathsf{E}}$ of (3.5) and the associated Levi-Civita connection ∇^{E}. This data gives the fibre metrics for all sorts of tensors defined on the total space E. We, however, are interested only in the lifted tensors such as are described in Sect. 3.1.

The reader will definitely observe a certain repetitiveness to our constructions in this section, rather similar to that seen in Sect. 3.3. However, the ideas here are important and the notation is confusing, so we do not skip anything. Additionally, a few of the cases we consider require considerations different from the majority of the other cases, so it is worth highlighting these differences.

While the results in this section are most important in the real analytic setting, we treat the smooth and real analytic cases simultaneously, as we have done in Chap. 3.

4.1.1 Fibre Norms for Horizontal Lifts of Functions

We let $r \in \{\infty, \omega\}$ and let $\pi_{\mathsf{E}}\colon \mathsf{E} \to \mathsf{M}$ be a C^r-vector bundle. For $f \in \mathrm{C}^r(\mathsf{M})$, we have $\pi_{\mathsf{E}}^* f \in \mathrm{C}^r(\mathsf{E})$. We can, therefore, think of the m-jet of $\pi_{\mathsf{E}}^* f$ as being characterised by $j_m f$, as well as by $j_m \pi_{\mathsf{E}}^* f$, and of comparing these two characterisations. Thus we have the two fibre norms

$$\|j_m f(x)\|^2_{\mathbb{G}_{\mathsf{M},m}} = \sum_{j=0}^{m} \frac{1}{(j!)^2} \|\nabla^{\mathsf{M},j} f(x)\|^2_{\mathbb{G}_{\mathsf{M}}}$$

and

$$\|j_m \pi_{\mathsf{E}}^* f(e)\|^2_{\mathbb{G}_{\mathsf{E},m}} = \sum_{j=0}^{m} \frac{1}{(j!)^2} \|\nabla^{\mathsf{E},j} \pi_{\mathsf{E}}^* f(e)\|^2_{\mathbb{G}_{\mathsf{E}}}. \tag{4.1}$$

These fibre norms can be related by virtue of Lemma 3.28. To do so, we make use of the following lemma.

Lemma 4.1 (Fibre Norms for Horizontal Lifts of Functions)

$$\|\pi_{\mathsf{E}}^* \nabla^{\mathsf{M},m} f(e)\|_{\mathbb{G}_{\mathsf{E}}} = \|\nabla^{\mathsf{M},m} f(\pi_{\mathsf{E}}(e))\|_{\mathbb{G}_{\mathsf{M}}}.$$

Proof We have the fibre metric $\mathbb{G}_{\mathsf{E}}^{-1}$ on $\mathsf{T}^*\mathsf{E}$ associated with the Riemannian metric $\mathbb{G}_{\mathsf{E}}$. The subbundles $\mathsf{H}^*\mathsf{E}$ and $\mathsf{V}^*\mathsf{E}$ are $\mathbb{G}_{\mathsf{E}}^{-1}$-orthogonal. We note that $T_e^*\pi_{\mathsf{E}}\colon \mathsf{T}^*_{\pi_{\mathsf{E}}(e)}\mathsf{M} \to \mathsf{H}_e^*\mathsf{E}$ is an isometry. Thus we have the formula

$$\|\pi_{\mathsf{E}}^* B\|_{\mathbb{G}_{\mathsf{E}}} = \|B\|_{\mathbb{G}_{\mathsf{M}}}, \qquad B \in \Gamma^0(\mathsf{T}^m(\mathsf{T}^*\mathsf{M})),$$

and the assertion of the lemma is merely a special case of this formula. □

We note that the fibre norm (4.4) makes use of the vector bundle mapping

$$B^m_{\nabla^{\mathsf{E}}} \in \mathrm{VB}^r(\mathsf{P}^{*m}\mathsf{E}; \mathsf{S}^{\leq m}(\mathsf{T}^*\mathsf{E}))$$

from Lemma 3.28. If instead we use the vector bundle mapping

$$A^m_{\nabla^{\mathsf{E}}} \in \mathrm{VB}^r(\mathsf{P}^{*m}\mathsf{E}; \mathsf{S}^{\leq m}(\pi_{\mathsf{E}}^*\mathsf{T}^*\mathsf{M}))$$

from Lemma 3.28, then we have the alternative fibre norm

$$\|j_m\pi_{\mathsf{E}}^* f(e)\|'^2_{\mathbb{G}_{\mathsf{E},m}} = \sum_{j=0}^{m} \frac{1}{(j!)^2}\|\pi_{\mathsf{E}}^*\nabla^{\mathsf{M},j} f(e)\|^2_{\mathbb{G}_{\mathsf{E}}} = \sum_{j=0}^{m} \frac{1}{(j!)^2}\|\nabla^{\mathsf{M},j} f(\pi_{\mathsf{E}}(e))\|^2_{\mathbb{G}_{\mathsf{M}}}.$$

The relationship between the fibre norms $\|\cdot\|_{\mathbb{G}_{\mathsf{E},m}}$ and $\|\cdot\|'_{\mathbb{G}_{\mathsf{E},m}}$ can be phrased as, "What is the relationship between the norms of the jet of the lift and the lift of the jet?" This is a question we will phrase below for other sorts of lifts, and will address comprehensively when we prove the continuity of the various lifting operations in Sect. 5.3.

4.1.2 Fibre Norms for Vertical Lifts of Sections

We let $r \in \{\infty, \omega\}$ and let $\pi_{\mathsf{E}}: \mathsf{E} \to \mathsf{M}$ be a C^r-vector bundle. For $\xi \in \Gamma^r(\mathsf{E})$, we have $\xi^{\mathrm{v}} \in \Gamma^r(\mathsf{TE})$. We can, therefore, think of the m-jet of ξ^{v} as being characterised by $j_m\xi$, as well as by $j_m\xi^{\mathrm{v}}$, and of comparing these two characterisations. Thus we have the two fibre norms

$$\|j_m\xi(x)\|^2_{\mathbb{G}_{\mathsf{M},\pi_{\mathsf{E}},m}} = \sum_{j=0}^{m} \frac{1}{(j!)^2}\|\nabla^{\mathsf{M},\pi_{\mathsf{E}},j}\xi(x)\|^2_{\mathbb{G}_{\mathsf{M},\pi_{\mathsf{E}}}}$$

and

$$\|j_m\xi^{\mathrm{v}}(e)\|^2_{\mathbb{G}_{\mathsf{E},m}} = \sum_{j=0}^{m} \frac{1}{(j!)^2}\|\nabla^{\mathsf{E},j}\xi^{\mathrm{v}}(e)\|^2_{\mathbb{G}_{\mathsf{E}}}. \tag{4.2}$$

These fibre norms can be related by virtue of Lemma 3.33. To do so, we make use of the following lemma.

Lemma 4.2 (Fibre Norms for Vertical Lifts of Sections)

$$\|(\nabla^{\mathsf{M},\pi_{\mathsf{E}},m}\xi)^{\mathrm{v}}(e)\|_{\mathbb{G}_{\mathsf{E}}} = \|\nabla^{\mathsf{M},\pi_{\mathsf{E}},m}\xi(\pi_{\mathsf{E}}(e))\|_{\mathbb{G}_{\mathsf{M},\pi_{\mathsf{E}}}}.$$

Proof The subbundles HE and VE are $\mathbb{G}_{\mathsf{E}}$-orthogonal and the subbundles $\mathsf{H}^*\mathsf{E}$ and $\mathsf{V}^*\mathsf{E}$ are $\mathbb{G}_{\mathsf{E}}^{-1}$-orthogonal. We note that the identification $\mathsf{V}_e\mathsf{E} \simeq \mathsf{E}_{\pi_{\mathsf{E}}(e)}$ is an isometry and that $T_e^*\pi_{\mathsf{E}} \colon \mathsf{T}^*_{\pi_{\mathsf{E}}(e)}\mathsf{M} \to \mathsf{H}^*_e\mathsf{E}$ is an isometry. Thus we have the formula

$$\|B^{\mathrm{v}}\|_{\mathbb{G}_{\mathsf{E}}} = \|B\|_{\mathbb{G}_{\mathsf{M},\pi_{\mathsf{E}}}}, \qquad B \in \Gamma^0(\mathsf{T}^m(\mathsf{T}^*\mathsf{M}) \otimes \mathsf{E}),$$

and the assertion of the lemma is merely a special case of this formula. □

We note that the fibre norm (4.2) makes use of the vector bundle mapping

$$B^m_{\nabla^{\mathsf{E}}} \in \mathrm{VB}^r(\mathsf{P}^{*m}\mathsf{E} \otimes \mathsf{VE}; \mathsf{S}^{\leq m}(\mathsf{T}^*\mathsf{E}) \otimes \mathsf{VE})$$

from Lemma 3.33. If instead we use the vector bundle mapping

$$A^m_{\nabla^{\mathsf{E}}} \in \mathrm{VB}^r(\mathsf{P}^{*m}\mathsf{E} \otimes \mathsf{VE}; \mathsf{S}^{\leq m}(\pi_{\mathsf{E}}^*\mathsf{T}^*\mathsf{M}) \otimes \mathsf{VE})$$

from Lemma 3.33, then we have the alternative fibre norm

$$\begin{aligned}\|j_m\xi^{\mathrm{v}}(e)\|'^2_{\mathbb{G}_{\mathsf{E},m}} &= \sum_{j=0}^m \frac{1}{(j!)^2}\|(\nabla^{\mathsf{M},\pi_{\mathsf{E}},j}\xi)^{\mathrm{v}}(e)\|^2_{\mathbb{G}_{\mathsf{E}}} \\ &= \sum_{j=0}^m \frac{1}{(j!)^2}\|\nabla^{\mathsf{M},\pi_{\mathsf{E}},j}\xi(\pi_{\mathsf{E}}(e))\|^2_{\mathbb{G}_{\mathsf{M},\pi_{\mathsf{E}}}}.\end{aligned}$$

Again, this points out the matter of the relationship between the norms of the jet of a lift versus the lift of the jet, and this matter will be considered in detail in the continuity results of Sect. 5.3.

4.1.3 Fibre Norms for Horizontal Lifts of Vector Fields

We let $r \in \{\infty, \omega\}$ and let $\pi_{\mathsf{E}} \colon \mathsf{E} \to \mathsf{M}$ be a C^r-vector bundle. For $X \in \Gamma^r(\mathsf{TM})$, we have $X^{\mathrm{h}} \in \Gamma^r(\mathsf{TE})$. We can, therefore, think of the m-jet of X^{h} as being characterised by j_mX, as well as by j_mX^{h}, and of comparing these two characterisations. Thus we have the two fibre norms

$$\|j_m X(x)\|^2_{\mathbb{G}_{\mathsf{M},m}} = \sum_{j=0}^{m} \frac{1}{(j!)^2} \|\nabla^{\mathsf{M},j} X(x)\|^2_{\mathbb{G}_{\mathsf{M}}}$$

and

$$\|j_m X^{\mathrm{h}}(e)\|^2_{\mathbb{G}_{\mathsf{E},m}} = \sum_{j=0}^{m} \frac{1}{(j!)^2} \|\nabla^{\mathsf{E},j} X^{\mathrm{h}}(e)\|^2_{\mathbb{G}_{\mathsf{E}}}. \tag{4.3}$$

These fibre norms can be related by virtue of Lemma 3.38. To do so, we make use of the following lemma.

Lemma 4.3 (Fibre Norms for Horizontal Lifts of Vector Fields)

$$\|(\nabla^{\mathsf{M},m} X)^{\mathrm{h}}(e)\|_{\mathbb{G}_{\mathsf{E}}} = \|\nabla^{\mathsf{M},m} X(\pi_{\mathsf{E}}(e))\|_{\mathbb{G}_{\mathsf{M}}}.$$

Proof The subbundles HE and VE are $\mathbb{G}_{\mathsf{E}}$-orthogonal. We note that the identification $\mathsf{H}_e\mathsf{E} \simeq \mathsf{T}_{\pi_{\mathsf{E}}(e)}\mathsf{M}$ is an isometry and that $T_e^*\pi_{\mathsf{E}} \colon \mathsf{T}^*_{\pi_{\mathsf{E}}(e)}\mathsf{M} \to \mathsf{H}^*_e\mathsf{E}$ is an isometry. Thus we have the formula

$$\|B^{\mathrm{h}}\|_{\mathbb{G}_{\mathsf{E}}} = \|B\|_{\mathbb{G}_{\mathsf{M}}}, \qquad B \in \Gamma^0(\mathrm{T}^m(\mathsf{T}^*\mathsf{M}) \otimes \mathsf{TM}),$$

and the assertion of the lemma is merely a special case of this formula. □

We note that the fibre norm (4.3) makes use of the vector bundle mapping

$$B^m_{\nabla^{\mathsf{E}}} \in \mathrm{VB}^r(\mathsf{P}^{*m}\mathsf{E} \otimes \mathsf{HE}; \mathsf{S}^{\leq m}(\mathsf{T}^*\mathsf{E}) \otimes \mathsf{HE})$$

from Lemma 3.38. If instead we use the vector bundle mapping

$$A^m_{\nabla^{\mathsf{E}}} \in \mathrm{VB}^r(\mathsf{P}^{*m}\mathsf{E} \otimes \mathsf{HE}; \mathsf{S}^{\leq m}(\pi_{\mathsf{E}}^*\mathsf{T}^*\mathsf{M}) \otimes \mathsf{HE})$$

from Lemma 3.38, then we have the alternative fibre norm

$$\|j_m X^{\mathrm{h}}(e)\|'^2_{\mathbb{G}_{\mathsf{E},m}} = \sum_{j=0}^{m} \frac{1}{(j!)^2} \|(\nabla^{\mathsf{M},j} X)^{\mathrm{h}}(e)\|^2_{\mathbb{G}_{\mathsf{E}}} = \sum_{j=0}^{m} \frac{1}{(j!)^2} \|\nabla^{\mathsf{M},j} X(\pi_{\mathsf{E}}(e))\|^2_{\mathbb{G}_{\mathsf{M}}}.$$

Again, this points out the matter of the relationship between the norms of the jet of a lift versus the lift of the jet, and this matter will be considered in detail in the continuity results of Sect. 5.3.

4.1.4 Fibre Norms for Vertical Lifts of Dual Sections

We let $r \in \{\infty, \omega\}$ and let $\pi_{\mathsf{E}} \colon \mathsf{E} \to \mathsf{M}$ be a C^r-vector bundle. For $\lambda \in \Gamma^r(\mathsf{E}^*)$, we have $\lambda^{\mathrm{v}} \in \Gamma^r(\mathsf{T}^*\mathsf{E})$. We can, therefore, think of the m-jet of λ^{v} as being characterised by $j_m\lambda$, as well as by $j_m\lambda^{\mathrm{v}}$, and of comparing these two characterisations. Thus we have fibre norms

$$\|j_m\lambda(x)\|^2_{\mathbb{G}_{\mathsf{M},\pi_{\mathsf{E}},m}} = \sum_{j=0}^{m} \frac{1}{(j!)^2} \|\nabla^{\mathsf{M},\pi_{\mathsf{E}},j}\lambda(x)\|^2_{\mathbb{G}_{\mathsf{M},\pi_{\mathsf{E}}}}$$

and

$$\|j_m\lambda^{\mathrm{v}}(e)\|^2_{\mathbb{G}_{\mathsf{E},m}} = \sum_{j=0}^{m} \frac{1}{(j!)^2} \|\nabla^{\mathsf{E},j}\lambda^{\mathrm{v}}(e)\|^2_{\mathbb{G}_{\mathsf{E}}}. \tag{4.4}$$

These fibre norms can be related by virtue of Lemma 3.43. To do so, we make use of the following lemma.

Lemma 4.4 (Fibre Norms for Vertical Lifts of Dual Sections)

$$\|(\nabla^{\mathsf{M},\pi_{\mathsf{E}},m}\lambda)^{\mathrm{v}}(e)\|_{\mathbb{G}_{\mathsf{E}}} = \|\nabla^{\mathsf{M},\pi_{\mathsf{E}},m}\lambda(\pi_{\mathsf{E}}(e))\|_{\mathbb{G}_{\mathsf{M},\pi_{\mathsf{E}}}}.$$

Proof The subbundles $\mathsf{H}^*\mathsf{E}$ and $\mathsf{V}^*\mathsf{E}$ are $\mathbb{G}_{\mathsf{E}}^{-1}$-orthogonal. We note that the identification $\mathsf{V}^*_e\mathsf{E} \simeq \mathsf{E}^*_{\pi_{\mathsf{E}}(e)}$ is an isometry and that $T^*_e\pi_{\mathsf{E}} \colon \mathsf{T}^*_{\pi_{\mathsf{E}}(e)}\mathsf{M} \to \mathsf{H}^*_e\mathsf{E}$ is an isometry. Thus we have the formula

$$\|B^{\mathrm{v}}\|_{\mathbb{G}_{\mathsf{E}}} = \|B\|_{\mathbb{G}_{\mathsf{M},\pi_{\mathsf{E}}}}, \qquad B \in \Gamma^0(\mathsf{T}^m(\mathsf{T}^*\mathsf{M}) \otimes \mathsf{E}^*),$$

and the assertion of the lemma is merely a special case of this formula. □

We note that the fibre norm (4.4) makes use of the vector bundle mapping

$$B^m_{\nabla\mathsf{E}} \in \mathrm{VB}^r(\mathsf{P}^{*m}\mathsf{E} \otimes \mathsf{V}^*\mathsf{E}; \mathsf{S}^{\leq m}(\mathsf{T}^*\mathsf{E}) \otimes \mathsf{V}^*\mathsf{E})$$

from Lemma 3.43. If instead we use the vector bundle mapping

$$A^m_{\nabla\mathsf{E}} \in \mathrm{VB}^r(\mathsf{P}^{*m}\mathsf{E} \otimes \mathsf{V}^*\mathsf{E}; \mathsf{S}^{\leq m}(\pi^*_{\mathsf{E}}\mathsf{T}^*\mathsf{M}) \otimes \mathsf{V}^*\mathsf{E})$$

from Lemma 3.43, then we have the alternative fibre norm

$$\begin{aligned}\|j_m\lambda^{\mathrm{v}}(e)\|'^2_{\mathbb{G}_{\mathsf{E},m}} &= \sum_{j=0}^{m} \frac{1}{(j!)^2} \|(\nabla^{\mathsf{M},\pi_{\mathsf{E}},j}\lambda)^{\mathrm{v}}(e)\|^2_{\mathbb{G}_{\mathsf{E}}}\\ &= \sum_{j=0}^{m} \frac{1}{(j!)^2} \|\nabla^{\mathsf{M},\pi_{\mathsf{E}},j}\lambda(\pi_{\mathsf{E}}(e))\|^2_{\mathbb{G}_{\mathsf{M},\pi_{\mathsf{E}}}}.\end{aligned}$$

Again, this points out the matter of the relationship between the norms of the jet of a lift versus the lift of the jet, and this matter will be considered in detail in the continuity results of Sect. 5.3.

4.1.5 Fibre Norms for Vertical Lifts of Endomorphisms

We let $r \in \{\infty, \omega\}$ and let $\pi_{\mathsf{E}} \colon \mathsf{E} \to \mathsf{M}$ be a C^r-vector bundle. For $L \in \Gamma^r(\mathrm{T}^1_1(\mathsf{E}))$, we have $L^{\mathrm{v}} \in \Gamma^r(\mathrm{T}^1_1(\mathsf{E}))$. We can, therefore, think of the m-jet of L^{v} as being characterised by $j_m L$, as well as by $j_m L^{\mathrm{v}}$, and of comparing these two characterisations. Thus we have the two fibre norms

$$\|j_m L(x)\|^2_{\mathbb{G}_{\mathsf{M},\pi_{\mathsf{E}},m}} = \sum_{j=0}^{m} \frac{1}{(j!)^2} \|\nabla^{\mathsf{M},\pi_{\mathsf{E}},j} L(x)\|^2_{\mathbb{G}_{\mathsf{M},\pi_{\mathsf{E}}}}$$

and

$$\|j_m L^{\mathrm{v}}(e)\|^2_{\mathbb{G}_{\mathsf{E},m}} = \sum_{j=0}^{m} \frac{1}{(j!)^2} \|\nabla^{\mathsf{E},j} L^{\mathrm{v}}(e)\|^2_{\mathbb{G}_{\mathsf{E}}}. \tag{4.5}$$

These fibre norms can be related by virtue of Lemma 3.48. To do so, we make use of the following lemma.

Lemma 4.5 (Fibre Norms for Vertical Lifts of Endomorphisms)

$$\|(\nabla^{\mathsf{M},\pi_{\mathsf{E}},m} L)^{\mathrm{v}}(e)\|_{\mathbb{G}_{\mathsf{E}}} = \|\nabla^{\mathsf{M},\pi_{\mathsf{E}},m} L(\pi_{\mathsf{E}}(e))\|_{\mathbb{G}_{\mathsf{M},\pi_{\mathsf{E}}}}.$$

Proof The subbundles $\mathsf{H}^*\mathsf{E}$ and $\mathsf{V}^*\mathsf{E}$ are $\mathbb{G}_{\mathsf{E}}^{-1}$-orthogonal. We note that the identifications $\mathsf{V}_e\mathsf{E} \simeq \mathsf{E}_{\pi_{\mathsf{E}}(e)}$ and $\mathsf{V}^*_e\mathsf{E} \simeq \mathsf{E}^*_{\pi_{\mathsf{E}}(e)}$ are isometries, and that $T^*_e\pi_{\mathsf{E}} \colon \mathsf{T}^*_{\pi_{\mathsf{E}}(e)}\mathsf{M} \to \mathsf{H}^*_e\mathsf{E}$ is an isometry. Thus we have the formula

$$\|B^{\mathrm{v}}\|_{\mathbb{G}_{\mathsf{E}}} = \|B\|_{\mathbb{G}_{\mathsf{M},\pi_{\mathsf{E}}}}, \qquad B \in \Gamma^0(\mathrm{T}^m(\mathsf{T}^*\mathsf{M}) \otimes \mathrm{T}^1_1(\mathsf{E})),$$

and the assertion of the lemma is merely a special case of this formula. □

We note that the fibre norm (4.5) makes use of the vector bundle mapping

$$B^m_{\nabla^{\mathsf{E}}} \in \mathrm{VB}^r(\mathsf{P}^{*m}\mathsf{E} \otimes \mathrm{T}^1_1(\mathsf{VE}); \mathrm{S}^{\leq m}(\mathsf{T}^*\mathsf{E}) \otimes \mathrm{T}^1_1(\mathsf{VE}))$$

from Lemma 3.48. If instead we use the vector bundle mapping

$$A^m_{\nabla^{\mathsf{E}}} \in \mathrm{VB}^r(\mathsf{P}^{*m}\mathsf{E} \otimes \mathrm{T}^1_1(\mathsf{VE}); \mathrm{S}^{\leq m}(\pi^*_{\mathsf{E}}\mathsf{T}^*\mathsf{M}) \otimes \mathrm{T}^1_1(\mathsf{VE}))$$

from Lemma 3.48, then we have the alternative fibre norm

$$\|j_m L^{\mathrm{v}}(e)\|'^2_{\mathbb{G}_{\mathsf{E},m}} = \sum_{j=0}^{m} \frac{1}{(j!)^2} \|(\nabla^{\mathsf{M},\pi_{\mathsf{E}},j} L)^{\mathrm{v}}(e)\|^2_{\mathbb{G}_{\mathsf{E}}}$$

$$= \sum_{j=0}^{m} \frac{1}{(j!)^2} \|\nabla^{\mathsf{M},\pi_{\mathsf{E}},j} L(\pi_{\mathsf{E}}(e))\|^2_{\mathbb{G}_{\mathsf{M},\pi_{\mathsf{E}}}}.$$

Again, this points out the matter of the relationship between the norms of the jet of a lift versus the lift of the jet, and this matter will be considered in detail in the continuity results of Sect. 5.3.

4.1.6 Fibre Norms for Vertical Evaluations of Dual Sections

We let $r \in \{\infty, \omega\}$ and let $\pi_{\mathsf{E}} \colon \mathsf{E} \to \mathsf{M}$ be a C^r-vector bundle. For $\lambda \in \Gamma^r(\mathsf{E}^*)$, we have $\lambda^{\mathrm{e}} \in \mathrm{C}^r(\mathsf{E})$. We can, therefore, think of the m-jet of λ^{e} as being characterised by $j_m\lambda$, as well as by $j_m\lambda^{\mathrm{e}}$, and of comparing these two characterisations. Thus we have the two fibre norms

$$\|j_m\lambda(x)\|^2_{\mathbb{G}_{\mathsf{M},\pi_{\mathsf{E}},m}} = \sum_{j=0}^{m} \frac{1}{(j!)^2} \|\nabla^{\mathsf{M},\pi_{\mathsf{E}},j}\lambda(x)\|^2_{\mathbb{G}_{\mathsf{M},\pi_{\mathsf{E}}}}$$

and

$$\|j_m\lambda^{\mathrm{e}}(e)\|^2_{\mathbb{G}_{\mathsf{E},m}} = \sum_{j=0}^{m} \frac{1}{(j!)^2} \|\nabla^{\mathsf{E},j}\lambda^{\mathrm{e}}(e)\|^2_{\mathbb{G}_{\mathsf{E}}}. \tag{4.6}$$

These fibre norms can be related by virtue of Lemma 3.53. To do so, we make use of the following lemma.

Lemma 4.6 (Fibre Norms for Vertical Evaluations of Dual Sections)

$$\|(\nabla^{\mathsf{M},\pi_{\mathsf{E}},m}\lambda)^{\mathrm{e}}(e)\|_{\mathbb{G}_{\mathsf{E}}} = \|\nabla^{\mathsf{M},\pi_{\mathsf{E}},m}\lambda(\pi_{\mathsf{E}}(e))(e)\|_{\mathbb{G}_{\mathsf{M},\pi_{\mathsf{E}}}}.$$

Proof The subbundles $\mathsf{H}^*\mathsf{E}$ and $\mathsf{V}^*\mathsf{E}$ are $\mathbb{G}_{\mathsf{E}}^{-1}$-orthogonal. We note that the identification $\mathsf{V}^*_e\mathsf{E} \simeq \mathsf{E}^*_{\pi_{\mathsf{E}}(e)}$ is an isometry, and that $T^*_e\pi_{\mathsf{E}} \colon \mathsf{T}^*_{\pi_{\mathsf{E}}(e)}\mathsf{M} \to \mathsf{H}^*_e\mathsf{E}$ is an isometry. Thus we have the formula

$$\|B^{\mathrm{e}}(e)\|_{\mathbb{G}_{\mathsf{E}}} = \|B(\pi_{\mathsf{E}}(e))(e)\|_{\mathbb{G}_{\mathsf{M},\pi_{\mathsf{E}}}}, \qquad B \in \Gamma^0(\mathrm{T}^m(\mathsf{T}^*\mathsf{M}) \otimes \mathsf{E}^*),$$

and the assertion of the lemma is merely a special case of this formula. □

We note that the fibre norm (4.6) makes use of the vector bundle mapping

$$B^m_{\nabla\mathsf{E}} \in \mathrm{VB}^r(\mathsf{P}^{*m}\mathsf{E}; \mathsf{S}^{\leq m}(\mathsf{T}^*\mathsf{E}))$$

from Lemma 3.53. If instead we use the vector bundle mapping

$$A^m_{\nabla\mathsf{E}} \in \mathrm{VB}^r(\mathsf{P}^{*m}\mathsf{E}; \mathsf{S}^{\leq m}(\pi^*_{\mathsf{E}}\mathsf{T}^*\mathsf{M}))$$

from Lemma 3.53, then we have the alternative fibre norm

$$\begin{aligned}\|j_m\lambda^{\mathrm{e}}(e)\|'^2_{\mathbb{G}_{\mathsf{E},m}} &= \sum_{j=0}^{m}\frac{1}{(j!)^2}\|(\nabla^{\mathsf{M},\pi_{\mathsf{E}},j}\lambda)^{\mathrm{e}}(e)\|^2_{\mathbb{G}_{\mathsf{E}}}\\ &= \sum_{j=0}^{m}\frac{1}{(j!)^2}\|\nabla^{\mathsf{M},\pi_{\mathsf{E}},j}\lambda(\pi_{\mathsf{E}}(e))(e)\|^2_{\mathbb{G}_{\mathsf{M},\pi_{\mathsf{E}}}}.\end{aligned}$$

Again, this points out the matter of the relationship between the norms of the jet of a lift versus the lift of the jet, and this matter will be considered in detail in the continuity results of Sect. 5.3.

4.1.7 Fibre Norms for Vertical Evaluations of Endomorphisms

We let $r \in \{\infty, \omega\}$ and let $\pi_{\mathsf{E}}\colon \mathsf{E} \to \mathsf{M}$ be a C^r-vector bundle. For $L \in \Gamma^r(\mathsf{T}^1_1(\mathsf{E}))$, we have $L^{\mathrm{e}} \in \Gamma^r(\mathsf{TE})$. We can, therefore, think of the m-jet of L^{e} as being characterised by j_mL, as well as by j_mL^{e}, and of comparing these two characterisations. Thus we have the two fibre norms

$$\|j_mL(x)\|^2_{\mathbb{G}_{\mathsf{M},\pi_{\mathsf{E}},m}} = \sum_{j=0}^{m}\frac{1}{(j!)^2}\|\nabla^{\mathsf{M},\pi_{\mathsf{E}},j}L(x)\|^2_{\mathbb{G}_{\mathsf{M},\pi_{\mathsf{E}}}}$$

and

$$\|j_mL^{\mathrm{e}}(e)\|^2_{\mathbb{G}_{\mathsf{E},m}} = \sum_{j=0}^{m}\frac{1}{(j!)^2}\|\nabla^{\mathsf{E},j}L^{\mathrm{e}}(e)\|^2_{\mathbb{G}_{\mathsf{E}}}. \tag{4.7}$$

These fibre norms can be related by virtue of Lemma 3.58. To do so, we make use of the following lemma.

Lemma 4.7 (Fibre Norms for Vertical Evaluations of Endomorphisms)

$$\|(\nabla^{\mathsf{M},\pi_{\mathsf{E}},m}L)^{\mathrm{e}}(e)\|_{\mathbb{G}_{\mathsf{E}}} = \|\nabla^{\mathsf{M},\pi_{\mathsf{E}},m}L(\pi_{\mathsf{E}}(e))(e)\|_{\mathbb{G}_{\mathsf{M},\pi_{\mathsf{E}}}}.$$

Proof The subbundles $\mathsf{H}^*\mathsf{E}$ and $\mathsf{V}^*\mathsf{E}$ are $\mathbb{G}_{\mathsf{E}}^{-1}$-orthogonal. We note that the identification $\mathsf{V}_e^*\mathsf{E} \simeq \mathsf{E}_{\pi_{\mathsf{E}}(e)}^*$ is an isometry and that $T_e^*\pi_{\mathsf{E}}\colon \mathsf{T}_{\pi_{\mathsf{E}}(e)}^*\mathsf{M} \to \mathsf{H}_e^*\mathsf{E}$ is an isometry. Thus we have the formula

$$\|B^{\mathrm{e}}(e)\|_{\mathbb{G}_{\mathsf{E}}} = \|B(\pi_{\mathsf{E}}(e))(e)\|_{\mathbb{G}_{\mathsf{M},\pi_{\mathsf{E}}}}, \qquad B \in \Gamma^0(\mathsf{T}^m(\mathsf{T}^*\mathsf{M}) \otimes \mathrm{T}_1^1(\mathsf{E})),$$

and the assertion of the lemma is merely a special case of this formula. □

We note that the fibre norm (4.7) makes use of the vector bundle mapping

$$B_{\nabla^{\mathsf{E}}}^m \in \mathrm{VB}^r(\mathsf{P}^{*m}\mathsf{E} \otimes \mathsf{VE}; \mathsf{S}^{\leq m}(\mathsf{T}^*\mathsf{E}) \otimes \mathsf{VE})$$

from Lemma 3.58. If instead we use the vector bundle mapping

$$A_{\nabla^{\mathsf{E}}}^m \in \mathrm{VB}^r(\mathsf{P}^{*m}\mathsf{E} \otimes \mathsf{VE}; \mathsf{S}^{\leq m}(\pi_{\mathsf{E}}^*\mathsf{T}^*\mathsf{M}) \otimes \mathsf{VE})$$

from Lemma 3.58, then we have the alternative fibre norm

$$\|j_m L^{\mathrm{e}}(e)\|_{\mathbb{G}_{\mathsf{E},m}}'^2 = \sum_{j=0}^m \frac{1}{(j!)^2}\|(\nabla^{\mathsf{M},\pi_{\mathsf{E}},j}L)^{\mathrm{e}}(e)\|_{\mathbb{G}_{\mathsf{E}}}^2 = \sum_{j=0}^m \frac{1}{(j!)^2}\|\nabla^{\mathsf{M},\pi_{\mathsf{E}},j}L(e)\|_{\mathbb{G}_{\mathsf{M},\pi_{\mathsf{E}}}}^2.$$

Again, this points out the matter of the relationship between the norms of the jet of a lift versus the lift of the jet, and this matter will be considered in detail in the continuity results of Sect. 5.3.

4.1.8 Fibre Norms for Pull-Backs of Functions

We let $r \in \{\infty, \omega\}$ and let M and N be C^r-manifolds, and let $\Phi \in \mathrm{C}^r(\mathsf{M}; \mathsf{N})$. For $f \in \mathrm{C}^r(\mathsf{N})$, we have $\Phi^* f \in \mathrm{C}^m(\mathsf{M})$. We can, therefore, think of the m-jet of $\Phi^* f$ as being characterised by $j_m f$, as well as by $j_m(\Phi^* f)$, and of comparing these two characterisations. Thus we have the two fibre norms

$$\|j_m f(x)\|_{\mathbb{G}_{\mathsf{N},m}}^2 = \sum_{j=0}^m \frac{1}{(j!)^2}\|\nabla^{\mathsf{N},j} f(x)\|_{\mathbb{G}_{\mathsf{N}}}^2$$

and

$$\|j_m \Phi^* f(e)\|_{\mathbb{G}_{\mathsf{M},m}}^2 = \sum_{j=0}^m \frac{1}{(j!)^2}\|\nabla^{\mathsf{M},j}\Phi^* f(e)\|_{\mathbb{G}_{\mathsf{M}}}^2. \tag{4.8}$$

These fibre norms can be related by virtue of Lemma 3.65. To make use of this relationship, we shall also need to relate the norms of the terms in these expressions.

In the preceding sections, this was easy to do since the Riemannian metric on E was related in a specific way to the Riemannian metric on M and the fibre metric in E. Here, this is not so simple since, if we choose a Riemannian metric $\mathbb{G}_{\mathsf{M}}$ on M and a Riemannian metric $\mathbb{G}_{\mathsf{N}}$ on N, these will be have no useful pointwise relationship. So, rather than getting an equality between certain norms, the best we can achieve (and all that we need) is a useful bound, and this is the content of the next lemma.

Lemma 4.8 (Fibre Norms for Pull-Backs of Functions) *For a compact set* $\mathcal{K} \subseteq \mathsf{M}$*:*

(i) there exists $C \in \mathbb{R}_{>0}$ *such that*

$$\|\Phi^* \nabla^{\mathsf{N},m} f(x)\|_{\mathbb{G}_{\mathsf{M}}} \leq C^m \|\nabla^{\mathsf{N},m} f(\Phi(x))\|_{\mathbb{G}_{\mathsf{N}}}, \qquad x \in \mathcal{K},\ m \in \mathbb{Z}_{\geq 0};$$

(ii) if Φ *is a submersion or an injective immersion, then* C *from part (i) can be chosen so that it also holds that*

$$\|\nabla^{\mathsf{N},m} f(\Phi(x))\|_{\mathbb{G}_{\mathsf{N}}} \leq C^m \|\Phi^* \nabla^{\mathsf{N},m} f(x)\|_{\mathbb{G}_{\mathsf{M}}}, \qquad x \in \mathcal{K},\ m \in \mathbb{Z}_{\geq 0}.$$

Proof The essential part of the proof is the following linear algebraic sublemma.

Sublemma 1 Let $(\mathsf{U}, \mathbb{G}_{\mathsf{U}})$ and $(\mathsf{V}, \mathbb{G}_{\mathsf{V}})$ be finite-dimensional $\mathbb{R}$-inner product spaces and let $\Phi \in \mathrm{Hom}_{\mathbb{R}}(\mathsf{U}; \mathsf{V})$. Then there exists $C \in \mathbb{R}_{>0}$ such that

$$\|\Phi^* A\|_{\mathbb{G}_{\mathsf{U}}} \leq C^k \|A\|_{\mathbb{G}_{\mathsf{V}}}$$

for every $A \in \mathrm{T}^k(\mathsf{V}^*)$, $k \in \mathbb{Z}_{\geq 0}$. If, additionally, Φ is a surjective or injective, then C can be chosen so that, additionally, it holds that

$$\|A\|_{\mathbb{G}_{\mathsf{V}}} \leq C^k \|\Phi^* A\|_{\mathbb{G}_{\mathsf{U}}}$$

for every $A \in \mathrm{T}^k(\mathsf{V}^*)$, $k \in \mathbb{Z}_{\geq 0}$.

Proof Let $(f_1, \dots, f_m)$ and $(e_1, \dots, e_n)$ be orthonormal bases for U and V with dual bases $(f^1, \dots, f^m)$ and $(e^1, \dots, e^n)$. Write

$$A = \sum_{j_1, \dots, j_k = 1}^{n} A_{j_1 \cdots j_k} e^{j_1} \otimes \dots \otimes e^{j_k}$$

and

$$\Phi = \sum_{j=1}^{n} \sum_{a=1}^{m} \Phi_a^j e_j \otimes f^a.$$

Then

$$\Phi^* A = \sum_{j_1,\dots,j_k=1}^{n} \sum_{a_1,\dots,a_k=1}^{m} \Phi^{j_1}_{a_1} \cdots \Phi^{j_k}_{a_k} A_{j_1\cdots j_k} f^{a_1} \otimes \dots \otimes f^{a_k}.$$

Denote

$$\|\Phi\|_\infty = \max\left\{\left|\Phi^j_a\right| \;\middle|\; a \in \{1,\dots,m\},\, j \in \{1,\dots,n\}\right\}.$$

We have

$$\begin{aligned}
\|\Phi^* A\|^2_{\mathbb{G}_\mathsf{U}} &= \sum_{a_1,\dots,a_k=1}^{m} \left(\sum_{j_1,\dots,j_k=1}^{n} \Phi^{j_1}_{a_1} \cdots \Phi^{j_k}_{a_k} A_{j_1\cdots j_k}\right)^2 \\
&\le \sum_{a_1,\dots,a_k=1}^{m} \left(\sum_{j_1,\dots,j_k=1}^{n} \left|\Phi^{j_1}_{a_1} \cdots \Phi^{j_k}_{a_k} A_{j_1\cdots j_k}\right|\right)^2 \\
&\le \sum_{a_1,\dots,a_k=1}^{m} \left(\sum_{j_1,\dots,j_k=1}^{n} \left|\Phi^{j_1}_{a_1} \cdots \Phi^{j_k}_{a_k}\right|^2\right)\left(\sum_{j_1,\dots,j_k=1}^{n} \left|A_{j_1\cdots j_k}\right|^2\right) \\
&\le (nm\|\Phi\|^2_\infty)^k \|A\|^2_{\mathbb{G}_\mathsf{V}},
\end{aligned}$$

using Cauchy–Schwartz. The first part of the result follows by taking $C = \sqrt{nm}\|\Phi\|_\infty$.

If Φ is surjective, let $\Psi \in \mathrm{Hom}_{\mathbb{R}}(\mathsf{V};\mathsf{U})$ be a right-inverse for Φ. Then, by the first part of the result, there exists $C \in \mathbb{R}_{>0}$ such that

$$\|A\|_{\mathbb{G}_\mathsf{V}} = \|(\Phi \circ \Psi)^* A\|_{\mathbb{G}_\mathsf{V}} = \|\Psi^*\Phi^* A\|_{\mathbb{G}_\mathsf{V}} \le C^k \|\Phi^* A\|_{\mathbb{G}_\mathsf{U}}$$

for every $A \in \mathrm{T}^k(\mathsf{V}^*)$, $k \in \mathbb{Z}_{\ge 0}$.

If Φ is injective, we choose the orthonormal basis $(e_1,\dots,e_n)$ so that $(e_1,\dots,e_m)$ is a basis for image(Φ). In this case we have

$$\Phi = \sum_{a,b=1}^{m} \Phi^b_a e_b \otimes f^a,$$

where the $m \times m$ matrix with components Φ^b_a, $a, b \in \{1,\dots,m\}$, is invertible, and

$$\Phi^* A = \sum_{b_1,\dots,b_k=1}^{m} \sum_{a_1,\dots,a_k=1}^{m} \Phi^{b_1}_{a_1} \cdots \Phi^{b_k}_{a_k} A_{b_1\cdots b_k} f^{a_1} \otimes \dots \otimes f^{a_k}.$$

Letting Ψ_a^b, $a, b \in \{1, \dots, m\}$, be defined by

$$\Psi_a^c \Phi_c^b = \begin{cases} 1, & a = b, \\ 0, & a \neq b, \end{cases}$$

we have

$$A = \sum_{b_1,\dots,b_k=1}^{m} \sum_{a_1,\dots,a_k=1}^{m} \Psi_{a_1}^{b_1} \cdots \Psi_{a_k}^{b_k} (\Phi^* A)_{b_1 \cdots b_k} e^{a_1} \otimes \dots \otimes e^{a_k},$$

and the conclusion in this case follows just as in the proof of the first part of the sublemma. ▽

(i) Let $x \in \mathcal{K}$ and take $C_x \in \mathbb{R}_{>0}$ as in the sublemma such that

$$\|\Phi^* \nabla^{\mathsf{N},m} f(x)\|_{\mathbb{G}_\mathsf{M}} \leq C_x^m \|\nabla^{\mathsf{N},m} f(\Phi(x))\|_{\mathbb{G}_\mathsf{N}}, \qquad m \in \mathbb{Z}_{\geq 0}.$$

By continuity, and noting the exact form of the constant C from the sublemma (i.e., depending on the size of the derivative of $T_x\Phi$), there exists a neighbourhood $\mathcal{U}_x$ of x such that

$$\|\Phi^* \nabla^{\mathsf{N},m} f(x')\|_{\mathbb{G}_\mathsf{M}} \leq (2C_x)^m \|\nabla^{\mathsf{N},m} f(\Phi(x'))\|_{\mathbb{G}_\mathsf{N}}, \qquad x' \in \mathcal{U}_x,\ m \in \mathbb{Z}_{>0}.$$

Then take $x_1, \dots, x_k \in \mathcal{K}$ such that $\mathcal{K} \subseteq \cup_{j=1}^k \mathcal{U}_{x_j}$. The first part of the lemma then follows by taking

$$C = \max\{2C_{x_1}, \dots, 2C_{x_k}\}.$$

(ii) Let $x \in \mathcal{K}$ and choose coordinate charts $(\mathcal{U}_x, \chi_x)$ and $(\mathcal{V}_x, \eta_x)$ about x and $\Phi(x)$ so that the local representative of Φ is linear [1, Theorems 3.5.2 and 3.5.7]. One then has two Riemannian metrics, $\mathbb{G}_\mathsf{M}$ and the Euclidean metric, on $\mathcal{U}_x$ and two Riemannian metrics, $\mathbb{G}_\mathsf{N}$ and the Euclidean metric, on $\mathcal{V}_x$. By the second assertion of sublemma and by Lemma 4.19 below, and by shrinking the chart domains $\mathcal{U}_x$ and $\mathcal{V}_x$ appropriately, one has

$$\|\nabla^{\mathsf{N},m} f(\Phi(x'))\|_{\mathbb{G}_\mathsf{N}} \leq (2C_x)^m \|\Phi^* \nabla^{\mathsf{N},m} f(x')\|_{\mathbb{G}_\mathsf{M}}, \qquad x' \in \mathcal{U}_x,\ m \in \mathbb{Z}_{>0}.$$

A compactness argument as in the proof of part (i) gives this part of the result. □

We note that the fibre norm (4.8) makes use of the vector bundle mapping

$$B_{\nabla^\mathsf{E}}^m \in \mathrm{VB}^r(\mathsf{T}_\Phi^{*m}\mathsf{M}; \mathsf{S}^{\leq m}(\mathsf{T}^*\mathsf{M}))$$

from Lemma 3.65. If instead we use the vector bundle mapping

$$A^m_{\nabla^{\mathsf{E}}} \in \mathrm{VB}^r(\mathsf{T}^{*m}_{\Phi}\mathsf{M}; \mathsf{S}^{\leq m}(\Phi^*\mathsf{T}^*\mathsf{N}))$$

from Lemma 3.65, then we have the alternative fibre norm

$$\|j_m\Phi^* f(e)\|'^2_{\mathbb{G}_{\mathsf{M},m}} = \sum_{j=0}^{m} \frac{1}{(j!)^2} \|\Phi^*\nabla^{\mathsf{N},j} f(e)\|^2_{\mathbb{G}_{\mathsf{M}}}.$$

The relationship between the fibre norms $\|\cdot\|_{\mathbb{G}_{\mathsf{M},m}}$ and $\|\cdot\|'_{\mathbb{G}_{\mathsf{M},m}}$ can be phrased as, "What is the relationship between the norms of the jet of the pull-back and the pull-back of the jet?"

4.2 Estimates Related to Jet Bundle Norms

In Sect. 3.3 we gave formulae relating derivatives of geometric objects to derivatives of their lifts, and vice versa. In Sect. 4.1 we defined fibre metrics associated with spaces of derivatives of lifted objects. In each of the multitude of constructions, there arose certain vector bundle mappings—denoted by A^m_s, B^m_s, C^m_s, and D^m_s, $m \in \mathbb{Z}_{\geq 0}$, $s \in \{0, 1, \dots, m\}$—that satisfied recursion relations. In Lemma 4.17, we will establish quite specific bounds for these vector bundle mappings. We will use these bounds in two ways.

1. In Sect. 4.3 we prove that the C^ω-topology defined by the seminorms $p^\omega_{\mathcal{K},\boldsymbol{a}}$, $\mathcal{K} \subseteq \mathsf{M}$ compact, $\boldsymbol{a} \in c_0(\mathbb{Z}_{\geq 0}; \mathbb{R}_{>0})$, is independent of the choices of metrics and connections used to define these seminorms. To be clear, this does not follow from Theorem 2.19 since in the proof of that theorem we relied in an essential way on Proposition 2.18, and our proof of that proposition relied in an essential way on the results of Lemma 4.22 and Theorem 4.24, which themselves rely crucially on Lemma 4.17. Thus our work in this section—and hence all of the work that leads up to it—lie at the core of the soundness of the approach of this book.
2. In Chap. 5 we will prove a variety of continuity results for basic operations in differential geometry. As we shall see, the bounds of Lemma 4.17 are used in these proofs in a routine way.

The main results in this section are important, but somewhat elaborate. Moreover, the bounds of Sect. 4.2.2 require, for the first time, real analyticity.

4.2.1 Algebraic Estimates

To work with the topologies we presented in Sect. 2.4, we will have to compute and estimate high-order derivatives of various sorts of tensors. In this section we collect the purely linear algebraic estimates that we shall need. All norms on tensor products are those induced by an inner product as in Lemma 2.16. For simplicity, therefore, we shall often omit any particular symbols attached to "$\|\cdot\|$" to connote which norm we are talking about; all vector spaces have a unique norm (given the data) that we shall use.

We start by giving the norm of the identity mapping.

Lemma 4.9 (Norm of the Identity Map) *If* V *is a finite-dimensional* $\mathbb{R}$*-vector space with inner product* $\mathbb{G}$*, then* $\|\mathrm{id}_{\mathsf{V}}\| = \sqrt{\dim_{\mathbb{R}}(\mathsf{V})}$.

Proof Let $(e_1, \dots, e_n)$ be an orthonormal basis for V with dual basis $(e^1, \dots, e^n)$ the dual basis. Write

$$\mathrm{id}_{\mathsf{V}} = \sum_{j=1}^{n}\sum_{k=1}^{n} \delta^k_j e_k \otimes e^j.$$

We have

$$\|A\|^2 = \sum_{j=1}^{n}\sum_{k=1}^{n} (\delta^k_j)^2 = n,$$

as claimed. □

Next we consider the norm of the tensor product of linear maps.

Lemma 4.10 (Norms of Tensor Products) *Let* U, V, W, *and* X *be finite-dimensional* $\mathbb{R}$*-vector spaces with inner products. Then, for* $A \in \mathrm{Hom}_{\mathbb{R}}(\mathsf{U};\mathsf{V})$ *and* $B \in \mathrm{Hom}_{\mathbb{R}}(\mathsf{W};\mathsf{X})$,

$$\|A \otimes B\| = \|A\|\|B\|.$$

Proof Let $(e_1, \dots, e_n)$, $(f_1, \dots, f_m)$, $(g_1, \dots, g_k)$, and $(h_1, \dots, h_l)$ be orthonormal bases for U, V, W, and X, respectively. Let $(e^1, \dots, e^n)$, $(f^1, \dots, f^m)$, $(g^1, \dots, g^k)$, and $(h^1, \dots, h^l)$ be the dual bases. Write

$$A = \sum_{j=1}^{n}\sum_{a=1}^{m} A^a_j f_a \otimes e^j, \quad B = \sum_{i=1}^{k}\sum_{b=1}^{l} B^b_i h_b \otimes g^i$$

so that

$$A \otimes B = \sum_{j=1}^{n} \sum_{i=1}^{k} \sum_{a=1}^{m} \sum_{b=1}^{l} A_j^a B_i^b (f_a \otimes h_b) \otimes (e^j \otimes g^i).$$

Then

$$\begin{aligned}\|A \otimes B\|^2 &= \sum_{j=1}^{n} \sum_{i=1}^{k} \sum_{a=1}^{m} \sum_{b=1}^{l} |A_j^a B_i^b|^2 \\ &= \left(\sum_{j=1}^{n} \sum_{a=1}^{m} |A_j^a|^2\right) \left(\sum_{i=1}^{k} \sum_{b=1}^{l} |B_i^b|^2\right) = \|A\|^2 \|B\|^2,\end{aligned}$$

as claimed. □

Our next estimate concerns the relationship between norms of tensors evaluated on arguments.

Lemma 4.11 (Norm of Tensor Evaluation) *Let* U *and* V *be finite-dimensional* $\mathbb{R}$-*vector spaces with inner products* $\mathbb{G}$ *and* $\mathbb{H}$, *respectively. Then*

$$\|L(u)\| \leq \|L\| \, \|u\|$$

for all linear mappings $L \in \mathrm{Hom}_{\mathbb{R}}(\mathsf{U}; \mathsf{V})$ *and for all* $u \in \mathsf{U}$.

Proof Let $(f_1, \ldots, f_m)$ and $(e_1, \ldots, e_n)$ be orthonormal bases for U and V with $(f^1, \ldots, f^m)$ and $(e^1, \ldots, e^n)$ their dual bases. For $L \in \mathrm{Hom}_{\mathbb{R}}(\mathsf{U}; \mathsf{V})$, write

$$L = \sum_{a=1}^{m} \sum_{j=1}^{n} L_a^j e_j \otimes f^a.$$

Then we compute, using Cauchy–Schwarz,

$$\begin{aligned}\|L(u)\|^2 &= \sum_{j=1}^{n} \left(\sum_{a=1}^{m} L_a^j u^a\right)^2 \leq \sum_{j=1}^{n} \left(\sum_{a=1}^{m} |L_a^j u^a|\right)^2 \\ &\leq \sum_{j=1}^{n} \left(\sum_{a=1}^{m} |L_a^j|^2\right) \left(\sum_{a=1}^{m} |u^a|^2\right) = \|L\|^2 \|u\|^2,\end{aligned}$$

giving the lemma. □

We shall also make use of a sort of "reverse inequality" related to the above.

Lemma 4.12 (Upper Bound for Norm of Linear Map) *Let* U *and* V *be finite-dimensional* $\mathbb{R}$*-vector spaces with inner products* $\mathbb{G}_{\mathsf{U}}$ *and* $\mathbb{G}_{\mathsf{V}}$*. For* $L \in \mathrm{Hom}_{\mathbb{R}}(\mathsf{U}; \mathsf{V})$,

$$\|L\| \leq \sqrt{\dim_{\mathbb{R}}(\mathsf{U})} \sup\{\|L(u)\| \mid \|u\| = 1\}.$$

Proof The result is true with equality and without the constant if one uses the induced norm for $\mathrm{Hom}_{\mathbb{R}}(\mathsf{U}; \mathsf{V})$, rather than the tensor norm as we do here. So the statement of the lemma is really about relating the induced norm with the tensor norm.

The tensor norm, in the case of linear mappings as we have here, is really the Frobenius norm, and as such it is computed as the ℓ^2-norm of the vector of the set of $\dim_{\mathbb{R}}(\mathsf{U})$ eigenvalues of $\sqrt{L^T \circ L}$. On the other hand, the induced norm is the ℓ^∞ norm of this same vector of eigenvalues of $\sqrt{L^T \circ L}$. These interpretations can be found in [6, page 7]. For this reason, an application of (1.4) gives the result. □

Another tensor estimate we shall find useful concerns symmetrisation.

Lemma 4.13 (Norms of Symmetrised Tensors) *Let* V *be a finite-dimensional* $\mathbb{R}$*-vector space and let* $\mathbb{G}$ *be an inner product on* V*. Then*

$$\|\mathrm{Sym}_k(A)\| \leq \|A\|$$

for every $A \in \mathrm{T}^k(\mathsf{V}^*)$ *and* $k \in \mathbb{Z}_{>0}$.

Proof The result follows from the following sublemma.

Sublemma 1 The map $\mathrm{Sym}_k : \mathrm{T}^k(\mathsf{V}^*) \to \mathrm{S}^k(\mathsf{V}^*)$ is the orthogonal projection.

Proof Let us simply denote by $\mathbb{G}$ the inner product on $\mathrm{T}^k(\mathsf{V}^*)$, defined as in Lemma 2.16. It suffices to show that $\mathbb{G}(A, S) = \mathbb{G}(\mathrm{Sym}_k(A), S)$ for every $A \in \mathrm{T}^k(\mathsf{V}^*)$ and $S \in \mathrm{S}^k(\mathsf{V}^*)$. It also suffices to show that this is true as A runs over a set of generators for $\mathrm{T}^k(\mathsf{V}^*)$ and S runs over a set of generators for $\mathrm{S}^k(\mathsf{V}^*)$.

Thus we let $(e_1, \dots, e_n)$ be an orthonormal basis for V with dual basis $(e^1, \dots, e^n)$. Then we have generators

$$e^{a_1} \otimes \dots \otimes e^{a_k}, \qquad a_1, \dots, a_k \in \{1, \dots, n\},$$

for $\mathrm{T}^k(\mathsf{V}^*)$ and

$$\mathrm{Sym}_k(e^{b_1} \otimes \dots \otimes e^{b_k}), \qquad b_1, \dots, b_k \in \{1, \dots, n\},$$

for $\mathsf{S}^k(\mathsf{V}^*)$. For $a_1, \dots, a_k, b_1, \dots, b_k \in \{1, \dots, n\}$, we wish to show that the inner product

$$\begin{aligned}
&\mathbb{G}(e^{a_1} \otimes \dots \otimes e^{a_k}, \mathrm{Sym}_k(e^{b_1} \otimes \dots \otimes e^{b_k})) \\
&\qquad = \frac{1}{k!} \sum_{\sigma \in \mathfrak{S}_k} \mathbb{G}(e^{a_1} \otimes \dots \otimes e^{a_k}, e^{b_{\sigma(1)}} \otimes \dots \otimes e^{b_{\sigma(k)}}) \\
&\qquad = \frac{1}{k!} \sum_{\sigma \in \mathfrak{S}_k} \mathbb{G}(e^{a_1}, e^{b_{\sigma(1)}}) \cdots \mathbb{G}(e^{a_k}, e^{b_{\sigma(k)}})
\end{aligned}$$

is equal to

$$\mathbb{G}(\mathrm{Sym}_k(e^{a_1} \otimes \dots \otimes e^{a_k}), \mathrm{Sym}_k(e^{b_1} \otimes \dots \otimes e^{b_k})).$$

Unless $\{a_1, \dots, a_k\}$ and $\{b_1, \dots, b_k\}$ agree as multisets (i.e., they agree as sets, and also multiplicities of members of the sets agree), we have

$$\begin{aligned}
0 = \mathbb{G}(e^{a_1} \otimes \dots \otimes e^{a_k}, \mathrm{Sym}_k(e^{b_1} \otimes \dots \otimes e^{b_k})) \\
= \mathbb{G}(\mathrm{Sym}_k(e^{a_1} \otimes \dots \otimes e^{a_k}), \mathrm{Sym}_k(e^{b_1} \otimes \dots \otimes e^{b_k})).
\end{aligned}$$

Thus we can suppose that $\{a_1, \dots, a_k\}$ and $\{b_1, \dots, b_k\}$ agree as multisets.

In this case, since

$$\mathrm{Sym}_k(e^{a_1} \otimes \dots \otimes e^{a_k}) = \mathrm{Sym}_k(e^{b_1} \otimes \dots \otimes e^{b_k}),$$

we can assume, without loss of generality, that $a_j = b_j$, $j \in \{1, \dots, k\}$. For $l \in \{1, \dots, n\}$, let $k_l^{\boldsymbol{a}} \in \mathbb{Z}_{\geq 0}$ be the number of occurrences of l in the list $(a_1, \dots, a_k)$. Let $\mathfrak{S}_k^{\boldsymbol{a}} \subseteq \mathfrak{S}_k$ be those permutations σ for which $a_j = a_{\sigma(j)}$, $j \in \{1, \dots, k\}$. Note that $\mathrm{card}(\mathfrak{S}_k^{\boldsymbol{a}}) = k_1^{\boldsymbol{a}}! \cdots k_n^{\boldsymbol{a}}!$ since $\mathfrak{S}_k^{\boldsymbol{a}}$ consists of compositions of permutations that permute all the 1's, all the 2's, etc., in the list $(a_1, \dots, a_k)$. With these bits of notation, we have

$$e^{a_1} \otimes \dots \otimes e^{a_k} = e^{a_{\sigma(1)}} \otimes \dots \otimes e^{a_{\sigma(k)}} \iff \sigma \in \mathfrak{S}_k^{\boldsymbol{a}}.$$

Therefore,

$$\mathbb{G}(e^{a_1} \otimes \dots \otimes e^{a_k}, e^{a_{\sigma(1)}} \otimes \dots \otimes e^{a_{\sigma(k)}}) = \begin{cases} 1, & \sigma \in \mathfrak{S}_k^{\boldsymbol{a}}, \\ 0, & \text{otherwise.} \end{cases}$$

We then have

$$\begin{aligned}\mathbb{G}(e^{a_1} \otimes \ldots \otimes e^{a_k}, \operatorname{Sym}_k(e^{a_1} \otimes \ldots \otimes e^{a_k}))\\ = \frac{k_1^{\boldsymbol{a}}! \cdots k_n^{\boldsymbol{a}}!}{k!} \mathbb{G}(e^{a_1} \otimes \ldots \otimes e^{a_k}, e^{a_1} \otimes \ldots \otimes e^{a_k}) = \frac{k_1^{\boldsymbol{a}}! \cdots k_n^{\boldsymbol{a}}!}{k!}.\end{aligned}$$

Next we calculate

$$\mathbb{G}(\operatorname{Sym}_k(e^{a_1} \otimes \ldots \otimes e^{a_k}), \operatorname{Sym}_k(e^{a_1} \otimes \ldots \otimes e^{a_k})).$$

Let $\sigma \in \mathfrak{S}_k$ and, for $l \in \{1, \ldots, n\}$, let $k_l^{\sigma(\boldsymbol{a})} \in \mathbb{Z}_{\geq 0}$ be the number of occurrences of l in the list $(a_{\sigma(1)}, \ldots, a_{\sigma(k)})$. Let $\mathfrak{S}_k^{\sigma(\boldsymbol{a})} \subseteq \mathfrak{S}_k$ be those permutations σ' for which $a_{\sigma(j)} = a_{\sigma'(j)}$, $j \in \{1, \ldots, k\}$. As above, $\operatorname{card}(\mathfrak{S}_k^{\sigma(\boldsymbol{a})}) = k_1^{\sigma(\boldsymbol{a})}! \cdots k_n^{\sigma(\boldsymbol{a})}!$. Also as above, we then have

$$\mathbb{G}(e^{\sigma(1)} \otimes \ldots \otimes e^{\sigma(k)}, \operatorname{Sym}_k(e^{a_1} \otimes \ldots \otimes e^{a_k})) = \frac{k_1^{\sigma(\boldsymbol{a})}! \cdots k_n^{\sigma(\boldsymbol{a})}!}{k!} = \frac{k_1^{\boldsymbol{a}}! \cdots k_n^{\boldsymbol{a}}!}{k!},$$

if $k_1^{\boldsymbol{a}}, \ldots, k_n^{\boldsymbol{a}}$ are as in the preceding paragraph. Therefore,

$$\begin{aligned}\mathbb{G}(\operatorname{Sym}_k(e^{a_1} \otimes \ldots \otimes e^{a_k}), \operatorname{Sym}_k(e^{a_1} \otimes \ldots \otimes e^{a_k}))\\ = \frac{1}{k!} \sum_{\sigma \in \mathfrak{S}_k} \mathbb{G}(e^{\sigma(1)} \otimes \ldots \otimes e^{\sigma(k)}, \operatorname{Sym}_k(e^{a_1} \otimes \ldots \otimes e^{a_k}))\\ = \frac{1}{k!} \sum_{\sigma \in \mathfrak{S}_k} \frac{k_1^{\boldsymbol{a}}! \cdots k_n^{\boldsymbol{a}}!}{k!} = \frac{k_1^{\boldsymbol{a}}! \cdots k_n^{\boldsymbol{a}}!}{k!},\end{aligned}$$

and so we have

$$\begin{aligned}\mathbb{G}(e^{a_1} \otimes \ldots \otimes e^{a_k}, \operatorname{Sym}_k(e^{b_1} \otimes \ldots \otimes e^{b_k}))\\ = \mathbb{G}(\operatorname{Sym}_k(e^{a_1} \otimes \ldots \otimes e^{a_k}), \operatorname{Sym}_k(e^{a_1} \otimes \ldots \otimes e^{a_k})),\end{aligned}$$

and the sublemma follows. ▽

Now, given $A \in \mathrm{T}^k(\mathsf{V}^*)$, we write $A = \operatorname{Sym}_k(A) + A_1$ where A_1 is orthogonal to $\mathrm{S}^k(\mathsf{V}^*)$. We then have $\|A\|^2 = \|\operatorname{Sym}_k(A)\|^2 + \|A_1\|^2$, from which the lemma follows. □

The sublemma from the preceding lemma is proved, differently, in [56, page 124].

We shall also require the norm of the inclusions $\Delta_{r,s}$ defined in (3.21).

Lemma 4.14 (Norm of Inclusion of Symmetric Tensors) *Let* V *be a finite-dimensional* $\mathbb{R}$*-vector space and let* $\mathbb{G}$ *be an inner product on* V*. Then*

$$\|\Delta_{r,s}(A)\| \leq \frac{(r+s)!}{r!s!}\|A\|$$

for every $A \in \mathrm{S}^{r+s}(\mathsf{V}^*)$ *and* $r, s \in \mathbb{Z}_{\geq 0}$.

Proof Let $(e_1, \ldots, e_n)$ be an orthonormal basis for V with $(e^1, \ldots, e^n)$ the dual basis. Write $A \in \mathrm{S}^{r+s}(\mathsf{V}^*)$ as

$$A = \sum_{j_1,\ldots,j_{r+s}=1}^{n} A_{j_1\cdots j_n} e^{j_1} \otimes \ldots \otimes e^{j_{r+s}},$$

for coefficients satisfying

$$A_{j_{\sigma(1)}\cdots j_{\sigma(r+s)}} = A_{j_1\cdots j_{r+s}}, \qquad \sigma \in \mathfrak{S}_{r+s}.$$

Then, by Lemma 3.20,

$$\begin{aligned}\Delta_{r,s}(A)(e_{j_1}, \cdots, e_{j_{r+s}}) = \sum_{\sigma\in\mathfrak{S}_{r,s}} A(e_{j_{\sigma(1)}}, \ldots, e_{j_{\sigma(r+s)}})\\ = \frac{(r+s)!}{r!s!}A(e_{j_1}, \ldots, e_{j_{r+s}}),\end{aligned}$$

which gives the result. □

Let us also determine the norm of various insertion operators that we shall use. We shall use notation that is specific to the manner in which we use these estimates, and this will seem unmotivated out of context. Let U, V, and W be finite-dimensional $\mathbb{R}$-vector spaces, let $m, s, r \in \mathbb{Z}_{>0}$ and $a \in \{0, 1, \ldots, r\}$, let $S \in \mathrm{T}^1_{r-a+2}(\mathsf{U})$, and let

$$A \in \mathrm{T}^{s+r-a+1}_{m+a+1}(\mathsf{U}) \otimes \mathsf{W} \otimes \mathsf{V}^*.$$

We then have the mapping

$$I^1_{A,S,j} \colon \mathrm{T}^s(\mathsf{U}^*) \otimes \mathsf{V} \to \mathrm{T}^{m+r+1}(\mathsf{U}^*) \otimes \mathsf{W}$$

defined by

$$I^1_{A,S,j}(\beta) = A(\mathrm{Ins}_j(\beta, S)).$$

Here we implicitly use the isomorphism

$$\kappa\colon \mathrm{T}^s_m(\mathsf{U}) \to \mathrm{Hom}_{\mathbb{R}}(\mathrm{T}^s(\mathsf{U}^*); \mathrm{T}^m(\mathsf{U}^*)),$$

for a finite-dimensional $\mathbb{R}$-vector space U and for $m, s \in \mathbb{Z}_{\geq 0}$, via

$$\kappa(v_1\otimes\ldots\otimes v_s\otimes\alpha^1\otimes\ldots\otimes\alpha^m)(\beta^1\otimes\ldots\otimes\beta^s) = \langle\beta^1; v_1\rangle\cdots\langle\beta^s; v_s\rangle\alpha^1\otimes\ldots\otimes\alpha^m,$$

for $v_a \in \mathsf{U}$, $a \in \{1,\ldots,s\}$, and $\alpha^j, \beta^b \in \mathsf{U}^*$, $b \in \{1,\ldots,s\}$, $j \in \{1,\ldots,m\}$. Thus, for additional finite-dimensional $\mathbb{R}$-vector spaces V and W, we have the identification

$$\mathrm{T}^m_s(\mathsf{U}) \otimes \mathsf{W} \otimes \mathsf{V}^* \simeq \mathrm{Hom}_{\mathbb{R}}(\mathrm{T}^s(\mathsf{U}^*) \otimes \mathsf{V}; \mathrm{T}^m(\mathsf{U}^*) \otimes \mathsf{W}).$$

We now have the following result.

Lemma 4.15 (Norm of Composition with Tensor Insertion I) *With the preceding notation,*

$$\|I^1_{A,S,j}\| \leq \|A\|\|S\|.$$

Proof Let $(f_1,\ldots,f_m)$ be an orthonormal basis for U with dual basis $(f^1,\ldots,f^m)$. Let $(e_1,\ldots,e_n)$ be an orthonormal basis for V with $(e^1,\ldots,e^n)$ the dual basis. Let $(g_1,\ldots,g_k)$ be an orthonormal basis for W with $(g^1,\ldots,g^k)$ the dual basis. Let us write

$$S = \sum_{a=1}^{m} \sum_{a_1,\ldots,a_{r-a+2}=1}^{m} S^a_{a_1\cdots a_{r-a+2}} f_a \otimes f^{a_1} \otimes\ldots\otimes f^{a_{r-a+2}}$$

and

$$A = \sum_{a_1,\ldots,a_{s+r-a+1}=1}^{m} \sum_{b_1,\ldots,b_{m+a+1}=1}^{m} \sum_{\alpha=1}^{k}\sum_{l=1}^{n} A^{a_a\cdots a_{s+r-a+1}\alpha}_{b_1\cdots b_{m+a+1}l} f^{b_1}\otimes\ldots\otimes f^{b_{m+a+1}}$$
$$\otimes f_{a_1}\otimes\ldots\otimes f_{a_{s+1}}\otimes g_\alpha \otimes e^l.$$

We then have, for $a_1,\ldots,a_s \in \{1,\ldots,m\}$, $\alpha \in \{1,\ldots,k\}$, and $l \in \{1,\ldots,n\}$,

$$\begin{aligned}
&\mathrm{Ins}_{S,j}(f^{a_1}\otimes\ldots\otimes f^{a_s}\otimes g_\alpha\otimes e^l)\\
&\quad= \mathrm{Ins}_j(f^{a_1}\otimes\ldots\otimes f^{a_j}\otimes\ldots\otimes f^{a_s}\otimes g_\alpha\otimes e^l, S)\\
&\quad= \sum_{b_1,\ldots,b_{r-a+2}=1}^{m} S^{a_j}_{b_1\cdots b_{r-a+2}} f^{a_1}\otimes\ldots\otimes f^{a_{j-1}}\otimes f^{b_1}\otimes\ldots\otimes f^{b_{r-a+2}}
\end{aligned}$$

$$\otimes f^{a_{j+1}} \otimes \ldots \otimes f^{a_s} \otimes g_\alpha \otimes e^l$$

$$= \sum_{c_1,\ldots,c_{j-1}=1}^{m} \sum_{c_{j+1},\ldots,c_s=1}^{m} \sum_{b_1,\ldots,b_{r-a+2}=1}^{n} \sum_{\beta=1}^{k} \sum_{p=1}^{n} S^{a_j}_{b_1\cdots b_{r-a+2}}$$

$$\times \delta^{a_1}_{c_1} \cdots \delta^{a_{j-1}}_{c_{j-1}} \delta^{a_{j+1}}_{c_{j+1}} \cdots \delta^{a_s}_{c_s} \delta^{\beta}_{\alpha} \delta^{l}_{p} f^{c_1} \otimes \ldots \otimes f^{c_{j-1}}$$

$$\otimes f^{b_1} \otimes \ldots \otimes f^{b_{r-a+2}} \otimes f^{c_{j+1}} \otimes \ldots \otimes f^{c_s} \otimes g_\beta \otimes e^p.$$

Thus

$$I^1_{A,S,j}(f^{a_1} \otimes \ldots \otimes f^{a_s} \otimes g_\alpha \otimes e^l)$$

$$= \sum_{b_1,\ldots,b_{r-a+2}=1}^{m} \sum_{d_1,\ldots,d_{m+a+1}=1}^{m} \sum_{\alpha=1}^{k} \sum_{l=1}^{n} A^{a_1\cdots a_{j-1} b_1 \cdots b_{r-a+2} a_{j+1} \cdots a_s \alpha}_{d_1 \cdots d_{m+a+1} l}$$

$$\times S^{a_j}_{b_1\cdots b_{r-a+2}} f^{d_1} \otimes \ldots \otimes f^{d_{m+a+1}} \otimes g_\alpha \otimes e^l.$$

Then we calculate, using Cauchy–Schwarz,

$$\|I^1_{A,S,j}\|^2 = \sum_{a_1,\ldots,a_s=1}^{m} \sum_{d_1,\ldots,d_{m+a+1}=1}^{m} \sum_{\alpha=1}^{k} \sum_{l=1}^{n}$$

$$\left(\sum_{b_1,\ldots,b_{r-a+2}=1}^{m} A^{a_1\cdots a_{j-1} b_1 \cdots b_{r-a+2} a_{j+1} \cdots a_s \alpha}_{d_1 \cdots d_{m+a+1} l} S^{a_j}_{b_1\cdots b_{r-a+2}} \right)^2$$

$$\leq \sum_{a_1,\ldots,a_s=1}^{m} \sum_{d_1,\ldots,d_{m+a+1}=1}^{m} \sum_{\alpha=1}^{k} \sum_{l=1}^{n}$$

$$\left(\sum_{b_1,\ldots,b_{r-a+2}=1}^{m} \left| A^{a_1\cdots a_{j-1} b_1 \cdots b_{r-a+2} a_{j+1} \cdots a_s \alpha}_{d_1 \cdots d_{m+a+1} l} S^{a_j}_{b_1\cdots b_{r-a+2}} \right| \right)^2$$

$$\leq \sum_{a_1,\ldots,a_s=1}^{m} \sum_{d_1,\ldots,d_{m+a+1}=1}^{m} \sum_{\alpha=1}^{k} \sum_{l=1}^{n}$$

$$\left(\sum_{b_1,\ldots,b_{r-a+2}=1}^{m} \left| A^{a_1\cdots a_{j-1} b_1 \cdots b_{r-a+2} a_{j+1} \cdots a_s}_{d_1 \cdots d_{m+a+1} l} \right|^2 \right) \left(\sum_{b_1,\ldots,b_{r-a+2}=1}^{m} \left| S^{a_j}_{b_1\cdots b_{r-a+2}} \right|^2 \right)$$

$$\leq \|A\|^2 \|S\|^2,$$

as claimed. □

Now we perform the same sort of estimate for a similar construction. We take U, V, and W as above, and m, s, r, and a as above. We also still take $S \in \mathrm{T}^1_{r-a+2}(\mathsf{U})$, but here we take

$$B \in \mathrm{T}^s_{m+a}(\mathsf{U}) \otimes \mathsf{W} \otimes \mathsf{V}^*.$$

We then have the mapping

$$I^2_{B,S,j} : \mathrm{T}^s(\mathsf{U}^*) \otimes \mathsf{V} \to \mathrm{T}^{m+r+1}(\mathsf{U}^*) \otimes \mathsf{W}$$

defined by

$$I^2_{B,S,j}(\beta) = \mathrm{Ins}_j(B(\beta), S)$$

We now have the following result, whose proof follows from direct computation, just as does Lemma 4.15.

Lemma 4.16 (Norm of Composition with Tensor Insertion II) *With the preceding notation,*

$$\|I^2_{B,S,j}\| \leq \|B\| \|S\|.$$

4.2.2 Tensor Field Estimates

We next turn to providing estimates for the tensors A^m_s, B^m_s, C^m_s, and D^m_s, $m \in \mathbb{Z}_{\geq 0}$, $s \in \{0, 1, \dots, m\}$, that appear in the lemmata from Sect. 3.3. In this section is where all of our seemingly pointless computations from Sects. 3.1 and 3.2, and our only slightly less seemingly pointless constructions from Sects. 3.3 and 4.1, bear fruit. We first develop a general estimate, and then show how this estimate can be made to apply to all of the specific tensors from Sect. 3.3.

We work with real analytic vector bundles $\pi_{\mathsf{E}} : \mathsf{E} \to \mathsf{M}$ and $\pi_{\mathsf{F}} : \mathsf{F} \to \mathsf{M}$. The rôle of $\pi_{\mathsf{E}} : \mathsf{E} \to \mathsf{M}$ in this discussion and that in Sect. 3.3 is different. One should think of E in Sect. 3.3 as being played by M here. This is because the lifted tensors in Sect. 3.3 are defined as having E as their base space. So here we rename this base space as M. As a consequence of this, one should think of (1) the rôle of M in the lemma below as being played by E in the lemmata of Sect. 3.3 (as we just said), (2) the rôle of ∇^{M} in the lemma below as being played by ∇^{E} in the lemmata of Sect. 3.3, and (3) the rôles of $\nabla^{\pi_{\mathsf{E}}}$ and $\nabla^{\pi_{\mathsf{F}}}$, and consequently $\nabla^{\pi_{\mathsf{E}} \otimes \pi_{\mathsf{F}}}$, in the lemma below as being played by the induced connection in an appropriate tensor bundle in the lemmata of Sect. 3.3. In our development here, we use the symbol $\nabla^{\mathsf{M}, \pi_{\mathsf{E}}}$, etc., to denote the connection induced in any of the myriad bundles formed by taking tensor products of TM, $\mathsf{T}^*\mathsf{M}$, E, and E^*, etc., cf. the constructions at the beginning of Sects. 2.3.1 and 3.2.3.

With this notational discussion out of the way, the main technical result we have is the following.

Lemma 4.17 (Bound for Families of Real Analytic Tensors Defined by Recursion) *Let $\pi_{\mathsf{E}}\colon \mathsf{E} \to \mathsf{M}$ and $\pi_{\mathsf{F}}\colon \mathsf{F} \to \mathsf{M}$ be real analytic vector bundles, let ∇^{M} be a real analytic affine connection on M, let $\nabla^{\pi_{\mathsf{E}}}$ and $\nabla^{\pi_{\mathsf{F}}}$ be real analytic linear connections in E and F, respectively. Let $\mathbb{G}_{\mathsf{M}}$ be a real analytic Riemannian metric on M, and let $\mathbb{G}_{\pi_{\mathsf{E}}}$ and $\mathbb{G}_{\pi_{\mathsf{F}}}$ be real analytic fibre metrics for E and F, respectively. Suppose that we are given the following data:*

(i) $\phi_m \in \Gamma^\omega(\mathrm{T}^m_m(\mathsf{TM}) \otimes \mathsf{F} \otimes \mathsf{E}^*)$, $m \in \mathbb{Z}_{\geq 0}$;
(ii) $\Phi^s_m \in \Gamma^\omega(\mathrm{End}(\mathrm{T}^s_{m+1}(\mathsf{TM}) \otimes \mathsf{F} \otimes \mathsf{F}^*))$, $m \in \mathbb{Z}_{\geq 0}$, $s \in \{0, 1, \dots, m\}$;
(iii) $\Psi^s_{jm} \in \Gamma^\omega(\mathrm{Hom}(\mathrm{T}^s_m(\mathsf{TM}) \otimes \mathsf{F} \otimes \mathsf{E}^*; \mathrm{T}^s_{m+1}(\mathsf{TM}) \otimes \mathsf{F} \otimes \mathsf{E}^*))$, $m \in \mathbb{Z}_{\geq 0}$, $s \in \{0, 1, \dots, m\}$, $j \in \{0, 1, \dots, m\}$;
(iv) $\Lambda^s_m \in \Gamma^\omega(\mathrm{Hom}(\mathrm{T}^{s-1}_m(\mathsf{TM}) \otimes \mathsf{F} \otimes \mathsf{E}^*; \mathrm{T}^s_{m+1}(\mathsf{TM}) \otimes \mathsf{F} \otimes \mathsf{E}^*))$, $m \in \mathbb{Z}_{\geq 0}$, $s \in \{1, \dots, m\}$;
(v) $A^m_s \in \Gamma^\omega(\mathrm{T}^s_m(\mathsf{TM}) \otimes \mathsf{F} \otimes \mathsf{E}^*)$, $m \in \mathbb{Z}_{\geq 0}$, $s \in \{0, 1, \dots, m\}$,

and that the data satisfies the recursion relations prescribed by $A^0_0 = \phi_0$ and

$$
\begin{aligned}
A^{m+1}_{m+1} &= \Phi^{m+1}_m \circ \phi_{m+1}, && m \in \mathbb{Z}_{\geq 0}\\
A^{m+1}_s &= \Phi^s_m \circ \nabla^{\mathsf{M},\pi_{\mathsf{E}}\otimes\pi_{\mathsf{F}}} A^m_s \\
&\quad + \sum_{j=0}^m \Psi^s_{jm} \circ A^m_s + \Lambda^s_m \circ A^m_{s-1}, && m \in \mathbb{Z}_{>0},\ s \in \{1, \dots, m\},\\
A^{m+1}_0 &= \Phi^0_m \circ \nabla^{\mathsf{M},\pi_{\mathsf{E}}\otimes\pi_{\mathsf{F}}} A^m_0 + \sum_{j=0}^m \Psi^0_{jm} \circ A^m_0, && m \in \mathbb{Z}_{\geq 0}.
\end{aligned}
$$

Suppose that the data are such that, for each compact $\mathcal{K} \subseteq \mathsf{M}$, there exist $C_1, \sigma_1 \in \mathbb{R}_{>0}$ satisfying

(i) $\|D^r_{\nabla^{\mathsf{M}},\nabla^{\pi_{\mathsf{E}}\otimes\pi_{\mathsf{F}}}} \phi_m(x)\|_{\mathbb{G}_{\mathsf{M},\pi_{\mathsf{E}}\otimes\pi_{\mathsf{F}}}} \leq C_1 \sigma_1^{-r} r!$, $m, r \in \mathbb{Z}_{\geq 0}$;
(ii) $\|D^r_{\nabla^{\mathsf{M}},\nabla^{\pi_{\mathsf{F}}}} \Phi^s_m(x) \circ A\|_{\mathbb{G}_{\mathsf{M},\pi_{\mathsf{E}}\otimes\pi_{\mathsf{F}}}} \leq C_1 \sigma_1^{-r} r! \|A\|_{\mathbb{G}_{\mathsf{M},\pi_{\mathsf{E}}\otimes\pi_{\mathsf{F}}}}$, $A \in \mathrm{T}^s_{m+1}(\mathsf{T}_x\mathsf{M} \otimes \mathsf{F}_x \otimes \mathsf{E}^*_x)$, $m, r \in \mathbb{Z}_{\geq 0}$, $s \in \{0, 1, \dots, m+1\}$;
(iii) $\|D^r_{\nabla^{\mathsf{M}},\nabla^{\pi_{\mathsf{E}}\otimes\pi_{\mathsf{F}}}} \Psi^s_{jm}(x) \circ A\|_{\mathbb{G}_{\mathsf{M},\pi_{\mathsf{E}}\otimes\pi_{\mathsf{F}}}} \leq C_1 \sigma_1^{-r} r! \|A\|_{\mathbb{G}_{\mathsf{M},\pi_{\mathsf{E}}\otimes\pi_{\mathsf{F}}}}$, $A \in \mathrm{T}^s_m(\mathsf{T}_x\mathsf{M} \otimes \mathsf{F}_x \otimes \mathsf{E}^*_x)$, $m, r \in \mathbb{Z}_{\geq 0}$, $s \in \{0, 1, \dots, m\}$, $j \in \{0, 1, \dots, m\}$;
(iv) $\|D^r_{\nabla^{\mathsf{M}},\nabla^{\pi_{\mathsf{E}}\otimes\pi_{\mathsf{F}}}} \Lambda^s_m(x) \circ A\|_{\mathbb{G}_{\mathsf{M},\pi_{\mathsf{E}}\otimes\pi_{\mathsf{F}}}} \leq C_1 \sigma_1^{-r} r! \|A\|_{\mathbb{G}_{\mathsf{M},\pi_{\mathsf{E}}\otimes\pi_{\mathsf{F}}}}$, $A \in \mathrm{T}^{s-1}_m(\mathsf{T}_x\mathsf{M} \otimes \mathsf{F}_x \otimes \mathsf{E}^*_x)$, $m, r \in \mathbb{Z}_{\geq 0}$, $s \in \{0, 1, \dots, m\}$.

for $x \in \mathcal{K}$.

Then, for $\mathcal{K} \subseteq \mathsf{M}$ compact, there exist $C, \sigma, \rho \in \mathbb{R}_{>0}$ such that

$$\|D^r_{\nabla^{\mathsf{M}},\nabla^{\pi_{\mathsf{E}}\otimes\pi_{\mathsf{F}}}} A^m_s(x)\|_{\mathbb{G}_{\mathsf{M},\pi_{\mathsf{E}}\otimes\pi_{\mathsf{F}}}} \leq C\sigma^{-m}\rho^{-(m+r-s)}(m+r-s)!$$

for $m, r \in \mathbb{Z}_{\geq 0}$, $s \in \{0, 1, \dots, m\}$, and $x \in \mathcal{K}$.

Proof We prove the lemma with a sort of meandering induction, covering various special cases of m and s, before giving a proof for the general case.

Before we embark on the proof, we organise some data that will arise in the estimate that we prove.

1. We take $\mathcal{K} \subseteq \mathsf{M}$ compact and define $C_1, \sigma_1 \in \mathbb{R}_{>0}$ as in the statement of the lemma. We shall assume, without loss of generality, that $C_1 > 1$ and $\sigma_1 < 1$. We shall also make use of Lemma 2.22 to further assume, without loss of generality, that

$$\|D^r_{\nabla^{\mathsf{M}},\nabla^{\pi_{\mathsf{E}}}} R_{\nabla^{\pi_{\mathsf{E}}}}(x)\|_{\mathbb{G}_{\mathsf{M},\pi_{\mathsf{E}}}} \leq \sigma_1^{-r} r!,$$
$$\|D^r_{\nabla^{\mathsf{M}}} T_{\nabla^{\mathsf{M}}}(x)\|_{\mathbb{G}_{\mathsf{M}}} \leq \sigma_1^{-r} r!, \qquad x \in \mathcal{K},\ r \in \mathbb{Z}_{\geq 0}.$$

2. Choose $\beta \geq 2$ so that

$$\sum_{k=0}^{\infty} \beta^{-k} < \infty,$$

and let $\alpha = \frac{\beta}{\beta-1} > 1$ denote the value of this sum. Let $\gamma = 6\alpha$.
3. We note that, for any $a, b, c \in \mathbb{Z}_{>0}$ with $b < c$, we have

$$\frac{(a+b)!}{b!} < \frac{(a+c)!}{c!}. \tag{4.9}$$

This is a direct computation:

$$\frac{(a+b)!}{b!} = (1+b)\cdots(a+b) < (1+c)\cdots(a+c) = \frac{(a+c)!}{c!}.$$

4. For $m \in \mathbb{Z}_{\geq 0}$ and $s \in \{0, 1, \dots, m\}$, we denote

$$C_{m,s} = \begin{cases} 1, & m = 0 \text{ or } s = 0, \\ \binom{m-1}{s-1}, & \text{otherwise.} \end{cases}$$

We note that

(a) $C_{m,m} = 1$, that
(b) $C_{m,s} \leq C_{m+1,s}$, that
(c) $C_{m,s} \leq C_{m+1,s+1}$, and that
(d) $mC_{m,s} \leq (m+1-s)C_{m+1,s}$.

The first and second of these assertions is obvious. For the third, for $m, s \in \mathbb{Z}_{>0}$ with $s \leq m$, we compute

$$C_{m,s} = \frac{(m-1)!}{(s-1)!(m-s)!} \leq \frac{m}{s}\frac{(m-1)!}{(s-1)!(m-s)!} = C_{m+1,s+1}.$$

For the fourth, for $m \in \mathbb{Z}_{>0}$ and $s \in \mathbb{Z}_{>0}$ satisfying $s \leq m$, we compute

$$\begin{aligned} mC_{m,s} &= m\frac{(m-1)!}{(s-1)!(m-s)!} \\ &\leq (m-s+1)\frac{m!}{(s-1)!(m+1-s)!} = (m-s+1)C_{m+1,s}. \end{aligned}$$

5. We shall have occasion below, and also elsewhere, to use a standard multinomial estimate. First let $\alpha_1, \dots, \alpha_n \in \mathbb{R}_{>0}$ and note that

$$(\alpha_1 + \dots + \alpha_n)^m = \sum_{m_1+\dots+m_n=m} \frac{m!}{m_1!\cdots m_n!}\alpha_1^{m_1}\cdots\alpha_n^{m_n},$$

which is well-known, but can easily be proved by induction on m. Taking $\alpha_1 = \dots = \alpha_n = 1$, we see that

$$\frac{m!}{m_1!\cdots m_n!} \leq n^m \tag{4.10}$$

whenever $m_1, \dots, m_n \in \mathbb{Z}_{\geq 0}$ sum to m.

Given all of this, we shall prove that

$$\|D^r_{\nabla^{\mathsf{M}},\nabla^{\pi_{\mathsf{E}}\otimes\pi_{\mathsf{F}}}} A^m_s(\|_{\mathbb{G}_{\mathsf{M},\pi_{\mathsf{E}}\otimes\pi_{\mathsf{F}}}} \leq C_1(2C_1\sigma_1^{-1}\gamma)^m C_{m,s}\left(\frac{\beta}{\sigma_1}\right)^{m+r-s}(m+r-s)! \tag{4.11}$$

for $m, r \in \mathbb{Z}_{\geq 0}$, $s \in \{0, 1, \dots, m\}$, and $x \in \mathcal{K}$.

Case $m = s = 0$:

Directly using the hypotheses, we have

$$\begin{aligned} &\|D^r_{\nabla^{\mathsf{M}},\nabla^{\pi_{\mathsf{E}}\otimes\pi_{\mathsf{F}}}} A^0_0(x)\|_{\mathbb{G}_{\mathsf{M},\pi_{\mathsf{E}}\otimes\pi_{\mathsf{F}}}} \\ &\quad= \|D^r_{\nabla^{\mathsf{M}},\nabla^{\pi_{\mathsf{E}}\otimes\pi_{\mathsf{F}}}} \phi_0(x)\|_{\mathbb{G}_{\mathsf{M},\pi_{\mathsf{E}}\otimes\pi_{\mathsf{F}}}} \\ &\quad\leq C_1\sigma_1^{-r}r! \leq C_1(2C_1\sigma_1^{-1}\gamma)^0 C_{0,0}\left(\frac{\beta}{\sigma_1}\right)^{0+r-0}(0+r-0)! \end{aligned}$$

for $r \in \mathbb{Z}_{\geq 0}$ and $x \in \mathcal{K}$. This gives (4.11) in this case.

Case $m \in \mathbb{Z}_{>0}$ and $s = m$:

By Lemma 3.15, we have

$$
\begin{aligned}
D^r_{\nabla^{\mathsf{M}},\nabla^{\pi_{\mathsf{E}}\otimes\pi_{\mathsf{F}}}} A^m_m &= D^r_{\nabla^{\mathsf{M}},\nabla^{\pi_{\mathsf{E}}\otimes\pi_{\mathsf{F}}}}(\Phi^m_{m-1}\circ\phi_m)\\
&= \sum_{a=0}^{r}\binom{r}{a} D^a_{\nabla^{\mathsf{M}},\nabla^{\pi_{\mathsf{E}}}}\Phi^m_{m-1}(D^{r-a}_{\nabla^{\mathsf{M}},\nabla^{\pi_{\mathsf{E}}\otimes\pi_{\mathsf{F}}}}\phi_m)
\end{aligned}
$$

for $m, r \in \mathbb{Z}_{\geq 0}$. Therefore, by Lemma 4.11, using the hypotheses, and by (4.9) above,

$$
\begin{aligned}
&\|D^r_{\nabla^{\mathsf{M}},\nabla^{\pi_{\mathsf{E}}\otimes\pi_{\mathsf{F}}}} A^m_m(x)\|_{\mathbb{G}_{\mathsf{M},\pi_{\mathsf{E}}\otimes\pi_{\mathsf{F}}}}\\
&\quad\leq \sum_{a=0}^{r}\frac{r!}{a!(r-a)!}(C_1\sigma_1^{-a}a!)(C_1\sigma_1^{-(m+r-a)}(r-a)!)\\
&\quad\leq C_1C_1\sigma_1^{-m}r!\sum_{a=0}^{r}\sigma_1^{-a}\left(\frac{\beta}{\sigma_1}\right)^{r-a} \leq C_1C_1\sigma_1^{-m}\left(\frac{\beta}{\sigma_1}\right)^{r} r!\sum_{a=0}^{r}\beta^{-a}\\
&\quad\leq C_1(2C_1\sigma_1^{-1}\gamma)^m C_{m,m}\left(\frac{\beta}{\sigma_1}\right)^{m+r-m}(m+r-m)!.
\end{aligned}
$$

As this holds for every $m \in \mathbb{Z}_{>0}$, $r \in \mathbb{Z}_{\geq 0}$, and $x \in \mathcal{K}$, this gives (4.11) in this case.

Case $m = 1$ and $s = 0$:

By Lemma 3.15 we have

$$
\begin{aligned}
D^r_{\nabla^{\mathsf{M}},\nabla^{\pi_{\mathsf{E}}\otimes\pi_{\mathsf{F}}}} A^1_0 &= \underbrace{\sum_{a=0}^{r}\binom{r}{a} D^a_{\nabla^{\mathsf{M}},\nabla^{\pi_{\mathsf{E}}\otimes\pi_{\mathsf{F}}}}\Phi^0_0(D^{r-a}_{\nabla^{\mathsf{M}},\nabla^{\pi_{\mathsf{E}}\otimes\pi_{\mathsf{F}}}}\nabla^{\mathsf{M},\pi_{\mathsf{E}}\otimes\pi_{\mathsf{F}}}A^0_0)}_{\text{term 1}}\\
&\quad+\underbrace{\sum_{a=0}^{r}\binom{r}{a} D^a_{\nabla^{\mathsf{M}},\nabla^{\pi_{\mathsf{E}}\otimes\pi_{\mathsf{F}}}}\Psi^0_{00}(D^{r-a}_{\nabla^{\mathsf{M}},\nabla^{\pi_{\mathsf{E}}\otimes\pi_{\mathsf{F}}}}A^0_0)}_{\text{term 2}}.
\end{aligned}
$$

If we specialise the computations below from the last case in the proof, we obtain

$$
\|D^r_{\nabla^{\mathsf{M}},\nabla^{\pi_{\mathsf{E}}\otimes\pi_{\mathsf{F}}}} A^1_0(x)\|_{\mathbb{G}_{\mathsf{M},\pi_{\mathsf{E}}\otimes\pi_{\mathsf{F}}}} \leq C_1(2C_1\sigma_1^{-1}\gamma)^1 C_{1,0}\left(\frac{\beta}{\sigma_1}\right)^{1+r-0}(1+r-0)!
$$

and this gives (4.11) in this case.

Case $m \in \mathbb{Z}_{>0}$ and $s = 0$:

We use induction on m, the desired estimate having been shown to be true for $m = 1$. By Lemma 3.15 we have

$$D^r_{\nabla^{\mathsf{M}},\nabla^{\pi_{\mathsf{E}}\otimes\pi_{\mathsf{F}}}} A^{m+1}_0 = \underbrace{\sum_{a=0}^{r}\binom{r}{a} D^a_{\nabla^{\mathsf{M}},\nabla^{\pi_{\mathsf{E}}}} \Phi^0_m (D^{r-a}_{\nabla^{\mathsf{M}},\nabla^{\pi_{\mathsf{E}}\otimes\pi_{\mathsf{F}}}} \nabla^{\mathsf{M},\pi_{\mathsf{E}}\otimes\pi_{\mathsf{F}}} A^m_0)}_{\text{term 1}}$$
$$+ \underbrace{\sum_{a=0}^{r}\binom{r}{a} D^a_{\nabla^{\mathsf{M}},\nabla^{\pi_{\mathsf{E}}\otimes\pi_{\mathsf{F}}}} \Psi^0_{0m} (D^{r-a}_{\nabla^{\mathsf{M}},\nabla^{\pi_{\mathsf{E}}\otimes\pi_{\mathsf{F}}}} A^m_0)}_{\text{term 2(a)}}$$
$$+ \underbrace{\sum_{j=1}^{m}\sum_{a=0}^{r}\binom{r}{a} D^a_{\nabla^{\mathsf{M}},\nabla^{\pi_{\mathsf{E}}\otimes\pi_{\mathsf{F}}}} \Psi^0_{jm} (D^{r-a}_{\nabla^{\mathsf{M}},\nabla^{\pi_{\mathsf{E}}\otimes\pi_{\mathsf{F}}}} A^m_0)}_{\text{term 2(b)}}.$$

As in the previous case, we can specialise the computations from the last case in the proof to give

$$\|D^r_{\nabla^{\mathsf{M}},\nabla^{\pi_{\mathsf{E}}\otimes\pi_{\mathsf{F}}}} A^{m+1}_0(x)\|_{\mathbb{G}_{\mathsf{M},\pi_{\mathsf{E}}\otimes\pi_{\mathsf{F}}}}$$
$$\leq C_1(2C_1\sigma_1^{-1}\gamma)^{m+1} C_{m+1,0}\left(\frac{\beta}{\sigma_1}\right)^{m+1+r-0} (m+1+r-0)!.$$

This proves (4.11) by induction in this case.

Case $m \in \mathbb{Z}_{>0}$ and $s \in \{1, \dots, m-1\}$:

We use induction first on m (the result having been proved for the case $m = 0$) and, for fixed m, by induction on s (the result having been proved for the case $s = 0$). By Lemma 3.15 we have

$$D^r_{\nabla^{\mathsf{M}},\nabla^{\pi_{\mathsf{E}}\otimes\pi_{\mathsf{F}}}} A^{m+1}_s = \underbrace{\sum_{a=0}^{r}\binom{r}{a} D^a_{\nabla^{\mathsf{M}},\nabla^{\pi_{\mathsf{E}}}} \Phi^s_m (D^{r-a}_{\nabla^{\mathsf{M}},\nabla^{\pi_{\mathsf{E}}\otimes\pi_{\mathsf{F}}}} \nabla^{\mathsf{M},\pi_{\mathsf{E}}\otimes\pi_{\mathsf{F}}} A^m_s)}_{\text{term 1}}$$
$$+ \underbrace{\sum_{a=0}^{r}\binom{r}{a} D^a_{\nabla^{\mathsf{M}},\nabla^{\pi_{\mathsf{E}}\otimes\pi_{\mathsf{F}}}} \Psi^s_{0m} (D^{r-a}_{\nabla^{\mathsf{M}},\nabla^{\pi_{\mathsf{E}}\otimes\pi_{\mathsf{F}}}} A^m_s)}_{\text{term 2(a)}}$$

$$+\underbrace{\sum_{j=1}^{m}\sum_{a=0}^{r}\binom{r}{a}D^{a}_{\nabla^{\mathsf{M}},\nabla^{\pi_{\mathsf{E}}\otimes\pi_{\mathsf{F}}}}\Psi^{s}_{jm}(D^{r-a}_{\nabla^{\mathsf{M}},\nabla^{\pi_{\mathsf{E}}\otimes\pi_{\mathsf{F}}}}A^{m}_{s})}_{\text{term 2(b)}}$$

$$+\underbrace{\sum_{a=0}^{r}\binom{r}{a}D^{a}_{\nabla^{\mathsf{M}},\nabla^{\pi_{\mathsf{E}}\otimes\pi_{\mathsf{F}}}}\Lambda^{s}_{m}(D^{r-a}_{\nabla^{\mathsf{M}},\nabla^{\pi_{\mathsf{E}}\otimes\pi_{\mathsf{F}}}}A^{m}_{s-1})}_{\text{term 3}}.$$

We shall evaluate the components of this expression one-by-one.

For term 1, by Lemma 3.21, we have

$$\begin{aligned}
&D^{r-a}_{\nabla^{\mathsf{M}},\nabla^{\pi_{\mathsf{E}}\otimes\pi_{\mathsf{F}}}}(\nabla^{\mathsf{M},\pi_{\mathsf{E}}\otimes\pi_{\mathsf{F}}}A^{m}_{s})\\
&\quad=\underbrace{D^{r-a+1}_{\nabla^{\mathsf{M}},\pi_{\mathsf{E}}\otimes\pi_{\mathsf{F}}}A^{m}_{s}}_{\text{(a)}}+\underbrace{\theta_{r-a,1}\otimes\mathrm{id}(D^{r-a-1}_{\nabla^{\mathsf{M}},\nabla^{\pi_{\mathsf{E}}\otimes\pi_{\mathsf{F}}}}R_{\nabla^{\pi_{\mathsf{E}}}}(A^{m}_{s}))}_{\text{(b)}}\\
&\qquad\qquad-\underbrace{\theta_{r-a,1}\otimes\mathrm{id}(D^{r-a-1}_{\nabla^{\mathsf{M}}}T_{\nabla^{\mathsf{M}}}(\nabla^{\mathsf{M},\pi_{\mathsf{E}}\otimes\pi_{\mathsf{F}}}A^{m}_{s}))}_{\text{(c)}}.
\end{aligned}$$

We examine each of the three terms on the right separately.

For term (a), by the induction hypothesis,

$$\begin{aligned}
&\|D^{r-a+1}_{\nabla^{\mathsf{M}},\pi_{\mathsf{E}}\otimes\pi_{\mathsf{F}}}A^{m}_{s}(x)\|_{\mathbb{G}_{\mathsf{M},\pi_{\mathsf{E}}\otimes\pi_{\mathsf{F}}}}\\
&\qquad\le C_1(2C_1\sigma_1^{-1}\gamma)^{m}C_{m,s}\left(\frac{\beta}{\sigma_1}\right)^{m+r-a-s+1}(m+r-a+1-s)!.
\end{aligned}$$

Therefore, using Lemmata 3.15 and 4.11 and observation (4.9), we estimate

$$\begin{aligned}
&\|\text{term 1(a)}(x)\|_{\mathbb{G}_{\mathsf{M},\pi_{\mathsf{E}}\otimes\pi_{\mathsf{F}}}}\\
&\quad\le\sum_{a=0}^{r}\frac{r!}{a!(r-a)!}(C_1\sigma_1^{-a}a!)\\
&\qquad\times\left(C_1(2C_1\sigma_1^{-1}\gamma)^{m}C_{m,s}\left(\frac{\beta}{\sigma_1}\right)^{m+r-a+1-s}(m+r-a+1-s)!\right)\\
&\quad\le C_1(2C_1\sigma_1^{-1})^{m+1}\gamma^{m}C_{m,s}(m+r+1-s)!\sum_{a=0}^{r}\sigma_1^{-a}\left(\frac{\beta}{\sigma_1}\right)^{m+r-a+1-s}
\end{aligned}$$

$$\le C_1(2C_1\sigma_1^{-1})^{m+1}\gamma^m C_{m,s}\left(\frac{\beta}{\sigma_1}\right)^{m+r+1-s}(m+r+1-s)!\sum_{a=0}^{r}\beta^{-a}$$
$$\le C_1(2C_1\sigma_1^{-1})^{m+1}\gamma^m \alpha C_{m,s}\left(\frac{\beta}{\sigma_1}\right)^{m+r+1-s}(m+r+1-s)!$$

for $x \in \mathcal{K}$.

Now we consider terms (b) and (c). As in observation 1 at the beginning of the proof, we have

$$\|D^{r-a-1}_{\nabla^{\mathsf{M}},\nabla^{\pi_{\mathsf{E}}\otimes\pi_{\mathsf{F}}}} R_{\nabla^{\pi_{\mathsf{E}}}}(x)\|_{\mathbb{G}_{\mathsf{M},\pi_{\mathsf{E}}\otimes\pi_{\mathsf{F}}}} \le \sigma_1^{-(r-a-1)}(r-a-1)!,$$
$$\|D^{r-a-1}_{\nabla^{\mathsf{M}}} T_{\nabla^{\mathsf{E}}}(x)\|_{\mathbb{G}_{\mathsf{M}}} \le \sigma_1^{-(r-a-1)}(r-a-1)!$$

for $x \in \mathcal{K}$. By the induction hypothesis,

$$\|A^m_s(x)\|_{\mathbb{G}_{\mathsf{M},\pi_{\mathsf{E}}\otimes\pi_{\mathsf{F}}}} \le C_1(2C_1\sigma_1^{-1}\gamma)^m C_{m,s}\left(\frac{\beta}{\sigma_1}\right)^{m-s}(m-s)!$$

and

$$\|\nabla^{\mathsf{M},\pi_{\mathsf{E}}} A^m_s(x)\|_{\mathbb{G}_{\mathsf{M},\pi_{\mathsf{E}}\otimes\pi_{\mathsf{F}}}} \le C_1(2C_1\sigma_1^{-1}\gamma)^m C_{m,s}\left(\frac{\beta}{\sigma_1}\right)^{m+1-s}(m+1-s)!$$

for $x \in \mathcal{K}$. By Lemmata 4.9, 4.13, and 4.14, by (4.10), and since $\beta \ge 2$,

$$\|\theta_{r-a,1}\otimes \mathrm{id}\|_{\mathbb{G}_{\mathsf{M},\pi_{\mathsf{E}}\otimes\pi_{\mathsf{F}}}} = \frac{1}{2}\|\Delta_{r-a-1,1}\|_{\mathbb{G}_{\mathsf{M},\pi_{\mathsf{E}}\otimes\pi_{\mathsf{F}}}}\|\Delta_{1,0}\|_{\mathbb{G}_{\mathsf{M},\pi_{\mathsf{E}}\otimes\pi_{\mathsf{F}}}}\|\mathrm{id}\|_{\mathbb{G}_{\mathsf{M},\pi_{\mathsf{E}}\otimes\pi_{\mathsf{F}}}}$$
$$\le \frac{1}{2}\frac{(r-a)!}{(r-a-1)!1!}\sqrt{m+s} \le \beta^{r-a} m.$$

Putting the preceding estimates together and using Lemma 4.11 and observation 4 gives

$$\|\text{term (b)}(x)\|_{\mathbb{G}_{\mathsf{M},\pi_{\mathsf{E}}\otimes\pi_{\mathsf{F}}}}$$
$$\le 2C_1(2C_1\sigma_1^{-1}\gamma)^m m C_{m,s}\left(\frac{\beta}{\sigma_1}\right)^{m+r-a-s}(r-a-1)!(m-s)!$$
$$\le C_1 2^{m+1}(C_1\sigma_1^{-1}\gamma)^m C_{m+1,s}\left(\frac{\beta}{\sigma_1}\right)^{m+r-a-s}(m+r-a-s)!$$

and, similarly,

$$\|\text{term (c)}(x)\|_{\mathbb{G}_{\mathsf{M},\pi_{\mathsf{E}}\otimes\pi_{\mathsf{F}}}}$$
$$\leq C_1 2^{m+1}(C_1\sigma_1^{-1}\gamma)^m C_{m+1,s}\left(\frac{\beta}{\sigma_1}\right)^{m+r-a+1-s}(m+r-a+1-s)!.$$

Now a computation like that for term 1(a) gives

$$\|\text{term 1(b)}(x)\|_{\mathbb{G}_{\mathsf{M},\pi_{\mathsf{E}}\otimes\pi_{\mathsf{F}}}}, \|\text{term 1(c)}(x)\|_{\mathbb{G}_{\mathsf{M},\pi_{\mathsf{E}}\otimes\pi_{\mathsf{F}}}}$$
$$\leq C_1(2C_1\sigma_1^{-1})^{m+1}\gamma^m\alpha C_{m+1,s}\left(\frac{\beta}{\sigma_1}\right)^{m+r+1-s}(m+r+1-s)!.$$

Let us now consider term 2(a). Here a similar analysis to the above gives

$$\|\text{term 2(a)}(x)\|_{\mathbb{G}_{\mathsf{M},\pi_{\mathsf{E}}\otimes\pi_{\mathsf{F}}}} \leq C_1(2C_1\sigma_1^{-1})^{m+1}\gamma^m\alpha C_{m,s}\left(\frac{\beta}{\sigma_1}\right)^{m+r-s}(m+r-s)!.$$

For term 2(b) we compute, along similar lines and using observation 4,

$$\|\text{term 2(b)}(x)\|_{\mathbb{G}_{\mathsf{M},\pi_{\mathsf{E}}\otimes\pi_{\mathsf{F}}}}$$
$$\leq C_1(2C_1\sigma_1^{-1})^{m+1}\gamma^m\alpha m C_{m,s}\left(\frac{\beta}{\sigma_1}\right)^{m+r-s}(m+r-s)!$$
$$\leq C_1(2C_1\sigma_1^{-1})^{m+1}\gamma^m\alpha(m-s+1)C_{m+1,s}\left(\frac{\beta}{\sigma_1}\right)^{m+r-s}(m+r-s)!$$
$$\leq C_1(2C_1\sigma_1^{-1})^{m+1}\gamma^m\alpha C_{m+1,s}\left(\frac{\beta}{\sigma_1}\right)^{m+r-s}(m+r+1-s)!.$$

An entirely similar analysis can be applied to term 3 to give

$$\|\text{term 3}(x)\|_{\mathbb{G}_{\mathsf{M},\pi_{\mathsf{E}}\otimes\pi_{\mathsf{F}}}}$$
$$\leq C_1(2C_1\sigma_1^{-1})^{m+1}\gamma^m C_{m,s-1}\alpha\left(\frac{\beta}{\sigma_1}\right)^{m+r-s+1}(m+r-s+1)!.$$

Adding these as in the previous case and using our observation 4 above, we have

$$\|D^r_{\nabla^{\mathsf{M}},\nabla^{\pi_{\mathsf{E}}\otimes\pi_{\mathsf{F}}}}A^{m+1}_s(x)\|_{\mathbb{G}_{\mathsf{M},\pi_{\mathsf{E}}\otimes\pi_{\mathsf{F}}}}$$
$$\leq C_1(2C_1\sigma_1^{-1}\gamma)^{m+1}C_{m+1,s}\left(\frac{\beta}{\sigma_1}\right)^{m+1+r-s}(m+1+r-s)!,$$

proving (4.11) by induction in this case.

We now note that a standard binomial estimate via (4.10) gives $C_{m,s} \leq 2^m$. The lemma now follows from (4.11) by taking

$$C = C_1, \quad \sigma = 4C_1\sigma_1^{-1}\gamma, \quad \rho = \frac{\beta}{\sigma_1}. \qquad \square$$

We now apply the lemma to the recursion relations that we proved in Lemmata 3.23, 3.24, 3.29, 3.30, 3.34, 3.35, 3.39, 3.40, 3.44, 3.45, 3.49, 3.50, 3.54, 3.55, 3.61, and 3.62. We first provide the correspondence between the data from the preceding lemmata with the data of Lemma 4.17.

1. Lemma 3.23: We have
   (a) $\mathsf{M} = \mathsf{E}$, $\mathsf{E} = \mathsf{F} = \mathbb{R}_{\mathsf{E}}$,
   (b) $\phi_m(\beta_m) = \beta_m$, $\beta_m \in \mathrm{T}^m(\mathsf{T}^*\mathsf{M})$, $m \in \mathbb{Z}_{\geq 0}$,
   (c) $\Phi^s_m(\alpha^{m+1}_s) = \alpha^{m+1}_s$, $\alpha^{m+1}_s \in \mathrm{T}^s_{m+1}(\mathsf{T}^*\mathsf{M}) \otimes \mathsf{F} \otimes \mathsf{E}^*)$, $m \in \mathbb{Z}_{\geq 0}$, $s \in \{0, 1, \dots, m+1\}$,
   (d) $\Psi^s_{jm}(\alpha^m_s)(\beta_s) = -\alpha^m_s \otimes \mathrm{id}_{\mathsf{T}^*\mathsf{M}}(\mathrm{Ins}_j(\beta_s, B_{\pi_{\mathsf{E}}}))$, $\alpha^m_s \in \mathrm{Hom}(\mathrm{T}^s(\mathsf{T}^*\mathsf{M}) \otimes \mathsf{E}; \mathrm{T}^m(\mathsf{T}^*\mathsf{M}) \otimes \mathsf{F})$, $\beta_s \in \mathrm{T}^s(\mathsf{T}^*\mathsf{M}) \otimes \mathsf{E}$, $m \in \mathbb{Z}_{>0}$, $s \in \{1, \dots, m\}$, $j \in \{1, \dots, s\}$,
   (e) $\Lambda^s_m(\alpha^m_{s-1}) = \alpha^m_{s-1} \otimes \mathrm{id}_{\mathsf{T}^*\mathsf{M}}$, $\alpha^m_{s-1} \in \mathrm{Hom}(\mathrm{T}^{s-1}(\mathsf{T}^*\mathsf{M}) \otimes \mathsf{E}; \mathrm{T}^m(\mathsf{T}^*\mathsf{M}) \otimes \mathsf{F})$, $m \in \mathbb{Z}_{>0}$, $s \in \{1, \dots, m\}$,
   (f) $\Psi^0_{jm} = 0$, $m \in \mathbb{Z}_{\geq 0}$, and
   (g) $\Lambda^0_m = 0$, $m \in \mathbb{Z}_{\geq 0}$.
2. Lemma 3.24: We have
   (a) $\mathsf{M} = \mathsf{E}$, $\mathsf{E} = \mathsf{F} = \mathbb{R}_{\mathsf{E}}$,
   (b) $\phi_m(\beta_m) = \beta_m$, $\beta_m \in \mathrm{T}^m(\mathsf{T}^*\mathsf{M})$, $m \in \mathbb{Z}_{\geq 0}$,
   (c) $\Phi^s_m(\alpha^{m+1}_s) = \alpha^{m+1}_s$, $\alpha^{m+1}_s \in \mathrm{T}^s_{m+1}(\mathsf{T}^*\mathsf{M}) \otimes \mathsf{F} \otimes \mathsf{E}^*$, $m \in \mathbb{Z}_{\geq 0}$, $s \in \{0, 1, \dots, m+1\}$,
   (d) $\Psi^s_{jm}(\alpha^m_s)(\beta_s) = \mathrm{Ins}_j(\alpha^m_s(\beta_s), B_{\pi_{\mathsf{E}}})$, $\alpha^m_s \in \mathrm{Hom}(\mathrm{T}^s(\mathsf{T}^*\mathsf{M}) \otimes \mathsf{E}; \mathrm{T}^m(\mathsf{T}^*\mathsf{M}) \otimes \mathsf{F})$, $\beta_s \in \mathrm{T}^s(\mathsf{T}^*\mathsf{M}) \otimes \mathsf{E}$, $m \in \mathbb{Z}_{>0}$, $s \in \{1, \dots, m\}$, $j \in \{1, \dots, m\}$, and
   (e) $\Lambda^s_m(\alpha^m_{s-1}) = \alpha^m_{s-1} \otimes \mathrm{id}_{\mathsf{T}^*\mathsf{M}}$, $\alpha^m_{s-1} \in \mathrm{Hom}(\mathrm{T}^{s-1}(\mathsf{T}^*\mathsf{M}) \otimes \mathsf{E}; \mathrm{T}^m(\mathsf{T}^*\mathsf{M}) \otimes \mathsf{F})$, $m \in \mathbb{Z}_{>0}$, $s \in \{0, \dots, m\}$.
3. Lemma 3.29: We have

$$\mathsf{M} = \mathsf{E}, \quad \mathsf{E} = \mathsf{F} = \mathsf{VE},$$

   and all other data derived from Lemma 3.29, similarly to the case of Lemma 3.23.

4. Lemma 3.30: We have

$$\mathsf{M} = \mathsf{E}, \quad \mathsf{E} = \mathsf{F} = \mathsf{VE},$$

and all other data derived from Lemma 3.30, similarly to the case of Lemma 3.24.
5. Lemma 3.34: We have

$$\mathsf{M} = \mathsf{E}, \quad \mathsf{E} = \mathsf{F} = \mathsf{HE},$$

and all other data derived from Lemma 3.34, similarly to the case of Lemma 3.23.
6. Lemma 3.35: We have

$$\mathsf{M} = \mathsf{E}, \quad \mathsf{E} = \mathsf{F} = \mathsf{HE},$$

and all other data derived from Lemma 3.35, similarly to the case of Lemma 3.24.
7. Lemma 3.39: We have

$$\mathsf{M} = \mathsf{E}, \quad \mathsf{E} = \mathsf{F} = \mathsf{V}^*\mathsf{E},$$

and all other data derived from Lemma 3.39, similarly to the case of Lemma 3.23.
8. Lemma 3.40: We have

$$\mathsf{M} = \mathsf{E}, \quad \mathsf{E} = \mathsf{F} = \mathsf{V}^*\mathsf{E},$$

and all other data derived from Lemma 3.39, similarly to the case of Lemma 3.24.
9. Lemma 3.44: We have

$$\mathsf{M} = \mathsf{E}, \quad \mathsf{E} = \mathsf{F} = \mathrm{T}^1_1(\mathsf{VE}),$$

and all other data derived from Lemma 3.44, similarly to the case of Lemma 3.23.
10. Lemma 3.45: We have

$$\mathsf{M} = \mathsf{E}, \quad \mathsf{E} = \mathsf{F} = \mathrm{T}^1_1(\mathsf{VE}),$$

and all other data derived from Lemma 3.45, similarly to the case of Lemma 3.24.

11. Lemma 3.49: We have

    (a) $\mathsf{M} = \mathsf{E}$, $\mathsf{E} = \mathbb{R}_{\mathsf{E}} \oplus \mathsf{V}^*\mathsf{E}$, $\mathsf{F} = \mathbb{R}_{\mathsf{E}} \oplus \mathbb{R}_{\mathsf{E}}$,
    (b) $\phi_m(\beta_m, \delta_m) = \beta_m$, $(\beta_m, \delta_m) \in \mathrm{T}^m(\mathsf{T}^*\mathsf{M}) \otimes \mathsf{E}$, $m \in \mathbb{Z}_{\geq 0}$,
    (c) $\Phi^s_m(\alpha^{m+1}_s, \gamma^{m+1}_s) = (\alpha^{m+1}_m, \gamma^{m+1}_s)$,
    $(\alpha^{m+1}_s, \gamma^{m+1}_s) \in \mathrm{T}^s_{m+1}(\mathsf{T}^*\mathsf{M}) \otimes \mathsf{F} \otimes \mathsf{E}^*$, $m \in \mathbb{Z}_{\geq 0}$, $s \in \{0, 1, \dots, m-1\}$,
    (d) $\Phi^m_m(\alpha^{m+1}_m, \gamma^{m+1}_m) = (0, 0)$, $(\alpha^{m+1}_m, \gamma^{m+1}_m) \in \mathrm{T}^m_{m+1}(\mathsf{T}^*\mathsf{M}) \otimes \mathsf{F} \otimes \mathsf{E}^*$, $m \in \mathbb{Z}_{\geq 0}$,
    (e) $\Psi^s_{jm}(\alpha^m_s, \gamma^m_s)(\beta_s, \delta_s) = (-\alpha^m_s \otimes \mathrm{id}_{\mathsf{T}^*\mathsf{M}}(\mathrm{Ins}_j(\beta_s, B_{\pi_{\mathsf{E}}}))$,
    $-\sum_{j=1}^{s+1} \gamma^m_s \otimes \mathrm{id}_{\mathsf{T}^*\mathsf{M}}(\mathrm{Ins}_j(\delta_s, B_{\pi_{\mathsf{E}}})))$, $(\alpha^m_s, \gamma^m_s) \in \mathrm{T}^s_m(\mathsf{T}^*\mathsf{M}) \otimes \mathsf{F} \otimes \mathsf{E}^*$,
    $(\beta_s, \delta_s) \in \mathrm{T}^s(\mathsf{T}^*\mathsf{M}) \otimes \mathsf{E}$, $m \geq 2$, $s \in \{1, \dots, m-1\}$, $j \in \{1, \dots, s\}$,
    (f) $\Psi^m_{jm}(\alpha^m_m, \gamma^m_m)(\beta_m, \delta_m) = (-\mathrm{Ins}_j(\beta_m, B_{\pi_{\mathsf{E}}}), \delta_m)$,
    $(\alpha^m_m, \gamma^m_m) \in \mathrm{T}^m_m(\mathsf{T}^*\mathsf{M}) \otimes \mathsf{F} \otimes \mathsf{E}^*$, $(\beta_m, \delta_m) \in \mathrm{T}^m(\mathsf{T}^*\mathsf{M}) \otimes \mathsf{E}$,
    $m \geq 2$, $j \in \{1, \dots, m\}$,
    (g) $\Psi^m_{00}(\alpha^m_0, \gamma^m_0)(\beta_0, \delta_0) = (0, -\gamma^m_0 \otimes \mathrm{id}_{\mathsf{T}^*\mathsf{M}}(\mathrm{Ins}_1(\delta_0, B_{\pi_{\mathsf{E}}})))$,
    $(\alpha^m_0, \gamma^m_0) \in \mathrm{T}^m_0(\mathsf{T}^*\mathsf{M}) \otimes \mathsf{F} \otimes \mathsf{E}^*$, $(\beta_0, \delta_0) \in \mathsf{E}$, $m \geq 2$,
    (h) $\Lambda^s_m(\alpha^m_s, \gamma^m_s) = (\alpha^m_{s-1} \otimes \mathrm{id}_{\mathsf{T}^*\mathsf{M}}, \gamma^m_{s-1} \otimes \mathrm{id}_{\mathsf{T}^*\mathsf{M}})$, $m \geq 2$, $s \in \{1, \dots, m\}$.

12. Lemma 3.50: We have

    (a) $\mathsf{M} = \mathsf{E}$, $\mathsf{E} = \mathbb{R}_{\mathsf{E}} \oplus \mathsf{V}^*\mathsf{E}$, $\mathsf{F} = \mathbb{R}_{\mathsf{E}} \oplus \mathbb{R}_{\mathsf{E}}$,
    (b) $\phi_m(\beta_m, \delta_m) = \beta_m$, $(\beta_m, \delta_m) \in \mathrm{T}^m(\mathsf{T}^*\mathsf{M}) \otimes \mathsf{E}$, $m \in \mathbb{Z}_{\geq 0}$,
    (c) $\Phi^s_m(\alpha^{m+1}_s, \gamma^{m+1}_s) = (\alpha^{m+1}_m, \gamma^{m+1}_s)$, $(\alpha^{m+1}_s, \gamma^{m+1}_s) \in \mathrm{T}^s_{m+1}(\mathsf{T}^*\mathsf{M}) \otimes \mathsf{F} \otimes \mathsf{E}^*$,
    $m \in \mathbb{Z}_{\geq 0}$, $s \in \{0, 1, \dots, m-1\}$,
    (d) $\Phi^m_m(\alpha^{m+1}_m, \gamma^{m+1}_m) = (0, 0)$, $(\alpha^{m+1}_m, \gamma^{m+1}_m) \in \mathrm{T}^m_{m+1}(\mathsf{T}^*\mathsf{M}) \otimes \mathsf{F} \otimes \mathsf{E}^*$, $m \in \mathbb{Z}_{\geq 0}$,
    (e) $\Psi^s_{jm}(\alpha^m_s, \gamma^m_s)(\beta_s, \delta_s) = (\mathrm{Ins}_j(\alpha^m_s(\beta_s), B_{\pi_{\mathsf{E}}})$
    $-\mathrm{Ins}_{m+1}(\alpha^m_s(\beta_s), B_{\pi_{\mathsf{E}}}{}^*), -\overline{B}^m_s)$,
    $(\alpha^m_s, \gamma^m_s) \in \mathrm{T}^s_m(\mathsf{T}^*\mathsf{M}) \otimes \mathsf{F} \otimes \mathsf{E}^*$, $(\beta_s, \delta_s) \in \mathrm{T}^s(\mathsf{T}^*\mathsf{M}) \otimes \mathsf{E}$,
    $m \geq 2$, $s \in \{1, \dots, m-1\}$, $j \in \{1, \dots, m\}$,
    (f) $\Psi^m_{jm}(\alpha^m_m, \gamma^m_m)(\beta_m, \delta_m) = (\mathrm{Ins}_j(\beta_m, B_{\pi_{\mathsf{E}}}) - \mathrm{Ins}_{m+1}(\beta_m, B_{\pi_{\mathsf{E}}}{}^*), -\delta_m)$,
    $(\alpha^m_m, \gamma^m_m) \in \mathrm{T}^m_m(\mathsf{T}^*\mathsf{M}) \otimes \mathsf{F} \otimes \mathsf{E}^*$, $(\beta_m, \delta_m) \in \mathrm{T}^m(\mathsf{T}^*\mathsf{M}) \otimes \mathsf{E}$,
    $m \geq 2$, $j \in \{1, \dots, m\}$,
    (g) $\Psi^m_{00}(\alpha^m_0, \gamma^m_0)(\beta_0, \delta_0) = (0, -\gamma^m_0 \otimes \mathrm{id}_{\mathsf{T}^*\mathsf{M}}(\mathrm{Ins}_1(\delta_0, B_{\pi_{\mathsf{E}}})))$,
    $(\alpha^m_0, \gamma^m_0) \in \mathrm{T}^m_0(\mathsf{T}^*\mathsf{M}) \otimes \mathsf{F} \otimes \mathsf{E}^*$, $(\beta_0, \delta_0) \in \mathsf{E}$, $m \geq 2$,
    (h) $\Lambda^s_m(\alpha^m_s, \gamma^m_s) = (\alpha^m_{s-1} \otimes \mathrm{id}_{\mathsf{T}^*\mathsf{M}}, \gamma^m_{s-1} \otimes \mathrm{id}_{\mathsf{T}^*\mathsf{M}})$, $m \geq 2$, $s \in \{1, \dots, m\}$.

13. Lemma 3.54: We have

$$\mathsf{M} = \mathsf{E}, \quad \mathsf{E} = \mathsf{VE} \oplus \mathrm{T}^1_1(\mathsf{VE}), \quad \mathsf{F} = \mathsf{VE} \oplus \mathsf{VE},$$

and all other data derived from Lemma 3.54, similarly to the case of Lemma 3.49.

14. Lemma 3.55: We have

$$\mathsf{M} = \mathsf{E}, \;\; \mathsf{E} = \mathsf{VE} \oplus \mathrm{T}^1_1(\mathsf{VE}), \;\; \mathsf{F} = \mathsf{VE} \oplus \mathsf{VE},$$

and all other data derived from Lemma 3.55, similarly to the case of Lemma 3.50.

15. Lemma 3.61: We have

$$\mathsf{M} = \mathsf{M}, \;\; \mathsf{E} = \Phi^*\mathsf{T}^*\mathsf{N}, \;\; \mathsf{F} = \mathsf{T}^*\mathsf{M},$$

and all other data derived from Lemma 3.61, similarly to the case of Lemma 3.23.

16. Lemma 3.62: We have

$$\mathsf{M} = \mathsf{M}, \;\; \mathsf{E} = \mathsf{T}^*\mathsf{M}, \;\; \mathsf{F} = \Phi^*\mathsf{T}^*\mathsf{N},$$

and all other data derived from Lemma 3.62, similarly to the case of Lemma 3.24.

17. Lemma 3.67: We have

$$\mathsf{M} = \mathsf{M}, \;\; \mathsf{E} = \mathsf{E}, \;\; \mathsf{F} = \mathsf{E},$$

and all other data derived from Lemma 3.67, similarly to the case of Lemma 3.23.

18. Lemma 3.68: We have

$$\mathsf{M} = \mathsf{M}, \;\; \mathsf{E} = \mathsf{E}, \;\; \mathsf{F} = \mathsf{E},$$

and all other data derived from Lemma 3.68, similarly to the case of Lemma 3.24.

Having now translated the lemmata of Sect. 3.3 to the general Lemma 4.17, we now need to show that the data of the lemmata of Sect. 3.3 satisfy the hypotheses of Lemma 4.17 in the real analytic case. As is easily seen, there are a few sorts of expressions that appear repeatedly, and we shall simply give estimates for these terms and leave to the reader the putting together of the pieces.

The following lemma gives the required bounds.

Lemma 4.18 (Specific Bounds for Terms Coming from Recursion) *Let $\pi_{\mathsf{E}}\colon \mathsf{E} \to \mathsf{M}$ be a real analytic vector bundle, let ∇^{M} be a real analytic affine connection on M, and let $\nabla^{\pi_{\mathsf{E}}}$ be a real analytic vector bundle connection in E. Let $\mathbb{G}_{\mathsf{M}}$ be a real analytic Riemannian metric on M and let $\mathbb{G}_{\pi_{\mathsf{E}}}$ be a real analytic fibre metric for E. Let $S \in \Gamma^{\omega}(\mathrm{T}^1_2(\mathsf{TM}))$. Let $\mathcal{K} \subseteq \mathsf{M}$ be compact and let n be the larger of the dimension of M and the fibre dimension of E and let $\sigma_0 = n^{-1}$. Then we have the following bounds:*

(i) $\|D^r_{\nabla^M,\nabla^{\pi_E}} \mathrm{id}_{T^m(T^*M)\otimes E}(x)\|_{\mathbb{G}_{M,\pi_E}} \le \sigma_0^{-(m+r+1)}$,
$x \in \mathcal{K}$, $m, r \in \mathbb{Z}_{\ge 0}$, $s \in \{0, 1, \dots, m\}$;

(ii) $\|D^r_{\nabla^M,\nabla^{\pi_E}} \mathrm{id}_{T^m_s(T^*M)\otimes E}(x)\|_{\mathbb{G}_{M,\pi_E}} \le \sigma_0^{-(2m+r+1)}$,
$x \in \mathcal{K}$, $m, r \in \mathbb{Z}_{\ge 0}$, $s \in \{0, 1, \dots, m\}$;

(iii) *if* $\Phi^s_m(\alpha^{m+1}_s) = \alpha^{m+1}_s$, $\alpha^{m+1}_s \in T^s_{m+1}(T^*M) \otimes E$, *then*

$$\|D^r_{\nabla^M,\nabla^{\pi_E}} \Phi^s_m \circ D^a_{\nabla^M,\nabla^{\pi_E}} A^{m+1}_s(x)\|_{\mathbb{G}_{M,\pi_E}} \le \|D^a_{\nabla^M,\nabla^{\pi_E}} A^{m+1}_s(x)\|_{\mathbb{G}_{M,\pi_E}},$$

for $x \in \mathcal{K}$, $m, r \in \mathbb{Z}_{\ge 0}$, $s \in \{0, 1, \dots, m\}$;

(iv) *if*

$$\Psi^s_{jm}(\alpha^m_s)(\beta_s) = (\alpha^m_s \otimes \mathrm{id}_{T^*M})(\mathrm{Ins}_j(\beta_s, S)),$$
$$\alpha^m_s \in \mathrm{Hom}(T^s(T^*M) \otimes E; T^m(T^*M) \otimes E), \ \beta_s \in T^s(T^*M) \otimes E,$$

then there exist $C_1, \sigma_1 \in \mathbb{R}_{>0}$ *such that*

$$\|D^r_{\nabla^M,\nabla^{\pi_E}} \Psi^s_{jm} \circ D^a_{\nabla^M,\nabla^{\pi_E}} A^m_s(x)\|_{\mathbb{G}_{M,\pi_E}} \le C_1\sigma_1^{-r} r! \|D^a_{\nabla^M,\nabla^{\pi_E}} A^m_s(x)\|_{\mathbb{G}_{M,\pi_E}},$$

for $x \in \mathcal{K}$, $m, r \in \mathbb{Z}_{\ge 0}$, $s \in \{0, 1, \dots, m\}$;

(v) *if*

$$\Psi^s_{jm}(\alpha^m_s)(\beta_s) = \mathrm{Ins}_j(\alpha^m_s(\beta_s), S),$$
$$\alpha^m_s \in \mathrm{Hom}(T^s(T^*M) \otimes E; T^m(T^*M) \otimes E), \ \beta_s \in T^s(T^*M) \otimes E,$$

then there exist $C_1, \sigma_1 \in \mathbb{R}_{>0}$ *such that*

$$\|D^r_{\nabla^M,\nabla^{\pi_E}} \Psi^s_{jm} \circ D^a_{\nabla^M,\nabla^{\pi_E}} A^m_s(x)\|_{\mathbb{G}_{M,\pi_E}} \le C_1\sigma_1^{-r} r! \|D^a_{\nabla^M,\nabla^{\pi_E}} A^m_s(x)\|_{\mathbb{G}_{M,\pi_E}},$$

for $x \in \mathcal{K}$, $m, r \in \mathbb{Z}_{\ge 0}$, $s \in \{0, 1, \dots, m\}$;

(vi) *if*

$$\Lambda^s_m(\alpha^m_{s-1}) = \alpha^m_{s-1} \otimes \mathrm{id}_{T^*M}, \qquad \alpha^m_{s-1} \in \mathrm{Hom}(T^{s-1}(T^*M) \otimes E; T^m(T^*M) \otimes E),$$

then

$$\|D^r_{\nabla^M,\nabla^{\pi_E}} \Lambda^s_m \circ D^a_{\nabla^M,\nabla^{\pi_E}} A^m_{s-1}(x)\|_{\mathbb{G}_{M,\pi_E}} \le \|D^a_{\nabla^M,\nabla^{\pi_E}} A^m_{s-1}(x)\|_{\mathbb{G}_{M,\pi_E}},$$

for $x \in \mathcal{K}$, $m, r \in \mathbb{Z}_{\ge 0}$, $s \in \{0, 1, \dots, m\}$.

Proof Parts (i) and (ii) follow from Lemma 4.9 along with the fact that the covariant derivative of the identity tensor is zero. Part (iii) is a tautology, but one that arises in the lemmata of Sect. 3.3.

For the next two parts of the proof, let $C_1, \sigma_1 \in \mathbb{R}_{>0}$ be such that

$$\|D^r_{\nabla^{\mathsf{M}}} S(x)\|_{\mathbb{G}_{\mathsf{M}}} \leq C_1 \sigma_1^{-r} r!, \qquad x \in \mathcal{K}, \tag{4.12}$$

this being possible by Lemma 2.22, and recalling the rôle of the factorials in the definition (2.14) of the fibre norms.

(iv) Let us define

$$\widehat{\Psi}^s_{jm}(\beta^m_s)(\alpha_s) = (\beta^m_s)(\mathrm{Ins}_j(\alpha_s, S)),$$
$$\beta^m_s \in \mathrm{Hom}(\mathsf{T}^s(\mathsf{T}^*\mathsf{M}) \otimes \mathsf{E}; \mathsf{T}^m(\mathsf{T}^*\mathsf{M}) \otimes \mathsf{E}),\ \alpha_s \in \mathsf{T}^s(\mathsf{T}^*\mathsf{M}) \otimes \mathsf{E}$$

and

$$\tau^s_m(\alpha^m_s) = \alpha^m_s \otimes \mathrm{id}_{\mathsf{T}^*\mathsf{M}}, \qquad \alpha^m_s \in \mathrm{Hom}(\mathsf{T}^s(\mathsf{T}^*\mathsf{M}) \otimes \mathsf{E}; \mathsf{T}^m(\mathsf{T}^*\mathsf{M}) \otimes \mathsf{E})$$

so that $\Psi^s_{jm} = \widehat{\Psi}^s_m \circ \tau^s_m$. Note that $\widehat{\Psi}^s_{jm} = \mathrm{Ins}_{S,j}$ so that, by Lemma 3.14,

$$D^r_{\nabla^{\mathsf{M}},\nabla^{\pi\mathsf{E}}} \widehat{\Psi}^s_{jm}(D^a_{\nabla^{\mathsf{M}},\nabla^{\pi\mathsf{E}}} B^m_s) = \mathrm{Ins}_{D^r_{\nabla^{\mathsf{M}},\nabla^{\pi\mathsf{E}}} S,j}(D^a_{\nabla^{\mathsf{M}},\nabla^{\pi\mathsf{E}}} B^m_s).$$

Since the covariant derivative of the identity tensor is zero,

$$D^a_{\nabla^{\mathsf{M}},\nabla^{\pi\mathsf{E}}}(A^m_s \otimes \mathrm{id}_{\mathsf{T}^*\mathsf{M}}) = (D^a_{\nabla^{\mathsf{M}},\nabla^{\pi\mathsf{E}}} A^m_s) \otimes \mathrm{id}_{\mathsf{T}^*\mathsf{M}}),$$

from which we deduce that $D^a_{\nabla^{\mathsf{M}},\nabla^{\pi\mathsf{E}}} \tau^s_m = \tau^s_{m+a}$. Thus

$$\begin{aligned} D^r_{\nabla^{\mathsf{M}},\nabla^{\pi\mathsf{E}}} \Psi^s_{jm} \circ D^a_{\nabla^{\mathsf{M}},\nabla^{\pi\mathsf{E}}} A^m_s &= D^r_{\nabla^{\mathsf{M}},\nabla^{\pi\mathsf{E}}}(\widehat{\Psi}^s_{jm} \circ \tau^s_m) \circ D^a_{\nabla^{\mathsf{M}},\nabla^{\pi\mathsf{E}}} A^m_s \\ &= \mathrm{Ins}_{D^r_{\nabla^{\mathsf{M}},\nabla^{\pi\mathsf{E}}} S,j}(D^a_{\nabla^{\mathsf{M}},\nabla^{\pi\mathsf{E}}} A^m_s \otimes \mathrm{id}_{\mathsf{T}^*\mathsf{M}}). \end{aligned}$$

By Lemmata 4.10 and 4.15, this part of the lemma follows immediately.

(v) Here we have $\Psi^s_{jm}(\alpha^m_s) = \mathrm{Ins}_{S,j} \circ \alpha^m_s$ and, following the arguments from the preceding part of the proof,

$$D^r_{\nabla^{\mathsf{M}},\nabla^{\pi\mathsf{E}}} \Psi^s_m \circ D^a_{\nabla^{\mathsf{M}},\nabla^{\pi\mathsf{E}}} A^m_s = \mathrm{Ins}_{D^r_{\nabla^{\mathsf{M}},\nabla^{\pi\mathsf{E}}} S,j}(D^a_{\nabla^{\mathsf{M}},\nabla^{\pi\mathsf{E}}} A^m_s),$$

and so this part of the lemma follows from Lemma 4.16.

(vi) This follows from Lemma 4.10 and the fact that the covariant derivative of the identity tensor is zero. □

4.3 Independence of Topologies on Connections and Metrics

The seminorms introduced in Sect. 2.4 for defining topologies for the space of real analytic sections of a vector bundle $\pi_{\mathsf{E}}\colon \mathsf{E} \to \mathsf{M}$ are made upon a choice of various objects, namely (1) an affine connection ∇^{M} on M, (2) a linear connection $\nabla^{\pi_{\mathsf{E}}}$ in E, (3) a Riemannian metric $\mathbb{G}_{\mathsf{M}}$ on M, and (4) a fibre metric $\mathbb{G}_{\pi_{\mathsf{E}}}$ for E. In order for these topologies to be useful, they should be independent of all of these choices. This is made more urgent by our very specific choice in Sect. 3.2.1 of a Riemannian metric $\mathbb{G}_{\mathsf{E}}$ on the total space E and its Levi-Civita connection. These choices were made because they made available to us the formulae of Sect. 3.2.3 for differentiation of lifts of tensors, which lifts were used in an essential way in Sects. 3.3 and 4.2, and use of which will be made in Chap. 5, as well as in the present chapter. However, as we pointed out in the preamble to Sect. 4.2, it does remain to be shown that the seminorm topology defined by the real analytic seminorms $p^{\omega}_{\mathcal{K},\boldsymbol{a}}$, $\mathcal{K} \subseteq \mathsf{M}$ compact, $\boldsymbol{a} \in \mathrm{c}_0(\mathbb{Z}_{\geq 0}; \mathbb{R}_{>0})$, is independent of the choices of metrics and connections used to define these seminorms.

We have at various points indicated that many of our constructions can be carried out in the smooth case. However, in the smooth case, it is relatively easy to see that the topology defined by the seminorms (2.17) is independent of metrics and connections. Indeed, these seminorms can just as well be defined by *any* fibre metrics on the jet bundles of $\pi_{\mathsf{E}}\colon \mathsf{E} \to \mathsf{M}$, and the resulting topology will be independent of these choices. This is not the case for the topology defined by the real analytic seminorms, and this is where the technical developments of Chap. 3 and the preceding sections of this chapter are important.

4.3.1 Comparison of Metric-Related Notions for Different Connections and Metrics

We first consider how various constructions involving Riemannian metrics and fibre metrics vary when one varies these metrics. The first result concerns fibre norms for tensor products induced by a fibre metric.

Lemma 4.19 (Comparison of Fibre Norms for Different Fibre Metrics) *Let* $\pi_{\mathsf{E}}\colon \mathsf{E} \to \mathsf{M}$ *be a smooth vector bundle and let* $\mathbb{G}_1$ *and* $\mathbb{G}_2$ *be smooth fibre metrics on* E. *Let* $\mathcal{K} \subseteq \mathsf{M}$ *be compact. Then there exist* $C, \sigma \in \mathbb{R}_{>0}$ *such that*

$$\frac{\sigma^{r+s}}{C}\|A(x)\|_{\mathbb{G}_2} \leq \|A(x)\|_{\mathbb{G}_1} \leq \frac{C}{\sigma^{r+s}}\|A(x)\|_{\mathbb{G}_2}$$

for all $A \in \Gamma^0(\mathrm{T}^r_s(\mathsf{E}))$, $r, s \in \mathbb{Z}_{\geq 0}$, *and* $x \in \mathcal{K}$.

Proof We begin by proving a linear algebra result. □

Sublemma 1 If $\mathbb{G}_1$ and $\mathbb{G}_2$ are inner products on a finite-dimensional $\mathbb{R}$-vector space V, then there exists $C \in \mathbb{R}_{>0}$ such that

$$C^{-1}\mathbb{G}_1(v, v) \leq \mathbb{G}_2(v, v) \leq C\mathbb{G}_1(v, v)$$

for all $v \in \mathsf{V}$.

Proof Let $\mathbb{G}_j^\flat \in \mathrm{Hom}_{\mathbb{R}}(\mathsf{V}; \mathsf{V}^*)$ and $\mathbb{G}_j^\sharp \in \mathrm{Hom}_{\mathbb{R}}(\mathsf{V}^*; \mathsf{V})$, $j \in \{1, 2\}$, be the induced linear maps. Note that

$$\mathbb{G}_1(\mathbb{G}_1^\sharp \circ \mathbb{G}_2^\flat(v_1), v_2) = \mathbb{G}_2(v_1, v_2) = \mathbb{G}_2(v_2, v_1) = \mathbb{G}_1(\mathbb{G}_1^\sharp \circ \mathbb{G}_2^\flat(v_2), v_1),$$

showing that $\mathbb{G}_1^\sharp \circ \mathbb{G}_2^\flat$ is $\mathbb{G}_1$-symmetric. Let $(e_1, \dots, e_n)$ be a $\mathbb{G}_1$-orthonormal basis for V that is also a basis of eigenvectors for $\mathbb{G}_1^\sharp \circ \mathbb{G}_2^\flat$. The matrix representatives of $\mathbb{G}_1$ and $\mathbb{G}_2$ are then

$$[\mathbb{G}_1] = \begin{bmatrix} 1 & 0 & \cdots & 0 \\ 0 & 1 & \cdots & 0 \\ \vdots & \vdots & \ddots & \vdots \\ 0 & 0 & \cdots & 1 \end{bmatrix}, \qquad [\mathbb{G}_2] = \begin{bmatrix} \lambda_1 & 0 & \cdots & 0 \\ 0 & \lambda_2 & \cdots & 0 \\ \vdots & \vdots & \ddots & \vdots \\ 0 & 0 & \cdots & \lambda_n \end{bmatrix},$$

where $\lambda_1, \dots, \lambda_n \in \mathbb{R}_{>0}$. Let us assume without loss of generality that

$$\lambda_1 \leq \cdots \leq \lambda_n.$$

Then taking $C = \max\{\lambda_n, \lambda_1^{-1}\}$ gives the result, as one can verify directly. ▽

Next we use the preceding sublemma to give the linear algebraic version of the lemma.

Sublemma 2 Let V be a finite-dimensional $\mathbb{R}$-vector space and let $\mathbb{G}_1$ and $\mathbb{G}_2$ be inner products on V. Then there exist $C, \sigma \in \mathbb{R}_{>0}$ such that

$$\frac{\sigma^{r+s}}{C}\|A\|_{\mathbb{G}_2} \leq \|A\|_{\mathbb{G}_1} \leq \frac{C}{\sigma^{r+s}}\|A\|_{\mathbb{G}_2}$$

for all $A \in \mathrm{T}^r_s(\mathsf{V})$, $r, s \in \mathbb{Z}_{\geq 0}$.

Proof As in the proof of Sublemma 1, let $(e_1, \dots, e_n)$ be a $\mathbb{G}_1$-orthonormal basis for V consisting of eigenvectors for $\mathbb{G}_1^\sharp \circ \mathbb{G}_2^\flat$. Let $\lambda_1, \dots, \lambda_n \in \mathbb{R}_{>0}$ be the corresponding eigenvalues, supposing that

$$\lambda_1 \leq \cdots \leq \lambda_n.$$

Note that $\mathbb{G}_2(e_j, e_k) = \delta_{jk}\lambda_j$, $j \in \{1, \dots, n\}$, (δ_{jk} being the Kronecker delta symbol) so that $(\hat{e}_1 \triangleq \lambda_1^{-1} e_1, \dots, \hat{e}_n \triangleq \lambda_n^{-1} e_n)$ is a $\mathbb{G}_2$-orthonormal basis. Denote by $(e^1, \dots, e^n)$ and $(\hat{e}^1, \dots, \hat{e}^n)$ be the dual bases. Note that $\hat{e}^j = \lambda_j e^j$, $j \in \{1, \dots, n\}$.

Now let $A \in \mathrm{T}^r_s(\mathsf{V})$ and write

$$A = \sum_{j_1,\dots,j_r=1}^{n} \sum_{k_1,\dots,k_s=1}^{n} A^{j_1\cdots j_r}_{k_1\cdots k_s} e_{j_1} \otimes \dots \otimes e_{j_r} \otimes e^{k_1} \otimes \dots \otimes e^{k_s}$$

and

$$A = \sum_{j_1,\dots,j_r=1}^{n} \sum_{k_1,\dots,k_s=1}^{n} \widehat{A}^{j_1\cdots j_r}_{k_1\cdots k_s} \hat{e}_{j_1} \otimes \dots \otimes \hat{e}_{j_r} \otimes \hat{e}^{k_1} \otimes \dots \otimes \hat{e}^{k_s}.$$

We necessarily have

$$\widehat{A}^{j_1\cdots j_r}_{k_1\cdots k_s} = \lambda_{j_1} \cdots \lambda_{j_r} \lambda_{k_1}^{-1} \cdots \lambda_{k_s}^{-1} A^{j_1\cdots j_r}_{k_1\cdots k_s}, \qquad j_1, \dots, j_r, k_1, \dots, k_s \in \{1, \dots, n\}.$$

We have

$$\|A\|_{\mathbb{G}_1} = \left(\sum_{j_1,\dots,j_r=1}^{n} \sum_{k_1,\dots,k_s=1}^{n} \left| A^{j_1\cdots j_r}_{k_1\cdots k_s} \right|^2 \right)^{1/2},$$

$$\|A\|_{\mathbb{G}_2} = \left(\sum_{j_1,\dots,j_r=1}^{n} \sum_{k_1,\dots,k_s=1}^{n} \left| \widehat{A}^{j_1\cdots j_r}_{k_1\cdots k_s} \right|^2 \right)^{1/2}.$$

Therefore, if we let $\sigma = \min\{\lambda_1, \lambda_n^{-1}\}$, we have

$$\|A\|_{\mathbb{G}_2} \leq \sigma^{-(r+s)} \|A\|_{\mathbb{G}_1}.$$

This gives one half of the estimate in the sublemma, and the other is established analogously. ▽

The lemma follows from the preceding sublemma since C and σ depend only on $\mathbb{G}_1$ and $\mathbb{G}_2$ through the largest and smallest eigenvalues of $\mathbb{G}_1^\sharp \circ \mathbb{G}_2^\flat$, which are uniformly bounded above and below on $\mathcal{K}$. □

Remark 4.20 (Generalisation to Tensor Products of Vector Bundles) For simplicity we have stated the preceding result for a single vector bundle $\pi_{\mathsf{E}} \colon \mathsf{E} \to \mathsf{M}$ with tensors coming from the tensor algebra of this vector bundle. A moment's consideration of the proof of the lemma will convince the reader that the result will hold for tensors that are tensor products of elements of the tensor algebra of any finite number of vector bundles, each equipped with two fibre metrics. We shall use this generalisation without mention. ○

The following comparison result for the distance function associated to two Riemannian metrics is often useful, although we do not make use of it for our developments here. This is certainly a known result, although we could not locate a proof.

Lemma 4.21 (Comparison of Distance Functions for Riemannian Metrics) *If $\mathbb{G}_1$ and $\mathbb{G}_2$ are C^∞-Riemannian metrics on a C^∞-manifold M with metrics d_1 and d_2, respectively, and if $\mathcal{K} \subseteq \mathsf{M}$ is compact, then there exists $C \in \mathbb{R}_{>0}$ such that*

$$C^{-1}\mathrm{d}_1(x_1, x_2) \leq \mathrm{d}_2(x_1, x_2) \leq C\mathrm{d}_1(x_1, x_2)$$

for every $x_1, x_2 \in \mathcal{K}$.

Proof First we give a local version of the result. Let $x \in \mathsf{M}$. Let $\mathcal{N}_1$ and $\mathcal{N}_2$ be geodesically convex neighbourhoods of x with respect to the Riemannian metrics $\mathbb{G}_1$ and $\mathbb{G}_2$, respectively [42, Proposition IV.3.4]. Thus every pair of points in $\mathcal{N}_1$ can be connected by a unique distance-minimising geodesic for $\mathbb{G}_1$ that remains in $\mathcal{N}_1$, and similarly with $\mathcal{N}_2$ and $\mathbb{G}_2$. By Sublemma 1 from the proof of Lemma 4.19, let $C_x \in \mathbb{R}_{>0}$ be such that

$$C_x^{-2}\mathbb{G}_1(v_x, v_x) < \mathbb{G}_2(v_x, v_x) < C_x^2\mathbb{G}_1(v_x, v_x), \qquad v_x \in \mathsf{T}_x\mathsf{M}.$$

By continuity of $\mathbb{G}_1$ and $\mathbb{G}_2$, we can choose $\mathcal{N}_1$ and $\mathcal{N}_2$ sufficiently small that

$$C_x^{-2}\mathbb{G}_1(v_y, v_y) < \mathbb{G}_2(v_y, v_y) < C_x^2\mathbb{G}_1(v_y, v_y), \qquad y \in \mathcal{N}_1 \cup \mathcal{N}_2.$$

Now define $\mathcal{U}_x = \mathcal{N}_1 \cap \mathcal{N}_2$. Then every pair of points in $\mathcal{U}_x$ can be connected with a unique distance-minimising geodesic of both $\mathbb{G}_1$ and $\mathbb{G}_2$ that remains in $\mathcal{N}_1 \cup \mathcal{N}_2$. Now let $x_1, x_2 \in \mathcal{U}_x$. Let $\gamma\colon [0, 1] \to \mathsf{M}$ be the unique distance-minimising $\mathbb{G}_1$-geodesic connecting x_1 and x_2. Then

$$\begin{aligned}\mathrm{d}_2(x_1, x_2) \leq \ell_{\mathbb{G}_2}(\gamma) &= \int_0^1 \sqrt{\mathbb{G}_2(\gamma'(t), \gamma'(t))}\,\mathrm{d}t \\ &\leq C_x \int_0^1 \sqrt{\mathbb{G}_1(\gamma'(t), \gamma'(t))}\,]dt \\ &= C_x\ell_{\mathbb{G}_1}(\gamma) = C_x\mathrm{d}_1(x_1, x_2).\end{aligned}$$

One similarly shows that $\mathrm{d}_1(x_1, x_2) \leq C_x\mathrm{d}_2(x_1, x_2)$.

We now prove the assertion of the lemma. Let $\mathcal{K} \subseteq \mathsf{M}$ be compact and, for each $x \in \mathcal{K}$, let $\mathcal{U}_x$ be a neighbourhood of x and let $C_x \in \mathbb{R}_{>0}$ be as in the preceding paragraph. Let $x_1, \dots, x_k \in \mathcal{K}$ be such that $\mathcal{K} \subseteq \cup_{j=1}^k \mathcal{U}_{x_j}$. Let

$$D_a = \sup\{\mathrm{d}_a(x, y) \mid x, y \in \mathcal{K}\}, \qquad a \in \{1, 2\}.$$

By the Lebesgue Number Lemma [11, Theorem 1.6.11], let $r_a \in \mathbb{R}_{>0}$ be such that, if $x_1, x_2 \in \mathcal{K}$ satisfy $\mathrm{d}_a(x_1, x_2) < r_a$, $a \in \{1, 2\}$, then there exists $j \in \{1, \dots, k\}$ such that $x_1, x_2 \in \mathcal{U}_{x_j}$. Let us denote

$$C = \max\left\{C_{x_1}, \dots, C_{x_k}, \frac{D_1}{r_2}, \frac{D_2}{r_1}\right\}.$$

Now let $x_1, x_2 \in \mathcal{K}$. If $\mathrm{d}_1(x_1, x_2) < r_1$, then let $j \in \{1, \dots, k\}$ be such that $x_1, x_2 \in \mathcal{U}_j$. Then

$$\mathrm{d}_2(x_1, x_2) \leq C\mathrm{d}_1(x_1, x_2).$$

If $\mathrm{d}_1(x_1, x_2) \geq r_1$, then

$$\frac{\mathrm{d}_2(x_1, x_2) r_1}{D_2} \leq \frac{\mathrm{d}_2(x_1, x_2) r_1}{\mathrm{d}_2(x_1, x_2)} \leq d_1(x_1, x_2)$$

This gives $\mathrm{d}_2(x_1, x_2) \leq C\mathrm{d}_1(x_1, x_2)$. Swapping the rôles of $\mathbb{G}_1$ and $\mathbb{G}_2$ gives $\mathrm{d}_1(x_1, x_2) \leq C\mathrm{d}_2(x_1, x_2)$ for an appropriate C, giving the lemma. □

Now we can compare fibre norms for jet bundles associated with different metrics and connections.

Lemma 4.22 (Comparison of Fibre Norms for Jet Bundles for Different Metrics and Connections) *Let $r \in \{\infty, \omega\}$ and let $\pi_{\mathsf{E}} \colon \mathsf{E} \to \mathsf{M}$ be a C^r-vector bundle. Consider C^r-affine connections ∇^{M} and $\overline{\nabla}^{\mathsf{M}}$ on M, and C^r-vector bundle connections $\nabla^{\pi_{\mathsf{E}}}$ and $\overline{\nabla}^{\pi_{\mathsf{E}}}$ in E. Consider C^r-Riemannian metrics $\mathbb{G}_{\mathsf{M}}$ and $\overline{\mathbb{G}}_{\mathsf{M}}$ for M, and C^r-fibre metrics $\mathbb{G}_{\pi_{\mathsf{E}}}$ and $\overline{\mathbb{G}}_{\pi_{\mathsf{E}}}$ for E. Let $\mathcal{K} \subseteq \mathsf{M}$ be compact. Then there exist $C, \sigma \in \mathbb{R}_{>0}$ such that*

$$\frac{\sigma^m}{C} \|j_m\xi(x)\|_{\overline{\mathbb{G}}_{\mathsf{M},\pi_{\mathsf{E}},m}} \leq \|j_m\xi(x)\|_{\mathbb{G}_{\mathsf{M},\pi_{\mathsf{E}},m}} \leq \frac{C}{\sigma^m} \|j_m\xi(x)\|_{\overline{\mathbb{G}}_{\mathsf{M},\pi_{\mathsf{E}},m}}$$

for all $\xi \in \Gamma^m(\mathsf{E})$, $m \in \mathbb{Z}_{\geq 0}$, and $x \in \mathcal{K}$.

Proof We first make some preliminary constructions that will be useful.

By Lemma 3.69, we have

$$\begin{aligned}
\xi(x) &= \widehat{A}_0^0 \xi(x), \\
(\mathrm{Sym}_1 \otimes \mathrm{id}_{\mathsf{E}}) \overline{\nabla}^{\pi_{\mathsf{E}}} \xi(x) &= \widehat{A}_1^1 (\nabla^{\pi_{\mathsf{E}}} \xi(x)) + \widehat{A}_0^1(\xi(x)), \\
&\ \vdots \\
(\mathrm{Sym}_m \otimes \mathrm{id}_{\mathsf{E}}) \circ \overline{\nabla}^{\mathsf{M},\pi_{\mathsf{E}},m} \xi(x) &= \sum_{s=0}^{m} \widehat{A}_s^m ((\mathrm{Sym}_s \otimes \mathrm{id}_{\mathsf{E}}) \circ \nabla^{\mathsf{M},\pi_{\mathsf{E}},s} \xi(x)).
\end{aligned} \tag{4.13}$$

In like manner, by Lemma 3.70, we have

$$\begin{aligned}
\xi(x) &= \widehat{B}_0^0 \xi(x), \\
(\mathrm{Sym}_1 \otimes \mathrm{id}_{\mathsf{E}}) \nabla^{\pi_{\mathsf{E}}} \xi(x) &= \widehat{B}_1^1 (\overline{\nabla}^{\pi_{\mathsf{E}}} \xi(x)) + \widehat{B}_0^1(\xi(x)), \\
&\ \vdots \\
(\mathrm{Sym}_m \otimes \mathrm{id}_{\mathsf{E}}) \circ \nabla^{\mathsf{M},\pi_{\mathsf{E}},m} \xi(x) &= \sum_{s=0}^{m} \widehat{B}_s^m ((\mathrm{Sym}_s \otimes \mathrm{id}_{\mathsf{E}}) \circ \overline{\nabla}^{\mathsf{M},\pi_{\mathsf{E}},s} \xi(x)).
\end{aligned} \tag{4.14}$$

By Lemma 4.11, we have

$$\|A_s^m(\beta_s)\|_{\overline{\mathbb{G}}_{\mathsf{M},\pi_{\mathsf{E}}}} \leq \|A_s^m\|_{\overline{\mathbb{G}}_{\mathsf{M},\pi_{\mathsf{E}}}} \|\beta_s\|_{\overline{\mathbb{G}}_{\mathsf{M},\pi_{\mathsf{E}}}}$$

for $\beta_s \in \mathsf{T}^s(\mathsf{T}^*\mathsf{M}) \otimes \mathsf{E}$, $m \in \mathbb{Z}_{>0}$, $s \in \{0, 1, \dots, m\}$. By Lemma 4.13,

$$\|\mathrm{Sym}_s(A)\|_{\overline{\mathbb{G}}_{\mathsf{M},\pi_{\mathsf{E}}}} \leq \|A\|_{\overline{\mathbb{G}}_{\mathsf{M},\pi_{\mathsf{E}}}}$$

for $A \in \mathsf{T}^s(\mathsf{T}^*\mathsf{E})$ and $s \in \mathbb{Z}_{>0}$. Thus, recalling (3.40),

$$\|\widehat{A}_s^m(\mathrm{Sym}_s(\beta_s))\|_{\overline{\mathbb{G}}_{\mathsf{M},\pi_{\mathsf{E}}}} = \|\mathrm{Sym}_m \circ A_s^m(\beta_s)\|_{\overline{\mathbb{G}}_{\mathsf{M},\pi_{\mathsf{E}}}} \leq \|A_s^m\|_{\overline{\mathbb{G}}_{\mathsf{E}}} \|\beta_s\|_{\overline{\mathbb{G}}_{\mathsf{M},\pi_{\mathsf{E}}}},$$

for $\beta_s \in \mathsf{T}^s(\pi_{\mathsf{E}}^*\mathsf{T}^*\mathsf{M}) \otimes \mathsf{E}$, $m \in \mathbb{Z}_{>0}$, $s \in \{1, \dots, m\}$.

By Lemmata 4.17 and 4.18 with $r = 0$, there exist $\sigma_1, \rho_1 \in \mathbb{R}_{>0}$ such that

$$\|A_s^k(x)\|_{\overline{\mathbb{G}}_{\mathsf{M},\pi_{\mathsf{E}}}} \leq \sigma_1^{-k} \rho_1^{-(k-s)} (k-s)!, \qquad k \in \mathbb{Z}_{\geq 0},\ s \in \{0, 1, \dots, k\},\ x \in \mathcal{K}.$$

Without loss of generality, we assume that $\sigma_1, \rho_1 \leq 1$. Thus, redefining $\sigma_1 = \sigma_1\rho_1$, we have

$$\begin{aligned}\|\widehat{A}^k_s((\mathrm{Sym}_s \otimes \mathrm{id}_{\mathsf{E}}) \circ \overline{\nabla}^{\mathsf{M},\pi_{\mathsf{E}},s}\xi(x)\|_{\overline{\mathbb{G}}_{\mathsf{M},\pi_{\mathsf{E}}}} \\ \leq C_1\sigma_1^{-k}(k-s)!\|(\mathrm{Sym}_s \otimes \mathrm{id}_{\mathsf{E}}) \circ \nabla^{\mathsf{M},\pi_{\mathsf{E}},s}\xi(x)\|_{\overline{\mathbb{G}}_{\mathsf{M},\pi_{\mathsf{E}}}}\end{aligned}$$

for $m \in \mathbb{Z}_{\geq 0}$, $k \in \{0, 1, \ldots, m\}$, $s \in \{0, 1, \ldots, k\}$, $x \in \mathcal{K}$.

By Lemma 4.19 (along with Remark 4.20), let $C_2, \sigma_2 \in \mathbb{R}_{>0}$ be such that

$$\frac{\sigma_2^{m+1}}{C_2}\|A\|_{\overline{\mathbb{G}}_{\mathsf{M},\pi_{\mathsf{E}},m}} \leq \|A\|_{\mathbb{G}_{\mathsf{M},\pi_{\mathsf{E}},m}} \leq \frac{C_2}{\sigma_2^{m+1}}\|A\|_{\overline{\mathbb{G}}_{\mathsf{M},\pi_{\mathsf{E}},m}}$$

for $A \in \Gamma^\infty(\mathsf{S}^k(\mathsf{T}^*\mathsf{M})\otimes\mathsf{E})$. We shall assume, without loss of generality, that $\sigma_2 \leq 1$.

Thus, by (1.4) and (4.13),

$$\begin{aligned}&\frac{\sigma_2^{m+1}}{C_2}\|j_m\xi(x)\|_{\overline{\mathbb{G}}_{\mathsf{M},\pi_{\mathsf{E}},m}} \\ &\quad\leq \sum_{k=0}^{m}\frac{1}{k!}\|(\mathrm{Sym}_k \otimes \mathrm{id}_{\mathsf{E}}) \circ \nabla^{\mathsf{M},\pi_{\mathsf{E}},k}\xi(x)\|_{\mathbb{G}_{\mathsf{M},\pi_{\mathsf{E}}}} \\ &\quad= \sum_{k=0}^{m}\frac{1}{k!}\left\|\sum_{s=0}^{k}\widehat{A}^k_s((\mathrm{Sym}_s \otimes \mathrm{id}_{\mathsf{E}}) \circ \nabla^{\mathsf{M},\pi_{\mathsf{E}},s}\xi(x))\right\|_{\mathbb{G}_{\mathsf{M},\pi_{\mathsf{E}}}} \\ &\quad\leq \sum_{k=0}^{m}\sum_{s=0}^{k}C_1\sigma_1^{-k}\frac{s!(k-s)!}{k!}\frac{1}{s!}\|(\mathrm{Sym}_s \otimes \mathrm{id}_{\mathsf{E}}) \circ \nabla^{\mathsf{M},\pi_{\mathsf{E}},s}\xi(x)\|_{\mathbb{G}_{\mathsf{M},\pi_{\mathsf{E}}}}\end{aligned}$$

for $x \in \mathcal{K}$ and $m \in \mathbb{Z}_{\geq 0}$. Now note that

$$\frac{s!(k-s)!}{k!} \leq 1, \quad C_1\sigma_1^{-k} \leq C_1\sigma_1^{-m},$$

for $s \in \{0, 1, \ldots, m\}$, $k \in \{0, 1, \ldots, s\}$, since $\sigma_1 \leq 1$. Then

$$\begin{aligned}\frac{\sigma_2^{m+1}}{C_2}\|j_m\xi(x)\|_{\overline{\mathbb{G}}_{\mathsf{M},\pi_{\mathsf{E}},m}} &\leq C_1\sigma_1^{-m}\sum_{k=0}^{m}\sum_{s=0}^{k}\frac{1}{s!}\|(\mathrm{Sym}_s \otimes \mathrm{id}_{\mathsf{E}}) \circ \nabla^{\mathsf{M},\pi_{\mathsf{E}},s}\xi(x)\|_{\mathbb{G}_{\mathsf{M},\pi_{\mathsf{E}}}} \\ &\leq C_1\sigma_1^{-m}\sum_{k=0}^{m}\sum_{s=0}^{m}\frac{1}{s!}\|(\mathrm{Sym}_s \otimes \mathrm{id}_{\mathsf{E}}) \circ \nabla^{\mathsf{M},s}\xi(x)\|_{\mathbb{G}_{\mathsf{M},\pi_{\mathsf{E}}}}\end{aligned}$$

$$= (m+1)C_1\sigma_1^{-m} \times \sum_{s=0}^{m} \frac{1}{s!} \|(\mathrm{Sym}_s \otimes \mathrm{id}_{\mathsf{E}}) \circ \nabla^{\mathsf{M},\pi_{\mathsf{E}},s}\xi(x)\|_{\mathbb{G}_{\mathsf{M},\pi_{\mathsf{E}}}},$$

which gives

$$\begin{aligned}\|j_m\xi(x)\|_{\overline{\mathbb{G}}_{\mathsf{M},\pi_{\mathsf{E}},m}} &\le (m+1)C_1C_2\sigma_2(\sigma_1\sigma_2)^{-m} \\ &\quad \times \sum_{s=0}^{m} \frac{1}{s!} \|(\mathrm{Sym}_s \otimes \mathrm{id}_{\mathsf{E}}) \circ \nabla^{\mathsf{M},\pi_{\mathsf{E}},s}\xi(x)\|_{\mathbb{G}_{\mathsf{M},\pi_{\mathsf{E}}}} \\ &\le \sqrt{m+1}(m+1)C_1C_2\sigma_2(\sigma_1\sigma_2)^{-m}\|j_m\xi(x)\|_{\mathbb{G}_{\mathsf{M},\pi_{\mathsf{E}},m}},\end{aligned}$$

making use of (1.4). Now let $\sigma < \sigma_1\sigma_2$ and note that

$$\lim_{m\to\infty} (m+1)^{3/2} \frac{(\sigma_1\sigma_2)^{-m}}{\sigma^{-m}} = 0.$$

Thus there exists $N \in \mathbb{Z}_{>0}$ such that

$$(m+1)^{3/2}C_1C_2\sigma_2(\sigma_1\sigma_2)^{-m} \le C_1C_2\sigma_2\sigma^{-m}, \qquad m \ge N.$$

Let

$$C = \max\left\{C_1C_2\sigma_2, 2^{3/2}C_1C_2\sigma_2\frac{\sigma}{\sigma_1\sigma_2}, 3^{3/2}C_1C_2\sigma_2\left(\frac{\sigma}{\sigma_1\sigma_2}\right)^2, \dots, (N+1)^{3/2}C_1C_2\sigma_2\left(\frac{\sigma}{\sigma_1\sigma_2}\right)^N\right\}.$$

We then immediately have $(m+1)^{3/2}C_1C_2\sigma_2\sigma_1^{-m} \le C\sigma^{-m}$ for all $m \in \mathbb{Z}_{\ge 0}$. We then have

$$\|j_m\xi(x)\|_{\overline{\mathbb{G}}_{\mathsf{M},\pi_{\mathsf{E}},m}} \le C\sigma^{-m}\|j_m\xi(x)\|_{\mathbb{G}_{\mathsf{M},\pi_{\mathsf{E}},m}}.$$

This gives one half of the desired pair of estimates.

For the other half of the estimate, we use (4.14), and Lemmata 4.17 and 4.18 in the computations above to arrive at the estimate

$$\|j_m\xi(x)\|_{\mathbb{G}_{\mathsf{M},\pi_{\mathsf{E}},m}} \le C\sigma^{-m}\|j_m\xi(x)\|_{\overline{\mathbb{G}}_{\mathsf{M},\pi_{\mathsf{E}},m}},$$

which gives the result. □

Remark 4.23 (Adaptation to the Smooth Case) An analogous result to the preceding result holds in the smooth case, and with a much easier proof. In the result, one can replace "$C\sigma^{-m}$" with a fixed constant "C_m" for each m; i.e., one does not need any uniformity of the estimates in m. For this reason, the proof is also far simpler, e.g., one need not keep track of all the factorial terms that give rise to the exponential component in the estimates. ◦

4.3.2 Local Descriptions of the Real Analytic Topology

We endeavour to make our presentation as unencumbered of coordinates as possible. While the intrinsic jet bundle characterisations of the seminorms are useful for general definitions and elegant proofs, concrete proofs often require local descriptions of the topologies. In this section we provide these local descriptions of the topologies. By proving that these local descriptions are equivalent to the intrinsic descriptions above, we also prove that these intrinsic descriptions of topologies do not depend on the choice of metrics or connections.

Let us develop the notation for working with local descriptions of topologies. Let $\mathcal{U} \subseteq \mathbb{R}^n$ be an open set. We define local seminorms for $\mathrm{C}^\omega(\mathcal{U}; \mathbb{R}^k)$ as follows. Let $\boldsymbol{\Phi} \in \mathrm{C}^\omega(\mathcal{U}; \mathbb{R}^k)$. For $\mathcal{K} \subseteq \mathcal{U}$ compact and for $\boldsymbol{a} \in \mathrm{c}_0(\mathbb{Z}_{\geq 0}; \mathbb{R}_{>0})$, denote

$$p'^{\omega}_{\mathcal{K},\boldsymbol{a}}(\boldsymbol{\Phi}) = \sup\left\{\frac{a_0 a_1 \cdots a_m}{I!}|\boldsymbol{D}^I \Phi^a(\boldsymbol{x})| \;\middle|\; \boldsymbol{x} \in \mathcal{K},\ a \in \{1, \dots, k\},\ I \in \mathbb{Z}^n_{\geq 0},\ |I| \leq m,\ m \in \mathbb{Z}_{\geq 0}\right\}.$$

These seminorms, defined for all compact $\mathcal{K} \subseteq \mathcal{U}$ and $\boldsymbol{a} \in \mathrm{c}_0(\mathbb{Z}_{\geq 0}; \mathbb{R}_{>0})$, define the ***local*** $\mathbf{C}^\omega$***-topology*** for $\mathrm{C}^\omega(\mathcal{U}; \mathbb{R}^k)$.

There are many possible variations of the seminorms that one can use, and these variations are equivalent to the seminorms above. For example, rather than using the ∞-vector norm, one might use the 2-vector norm. In doing so, one uses (1.4) to give

$$\begin{aligned}\sup\{|\boldsymbol{D}^I \Phi^a(\boldsymbol{x})| \mid I \in \mathbb{Z}^n_{\geq 0},\ |I| = m,\ a \in \{1, \dots, k\}\} &\leq \|\boldsymbol{D}^m \boldsymbol{\Phi}(\boldsymbol{x})\| \\ &\leq \sqrt{kn^m} \sup\{|\boldsymbol{D}^I \Phi^a(\boldsymbol{x})| \mid I \in \mathbb{Z}^n_{\geq 0},\ |I| = m,\ a \in \{1, \dots, k\}\}.\end{aligned}$$

If we define

$$b_0 = 2\sqrt{k}a_0,\ b_j = 2\sqrt{n}a_j, \qquad j \in \mathbb{Z}_{>0},$$

then, noting that $n^j \leq n^m$ for $j \in \{0, 1, \dots, m\}$ and that $m + 1 \leq 2^m$ for $m \in \mathbb{Z}_{\geq 0}$, we have

$$p'^{\omega}_{\mathcal{K},\boldsymbol{a}}(\boldsymbol{\Phi}) \leq \sup \left\{ \frac{a_0 a_1 \cdots a_m}{I!} \|\boldsymbol{D}^{|I|}\boldsymbol{\Phi}(\boldsymbol{x})\| \,\middle|\, \boldsymbol{x} \in \mathcal{K},\ I \in \mathbb{Z}^n_{\geq 0},\ |I| \leq m,\ m \in \mathbb{Z}_{\geq 0} \right\} \leq p'^{\omega}_{\mathcal{K},\boldsymbol{b}}(\boldsymbol{\Phi}),$$

and this gives equivalence of the topologies using the ∞- and 2-norms. Another variation in the seminorms is that one might scale the derivatives by $\frac{1}{|I|!}$ rather than $\frac{1}{I!}$. In this case, we use the standard multinomial estimate (4.10) to give

$$\frac{|I|!}{I!} \leq n^m.$$

Thus, if we take

$$b_0 = a_0,\ b_j = n a_j, \qquad j \in \mathbb{Z}_{>0},$$

we have

$$p'^{\omega}_{\mathcal{K},\boldsymbol{b}}(\boldsymbol{\Phi}) \leq \sup \left\{ \frac{a_0 a_1 \cdots a_m}{|I|!} |\boldsymbol{D}^I \Phi^a(\boldsymbol{x})| \,\middle|\, \boldsymbol{x} \in \mathcal{K},\ a \in \{1, \dots, k\},\ I \in \mathbb{Z}^n_{\geq 0},\ |I| \leq m,\ m \in \mathbb{Z}_{\geq 0} \right\} \leq p'^{\omega}_{\mathcal{K},\boldsymbol{a}}(\boldsymbol{\Phi}).$$

This gives the equivalence of the topologies defined using the scaling factor $\frac{1}{|I|!}$ for derivatives in place of $\frac{1}{I!}$. One can also combine the previous modifications. Indeed, if we use the 2-norm and the scaling factor $\frac{1}{|I!|}$, then one readily sees that we recover the intrinsic seminorms on the trivial vector bundle $\mathbb{R}^k_{\mathcal{U}}$ of Sect. 2.4 using (1) the Euclidean inner product for the Riemannian metric on $\mathcal{U}$ and for the fibre metric on $\mathbb{R}^k$ and (2) standard differentiation as covariant differentiation. We shall use this observation in the proof of Theorem 4.24 below.

We wish to show that these local topologies can be used to define a topology for $\Gamma^{\omega}(\mathsf{E})$ that is equivalent to the intrinsic topologies defined in Sect. 2.4 using jet bundles, connections, and metrics. To state the result, let us indicate some notation. Let $(\mathcal{V}, \boldsymbol{\nu})$ be a vector bundle chart for $\pi_{\mathsf{E}}\colon \mathsf{E} \to \mathsf{M}$ with $(\mathcal{U}, \chi)$ the induced chart for M. Suppose that $\boldsymbol{\nu}(\mathcal{V}) = \chi(\mathcal{U}) \times \mathbb{R}^k$. Given a section ξ, we define $\boldsymbol{\nu}_*(\xi)\colon \chi(\mathcal{U}) \to \mathbb{R}^k$ by requiring that

$$\boldsymbol{\nu} \circ \xi \circ \chi^{-1}(\boldsymbol{x}) = (\boldsymbol{x}, \boldsymbol{\nu}_*(\xi)(\boldsymbol{x})).$$

With this notation, we have the following result.

Theorem 4.24 (Agreement of Intrinsic and Local Topologies) *Let $\pi_{\mathsf{E}}\colon \mathsf{E} \to \mathsf{M}$ be a C^{ω}-vector bundle. Let $\mathbb{G}_{\mathsf{M}}$ be a Riemannian metric on M, let $\mathbb{G}_{\pi_{\mathsf{E}}}$ be a fibre metric on E, let ∇^{M} be an affine connection on M, and let $\nabla^{\pi_{\mathsf{E}}}$ be a vector bundle connection on E, with all of these being of class C^{ω}. Then the following two collections of seminorms for $\Gamma^{\omega}(\mathsf{E})$ define the same topology:*

(i) $p^{\omega}_{\mathcal{K},\boldsymbol{a}}$*,* $\boldsymbol{a} \in c_0(\mathbb{Z}_{\geq 0}; \mathbb{R}_{>0})$*,* $\mathcal{K} \subseteq \mathsf{M}$ *compact;*

(ii) $p'^{\omega}_{\mathcal{K},\boldsymbol{a}} \circ \boldsymbol{\nu}_*$*,* $\boldsymbol{a} \in c_0(\mathbb{Z}_{\geq 0}; \mathbb{R}_{>0})$*,* $(\mathcal{V}, \boldsymbol{\nu})$ *is a vector bundle chart for* E *with* $(\mathcal{U}, \boldsymbol{\chi})$ *the induced chart for* M*,* $\mathcal{K} \subseteq \boldsymbol{\chi}(\mathcal{U})$ *compact.*

Proof As alluded to in the discussion above, it suffices to use the norm

$$\|\boldsymbol{D}^m \boldsymbol{\Phi}(\boldsymbol{x})\|_2 = \left(\sum_{\substack{I \in \mathbb{Z}^n_{\geq 0} \\ |I| = m}} \sum_{a=1}^{k} |\boldsymbol{D}^I \Phi^a(\boldsymbol{x})|^2 \right)^{1/2}$$

for derivatives of $\mathbb{R}^k$-valued functions on $\mathcal{U} \subseteq \mathbb{R}^n$. If we denote

$$j_m \boldsymbol{\Phi}(\boldsymbol{x}) = (\boldsymbol{\Phi}(\boldsymbol{x}), \boldsymbol{D}\boldsymbol{\Phi}(\boldsymbol{x}), \ldots, \boldsymbol{D}^m \boldsymbol{\Phi}(\boldsymbol{x})),$$

then we define

$$\|j_m \boldsymbol{\Phi}(\boldsymbol{x})\|^2_{2,m} = \sum_{j=0}^{m} \frac{1}{(j!)^2} \|\boldsymbol{D}^j \boldsymbol{\Phi}(\boldsymbol{x})\|^2_2,$$

this norm agreeing with the fibre norms used in Sect. 2.3.2 with the flat connections and with the Euclidean inner products. We use these norms to define seminorms that we denote by q' in place of the local seminorms p' as above.

We might like to use Lemma 4.22 in this proof. However, we cannot do so without a moment's thought. The reason for this is that the proof of Lemma 4.22 makes reference to Lemma 4.17, which itself can be applied only by virtue of Lemma 4.18. The proof of this latter lemma relies on the bound (4.12), which is deduced from Lemma 2.22. The proof of Lemma 2.22, we note, makes use of Lemma 4.22. To intrude on the potential circular logic, we must give a proof of this part of the theorem that does not rely on Lemma 4.22 as it is stated. In fact, the only part of the chain of results that we need to prove independently is the bound (4.12). In particular, if we can show that Lemma 4.22 holds in the current situation where

1. $\mathsf{M} = \mathcal{U} \subseteq \mathbb{R}^n$ and $\mathsf{E} = \mathbb{R}^k_{\mathcal{U}}$,
2. $\overline{\mathbb{G}}_{\mathcal{U}}$ and $\overline{\mathbb{G}}_{\pi_{\mathsf{E}}}$ are the Euclidean inner products, and
3. $\overline{\nabla}^{\mathsf{M}}$ and $\overline{\nabla}^{\pi_{\mathsf{E}}}$ are the flat connections,

this will be enough to make use of Lemma 4.22.

To this end, let $(\mathcal{V}, \boldsymbol{\nu})$ be a vector bundle chart for E with $(\mathcal{U}, \boldsymbol{\chi})$ the chart for M. Theorem 1.9 gives $C_1, \sigma_1 \in \mathbb{R}_{>0}$ such that

$$\|D^r_{\overline{\nabla}^{\mathcal{U}},\overline{\nabla}^{\pi_\mathsf{E}}} S_{\mathcal{U}}(\boldsymbol{x})\|_2, \ \|D^r_{\overline{\nabla}^{\mathcal{U}},\overline{\nabla}^{\pi_\mathsf{E}}} S_{\pi_\mathsf{E}}(\boldsymbol{x})\|_2 \le C_1\sigma_1^{-r} r!, \qquad \boldsymbol{x} \in \mathcal{K}. \tag{4.15}$$

This gives the bound (4.12) in this case, and so we can use Lemma 4.18, and then Lemma 4.17, and then the computation of Lemma 4.22 (all in our current local setting) to give

$$\frac{\sigma^m}{C}\|j_m\xi\|_{\mathbb{G}_{\mathsf{M},\pi_\mathsf{E}},m} \le \|j_m(\boldsymbol{\nu}_*(\xi))(\boldsymbol{\chi}(x))\|_{2,m} \le \frac{C}{\sigma^m}\|j_m\xi\|_{\mathbb{G}_{\mathsf{M},\pi_\mathsf{E}},m}.$$

Now, having established Lemma 4.22 in the case of interest, we proceed with the proof, making use of this fact.

Let $\mathcal{K} \subseteq \boldsymbol{\chi}(\mathcal{U})$ be compact and let $\boldsymbol{a} \in c_0(\mathbb{Z}_{\ge 0}; \mathbb{R}_{>0})$. As per our appropriate version of Lemma 4.22, there exist $C, \sigma \in \mathbb{R}_{>0}$ such that

$$\|j_m(\boldsymbol{\nu}_*(\xi))(\boldsymbol{\chi}(x))\|_{2,m} \le \frac{C}{\sigma^m}\|j_m\xi(x)\|_{\mathbb{G}_{\mathsf{M},\pi_\mathsf{E}},m}$$

for every $\xi \in \Gamma^\omega(\mathsf{E})$, $x \in \boldsymbol{\chi}^{-1}(\mathcal{K})$, and $m \in \mathbb{Z}_{\ge 0}$. Then

$$a_0a_1\cdots a_m\|j_m(\boldsymbol{\nu}_*(\xi))(\boldsymbol{\chi}(x))\|_{2,m} \le \frac{Ca_0a_1\cdots a_m}{\sigma^m}\|j_m\xi(x)\|_{\mathbb{G}_{\mathsf{M},\pi_\mathsf{E}},m}$$

for every $\xi \in \Gamma^\omega(\mathsf{E})$, $x \in \boldsymbol{\chi}^{-1}(\mathcal{K})$, and $m \in \mathbb{Z}_{\ge 0}$. Define $\boldsymbol{b} \in c_0(\mathbb{Z}_{\ge 0}; \mathbb{R}_{>0})$ by

$$b_0 = Ca_0, \ b_j = \frac{a_j}{\sigma}, \qquad j \in \mathbb{Z}_{>0}.$$

Then, taking supremums of the preceding inequality gives

$$q'^{\omega}_{\mathcal{K},\boldsymbol{a}} \circ \boldsymbol{\nu}_*(\xi) \le p^\omega_{\boldsymbol{\chi}^{-1}(\mathcal{K}),\boldsymbol{b}}(\xi)$$

for $\xi \in \Gamma^\omega(\mathsf{E})$.

Now let $\mathcal{K} \subseteq \mathsf{M}$ be compact and let $\boldsymbol{a} \in c_0(\mathbb{Z}_{\ge 0}; \mathbb{R}_{>0})$. Let $x \in \mathcal{K}$ and let $(\mathcal{V}_x, \boldsymbol{\nu}_x)$ be a vector bundle chart for E with $(\mathcal{U}_x, \boldsymbol{\chi}_x)$ the chart for M with $x \in \mathcal{U}_x$. We suppose that $\mathcal{U}_x$ is precompact and that, by our appropriate version of Lemma 4.22, there exist $C_x, \sigma_x \in \mathbb{R}_{>0}$ such that

$$\|j_m\xi(y)\|_{\mathbb{G}_{\mathsf{M},\pi_\mathsf{E}},m} \le \frac{C_x}{\sigma_x^m}\|j_m(\boldsymbol{\nu}_*\xi)(y)\|_{m,2}$$

for $\xi \in \Gamma^\omega(\mathsf{E})$, $y \in \mathrm{cl}(\mathcal{U}_x)$, $m \in \mathbb{Z}_{\geq 0}$. Therefore,

$$a_0 a_1 \cdots a_m \| j_m \xi(y) \|_{\mathbb{G}_{\pi_{\mathsf{E}},m}} \leq \frac{C_x a_0 a_1 \cdots a_m}{\sigma_x^m} \| j_m(\boldsymbol{\nu}_* \xi)(y) \|_{m,2}$$

for $\xi \in \Gamma^\omega(\mathsf{E})$, $y \in \mathrm{cl}(\mathcal{U}_x)$, $m \in \mathbb{Z}_{\geq 0}$. Compactness of $\mathcal{K}$ gives $x_1, \dots, x_s \in \mathcal{K}$ such that $\mathcal{K} \subseteq \cup_{j=1}^s \mathcal{U}_{x_j}$ and we then take

$$C = \max\{C_{x_1}, \dots, C_{x_s}\}, \quad \sigma = \min\{\sigma_{x_1}, \dots, \sigma_{x_s}\}.$$

We define $\boldsymbol{b} \in c_0(\mathbb{Z}_{\geq 0}; \mathbb{R}_{>0})$ by

$$b_0 = C a_0, \ b_j = \frac{a_j}{\sigma}, \qquad j \in \mathbb{Z}_{>0}.$$

We then arrive at the inequality

$$p^\omega_{\mathcal{K},\boldsymbol{a}}(\xi) \leq q'^\omega_{\mathrm{cl}(\mathcal{U}_{x_1}),\boldsymbol{b}} \circ \boldsymbol{\nu}_{x_1 *}(\xi) + \dots + q'^\omega_{\mathrm{cl}(\mathcal{U}_{x_s}),\boldsymbol{b}} \circ \boldsymbol{\nu}_{x_s *}(\xi)$$

which is valid for $\xi \in \Gamma^\omega(\mathsf{E})$. □

Since there is some slightly tormented logic necessitated by our invocation of local estimates, it is perhaps worth making it clear how the pertinent results are logically interconnected. This we do in the following diagram:

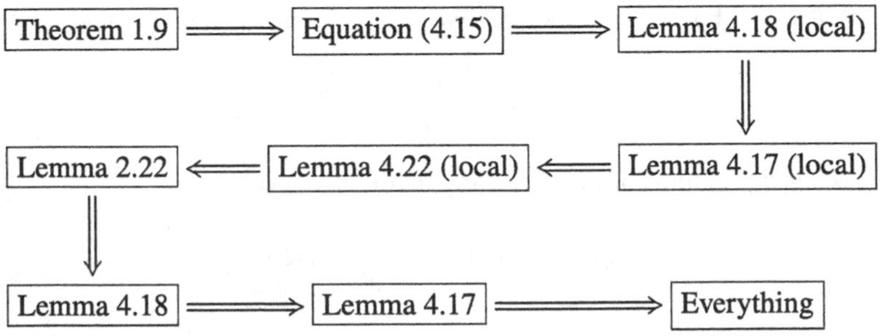

(4.16)

Remark 4.25 (Adaptation to the Smooth Case) An analogous version of the preceding theorem holds in the smooth case. The proof is somewhat simpler in the smooth case, unlike in the proof of Lemma 4.22 where the smooth case is significantly simpler than the real analytic case. Note also that, in the smooth case, one does not need the local estimates for derivatives of real analytic functions, so this also significantly simplifies the logic. ○

An immediate consequence of the theorem is the following.

Corollary 4.26 (Independence of Topologies on Connections and Metrics) *Let $\pi_{\mathsf{E}}\colon \mathsf{E} \to \mathsf{M}$ be a real analytic vector bundle. Let $\mathbb{G}_{\mathsf{M}}$ and $\overline{\mathbb{G}}_{\mathsf{M}}$ be C^{ω}-Riemannian metrics on M, let $\mathbb{G}_{\pi_{\mathsf{E}}}$ and $\overline{\mathbb{G}}_{\pi_{\mathsf{E}}}$ be C^{ω}-fibre metrics on E, let ∇^{M} and $\overline{\nabla}^{\mathsf{M}}$ be C^{ω}-affine connections on M, and let $\nabla^{\pi_{\mathsf{E}}}$ and $\overline{\nabla}^{\pi_{\mathsf{E}}}$ be C^{ω}-linear connections on E, giving rise to seminorms $p^{\omega}_{\mathcal{K},\boldsymbol{a}}$ and $\overline{p}^{\omega}_{\mathcal{K},\boldsymbol{a}}$, $\mathcal{K} \subseteq \mathsf{M}$ compact, $\boldsymbol{a} \in \mathrm{c}_0(\mathbb{Z}_{\geq 0}; \mathbb{R}_{>0})$.*

Then the topologies defined by the two families of seminorms

$$p^{\omega}_{\mathcal{K},\boldsymbol{a}}, \overline{p}^{\omega}_{\mathcal{K},\boldsymbol{a}}, \qquad \mathcal{K} \subseteq \mathsf{M} \text{ compact},\ \boldsymbol{a} \in \mathrm{c}_0(\mathbb{Z}_{\geq 0}; \mathbb{R}_{>0}),$$

agree.

Chapter 5
Continuity of Some Standard Geometric Operations

In this chapter we put to use the developments of the preceding two chapters to prove the continuity of a number of standard algebraic and differential operations on real analytic manifolds. The reader will notice as they go through the proofs that there are definite themes that emerge from the various proofs of continuity. Let us highlight these here, since our belief is that these themes are important in and of themselves, and justify the rather complex developments of Chaps. 3 and 4. We particularly draw attention to the following recurrent ideas.

1. We take full advantage of the algebraic bounds from Sect. 4.2.1 that were nominally developed to prove the bounds of Sect. 4.2.2. This gives independent interest to these algebraic bounds.
2. We make repeated use of the estimate from Lemma 4.17. As we discuss in the preamble to Sect. 4.2, the fact that Lemma 4.17 lies at the core of so much of what we do gives some motivation for the quite tedious geometric constructions from Chap. 3.
3. We shall prove in many cases that certain linear mappings between spaces of sections of real analytic vector bundles are continuous and open onto their image, i.e., homeomorphisms onto their image. One might hope to do this with a general Open Mapping Theorem. Indeed, since the space of real analytic sections of a vector bundle is both webbed and ultrabornological, one is perhaps in a position to use the Open Mapping Theorem of De Wilde [16]. However, since the images of our mappings are not necessarily ultrabornological (even closed subspaces of ultrabornological spaces may not be ultrabornological), we typically prove the openness by a direct argument, by virtue of our having given in Sect. 3.3 relations between iterated covariant derivatives going "both ways." Moreover, the use of seminorms to prove these results is in keeping with the general tenor of this work.

A. D. Lewis, *Geometric Analysis on Real Analytic Manifolds*, Lecture Notes in Mathematics 2333, https://doi.org/10.1007/978-3-031-37913-0_5

As we have indicated as we have been going along, the results in this chapter are applicable to (and easier in) the smooth case. At various points, we shall indicate the modifications required to adapt our approach to the smooth case.

5.1 Continuity of Algebraic Operations and Constructions

In this section we consider topological aspects of two sorts of operations on spaces of sections. First we consider the standard constructions of linear algebra: addition and "multiplication." Then we consider the relationships between direct sums and tensor products of, on the one hand, vector bundles, and, on the other hand, spaces of sections.

5.1.1 Continuity of Linear Algebraic Operations

We begin with a consideration of continuity of standard algebraic operations with vector bundles.

Theorem 5.1 (Continuity of Algebraic Operations) *Let* $\pi_{\mathsf{E}}\colon \mathsf{E} \to \mathsf{M}$ *and* $\pi_{\mathsf{F}}\colon \mathsf{F} \to \mathsf{M}$ *be* C^{ω}*-vector bundles. Then the following mappings are continuous:*

(i) $\Gamma^{\omega}(\mathsf{E}) \oplus \Gamma^{\omega}(\mathsf{E}) \ni (\xi, \eta) \mapsto \xi + \eta \in \Gamma^{\omega}(\mathsf{E})$;
(ii) $\Gamma^{\omega}(\mathsf{F} \otimes \mathsf{E}^*) \times \Gamma^{\omega}(\mathsf{E}) \ni (L, \xi) \mapsto L \circ \xi \in \Gamma^{\omega}(\mathsf{F})$.

Also, fixing a vector bundle mapping $L \in \Gamma^{\omega}(\mathsf{F} \otimes \mathsf{E}^*)$*, the following statements hold:*

(iii) if L *is injective, then the mapping* $\Gamma^{\omega}(\mathsf{E}) \ni \xi \mapsto L \circ \xi \in \Gamma^{\omega}(\mathsf{F})$ *is a topological monomorphism;*
(iv) if L *is surjective, then the mapping* $\Gamma^{\omega}(\mathsf{E}) \ni \xi \mapsto L \circ \xi \in \Gamma^{\omega}(\mathsf{F})$ *is a topological epimorphism.*

Proof We suppose that we have a real analytic affine connection ∇^{M} on M, and real analytic vector bundle connections $\nabla^{\pi_{\mathsf{E}}}$ and $\nabla^{\pi_{\mathsf{F}}}$ in E and F, respectively. We suppose that we have a real analytic Riemannian metric $\mathbb{G}_{\mathsf{M}}$ on M, and real analytic fibre metrics $\mathbb{G}_{\pi_{\mathsf{E}}}$ and $\mathbb{G}_{\pi_{\mathsf{F}}}$ on E and F, respectively. This gives the seminorms $p^{\omega}_{\mathcal{K},\boldsymbol{a}}$ and $q^{\omega}_{\mathcal{K},\boldsymbol{a}}$, $\mathcal{K} \subseteq \mathsf{M}$ compact, $\boldsymbol{a} \in \mathrm{c}_0(\mathbb{Z}_{\geq 0}; \mathbb{R}_{>0})$, for $\Gamma^{\omega}(\mathsf{E})$ and $\Gamma^{\omega}(\mathsf{F})$, respectively. We denote the induced seminorms for $\Gamma^{\omega}(\mathsf{F}\otimes\mathsf{E}^*)$ by $q^{\omega}_{\mathcal{K},\boldsymbol{a}} \otimes p^{\omega}_{\mathcal{K},\boldsymbol{a}}$, $\mathcal{K} \subseteq \mathsf{M}$ compact, $\boldsymbol{a} \in \mathrm{c}_0(\mathbb{Z}_{\geq 0}; \mathbb{R}_{>0})$.

(i) The fibre norms from Sect. 2.3.2 satisfy the triangle inequality, and this readily gives

$$p^{\omega}_{\mathcal{K},\boldsymbol{a}}(\xi + \eta) \leq p^{\omega}_{\mathcal{K},\boldsymbol{a}}(\xi) + p^{\omega}_{\mathcal{K},\boldsymbol{a}}(\eta),$$

which immediately gives this part of the result.

(ii) Let us make some preliminary computations from which this part of the theorem will follow easily.

First, by Lemma 4.11, we have

$$\|L \circ \xi(x)\|_{\mathbb{G}_{M,\pi_F}} \le \|L(x)\|_{\mathbb{G}_{M,\pi_F \otimes \pi_E}} \|\xi(x)\|_{\mathbb{G}_{M,\pi_E}}. \tag{5.1}$$

Next, by Lemmata 3.15, 4.11, and 4.13, we have

$$\|D^k_{\nabla^M, \nabla^{\pi_F}}(L \circ \xi(x))\|_{\mathbb{G}_{M,\pi_F}} \le \sum_{j=0}^{k} \binom{k}{j} \|D^j_{\nabla^M, \nabla^{\pi_F \otimes \pi_E}} L(x)\|_{\mathbb{G}_{M,\pi_F \otimes \pi_E}} \|D^{k-j}_{\nabla^M, \nabla^{\pi_E}} \xi(x)\|_{\mathbb{G}_{M,\pi_E}}$$

for $k \in \mathbb{Z}_{>0}$. By (1.4) (twice) we have

$$\begin{aligned}
\|j_m(L \circ \xi)(x)\|_{\mathbb{G}_{M,\pi_F,m}}
&\le \sum_{k=0}^{m} \frac{1}{k!} \|D^k_{\nabla^M, \nabla^{\pi_F}}(L \circ \xi(x))\|_{\mathbb{G}_{M,\pi_F}} \\
&\le \sum_{k=0}^{m} \frac{1}{k!} \sum_{j=0}^{k} \binom{k}{j} \|D^j_{\nabla^M, \nabla^{\pi_F \otimes \pi_E}} L(x)\|_{\mathbb{G}_{M,\pi_F \otimes \pi_E}} \|D^{k-j}_{\nabla^M, \nabla^{\pi_E}} \xi(x)\|_{\mathbb{G}_{M,\pi_E}} \\
&= \sum_{k=0}^{m} \sum_{j=0}^{k} \frac{\|D^j_{\nabla^M, \nabla^{\pi_F \otimes \pi_E}} L(x)\|_{\mathbb{G}_{M,\pi_F \otimes \pi_E}}}{j!} \frac{\|D^{k-j}_{\nabla^M, \nabla^{\pi_E}} \xi(x)\|_{\mathbb{G}_{M,\pi_E}}}{(k-j)!} \\
&\le (m+1)^2 \sup \left\{ \frac{\|D^j_{\nabla^M, \nabla^{\pi_F \otimes \pi_E}} L(x)\|_{\mathbb{G}_{M,\pi_F \otimes \pi_E}}}{j!} \;\middle|\; j \le m \right\} \\
&\quad \times \sup \left\{ \frac{\|D^{k-j}_{\nabla^M, \nabla^{\pi_E}} \xi(x)\|_{\mathbb{G}_{M,\pi_E}}}{(k-j)!} \;\middle|\; j \le m \right\} \\
&\le (m+1)^{5/2} \|j_m L(x)\|_{\mathbb{G}_{M,\pi_F \otimes \pi_E,m}} \|j_m \xi(x)\|_{\mathbb{G}_{M,\pi_E,m}}.
\end{aligned}$$

Noting that $(m+1)^{5/2} \le 3^{m+1}$, $m \in \mathbb{Z}_{>0}$, we finally get

$$\|j_m(L \circ \xi)(x)\|_{\mathbb{G}_{M,\pi_F,m}} \le 3^{m+1} \|j_m L(x)\|_{\mathbb{G}_{M,\pi_F \otimes \pi_E,m}} \|j_m \xi(x)\|_{\mathbb{G}_{M,\pi_E,m}}. \tag{5.2}$$

Let $\mathcal{K} \subseteq \mathsf{M}$ be compact and let $\boldsymbol{a} \in c_0(\mathbb{Z}_{\geq 0}; \mathbb{R}_{>0})$. Define $\boldsymbol{a}' \in c_0(\mathbb{Z}_{\geq 0}; \mathbb{R}_{>0})$ by $a'_j = \sqrt{3}a_j$, $j \in \mathbb{Z}_{\geq 0}$. We then have

$$q^{\omega}_{\mathcal{K},\boldsymbol{a}}(L \circ \xi) \leq q^{\omega}_{\mathcal{K},\boldsymbol{a}'} \otimes p^{\omega}_{\mathcal{K},\boldsymbol{a}'}(L) p^{\omega}_{\mathcal{K},\boldsymbol{a}'}(\xi) = (q^{\omega}_{\mathcal{K},\boldsymbol{a}'} \otimes p^{\omega}_{\mathcal{K},\boldsymbol{a}'}) \otimes p^{\omega}_{\mathcal{K},\boldsymbol{a}'}(L \otimes \xi).$$

By Jarchow [37, Theorem 15.1.2], this gives continuity of the bilinear map $(L, \xi) \mapsto L \circ \xi$.

To prove the last two parts of the theorem, we first prove a couple of technical lemmata.

Lemma 1 *Let* U *and* V *be locally convex topological vector spaces, and let* $L \in \mathrm{L}(\mathsf{U}; \mathsf{V})$. *If, for every continuous seminorm* q *for* U, *there exists a continuous seminorm* p *for* V *such that*

$$q(u) \leq p \circ L(u), \qquad u \in \mathsf{U},$$

then L *is a topological homomorphism.*

Proof Let us denote by $\mathscr{Q}$ the set of continuous seminorms for U and by $\mathscr{P}$ the set of continuous seminorms for V. Our hypothesis is that, for each $q \in \mathscr{Q}$, there exists $p_q \in \mathscr{P}$ such that $q(u) \leq p_q \circ L(u)$ for all $u \in \mathsf{U}$. We fix such a choice of p_q for each $q \in \mathscr{Q}$.

We first prove that there are 0-bases $\mathscr{B}_{\mathsf{U}}$ for U and $\mathscr{B}_{\mathsf{V}}$ for V such that, for each $\mathcal{B} \in \mathscr{B}_{\mathsf{U}}$, there exists $\mathcal{C} \in \mathscr{B}_{\mathsf{V}}$ such that

$$\mathcal{C} \cap \mathrm{image}(L) \subseteq L(\mathcal{B}).$$

To see this, first let $q \in \mathscr{Q}$ and let $p_q \in \mathscr{P}$ be as hypothesised. Then

$$p_q \circ L(u) < 1 \implies q(u) < 1 \implies L(u) \in L(q^{-1}([0, 1))).$$

Thus

$$p_q^{-1}([0, 1)) \cap \mathrm{image}(L) \subseteq L(q^{-1}([0, 1))).$$

Now let $\mathscr{B}_{\mathsf{U}}$ be the collection of all 0-neighbourhoods of the form

$$\mathcal{B} = \bigcap_{j=1}^{k} q_j^{-1}([0, 1)), \qquad q_j \in \mathscr{Q},\ j \in \{1, \dots, k\},\ k \in \mathbb{Z}_{>0}.$$

This is a 0-base for U. Then, by our above computations,

$$\left(\bigcap_{j=1}^{k} p_{q_j}^{-1}([0, 1))\right) \cap \mathrm{image}(L) \subseteq L\left(\bigcap_{j=1}^{k} q_j^{-1}([0, 1))\right).$$

Thus, the 0-base

$$\bigcap_{j=1}^{k} p_j^{-1}([0,1)), \qquad p_j \in \mathscr{P},\ j \in \{1,\dots,k\},\ k \in \mathbb{Z}_{>0},$$

for V has the desired property.

Now let $\mathcal{O} \subseteq \mathsf{V}$ be open and let $u \in \mathcal{O}$. Let $\mathcal{B} \in \mathscr{B}_{\mathsf{U}}$ be such that $u + \mathcal{B} \subseteq \mathcal{O}$ and let $\mathcal{C} \in \mathscr{B}_{\mathsf{V}}$ be such that $\mathcal{C} \cap \operatorname{image}(L) \subseteq L(\mathcal{B})$. Then

$$L(u) + \mathcal{C} \cap \operatorname{image}(L) \subseteq L(u) + L(\mathcal{B}) = L(u + \mathcal{B}) \subseteq L(\mathcal{O}).$$

Thus $L(u) + \mathcal{C} \cap \operatorname{image}(L)$ is a neighbourhood of $L(u)$ in $L(\mathcal{O})$ which shows that $L(\mathcal{O})$ is open in $\operatorname{image}(L)$. ▽

Lemma 2 *If $\pi_{\mathsf{E}}\colon \mathsf{E} \to \mathsf{M}$ and $\pi_{\mathsf{F}}\colon \mathsf{F} \to \mathsf{M}$ are C^{ω}-vector bundles, and if $L \in \Gamma^{\omega}(\mathsf{F} \otimes \mathsf{E}^*)$, then the following statements hold:*

(i) if L is injective, then there exists a left-inverse $L' \in \Gamma^{\omega}(\mathsf{E} \otimes \mathsf{F}^*)$;
(ii) if L is surjective, then the mapping $\Gamma^{\omega}(\mathsf{E}) \ni \xi \mapsto L \circ \xi \in \Gamma^{\omega}(\mathsf{F})$ is surjective.

Proof

(i) First we note that $\operatorname{image}(L)$ is a C^{ω}-subbundle of F and that L is a C^{ω}-vector bundle isomorphism onto $\operatorname{image}(L)$, cf. [1, Proposition 3.4.18]. Let $\mathsf{G} \subseteq \mathsf{F}$ be the $\mathbb{G}_{\pi_{\mathsf{F}}}$-orthogonal complement to $\operatorname{image}(L)$ which is then itself a C^{ω}-subbundle of F. Clearly, $\mathsf{F} = \operatorname{image}(L) \oplus \mathsf{G}$. Let

$$\begin{aligned} L'\colon \operatorname{image}(L) \oplus \mathsf{G} &\to \mathsf{E} \\ (L(e), g) &\mapsto e, \end{aligned}$$

and note that L' is obviously a left-inverse of L. It is also of class C^{ω} since the projection from F to the summand $\operatorname{image}(L)$ is of class C^{ω}.

(ii) Note that $\ker(L)$ is a C^{ω}-subbundle of E, cf. [1, Proposition 3.4.18]. Using a real analytic fibre metric as in the preceding part of the proof, let G be a real analytic subbundle of E for which $\mathsf{E} = \ker(L) \oplus \mathsf{G}$. Then $L|\mathsf{G}$ is a vector bundle isomorphism onto F. Thus the mapping $\Gamma^{\omega}(\mathsf{G}) \ni \xi \mapsto L \circ \xi \in \Gamma^{\omega}(\mathsf{F})$ is a vector space isomorphism. ▽

(iii) By Lemma 2(i), we suppose that there is a C^{ω}-vector bundle mapping L' that is a left-inverse for L. Then, from the first part of the proof, for a compact $\mathcal{K} \subseteq \mathsf{M}$ and for $\boldsymbol{a} \in c_0(\mathbb{Z}_{\geq 0}; \mathbb{R}_{>0})$, let $C \in \mathbb{R}_{>0}$ and $\boldsymbol{a}' \in c_0(\mathbb{Z}_{\geq 0}; \mathbb{R}_{>0})$ be such that

$$p^{\omega}_{\mathcal{K},\boldsymbol{a}}(L' \circ \eta) \leq C q^{\omega}_{\mathcal{K},\boldsymbol{a}'}(\eta), \quad \eta \in \Gamma^{\omega}(\mathsf{F}).$$

We then have, for $\xi \in \Gamma^\omega(\mathsf{E})$,

$$p^\omega_{\mathcal{K},\boldsymbol{a}}(\xi) = p^\omega_{\mathcal{K},\boldsymbol{a}}(L' \circ L \circ \xi) \leq C q^\omega_{\mathcal{K},\boldsymbol{a}'}(L \circ \xi).$$

By Lemma 1, this suffices to establish that L is open onto its image.

(iv) By Lemma 2(ii) and since the set of real analytic sections of a vector bundle, with the C^ω-topology, is a webbed and ultrabornological locally convex topological vector space (Proposition 2.14), this part of the result follows from the De Wilde Open Mapping Theorem. □

Remarks 5.2 (Adaptation to the Smooth Case) The preceding proof is easily adapted to the smooth case. Indeed, the proof is a little easier since one does not need to carefully keep track of the growth in m of the coefficient of the norm of the m-jet. ∘

The theorem admits the following corollary, where we use the topological notions of monomorphism and epimorphism from Definition 1.7 for exactness of sequences of locally convex topological vector spaces.

Corollary 5.3 (Short Exact Sequences of Spaces of Sections Induced by Short Exact Sequences of Vector Bundles) *Let* $\pi_{\mathsf{E}}\colon \mathsf{E} \to \mathsf{M}$, $\pi_{\mathsf{F}}\colon \mathsf{F} \to \mathsf{M}$, *and* $\pi_{\mathsf{G}}\colon \mathsf{G} \to \mathsf{M}$ *be real analytic vector bundles and suppose that we have a short exact sequence*

$$0 \longrightarrow \mathsf{E} \xrightarrow{\ \Psi\ } \mathsf{F} \xrightarrow{\ \Phi\ } \mathsf{G} \longrightarrow 0$$

of real analytic vector bundle mappings. Then the sequence

$$0 \longrightarrow \Gamma^r(\mathsf{E}) \xrightarrow{\ \xi \mapsto \Psi \circ \xi\ } \Gamma^\omega(\mathsf{F}) \xrightarrow{\ \eta \mapsto \Phi \circ \eta\ } \Gamma^\omega(\mathsf{G}) \longrightarrow 0$$

is short exact in the category of locally convex topological vector spaces.

A consequence of the corollary, and one that one might use without thinking about it, is that isomorphic vector bundles give rise to isomorphic (as locally convex topological vector spaces) spaces of sections.

5.1.2 Direct Sums and Tensor Products of Spaces of Sections

Given real analytic vector bundles $\pi_{\mathsf{E}}\colon \mathsf{E} \to \mathsf{M}$ and $\pi_{\mathsf{F}}\colon \mathsf{F} \to \mathsf{M}$, we can build two real analytic vector bundles, the direct sum $\mathsf{E} \oplus \mathsf{F}$ and the tensor product $\mathsf{E} \otimes \mathsf{F}$. We consider in this section topological matters associated with the spaces of sections of these vector bundles.

First we consider direct sums, where the situation is simple.

Theorem 5.4 (Topology for Space of Sections of Direct Sums) *If* $\pi_{\mathsf{E}} \colon \mathsf{E} \to \mathsf{M}$ *and* $\pi_{\mathsf{F}} \colon \mathsf{F} \to \mathsf{M}$ *are real analytic vector bundles, then the mapping*

$$\Gamma^{\omega}(\mathsf{E} \oplus \mathsf{F}) \ni (x \mapsto \xi(x) \oplus \eta(x)) \mapsto \xi \oplus \eta \in \Gamma^{\omega}(\mathsf{E}) \oplus \Gamma^{\omega}(\mathsf{F}),$$
$$\xi \in \Gamma^{\omega}(\mathsf{E}),\ \eta \in \Gamma^{\omega}(\mathsf{F}),$$

is an isomorphism of locally convex topological vector spaces. Moreover, the projections onto the components of the direct sum on the right are topological epimorphisms and the inclusions of the components of the direct sum into the direct sum on the right are topological monomorphisms.

Proof Denote by $p^{\omega}_{\mathcal{K},\boldsymbol{a}}$ and $q^{\omega}_{\mathcal{K},\boldsymbol{a}}$, $\mathcal{K} \subseteq \mathsf{M}$ compact, $\boldsymbol{a} \in c_0(\mathbb{Z}_{\geq 0}; \mathbb{R}_{>0})$, the seminorms for $\Gamma^{\omega}(\mathsf{E})$ and $\Gamma^{\omega}(\mathsf{F})$, respectively.

The mapping given in the statement of the theorem is certainly an algebraic isomorphism. Note that fibre metrics $\mathbb{G}_{\mathsf{E}}$ and $\mathbb{G}_{\mathsf{F}}$ give rise to the fibre metric

$$\mathbb{G}_{\mathsf{E}\oplus\mathsf{F}}(e_x \oplus f_x, e'_x \oplus f'_x) = \mathbb{G}_{\mathsf{E}}(e_x, e'_x) + \mathbb{G}_{\mathsf{F}}(f_x, f'_x)$$

for $\mathsf{E} \oplus \mathsf{F}$. The corresponding set of seminorms for $\Gamma^{\omega}(\mathsf{E} \oplus \mathsf{E})$ we denote by $r^{\omega}_{\mathcal{K},\boldsymbol{a}}$, $\mathcal{K} \subseteq \mathsf{M}$ compact, $a \in c_0(\mathbb{Z}_{\geq 0}; \mathbb{R}_{>0})$, for $\mathsf{E} \oplus \mathsf{F}$. A set of seminorms defining the topology of $\Gamma^{\omega}(\mathsf{E}) \oplus \Gamma^{\omega}(\mathsf{F})$ can be taken to be

$$\xi \oplus \eta \mapsto \max\{p^{\omega}_{\mathcal{K},\boldsymbol{a}}(\xi), q^{\omega}_{\mathcal{L},\boldsymbol{b}}(\eta)\}, \quad \mathcal{K}, \mathcal{L} \subseteq \mathsf{M} \text{ compact}, \ \boldsymbol{a}, \boldsymbol{b} \in c_0(\mathbb{Z}_{\geq 0}; \mathbb{R}_{>0}).$$

Let $\mathcal{K} \subseteq \mathsf{M}$ be compact and let $\boldsymbol{a} \in c_0(\mathbb{Z}_{\geq 0}; \mathbb{R}_{>0})$. The inequality (which makes use of (1.4))

$$\max\{p^{\omega}_{\mathcal{K},\boldsymbol{a}}(\xi), q^{\omega}_{\mathcal{L},\boldsymbol{b}}(\eta)\} \leq r^{\omega}_{\mathcal{K}\cup\mathcal{L},\boldsymbol{c}}(x \mapsto \xi(x) \oplus \eta(x)), \quad \xi \in \Gamma^{\omega}(\mathsf{E}),\ \eta \in \Gamma^{\omega}(\mathsf{F}),$$

with $\boldsymbol{c} \in c_0(\mathbb{Z}_{\geq 0}; \mathbb{R}_{>0})$ defined by $c_m = \max\{a_m, b_m\}$, $m \in \mathbb{Z}_{\geq 0}$, establishes the continuity of the mapping in the statement of the theorem.

For the continuity of the inverse, we let $\mathcal{K} \subseteq \mathsf{M}$ be compact and let $\boldsymbol{a} \in c_0(\mathbb{Z}_{\geq 0}, \mathbb{R}_{>0})$, and make use of the inequality

$$r^{\omega}_{\mathcal{K},\boldsymbol{a}}(x \mapsto \xi(x) \oplus \eta(x)) \leq \sqrt{2} \max\{p^{\omega}_{\mathcal{K},\boldsymbol{a}}(\xi), q^{\omega}_{\mathcal{K},\boldsymbol{a}}(\eta)\}, \quad \xi \in \Gamma^{\omega}(\mathsf{E}),\ \eta \in \Gamma^{\omega}(\mathsf{F}),$$

which again uses (1.4).

The final assertion of the theorem follows from general facts about finite direct sums [65, §2]. □

For tensor products, the situation is more complicated, and is worked out in the smooth case in [57, Chapter 12]. The first thing one must do is understand the algebraic setting properly, something which is interesting in its own right. In the next result we are working with the $\mathrm{C}^\omega(\mathsf{M})$-module structure of the set of sections of a vector bundle, and we use $\otimes_{\mathrm{C}^\omega(\mathsf{M})}$ to denote the tensor product of $\mathrm{C}^\omega(\mathsf{M})$-modules, as opposed to $\otimes$ which denotes the tensor product of $\mathbb{R}$-vector spaces.

Proposition 5.5 (Sections of a Tensor Product of Vector Bundles) *If* $\pi_{\mathsf{E}}\colon \mathsf{E} \to \mathsf{M}$ *and* $\pi_{\mathsf{F}}\colon \mathsf{F} \to \mathsf{M}$ *are real analytic vector bundles, then the bilinear mapping*

$$\Gamma^\omega(\mathsf{E}) \times \Gamma^\omega(\mathsf{F}) \ni (\xi, \eta) \mapsto (x \mapsto \xi(x) \otimes \eta(x)) \in \Gamma^\omega(\mathsf{E} \otimes \mathsf{F})$$

defines (by the universal property of tensor products)[1] *a* $\mathrm{C}^\omega(\mathsf{M})$*-module isomorphism*

$$\Gamma^\omega(\mathsf{E}) \otimes_{\mathrm{C}^\omega(\mathsf{M})} \Gamma^\omega(\mathsf{M}) \simeq \Gamma^\omega(\mathsf{E} \otimes \mathsf{F}).$$

Proof We first consider the case of trivial vector bundles, say $\mathsf{E} = \mathbb{R}^k_{\mathsf{M}}$ and $\mathsf{F} = \mathbb{R}^l_{\mathsf{M}}$. Let $(\boldsymbol{e}_1, \dots, \boldsymbol{e}_k)$ and $(\boldsymbol{f}_1, \dots, \boldsymbol{f}_l)$ be the standard basis for $\mathbb{R}^k$ and $\mathbb{R}^l$, respectively. We then have the sections $\xi_j(x) = (x, \boldsymbol{e}_j)$, $j \in \{1, \dots, k\}$, and $\eta_a(x) = (x, \boldsymbol{f}_a)$, $a \in \{1, \dots, l\}$, which are bases for the $\mathrm{C}^\omega(\mathsf{M})$-modules $\Gamma^\omega(\mathbb{R}^k_{\mathsf{M}})$ and $\Gamma^\omega(\mathbb{R}^l_{\mathsf{M}})$, respectively. One then readily verifies that $\xi_j \otimes \eta_a \in \Gamma^\omega((\mathbb{R}^k \otimes \mathbb{R}^l)_{\mathsf{M}})$ defined by

$$\xi_j \otimes \eta_a(x) = (x, \boldsymbol{e}_j \otimes \boldsymbol{f}_a), \qquad j \in \{1, \dots, k\},\ a \in \{1, \dots, l\},$$

give a basis of $\Gamma^\omega((\mathbb{R}^k \otimes \mathbb{R}^l)_{\mathsf{M}})$ as a $\mathrm{C}^\omega(\mathsf{M})$-module. This establishes the proposition in this case of trivial bundles.

If E and F are not necessarily trivial, then Corollary 2.5 gives E as a subbundle of $\mathbb{R}^k_{\mathsf{M}}$ for a suitable $k \in \mathbb{Z}_{>0}$ and F as a subbundle of $\mathbb{R}^l_{\mathsf{M}}$ for a suitable $l \in \mathbb{Z}_{>0}$. Let $\mathsf{E}^\perp$ and $\mathsf{F}^\perp$ denote the orthogonal complement subbundles using the standard Euclidean fibre metric (for example). We then have the commuting diagram

$$\begin{array}{ccc}
\Gamma^\omega(\mathsf{E} \oplus \mathsf{E}^\perp) \otimes_{\mathrm{C}^\omega(\mathsf{M})} \Gamma^\omega(\mathsf{F} \oplus \mathsf{F}^\perp) & \longrightarrow & \Gamma^\omega((\mathsf{E} \oplus \mathsf{E}^\perp) \otimes (\mathsf{F} \oplus \mathsf{F}^\perp)) \\
{\scriptstyle p}\downarrow\uparrow{\scriptstyle i} & & {\scriptstyle i}\uparrow\downarrow{\scriptstyle p} \\
\Gamma^\omega(\mathsf{E}) \otimes_{\mathrm{C}^\omega(\mathsf{M})} \Gamma^\omega(\mathsf{F}) & \longrightarrow & \Gamma^\omega(\mathsf{E} \otimes \mathsf{F})
\end{array}$$

Here the horizontal arrows are those induced by the map from the statement of the proposition; in particular the upper horizontal arrow is an isomorphism of $\mathrm{C}^\omega(\mathsf{M})$-modules. Note that the vector bundle isomorphism

$$(\mathsf{E} \oplus \mathsf{E}^\perp) \otimes (\mathsf{F} \oplus \mathsf{F}^\perp) \simeq (\mathsf{E} \otimes \mathsf{F}) \oplus (\mathsf{E} \otimes \mathsf{F}^\perp) \oplus (\mathsf{E}^\perp \otimes \mathsf{F}) \oplus (\mathsf{E}^\perp \otimes \mathsf{F}^\perp)$$

---

[1] By this we mean that the universal property of tensor products defines a module homomorphism

$$\Gamma^\omega(\mathsf{E}) \otimes_{\mathrm{C}^\omega(\mathsf{M})} \Gamma^\omega(\mathsf{M}) \to \Gamma^\omega(\mathsf{E} \otimes \mathsf{F}).$$

Thus the assertion of the proposition is that this homomorphism is an isomorphism.

cf. [35, Theorem IV.5.9] gives rise to the $C^\omega(M)$-module isomorphism

$$\Gamma^\omega((E\oplus E^\perp)\otimes(F\oplus F^\perp)) \simeq \Gamma^\omega(E\otimes F)\oplus\Gamma^\omega(E\otimes F^\perp)\oplus\Gamma^\omega(E^\perp\otimes F)\oplus\Gamma^\omega(E^\perp\otimes F^\perp)$$

by Theorem 5.4. By Theorem 5.4 and [35, Theorem IV.5.9] (now for $C^\omega(M)$-modules) we have

$$\begin{aligned}\Gamma^\omega(E \oplus E^\perp) \otimes_{C^\omega(M)} \Gamma^\omega(F \oplus F^\perp)&\\ \simeq (\Gamma^\omega(E) \otimes_{C^\omega(M)} \Gamma^\omega(F)) &\oplus (\Gamma^\omega(E) \otimes_{C^\omega(M)} \Gamma^\omega(F^\perp))\\ &\oplus (\Gamma^\omega(E^\perp) \otimes_{C^\omega(M)} \Gamma^\omega(F)) \oplus (\Gamma^\omega(E^\perp) \otimes_{C^\omega(M)} \Gamma^\omega(F^\perp)).\end{aligned}$$

The vertical arrows in the diagram above are then the associated inclusions and projections arising from the preceding two equations. Moreover, the arrows induced by inclusion are injective while those induced by projection are surjective. With all of this in place, we leave to the reader the routine and tedious verification that the diagram makes sense and commutes in the category of $C^\omega(M)$-modules. We also leave to the reader the routine verification that (1) by considering the vertical inclusions, one shows that the bottom horizontal arrow is injective and (2) by consider the vertical projections, the bottom horizontal arrow is surjective. □

To make the preceding algebraic construction have sense topologically, we need to consider the relationship between $\Gamma^\omega(E) \otimes_{C^\omega(M)} \Gamma^\omega(F)$ and $\Gamma^\omega(E) \otimes \Gamma^\omega(F)$. For the latter tensor product, we have well-defined locally convex topologies, and it is the projective tensor topology of which we shall make use. To connect the two tensor products, we make a few constructions. It is convenient to make these constructions in a fairly general setting, at least for the algebraic part of the construction.

First we have the following definition.

Definition 5.6 (Balanced Bilinear Map) Let R be a commutative unit ring, and let A, B, and C be R-modules. A $\mathbb{Z}$-bilinear mapping $\beta\colon A \times B \to C$ is ***R-balanced*** if

$$\beta(ra, b) = \beta(a, rb), \qquad r \in R,\ a \in A,\ b \in B. \qquad \circ$$

The following lemma explains the importance of balanced mappings.

Lemma 5.7 (Universal Property of Tensor Product Over Different Rings) *Let* R *be a commutative unit ring, let* $S \subseteq R$ *be a subring for which* $1 \in S$*, and let* A *and* B *be* R*-modules. Then there exists an* S*-module* X *and an* R*-balanced,* S*-bilinear map* $\gamma\colon A \times B \to X$ *such that, for every* S*-module* C *and every* R*-balanced,* S*-bilinear map* $\beta\colon A \times B \to C$*, there exists a unique* $\alpha \in \mathrm{Hom}_S(X; C)$ *such that the*

diagram

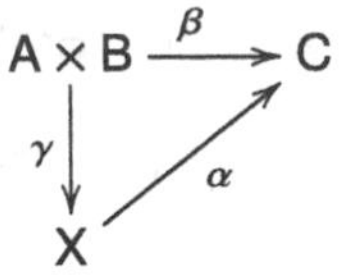

commutes.

Proof The construction of the S-module X, and the verification of its uniqueness, is the same as the standard construction of the tensor product $A \otimes_R B$, e.g., [35, §5], except that we work with S-modules rather than $\mathbb{Z}$-modules. Similarly, the universal property as stated for R-balanced, S-bilinear maps is verified as in the usual construction. □

As indicated in our outline of the proof, the S-module X one gets from the previous lemma is the same as the tensor product $A \otimes_R B$, except that it has the structure of an S-module and not simply a $\mathbb{Z}$-module. In the case $R = S$, the previous lemma yields the usual S-module tensor product $A \otimes_S B$ that is universal for S-bilinear mappings.

We denote the (unique up to S-module isomorphism) module X whose existence is asserted in the lemma by $A \otimes_R B$, with the rôle of S suppressed in the notation. We also let

$$\otimes_R : A \times B \to A \otimes_R B$$

be the associated R-balanced, S-bilinear map.

The rôle of S in the constructions is further clarified by the following result.

Lemma 5.8 (Tensor Products of Modules Over Different Rings) *Let* R *be a commutative unit ring, let* $S \subseteq R$ *be a subring for which* $1 \in S$*, and let* A, B, *and* C *be* R*-modules. Let* J *be the* $\mathbb{Z}$*-submodule of* $A \otimes_S B$ *generated by elements of the form*

$$(ra) \otimes_S b - a \otimes_S (rb), \qquad r \in R,\ a \in A,\ b \in B.$$

Then the mapping from $A \otimes_R B$ *to* $(A \otimes_S B)/J$ *defined by*

$$a \otimes_R b \mapsto a \otimes_S b + J$$

is an isomorphism of S*-modules.*

Proof First let us check that the statement makes sense in that the domain and codomain of the asserted isomorphism are S-modules. Since $A \otimes_R B$ is an R-module, it is certainly an S-module. To show that $(A \otimes_S B)/J$ has an S-module structure, we need only show that J is an S-submodule. To see this, we claim that the generators for J remain in J upon multiplication with S. Indeed,

$$s((ra)\otimes_S b - a\otimes_S (rb)) = r(sa)\otimes_S b - (sa)\otimes_S (rb), \ \ r \in \mathsf{R},\ s \in \mathsf{S},\ a \in \mathsf{A},\ b \in \mathsf{B}.$$

Now let

$$\otimes_R : \mathsf{A} \times \mathsf{B} \to \mathsf{A} \otimes_R \mathsf{B}, \quad \otimes_S : \mathsf{A} \times \mathsf{B} \to \mathsf{A} \otimes_S \mathsf{B}$$

be the canonical bilinear maps, the first using the R-module structure of A and B, and the second using the S-module structure. The universal property of the S-module tensor product gives a unique S-module homomorphism λ as in the diagram

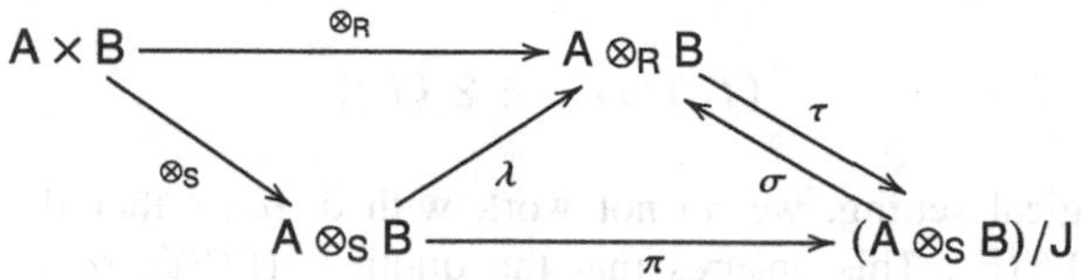

The diagram also introduces us to the projection π, which is an homomorphism of S-modules. Note that

$$\lambda((ra) \otimes_S b) = (ra) \otimes_R b = a \otimes_R (rb) = \lambda(a \otimes_S (rb)),$$

and so $\mathsf{J} \subseteq \ker(\lambda)$. Then the S-module homomorphism σ is that induced by the universal property of quotients [35, Theorem IV.1.7]. To define τ, let $r \in \mathsf{R}$, $a \in \mathsf{A}$, and $b \in \mathsf{B}$, so that $(ra) \otimes_S b - a \otimes_S (rb) \in \mathsf{J}$. Thus

$$\pi((ra) \otimes_S b) = \pi(a \otimes_S (rb)) \quad \Longrightarrow \quad \pi \circ \otimes_S (ra, b) = \pi \circ \otimes_S (a, rb);$$

that is, $\pi \circ \otimes_S$ is R-balanced and S-bilinear. By Lemma 5.7, there exists an S-module homomorphism $\tau : \mathsf{A} \otimes_R \mathsf{S} \to (\mathsf{A} \otimes_S \mathsf{B})/\mathsf{J}$ such that $\tau \circ \otimes_R = \pi \circ \otimes_S$. We have

$$\otimes_R = \lambda \circ \otimes_S = \sigma \circ \tau \circ \otimes_S = \sigma \circ \tau \circ \otimes_R,$$

and the uniqueness assertion of Lemma 5.7 in the special case $\mathsf{R} = \mathsf{S}$ gives $\sigma \circ \tau$ as the identity mapping of S-modules on $(\mathsf{A} \otimes_S \mathsf{B})/\mathsf{J}$. Again by the uniqueness assertion of Lemma 5.7 in the case that $\mathsf{R} = \mathsf{S}$,

$$\pi \circ \otimes_S = \tau \circ \otimes_R = \tau \circ \lambda \circ \otimes_S = \tau \circ \sigma \circ \pi \circ \otimes_S \implies \pi = \tau \circ \sigma \circ \pi.$$

Since π is surjective, it is right-invertible and so $\tau \circ \sigma$ is the identity mapping on the S-module $(\mathsf{A} \otimes_S \mathsf{B})/\mathsf{J}$. Therefore, τ is an S-module isomorphism. □

Now we work our way back from algebra to topology. One can do this in the setting of topological modules over topological rings [57, Chapter 10]. However, we do not undertake this level of generality for the restricted use we shall make of these developments. In any case, a slightly imaginative reader can see how a general theory will work based on our specific application.

We let $\pi_{\mathsf{E}} \colon \mathsf{E} \to \mathsf{M}$ and $\pi_{\mathsf{F}} \colon \mathsf{F} \to \mathsf{M}$ be real analytic vector bundles. The preceding constructions can be applied to the case of $\mathsf{S} = \mathbb{R}$, $\mathsf{R} = \mathrm{C}^{\omega}(\mathsf{M})$, $\mathsf{A} = \Gamma^{\omega}(\mathsf{E})$, and $\mathsf{B} = \Gamma^{\omega}(\mathsf{F})$. Recalling Proposition 5.5, we thus have $\mathbb{R}$-vector space isomorphisms

$$\Gamma^{\omega}(\mathsf{E} \otimes \mathsf{F}) \simeq \Gamma^{\omega}(\mathsf{E}) \otimes_{\mathrm{C}^{\omega}(\mathsf{M})} \Gamma^{\omega}(\mathsf{F}) \simeq (\Gamma^{\omega}(\mathsf{E}) \otimes \Gamma^{\omega}(\mathsf{F}))/\mathsf{J},$$

where J is the subspace of $\Gamma^{\omega}(\mathsf{E}) \otimes \Gamma^{\omega}(\mathsf{F})$ generated by elements of the form

$$(f\xi) \otimes \eta - \xi \otimes (f\eta).$$

In the topological setting, we do not work with J, but rather the closure $\mathrm{cl}(\mathsf{J})$ in $\Gamma^{\omega}(\mathsf{E}) \otimes \Gamma^{\omega}(\mathsf{F})$. This ensures that the quotient $(\Gamma^{\omega}(\mathsf{E}) \otimes \Gamma^{\omega}(\mathsf{F}))/\,\mathrm{cl}(\mathsf{J})$ is Hausdorff [37, Proposition 4.4.2]. We now note that we have the diagram

$$\Gamma^{\omega}(\mathsf{E}) \times \Gamma^{\omega}(\mathsf{F}) \xrightarrow{\ \otimes\ } \Gamma^{\omega}(\mathsf{E}) \otimes_{\pi} \Gamma^{\omega}(\mathsf{F}) \longrightarrow \cdots$$

$$\cdots \longrightarrow (\Gamma^{\omega}(\mathsf{E}) \otimes_{\pi} \Gamma^{\omega}(\mathsf{F}))/\mathsf{J} \longrightarrow (\Gamma^{\omega}(\mathsf{E}) \otimes_{\pi} \Gamma^{\omega}(\mathsf{F}))/\mathrm{cl}(\mathsf{J})$$

where $\otimes_{\pi}$ means that we are considering the projective tensor topology on the algebraic tensor product as in Sect. 1.8.5. The unnamed horizontal arrows are the quotient mappings, which are thus continuous. The composition of the three mappings gives a mapping

$$\otimes^{\pi}_{\mathrm{C}^{\omega}(\mathsf{M})} \colon \Gamma^{\omega}(\mathsf{E}) \times \Gamma^{\omega}(\mathsf{F}) \to \Gamma^{\omega}(\mathsf{E}) \otimes^{\pi}_{\mathrm{C}^{\omega}(\mathsf{M})} \Gamma^{\omega}(\mathsf{F}) \triangleq (\Gamma^{\omega}(\mathsf{E}) \otimes_{\pi} \Gamma^{\omega}(\mathsf{F}))/\,\mathrm{cl}(\mathsf{J}).$$

We call $\Gamma^{\omega}(\mathsf{E}) \otimes^{\pi}_{\mathrm{C}^{\omega}(\mathsf{M})} \Gamma^{\omega}(\mathsf{F})$ the ***balanced projective tensor product*** of $\Gamma^{\omega}(\mathsf{E})$ and $\Gamma^{\omega}(\mathsf{F})$. For a locally convex topological vector space V, let us denote by

$$\mathrm{L}^{\mathrm{C}^{\omega}(\mathsf{M})}(\Gamma^{\omega}(\mathsf{E}), \Gamma^{\omega}(\mathsf{F}); \mathsf{V})$$

the set of continuous $\mathbb{R}$-bilinear mappings that are $\mathrm{C}^{\omega}(\mathsf{M})$-balanced. We then have the following result.

Lemma 5.9 (Universal Property of Balanced Projective Tensor Product) *Let $\pi_{\mathsf{E}} \colon \mathsf{E} \to \mathsf{M}$ and $\pi_{\mathsf{F}} \colon \mathsf{F} \to \mathsf{M}$ be real analytic vector bundles. Then, for any locally convex topological vector space $(\mathsf{V}, \mathscr{O})$ and any $\beta \in \mathrm{L}^{\mathrm{C}^{\omega}(\mathsf{M})}(\Gamma^{\omega}(\mathsf{E}), \Gamma^{\omega}(\mathsf{F}); \mathsf{V})$,* *there exists a unique*

$$\phi_\beta \in \mathrm{L}(\Gamma^\omega(\mathsf{E}) \otimes^\pi_{\mathrm{C}^\omega(\mathsf{M})} \Gamma^\omega(\mathsf{F}); \mathsf{V})$$

for which the diagram

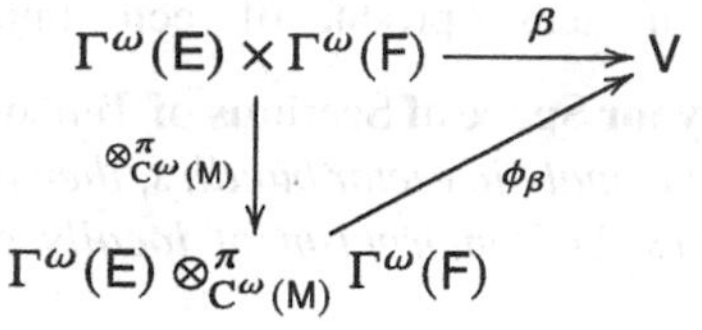

commutes.

Proof Let $(\mathsf{V}, \mathscr{O})$ be a locally convex topological vector space and let $\beta \in \mathrm{L}^{\mathrm{C}^\omega(\mathsf{M})}(\Gamma^\omega(\mathsf{E}), \Gamma^\omega(\mathsf{F}); \mathsf{V})$. The diagram

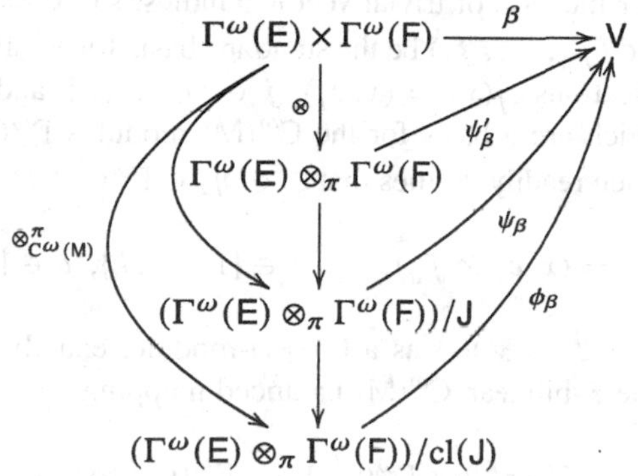

introduces the players in the proof. The proof, then, consists of understanding why the players are located where they are and what are their properties.

By Jarchow [37, Theorem 15.1.2] there exists a unique

$$\psi'_\beta \in \mathrm{L}(\Gamma^\omega(\mathsf{E}) \otimes_\pi \Gamma^\omega(\mathsf{F}); \mathsf{V})$$

such that the upper triangle in the diagram commutes. Since β is $\mathrm{C}^\omega(\mathsf{M})$-balanced and $\mathbb{R}$-bilinear, $\mathsf{J} \subseteq \ker(\psi'_\beta)$ (cf. the proof of Lemma 5.8). Thus, by universal properties of quotients [37, Proposition 4.1.2], we have the mapping ψ_β so that the upper two triangles in the diagram commute. As a consequence of the continuity of β and the fact that (1) the projective tensor topology is the final topology for $\Gamma^\omega(\mathsf{E}) \otimes \Gamma^\omega(\mathsf{F})$ associated with $\otimes$ (depending on one's source, this can be the definition of the projective tensor topology) and (2) the topology of $(\Gamma^\omega(\mathsf{E}) \otimes_\pi \Gamma^\omega(\mathsf{F}))/\mathsf{J}$ is the final topology associated with the quotient [37, Proposition 4.1.2], the universal property of final topologies ensures that ψ_β is continuous. Since

$\mathsf{J} \subseteq \ker(\psi'_\beta)$ and since ψ'_β is continuous, $\mathrm{cl}(\mathsf{J}) \subseteq \ker(\psi'_\beta)$. By universal properties of quotients, this gives rise to the continuous linear mapping ϕ_β. This mapping is unique by the uniqueness of the continuous linear mappings arising from the various universal properties for tensor products and quotients. □

Now let us see how to put all of this together to arrive at a description of the topology for the sections of a tensor product of vector bundles.

Theorem 5.10 (Topology for Space of Sections of Tensor Products) *If $\pi_{\mathsf{E}} : \mathsf{E} \to \mathsf{M}$ and $\pi_{\mathsf{F}} : \mathsf{F} \to \mathsf{M}$ are real analytic vector bundles, then the algebraic isomorphism of Proposition 5.5 induces an isomorphism of locally convex topological vector spaces*

$$\Gamma^\omega(\mathsf{E}) \otimes^\pi_{\mathrm{C}^\omega(\mathsf{M})} \Gamma^\omega(\mathsf{F}) \simeq \Gamma^\omega(\mathsf{E} \otimes \mathsf{F}).$$

Proof The proof bears some similarities to that for Proposition 5.5 and, as in that proof, we shall leave to the reader the verification of various routine details.

First we consider the case of trivial vector bundles, say $\mathsf{E} = \mathbb{R}^k_{\mathsf{M}}$ and $\mathsf{F} = \mathbb{R}^l_{\mathsf{M}}$. Let $(\boldsymbol{e}_1, \dots, \boldsymbol{e}_k)$ and $(\boldsymbol{f}_1, \dots, \boldsymbol{f}_l)$ be the standard basis for $\mathbb{R}^k$ and $\mathbb{R}^l$, respectively. We then have the sections $\xi_j(x) = (x, \boldsymbol{e}_j)$, $j \in \{1, \dots, k\}$, and $\eta_a(x) = (x, \boldsymbol{f}_a)$, $a \in \{1, \dots, l\}$, which are a basis for the $\mathrm{C}^\omega(\mathsf{M})$-modules $\Gamma^\omega(\mathbb{R}^k_{\mathsf{M}})$ and $\Gamma^\omega(\mathbb{R}^l_{\mathsf{M}})$, respectively. One then readily verifies that $\xi_j \otimes \eta_a \in \Gamma^\omega((\mathbb{R}^k \otimes \mathbb{R}^l)_{\mathsf{M}})$ defined by

$$\xi_j \otimes \eta_a(x) = (x, \boldsymbol{e}_j \otimes \boldsymbol{f}_a), \qquad j \in \{1, \dots, k\},\ a \in \{1, \dots, l\},$$

give a basis of $\Gamma^\omega((\mathbb{R}^k \otimes \mathbb{R}^l)_{\mathsf{M}})$ as a $\mathrm{C}^\omega(\mathsf{M})$-module, exactly as in the proof of Proposition 5.5. The $\mathbb{R}$-bilinear, $\mathrm{C}^\omega(\mathsf{M})$-balanced mapping

$$\begin{aligned} \gamma\colon \Gamma^\omega(\mathbb{R}^k_{\mathsf{M}}) \times \Gamma^\omega(\mathbb{R}^l_{\mathsf{M}}) &\to \Gamma^\omega((\mathbb{R}^k \otimes \mathbb{R}^l)_{\mathsf{M}}) \\ (\xi, \eta) &\mapsto (x \mapsto \xi(x) \otimes \eta(x)) \end{aligned}$$

is then expanded in bases as

$$\left(\sum_{j=1}^k f^j \xi_j, \sum_{a=1}^l g^a \eta_a\right) \mapsto \sum_{j=1}^k \sum_{a=1}^l (x \mapsto f^j(x) g^a(x) \xi_j(x) \otimes \eta_a(x)).$$

The continuity of this mapping follows from Theorem 5.1. By Lemma 5.9, there is an induced continuous linear mapping

$$\alpha\colon \Gamma^\omega(\mathbb{R}^k_{\mathsf{M}}) \otimes^\pi_{\mathrm{C}^\omega(\mathsf{M})} \Gamma^\omega(\mathbb{R}^l_{\mathsf{M}}) \to \Gamma^\omega((\mathbb{R}^k \otimes \mathbb{R}^l)_{\mathsf{M}})$$

such that the diagram

$$\begin{array}{ccc} \Gamma^\omega(\mathbb{R}^k_{\mathsf{M}}) \times \Gamma^\omega(\mathbb{R}^l_{\mathsf{M}}) & \xrightarrow{\ \gamma\ } & \Gamma^\omega((\mathbb{R}^k \otimes \mathbb{R}^l)_{\mathsf{M}}) \\ {\scriptstyle \otimes^\pi_{C^\omega(\mathsf{M})}} \big\downarrow & \overset{\beta}{\nearrow} \ \underset{\alpha}{\swarrow} & \\ \Gamma^\omega(\mathbb{R}^k_{\mathsf{M}}) \otimes^\pi_{C^\omega(\mathsf{M})} \Gamma^\omega(\mathbb{R}^l_{\mathsf{M}}) & & \end{array}$$

commutes. The mapping

$$\beta\colon \Gamma^\omega((\mathbb{R}^k \otimes \mathbb{R}^l)_{\mathsf{M}}) \to \Gamma^\omega(\mathbb{R}^k_{\mathsf{M}}) \otimes^\pi_{C^\omega(\mathsf{M})} \Gamma^\omega(\mathbb{R}^l_{\mathsf{M}})$$

in the diagram is defined by

$$\beta\left(x \mapsto \sum_{j=1}^{k}\sum_{a=1}^{l} A^{ja}(x)\xi_j(x) \otimes \eta_a(a)\right) = \sum_{j=1}^{k}\sum_{a=1}^{l} A^{ja}\xi_j \otimes^\pi_{C^\omega(\mathsf{M})} \eta_a.$$

Note that β is the composition of the algebraic isomorphism

$$\Gamma^\omega((\mathbb{R}^k \otimes \mathbb{R}^l)_{\mathsf{M}}) \ni \left(x \mapsto \sum_{j=1}^{k}\sum_{a=1}^{l} A^{ja}(x)\xi_j(x) \otimes \eta_a(x)\right)$$
$$\mapsto \sum_{j=1}^{k}\sum_{a=1}^{l} A^{ja}\xi \otimes_{C^\omega(\mathsf{M})} \eta \in \Gamma^\omega(\mathbb{R}^k_{\mathsf{M}}) \otimes_{C^\omega(\mathsf{M})} \Gamma^\omega(\mathbb{R}^l_{\mathsf{M}})$$

with the quotient by $\mathrm{cl}(\mathsf{J})$, and so is continuous by Theorem 5.1 and the definition of the quotient topology. A direct computation using the definitions of the symbols involved shows that α and β are inverses of one another.

The above establishes the theorem for trivial bundles.

If E and F are not necessarily trivial, then Corollary 2.5 gives E as a subbundle of $\mathbb{R}^k_{\mathsf{M}}$ for a suitable $k \in \mathbb{Z}_{>0}$ and F as a subbundle of $\mathbb{R}^l_{\mathsf{M}}$ for a suitable $l \in \mathbb{Z}_{>0}$. Let $\mathsf{E}^\perp$ and $\mathsf{F}^\perp$ denote the orthogonal complement subbundles using the standard Euclidean fibre metric (for example). Similarly to what we saw in the proof of Proposition 5.5 (now working with tensor products over $\mathbb{R}$ rather than $C^\omega(\mathsf{M})$), we have the algebraic isomorphism

$$\begin{aligned} &\Gamma^\omega(\mathsf{E} \oplus \mathsf{E}^\perp) \otimes \Gamma^\omega(\mathsf{F} \oplus \mathsf{F}^\perp) \\ &\qquad \simeq (\Gamma^\omega(\mathsf{E}) \otimes \Gamma^\omega(\mathsf{F})) \oplus (\Gamma^\omega(\mathsf{E}) \otimes \Gamma^\omega(\mathsf{F}^\perp)) \\ &\qquad\quad \oplus (\Gamma^\omega(\mathsf{E}^\perp) \otimes \Gamma^\omega(\mathsf{F})) \oplus (\Gamma^\omega(\mathsf{E}^\perp) \otimes \Gamma^\omega(\mathsf{F}^\perp)). \end{aligned} \tag{5.3}$$

We claim that this algebraic isomorphism is a topological isomorphism with the tensor products over $\mathbb{R}$ equipped with the projective tensor topology. To see this, we note that the algebraic isomorphism (5.3) is the linear mapping associated with the bilinear mapping

$$\begin{aligned}\Gamma^\omega(\mathsf{E}\oplus\mathsf{E}^\perp)\times\Gamma^\omega(\mathsf{F}\oplus\mathsf{F}^\perp)&\ni(\xi\oplus\xi^\perp,\eta\oplus\eta^\perp)\\&\mapsto(\xi\otimes\eta)\oplus(\xi\otimes\eta^\perp)\oplus(\xi^\perp\otimes\eta)\oplus(\xi^\perp\otimes\eta^\perp)\\&\in(\Gamma^\omega(\mathsf{E})\otimes\Gamma^\omega(\mathsf{F}))\oplus(\Gamma^\omega(\mathsf{E})\otimes\Gamma^\omega(\mathsf{F}^\perp))\\&\qquad\oplus(\Gamma^\omega(\mathsf{E}^\perp)\otimes\Gamma^\omega(\mathsf{F}))\oplus(\Gamma^\omega(\mathsf{E}^\perp)\otimes\Gamma^\omega(\mathsf{F}^\perp)).\end{aligned}$$

This bilinear mapping is continuous by Theorem 5.1. Thus the induced linear mapping is continuous by the universal property of the projective tensor topology [37, Theorem 15.1.2]. Therefore, the induced linear mapping is a topological isomorphism by the De Wilde Open Mapping Theorem, using the fact that the space of sections of a real analytic vector bundle is webbed and ultrabornological by Proposition 2.14.

We next claim that the algebraic isomorphism (5.3) induces a topological isomorphism

$$\begin{aligned}&\Gamma^\omega(\mathsf{E}\oplus\mathsf{E}^\perp)\otimes^\pi_{C^\omega(\mathsf{M})}\Gamma^\omega(\mathsf{F}\oplus\mathsf{F}^\perp)\\&\qquad\simeq(\Gamma^\omega(\mathsf{E})\otimes^\pi_{C^\omega(\mathsf{M})}\Gamma^\omega(\mathsf{F}))\oplus(\Gamma^\omega(\mathsf{E})\otimes^\pi_{C^\omega(\mathsf{M})}\Gamma^\omega(\mathsf{F}^\perp))\\&\qquad\oplus(\Gamma^\omega(\mathsf{E}^\perp)\otimes^\pi_{C^\omega(\mathsf{M})}\Gamma^\omega(\mathsf{F}))\oplus(\Gamma^\omega(\mathsf{E}^\perp)\otimes^\pi_{C^\omega(\mathsf{M})}\Gamma^\omega(\mathsf{F}^\perp)).\end{aligned}\tag{5.4}$$

To see this, we note that each algebraic tensor product $\otimes$ induces the topological tensor product $\otimes^\pi_{C^\omega(\mathsf{M})}$ by a quotient by an appropriate subspace, denote by $\mathrm{cl}(\mathsf{J})$ in our general development. Here we have a variety of J's, so we need to name them:

$$\begin{aligned}\Gamma^\omega(\mathsf{E}\oplus\mathsf{E}^\perp)\otimes\Gamma^\omega(\mathsf{F}\oplus\mathsf{F}^\perp)&\rightsquigarrow\mathsf{J}_0\\\Gamma^\omega(\mathsf{E})\otimes\Gamma^\omega(\mathsf{F})&\rightsquigarrow\mathsf{J}_1\\\Gamma^\omega(\mathsf{E})\otimes\Gamma^\omega(\mathsf{F}^\perp)&\rightsquigarrow\mathsf{J}_2\\\Gamma^\omega(\mathsf{E}^\perp)\otimes\Gamma^\omega(\mathsf{F})&\rightsquigarrow\mathsf{J}_3\\\Gamma^\omega(\mathsf{E}^\perp)\otimes\Gamma^\omega(\mathsf{F}^\perp)&\rightsquigarrow\mathsf{J}_4.\end{aligned}$$

Under the algebraic isomorphism (5.3) we have

$$\mathsf{J}_0\simeq\mathsf{J}_1\oplus\mathsf{J}_2\oplus\mathsf{J}_3\oplus\mathsf{J}_4,$$

as can be directly verified from the definitions. Continuity of the algebraic isomorphism (5.3) and Theorem 5.4 then give

$$\mathrm{cl}(J_0) \simeq \mathrm{cl}(J_1) \oplus \mathrm{cl}(J_2) \oplus \mathrm{cl}(J_3) \oplus \mathrm{cl}(J_4).$$

Now we observe that a direct sum of quotients is isomorphic to the quotient of the direct sums; this is a consequence of the transitivity of final topologies, cf. [34, Proposition 2.12.2]. Thus we have a topological isomorphism

$$\begin{aligned}&\frac{\Gamma^\omega(\mathsf{E} \oplus \mathsf{E}^\perp) \otimes \Gamma^\omega(\mathsf{F} \oplus \mathsf{F}^\perp)}{\mathrm{cl}(J_0)}\\ &\qquad\simeq \frac{\Gamma^\omega(\mathsf{E}) \otimes \Gamma^\omega(\mathsf{F})}{\mathrm{cl}(J_1)} \oplus \frac{\Gamma^\omega(\mathsf{E}) \otimes \Gamma^\omega(\mathsf{F}^\perp)}{\mathrm{cl}(J_2)}\\ &\qquad\qquad \oplus \frac{\Gamma^\omega(\mathsf{E}^\perp) \otimes \Gamma^\omega(\mathsf{F})}{\mathrm{cl}(J_3)} \oplus \frac{\Gamma^\omega(\mathsf{E}^\perp) \otimes \Gamma^\omega(\mathsf{F}^\perp)}{\mathrm{cl}(J_4)},\end{aligned}$$

which is exactly the desired conclusion (5.4).

Next we note that, just as in the proof of Proposition 5.5, we have the algebraic isomorphism

$$\Gamma^\omega((\mathsf{E}\oplus\mathsf{E}^\perp)\otimes(\mathsf{F}\oplus\mathsf{F}^\perp)) \simeq \Gamma^\omega(\mathsf{E}\otimes\mathsf{F})\oplus\Gamma^\omega(\mathsf{E}\otimes\mathsf{F}^\perp)\oplus\Gamma^\omega(\mathsf{E}^\perp\otimes\mathsf{F})\oplus\Gamma^\omega(\mathsf{E}^\perp\otimes\mathsf{F}^\perp),$$

which is a topological isomorphism by Corollary 5.3 since it is the mapping on sections induced by a vector bundle isomorphism.

Finally, the first part of the proof gives a topological isomorphism

$$\Phi\colon \Gamma^\omega(\mathsf{E} \oplus \mathsf{E}^\perp) \otimes^\pi_{C^\omega(\mathsf{M})} \Gamma^\omega(\mathsf{F} \oplus \mathsf{F}^\perp) \to \Gamma^\omega((\mathsf{E} \oplus \mathsf{E}^\perp) \otimes (\mathsf{F} \oplus \mathsf{F}^\perp))$$

Using the definitions of the topological isomorphisms (5.3) and (5.4), one can show that Φ can be represented as a 4×4 "matrix" relative to the two direct sum decompositions, with the matrix having the form

$$\Phi = \begin{bmatrix} \Phi_1 & 0 & 0 & 0 \\ 0 & \Phi_2 & 0 & 0 \\ 0 & 0 & \Phi_3 & 0 \\ 0 & 0 & 0 & \Phi_4 \end{bmatrix}.$$

The mapping $\Phi_1\colon \Gamma^\omega(\mathsf{E}) \otimes^\pi_{C^\omega(\mathsf{M})} \Gamma^\omega(\mathsf{F}) \to \Gamma^\omega(\mathsf{E} \otimes \mathsf{F})$ is the asserted mapping in the statement of the theorem, and it is a topological isomorphism since Φ is, and since the projections and inclusions on the first sums are epimorphisms and monomorphisms, respectively, by Theorem 5.4. □

5.2 Continuity of Operations Involving Differentiation

In this section we consider the continuity of operations involving differentiation. First we consider a general version of the assertion that "differentiation is continuous," and following this we give a collection of consequently elementary results derived from this general fact.

Theorem 5.11 (Prolongation of Sections Is Continuous) *Let $\pi_{\mathsf{E}}\colon \mathsf{E} \to \mathsf{M}$ be a C^{ω}-vector bundle. If $k \in \mathbb{Z}_{\geq 0}$, then the map*

$$J_k^{\omega}\colon \Gamma^{\omega}(\mathsf{E}) \to \Gamma^{\omega}(\mathsf{J}^k\mathsf{E})$$
$$\xi \mapsto j_k\xi$$

is a topological homomorphism.

Proof We let ∇^{M} be a torsion-free (for simplicity) C^{ω}-affine connection on M, $\nabla^{\pi_{\mathsf{E}}}$ be a C^{ω}-vector bundle connection in E, $\mathbb{G}_{\mathsf{M}}$ be a C^{ω}-Riemannian metric on M, and $\mathbb{G}_{\pi_{\mathsf{E}}}$ be a C^{ω}-vector bundle connection in E. We denote the associated seminorms for $\Gamma^{\omega}(\mathsf{E})$ by $p^{\omega}_{\mathcal{K},\boldsymbol{a}}$ and for $\Gamma^{\omega}(\mathsf{J}^k\mathsf{E})$ by $p^{k,\omega}_{\mathcal{K},\boldsymbol{a}}$, for $\mathcal{K} \subseteq \mathsf{M}$ compact and $\boldsymbol{a} \in \mathrm{c}_0(\mathbb{Z}_{\geq 0}; \mathbb{R}_{>0})$.

We recall from (3.20) that we have the vector bundle mapping $\iota_{\mathsf{E},m,k}\colon \mathsf{J}^{m+k}\mathsf{E} \to \mathsf{J}^m\mathsf{J}^k\mathsf{E}$ defined by the requirement that $\iota_{\mathsf{E},m,k} \circ j_{m+k}\xi = j_m j_k\xi$. The representation $\hat{\iota}_{\mathsf{E},m,k}$ of this vector bundle mapping with respect to the decompositions of jet bundles as in Sect. 2.3.1 is given by Lemma 3.22. We first perform estimates regarding this vector bundle mapping.

Let $j, l \in \mathbb{Z}_{>0}$. By Lemmata 4.9, 4.13, and 4.14, and by (4.10),

$$\|\theta_{j,l} \otimes \mathrm{id}_{\mathsf{E}}\|_{\mathbb{G}_{\mathsf{M},\pi_{\mathsf{E}}}} \leq \frac{1}{2}\|\Delta_{j-1,1} \otimes \Delta_{1,l-1}\|_{\mathbb{G}_{\mathsf{M}}}\|\mathrm{id}_{\mathsf{E}}\|_{\mathbb{G}_{\pi_{\mathsf{E}}}} \leq \frac{1}{2}\sqrt{d}\,jl,$$

where d is the fibre dimension of E. By Lemma 2.22, let $C_1, \sigma_1 \in \mathbb{R}_{>0}$ be such that

$$\|D^r_{\nabla^{\mathsf{M}},\nabla^{\pi_{\mathsf{E}}}} R_{\nabla^{\pi_{\mathsf{E}}}}(x)\|_{\mathbb{G}_{\mathsf{M},\pi_{\mathsf{E}}}} \leq C_1\sigma_1^{-r} r!, \qquad r \in \mathbb{Z}_{\geq 0},\ x \in \mathcal{K}.$$

Without loss of generality, we can assume that $C_1 \geq 1$ and $\sigma_1 \leq 1$. For $A_{j+l} \in \Gamma^{\omega}(\mathsf{S}^{j+l}(\mathsf{T}^*_x\mathsf{M}) \otimes \mathsf{E}_x)$ and $A_{l-1} \in \Gamma^{\omega}(\mathsf{S}^{l-l}(\mathsf{T}^*_x\mathsf{M}) \otimes \mathsf{E}_x)$, we now have

$$\|A_{j+l}(x) + \theta_{j,l} \otimes \mathrm{id}_{\mathsf{E}}(D^{j-1}_{\nabla^{\mathsf{M}},\nabla^{\pi_{\mathsf{E}}}} R_{\nabla^{\pi_{\mathsf{E}}}}(A_{l-1}))(x)\|_{\mathbb{G}_{\mathsf{M},\pi_{\mathsf{E}}}}$$
$$\leq \|A_{j+l}(x)\|_{\mathbb{G}_{\mathsf{M},\pi_{\mathsf{E}}}} + \frac{1}{2}\sqrt{d}C_1 jl\sigma_1^{j-1}(j-1)!\|A_{l-1}(x)\|_{\mathbb{G}_{\mathsf{M},\pi_{\mathsf{E}}}}.$$

Then, using (1.4), (4.9), and (4.10),

$$\sum_{l=0}^{m}\frac{1}{l!}\sum_{j=0}^{k}\|A_{j+l}(x)+\theta_{j,l}\otimes\mathrm{id}_{\mathsf{E}}(D^{j-1}_{\nabla^{\mathsf{M}},\nabla^{\pi_{\mathsf{E}}}}R_{\nabla^{\pi_{\mathsf{E}}}}(A_{l-1}))(x)\|_{\mathbb{G}_{\mathsf{M},\pi_{\mathsf{E}}}}$$

$$\leq\frac{1}{2}\sqrt{d}C_1\sigma_1^{-m}\sum_{l=0}^{m}\frac{1}{l!}\sum_{j=0}^{k}(\|A_{j+l}(x)\|_{\mathbb{G}_{\mathsf{M},\pi_{\mathsf{E}}}}+lj!\|A_{l-1}(x)\|_{\mathbb{G}_{\mathsf{M},\pi_{\mathsf{E}}}})$$

$$\leq\frac{1}{2}\sqrt{d}C_1\sigma_1^{-m}\sum_{l=0}^{m}\sum_{j=0}^{k}\frac{j!(l+j)!}{(l-1)!(j+l)!}(\|A_{j+l}(x)\|_{\mathbb{G}_{\mathsf{M},\pi_{\mathsf{E}}}}+\|A_{l-1}(x)\|_{\mathbb{G}_{\mathsf{M},\pi_{\mathsf{E}}}})$$

$$\leq\frac{1}{2}\sqrt{d}C_1\sigma_1^{-m}k!m(m+1)\cdots(m+k)$$

$$\times\left(\sum_{l=0}^{m}\sum_{j=0}^{k}\frac{1}{(j+l)!}\|A_{j+l}(x)\|_{\mathbb{G}_{\mathsf{M},\pi_{\mathsf{E}}}}+k\sum_{l=1}^{m}\frac{1}{(l-1)!}\|A_{l-1}(x)\|_{\mathbb{G}_{\mathsf{M},\pi_{\mathsf{E}}}}\right)$$

$$\leq\frac{1}{2}\sqrt{d}C_1\sigma_1^{-m}k!(m+k)^{k+1}$$

$$\times\left((m+1)\sum_{j=0}^{m+k}\frac{1}{j!}\|A_j(x)\|_{\mathbb{G}_{\mathsf{M},\pi_{\mathsf{E}}}}+k\sum_{j=0}^{m+k}\frac{1}{j!}\|A_j(x)\|_{\mathbb{G}_{\mathsf{M},\pi_{\mathsf{E}}}}\right)$$

$$\leq\frac{1}{2}\sqrt{d}C_1\sigma_1^{-m}k!(m+k)^{k+1}(m+1+k)\sum_{j=0}^{m+k}\frac{1}{j!}\|A_j(x)\|_{\mathbb{G}_{\mathsf{M},\pi_{\mathsf{E}}}}.$$

(In this computation, we take the convention that $A_{-1}=0$.) For $\hat{\sigma}\in(0,\sigma_1)$ we have

$$\lim_{m\to 0}\frac{1}{2}\hat{\sigma}^m\sqrt{d}C_1\sigma_1^{-m}k!(m+k)^{k+1}(m+1+k)=0.$$

Thus let $\hat{N}\in\mathbb{Z}_{>0}$ be such that

$$\frac{1}{2}\hat{\sigma}^m\sqrt{d}C_1\sigma_1^{-m}k!(m+k)^{k+1}(m+1+k)<1,\qquad m\geq\hat{N},$$

and define

$$\hat{C}=\max\left\{\frac{1}{2}\sqrt{d}C_1\sigma_1^{-m}k!(m+k)^{k+1}(m+1+k)\;\middle|\;m\in\{0,1,\dots,N\}\right\}.$$

Then, for any $m \in \mathbb{Z}_{\geq 0}$,

$$\sum_{l=0}^{m} \frac{1}{l!} \sum_{j=0}^{k} \|A_{j+l}(x) + \theta_{j,l} \otimes \mathrm{id}_{\mathsf{E}}(D^{j-1}_{\nabla^{\mathsf{M}},\nabla^{\pi_{\mathsf{E}}}} R_{\nabla^{\pi_{\mathsf{E}}}}(A_{l-1}))(x)\|_{\mathbb{G}_{\mathsf{M},\pi_{\mathsf{E}}}} \\ \leq \hat{C}\hat{\sigma}^{-m} \sum_{l=0}^{m+k} \frac{1}{j!} \|A_j\|_{\mathbb{G}_{\mathsf{M},\pi_{\mathsf{E}}}}.$$

In particular

$$\sum_{l=0}^{m} \frac{1}{l!} \sum_{j=0}^{k} \|D^{l,j}_{\nabla^{\mathsf{M}},\nabla^{\pi_{\mathsf{E}}}}(D^{j}_{\nabla^{\mathsf{M}},\nabla^{\pi_{\mathsf{E}}}}(\xi))\|_{\mathbb{G}_{\mathsf{M},\pi_{\mathsf{E}}}} \leq \hat{C}\hat{\sigma}^{-m} \sum_{l=0}^{m+k} \frac{1}{j!} \|D^{j}_{\nabla^{\mathsf{M}},\nabla^{\pi_{\mathsf{E}}}}(\xi)\|_{\mathbb{G}_{\mathsf{M},\pi_{\mathsf{E}}}}.$$

By (1.4), we have

$$\|j_m j_k \xi(x)\|_{\mathbb{G}_{\mathsf{M},\pi_{\mathsf{E}},k,m}} \leq \sqrt{k+m+1}\hat{C}\hat{\sigma}^{-m} \|j_{m+k}\xi(x)\|_{\mathbb{G}_{\mathsf{M},\pi_{\mathsf{E}},m+k}}.$$

Note that, for $\sigma < \hat{\sigma}$, we have

$$\lim_{m\to\infty} \sigma^m \sqrt{k+m+1}\hat{C}\hat{\sigma}^{-m} = 0.$$

Now let $N \in \mathbb{Z}_{>0}$ be large enough that

$$\sigma^m \sqrt{k+m+1}\hat{C}\hat{\sigma}^{-m} < 1, \qquad m \geq N,$$

and let

$$C = \max\left\{ \sqrt{k+m+1}\hat{C}\hat{\sigma}^{-m} \;\middle|\; m \in \{0, 1, \dots, N\} \right\}.$$

Then, for any $m \in \mathbb{Z}_{\geq 0}$, we have

$$\|j_{m+k}\xi(x)\|_{\mathbb{G}_{\mathsf{M},\pi_{\mathsf{E}},k,m}} \leq C\sigma^{-m} \|j_{m+k}\xi(x)\|_{\mathbb{G}_{\mathsf{M},\pi_{\mathsf{E}},m+k}}.$$

Now let $\mathcal{K} \subseteq \mathsf{M}$ be compact and let $\boldsymbol{a} \in c_0(\mathbb{Z}_{\geq 0}; \mathbb{R}_{>0})$. Define $\boldsymbol{a}' \in c_0(\mathbb{Z}_{\geq 0}; \mathbb{R}_{>0})$ by $a'_0 = a_0$, $a'_j = C$, $j \in \{1, \dots, k\}$, and $a'_j = \sigma^{-1} a_{j-k}$, $j \in \{k+1, k+2, \dots\}$. Then

$$\begin{aligned} a_0 a_1 \cdots a_m \|j_m j_k \xi(x)\|_{\mathbb{G}_{\mathsf{M},\pi_{\mathsf{E}},k,m}} &\leq C\sigma^{-m} a_0 a_1 \cdots a_m \|j_{k+m}\xi(x)\|_{\mathbb{G}_{\mathsf{M},\pi_{\mathsf{E}},m+k}} \\ &\leq a_0 C^k (\sigma^{-1} a_1) \cdots (\sigma^{-1} a_m) \|j_{k+m}\xi(x)\|_{\mathbb{G}_{\mathsf{M},\pi_{\mathsf{E}},m+k}} \\ &= a'_0 a'_1 \cdots a'_{k+m} \|j_{k+m}\xi(x)\|_{\mathbb{G}_{\mathsf{M},\pi_{\mathsf{E}},m+k}}, \end{aligned}$$

since $C \geq 1$. We then immediately have

$$p^{k,\omega}_{\mathcal{K},a}(j_k\xi) \leq p^{\omega}_{\mathcal{K},a'}(\xi),$$

which gives continuity of J^{ω}_k.

Consider the surjective vector bundle mapping $\pi_{\mathsf{M},k} \colon \mathsf{J}^k\mathsf{E} \to \mathsf{E}$ and the induced mapping on sections,

$$\begin{aligned}\Pi_{\mathsf{E},k} \colon \Gamma^{\omega}(\mathsf{J}^k\mathsf{E}) &\to \Gamma^{\omega}(\mathsf{E})\\ \Xi &\mapsto \pi_{\mathsf{E},k} \circ \Xi.\end{aligned}$$

Clearly, $\Pi_{\mathsf{E},k} \circ J^{\omega}_k(\xi) = \xi$. Let $\mathcal{K} \subseteq \mathsf{M}$ be compact and let $a \in c_0(\mathbb{Z}_{\geq 0}; \mathbb{R}_{>0})$. Following the proof of Theorem 5.1, let $C \in \mathbb{R}_{>0}$ and $a' \in c_0(\mathbb{Z}_{\geq 0}; \mathbb{R}_{>0})$ be such that

$$p^{\omega}_{\mathcal{K},a}(\Pi_{\mathsf{E},k} \circ \Xi) \leq C p^{k,\omega}_{\mathcal{K},a'}(\Xi), \qquad \Xi \in \Gamma^{\omega}(\mathsf{J}^k\mathsf{E}).$$

Then, for $\xi \in \Gamma^{\omega}(\mathsf{E})$, we have

$$p^{\omega}_{\mathcal{K},a}(\xi) = p^{\omega}_{\mathcal{K},a}(\Pi_{\mathsf{E},k} \circ j_k\xi) \leq C p^{k,\omega}_{\mathcal{K},a'}(j_k\xi).$$

Openness of J^{ω}_k onto its image now follows from Lemma 1 from the proof of Theorem 5.1. □

We comment that the proof of the theorem is a little easier in local coordinates since the inclusion $\iota_{\mathsf{E},k,m}$ is simpler in local coordinates than the expression that comes from Lemma 3.22. We use the connection formulation to be consistent with our geometric approach.

We can now prove a collection of results regarding standard operations involving differentiation, derived from the preceding result about basic prolongation.

Corollary 5.12 (Continuity of Differential) *Let* M *be a* C^{ω}*-manifold. Then the mapping*

$$\begin{aligned}\mathrm{d} \colon \mathrm{C}^{\omega}(\mathsf{M}) &\to \Gamma^{\omega}(\mathsf{T}^*\mathsf{M})\\ f &\mapsto \mathrm{d}f\end{aligned}$$

is a topological homomorphism.

Proof Note that $\mathsf{J}^1(\mathsf{M}; \mathbb{R}) \simeq \mathbb{R}_{\mathsf{M}} \oplus \mathsf{T}^*\mathsf{M}$ and that, under this identification, $j_1 f = f \oplus \mathrm{d}f$. Thus $\mathrm{d}f = \mathrm{pr}_2 \circ j_1 f$, where $\mathrm{pr}_2 \colon \mathsf{J}^1(\mathsf{M}; \mathbb{R}) \to \mathsf{T}^*\mathsf{M}$ is the C^{ω}-vector bundle mapping of projection onto the second factor. The result then immediately follows from Theorems 5.4 and 5.11. □

Corollary 5.13 (Continuity of Lie Derivative) *Let* M *be a* C^{ω}*-manifold. Then the map*

$$\begin{aligned}\mathscr{L}\colon \Gamma^{\omega}(\mathsf{TM}) \times \mathrm{C}^{\omega}(\mathsf{M}) &\to \mathrm{C}^{\omega}(\mathsf{M})\\ (X, f) &\mapsto \mathscr{L}_X f\end{aligned}$$

is continuous.

Proof We think of X as being a C^{ω}-vector bundle mapping via

$$\begin{aligned}X\colon \mathsf{T}^*\mathsf{M} &\to \mathbb{R}_{\mathsf{M}}\\ \alpha_x &\mapsto \langle \alpha_x; X(x)\rangle.\end{aligned}$$

Then the bilinear mapping of the corollary is given by the composition

$$(X, f) \mapsto (X, \mathrm{d}f) \mapsto X(\mathrm{d}f).$$

The left mapping is continuous since it is the product of the continuous mappings id and d. The right mapping is continuous by Theorem 5.1(ii), and so the corollary follows. □

Corollary 5.14 (Continuity of Covariant Derivative) *Let* $\pi_{\mathsf{E}}\colon \mathsf{E} \to \mathsf{M}$ *be a* C^{ω}*-vector bundle with a* C^{ω}*-vector bundle connection* $\nabla^{\pi_{\mathsf{E}}}$*. Then the map*

$$\begin{aligned}\nabla^{\pi_{\mathsf{E}}}\colon \Gamma^{\omega}(\mathsf{TM}) \times \Gamma^{\omega}(\mathsf{E}) &\to \Gamma^{\omega}(\mathsf{E})\\ (X, \xi) &\mapsto \nabla_X^{\pi_{\mathsf{E}}} \xi\end{aligned}$$

is continuous.

Proof As in the proof of Lemma 2.15, we have a C^{ω}-vector bundle mapping $S_{\nabla^{\pi_{\mathsf{E}}}}\colon \mathsf{E} \to \mathsf{J}^1\mathsf{E}$ over id_{M} that determines the connection $\nabla^{\pi_{\mathsf{E}}}$ by

$$\nabla^{\pi_{\mathsf{E}}}\xi(x) = j_1\xi(x) - S_{\nabla^{\pi_{\mathsf{E}}}}(\xi(x)).$$

The mapping $\xi \mapsto \nabla^{\pi_{\mathsf{E}}}\xi$ is continuous by Theorems 5.1 and 5.11. We note that $\nabla^{\pi_{\mathsf{E}}}\xi$ is to be thought of as a C^{ω}-vector bundle mapping by

$$\begin{aligned}\nabla^{\pi_{\mathsf{E}}}\xi\colon \mathsf{TM} &\to \mathsf{E}\\ X &\mapsto \nabla_X^{\pi_{\mathsf{E}}}\xi.\end{aligned}$$

The bilinear mapping of the lemma is then given by the composition

$$(X, \xi) \mapsto (X, \nabla^{\pi_{\mathsf{E}}}\xi) \mapsto \nabla^{\pi_{\mathsf{E}}}\xi(X).$$

The left mapping is continuous since it is the product of the continuous mappings id and $\xi \mapsto \nabla^{\pi_{\mathsf{E}}}\xi$. The right mapping is continuous by Theorem 5.1(ii), and so the lemma follows. □

Corollary 5.15 (Continuity of Lie Bracket) *Let* M *be a* C^{ω}*-manifold. Then the map*

$$[\cdot,\cdot]\colon \Gamma^{\omega}(\mathsf{TM}) \times \Gamma^{\omega}(\mathsf{TM}) \to \Gamma^{\omega}(\mathsf{TM})$$
$$(X, Y) \mapsto [X, Y]$$

is continuous.

Proof Let $\mathbb{G}_{\mathsf{M}}$ be a real analytic Riemannian metric on M and let ∇^{M} be the associated Levi-Civita connection. Since

$$[X, Y] = \nabla^{\mathsf{M}}_X Y - \nabla^{\mathsf{M}}_Y X,$$

the result follows from Corollary 5.14. □

Corollary 5.16 (Continuity of Linear Partial Differential Operators) *Let* $\pi_{\mathsf{E}}\colon \mathsf{E} \to \mathsf{M}$ *and* $\pi_{\mathsf{F}}\colon \mathsf{F} \to \mathsf{M}$ *be* C^{ω}*-vector bundles and let* $\Phi \in \mathrm{VB}^{\omega}(\mathsf{J}^k\mathsf{E}; \mathsf{F})$. *Then the kth-order linear partial differential operator* $D_{\Phi}\colon \Gamma^{\omega}(\mathsf{E}) \to \Gamma^{\omega}(\mathsf{F})$ *defined by* $D_{\Phi}(\xi)(x) = \Phi(j_k\xi(x))$, $x \in \mathsf{M}$, *is continuous.*

Proof The operator D_{Φ} is the composition of the continuous mappings $\xi \mapsto j_k\xi$ $\Gamma^{\omega}(\mathsf{E})$ to $\Gamma^{\omega}(\mathsf{J}^k\mathsf{E})$ and $\Xi \mapsto \Phi \circ \Xi$ from $\Gamma^{\omega}(\mathsf{J}^k\mathsf{E})$ to $\Gamma^{\omega}(\mathsf{F})$. □

The reader can no doubt imagine many extensions of results such as the ones we give, and we leave these for the reader to figure out as they need them.

5.3 Continuity of Lifting Operations

In Sects. 3.1.1–3.1.4 we introduced a variety of constructions for lifting objects from the base space of a vector bundle to the total space. In Sect. 3.3 we considered how to differentiate these constructions in multiple ways, and how to relate these multiple differentiations. In Sects. 4.1.1–4.1.7 we described fibre norms to give norms for these lifted objects. In this section, we put this all together to prove results that give substantial motivation for all of these quite elaborate constructions. That is, we show that these lift operations are topological homomorphisms. Many of the proofs are similar to one another, so we only give representative proofs.

We begin by considering horizontal lifts of functions. We note that continuity of the mapping in the next theorem follows from Theorem 5.26, but openness onto its image does not since the vector bundle projection is not proper. In any case, we give an independent proof of continuity, as it is a model for the proof of subsequent statements for which we will not give detailed proofs.

Theorem 5.17 (Horizontal Lift of Functions Is a Topological Monomorphism) *Let $\pi_{\mathsf{E}}\colon \mathsf{E} \to \mathsf{M}$ be a C^{ω}-vector bundle. Then the mapping*

$$\mathrm{C}^{\omega}(\mathsf{M}) \ni f \mapsto \pi^* f \in \mathrm{C}^{\omega}(\mathsf{E})$$

is a topological monomorphism.

Proof It is clear that the asserted map is injective, so we focus on its topological attributes.

We let $\mathbb{G}_{\mathsf{M}}$ be a C^{ω}-Riemannian metric on M, $\mathbb{G}_{\pi}$ be a C^{ω}-vector bundle connection in E, ∇^{M} be the Levi-Civita connection for $\mathbb{G}_{\mathsf{M}}$, and ∇^{π} be a C^{ω}-vector bundle connection in E. Corresponding to this, we have a Riemannian metric $\mathbb{G}_{\mathsf{E}}$ on E with its Levi-Civita connection ∇^{E}, as in Sect. 3.2.1. We denote the associated seminorms for $\mathrm{C}^{\omega}(\mathsf{M})$ and $\mathrm{C}^{\omega}(\mathsf{E})$ by $p^{\omega}_{\mathcal{K},\boldsymbol{a}}$ and $q^{\omega}_{\mathcal{L},\boldsymbol{a}}$ for $\mathcal{K} \subseteq \mathsf{M}$ and $\mathcal{L} \subseteq \mathsf{E}$ compact, and for $\boldsymbol{a} \in \mathrm{c}_0(\mathbb{Z}_{\geq 0}; \mathbb{R}_{>0})$.

Let us make some preliminary computations.

By Lemma 3.25, we have

$$\mathrm{Sym}_m \circ \nabla^{\mathsf{E},m} \pi_{\mathsf{E}}^* f(e) = \sum_{s=0}^{m} \widehat{A}^m_s (\mathrm{Sym}_{s+1} \circ \pi_{\mathsf{E}}^* \nabla^{\mathsf{M},s} f(e)). \tag{5.5}$$

By Lemma 4.11, we have

$$\|A^m_s(\beta_s)\|_{\mathbb{G}_{\mathsf{E}}} \leq \|A^m_s\|_{\mathbb{G}_{\mathsf{E}}} \|\beta_s\|_{\mathbb{G}_{\mathsf{E}}}$$

for $\beta_s \in \mathrm{T}^s(\mathsf{T}^*_e\mathsf{E})$, $m \in \mathbb{Z}_{>0}$, and $s \in \{0, 1, \dots, m\}$. By Lemma 4.13,

$$\|\mathrm{Sym}_s(A)\|_{\mathbb{G}_{\mathsf{E}}} \leq \|A\|_{\mathbb{G}_{\mathsf{E}}}$$

for $A \in \mathrm{T}^s(\mathsf{T}^*\mathsf{N})$ and $s \in \mathbb{Z}_{>0}$. Thus, recalling (3.29),

$$\|\widehat{A}^m_s(\mathrm{Sym}_s(\beta_s))\|_{\mathbb{G}_{\mathsf{E}}} = \|\mathrm{Sym}_m \circ A^m_s(\beta_s)\|_{\mathbb{G}_{\mathsf{E}}} \leq \|A^m_s\|_{\mathbb{G}_{\mathsf{E}}} \|\beta_s\|_{\mathbb{G}_{\mathsf{E}}},$$

for $\beta_s \in \mathrm{T}^s(\pi_{\mathsf{E}}^*\mathsf{T}^*\mathsf{M})$, $m \in \mathbb{Z}_{>0}$, $s \in \{1, \dots, m\}$.

Let $\mathcal{L} \subseteq \mathsf{E}$ be compact. By Lemmata 4.17 and 4.18 with $r = 0$, there exist $C_1, \sigma_1, \rho_1 \in \mathbb{R}_{>0}$ such that

$$\|A^k_s(e)\|_{\mathbb{G}_{\mathsf{E}}} \leq C_1 \sigma_1^{-k} \rho_1^{-(k-s)} (k-s)!, \qquad k \in \mathbb{Z}_{>0},\ s \in \{0, 1, \dots, k-1\},\ e \in \mathcal{L}.$$

Without loss of generality, we assume that $C_1 \geq 1$ and $\sigma_1, \rho_1 \leq 1$. Thus, using Lemma 4.1 and abbreviating $\sigma_2 = \sigma_1 \rho_1$, we have

$$\|\widehat{A}^k_s(\pi_{\mathsf{E}}^* \mathrm{Sym}_s \circ \nabla^{\mathsf{M},s} f(e))\|_{\mathbb{G}_{\mathsf{E}}} \leq C_1 \sigma_2^{-k} (k-s)! \|\mathrm{Sym}_s \circ \nabla^{\mathsf{M},s} \mathrm{d} f(\pi_{\mathsf{E}}(e))\|_{\mathbb{G}_{\mathsf{M}}}$$

for $k \in \mathbb{Z}_{\geq 0}$, $s \in \{0, 1, \ldots, k\}$, $e \in \mathcal{L}$. Thus, by (1.4), (5.5), and Lemma 4.8,

$$\begin{aligned}
\| j_m(\pi_{\mathsf{E}}^* f)(e)\|_{\mathbb{G}_{\mathsf{E},m}} &\leq \sum_{k=0}^{m} \frac{1}{k!} \|\mathrm{Sym}_k \circ \nabla^{\mathsf{E},k} \pi_{\mathsf{E}}^* f(e)\|_{\mathbb{G}_{\mathsf{E}}} \\
&= \sum_{k=0}^{m} \frac{1}{k!} \left\| \sum_{s=0}^{k} \widehat{A}_s^k (\pi_{\mathsf{E}}^* \mathrm{Sym}_s \circ \nabla^{\mathsf{M},s} f(e)) \right\|_{\mathbb{G}_{\mathsf{E}}} \\
&\leq \sum_{k=0}^{m} \sum_{s=0}^{k} C_1 \sigma_2^{-k} \frac{s!(k-s)!}{k!} \frac{1}{s!} \|\mathrm{Sym}_s \circ \nabla^{\mathsf{M},s} f(\pi_{\mathsf{E}}(e))\|_{\mathbb{G}_{\mathsf{M}}}
\end{aligned}$$

for $e \in \mathcal{L}$ and $m \in \mathbb{Z}_{\geq 0}$. Now note that

$$\frac{s!(k-s)!}{k!} \leq 1, \quad C_1 \sigma_2^{-k} \leq C_1 \sigma_2^{-m},$$

for $s \in \{0, 1, \ldots, m\}$, $k \in \{0, 1, \ldots, s\}$, since $\sigma_2 \leq 1$. Then

$$\begin{aligned}
\| j_m(\pi_{\mathsf{E}}^* f)(e)\|_{\mathbb{G}_{\mathsf{E},m}} &\leq C_1 \sigma_2^{-m} \sum_{k=0}^{m} \sum_{s=0}^{k} \frac{1}{s!} \|\mathrm{Sym}_s \circ \nabla^{\mathsf{M},s} f(\pi_{\mathsf{E}}(e))\|_{\mathbb{G}_{\mathsf{M}}} \\
&\leq C_1 \sigma_2^{-m} \sum_{k=0}^{m} \sum_{s=0}^{m} \frac{1}{s!} \|\mathrm{Sym}_s \circ \nabla^{\mathsf{M},s} f(\pi_{\mathsf{E}}(e))\|_{\mathbb{G}_{\mathsf{M}}} \\
&= (m+1) C_1 \sigma_2^{-m} \sum_{s=0}^{m} \frac{1}{s!} \|\mathrm{Sym}_s \circ \nabla^{\mathsf{M},s} f(\pi_{\mathsf{E}}(e))\|_{\mathbb{G}_{\mathsf{M}}}.
\end{aligned}$$

Now let $\sigma < \sigma_2$ and note that

$$\lim_{m \to \infty} (m+1) \frac{\sigma_2^{-m}}{\sigma^{-m}} = 0.$$

Thus there exists $N \in \mathbb{Z}_{>0}$ such that

$$(m+1) C_1 \sigma_2^{-m} \leq C_1 \sigma^{-m}, \qquad m \geq N.$$

Let

$$C = \max \left\{ C_1, C_1 \frac{\sigma}{\sigma_2}, 2C_1 \left(\frac{\sigma}{\sigma_2}\right)^2, \ldots, N C_1 \left(\frac{\sigma}{\sigma_2}\right)^N \right\}.$$

We then immediately have $(m+1)C_1\sigma_2^{-m} \leq C\sigma^{-m}$ for all $m \in \mathbb{Z}_{\geq 0}$. We then have, by (1.4),

$$\begin{aligned}\|j_m(\pi_{\mathsf{E}}^* f)(e)\|_{\mathbb{G}_{\mathsf{E},m}} &\leq C\sigma^{-m}\sum_{s=0}^{m}\frac{1}{s!}\|\mathrm{Sym}_s \circ \nabla^{\mathsf{M},s} f(\pi_{\mathsf{E}}(e))\|_{\mathbb{G}_{\mathsf{M}}} \\ &\leq C\sqrt{m+1}\sigma^{-m}\|j_m f(\pi_{\mathsf{E}}(e))\|_{\mathbb{G}_{\mathsf{M},m}}.\end{aligned}$$

By modifying C and σ guided by what we did just preceding, we get

$$\|j_m(\pi_{\mathsf{E}}^* f)(e)\|_{\mathbb{G}_{\mathsf{E},m}} \leq C\sigma^{-m}\|j_m f(\pi_{\mathsf{E}}(e))\|_{\mathbb{G}_{\mathsf{M},m}}.$$

Now let $\boldsymbol{a} \in c_0(\mathbb{Z}_{\geq 0}; \mathbb{R}_{>0})$ and define $\boldsymbol{a}' \in c_0(\mathbb{Z}_{\geq 0}; \mathbb{R}_{>0})$ be defined by $a_0' = Ca_0$ and $a_j' = a_j\sigma^{-1}$, $j \in \mathbb{Z}_{>0}$. Then we have

$$q^{\omega}_{\mathcal{L},\boldsymbol{a}}(\pi_{\mathsf{E}}^* f) \leq p^{\omega}_{\pi_{\mathsf{E}}(\mathcal{L}),\boldsymbol{a}'}(f),$$

giving continuity in this case.

Now we show that π_{E}^* is open onto its image. The idea here is to make some preliminary observations to put ourselves in a position to be able to say, "Now proceed as above."

By Lemma 3.26, we have

$$\mathrm{Sym}_m \circ \pi_{\mathsf{E}}^* \nabla^{\mathsf{M},m} f(e) = \sum_{s=0}^{m} \widehat{B}_s^m(\mathrm{Sym}_s \circ \nabla^{\mathsf{E},s}\pi_{\mathsf{E}}^* f(e)). \tag{5.6}$$

For a compact $\mathcal{L} \subseteq \mathsf{E}$ we can proceed as above to give a bound

$$\|j_m f(\pi_{\mathsf{E}}(e))\|_{\mathbb{G}_{\mathsf{M},m}} \leq C\sigma^{-m}\|j_m(\pi_{\mathsf{E}}^* f)(e)\|_{\mathbb{G}_{\mathsf{E},m}}, \qquad e \in \mathcal{L}.$$

We need to choose the compact set $\mathcal{L}$ in a specific way. We let $\mathcal{K} \subseteq \mathsf{M}$ be compact and choose a continuous section $\xi \in \Gamma^0(\mathsf{E})$, and then take $\mathcal{L} = \xi(\mathcal{K})$. Then we have the estimate

$$\|j_m f(x)\|_{\mathbb{G}_{\mathsf{M},m}} \leq C\sigma^{-m}\|j_m(\pi_{\mathsf{E}}^* f)(\xi(x))\|_{\mathbb{G}_{\mathsf{E},m}}, \qquad x \in \mathcal{K}.$$

Now, as advertised, we can proceed as above for continuity to give the bound

$$p^{\omega}_{\mathcal{K},\boldsymbol{a}}(f) \leq q^{\omega}_{\xi(\mathcal{K}),\boldsymbol{a}'}(\pi_{\mathsf{E}}^* f),$$

and from this we conclude that $f \mapsto \pi_{\mathsf{E}}^* f$ is indeed open onto its image by Lemma 1 from the proof of Theorem 5.1. □

Now we consider vertical lifts of sections.

Theorem 5.18 (Vertical Lift of Sections Is a Topological Monomorphism) *Let $\pi_{\mathsf{E}} : \mathsf{E} \to \mathsf{M}$ be a C^{ω}-vector bundle. Then the mapping*

$$\Gamma^{\omega}(\mathsf{E}) \ni \xi \mapsto \xi^{\mathrm{v}} \in \Gamma^{\omega}(\mathsf{TE})$$

is a topological monomorphism.

Proof This follows in the same manner as Theorem 5.17, using Lemmata 3.31, 3.32, and 4.2. □

One has the similar result for vertical lifts of endomorphisms.

Theorem 5.19 (Vertical Lift of Endomorphisms Is a Topological Monomorphism) *Let $\pi_{\mathsf{E}} : \mathsf{E} \to \mathsf{M}$ be a C^{ω}-vector bundle. Then the mapping*

$$\Gamma^{\omega}(\mathrm{End}(\mathsf{E})) \ni L \mapsto L^{\mathrm{v}} \in \Gamma^{\omega}(\mathrm{End}(\mathsf{TE}))$$

is a topological monomorphism.

Proof This follows in the same manner as Theorem 5.17, using Lemmata 3.46, 3.47, and 4.5. □

Now we consider horizontal lifts of vector fields.

Theorem 5.20 (Horizontal Lift of Vector Fields Is a Topological Monomorphism) *Let $\pi_{\mathsf{E}} : \mathsf{E} \to \mathsf{M}$ be a C^{ω}-vector bundle. Then the mapping*

$$\Gamma^{\omega}(\mathsf{TM}) \ni X \mapsto X^{\mathrm{h}} \in \Gamma^{\omega}(\mathsf{TE})$$

is a topological monomorphism.

Proof This follows in the same manner as Theorem 5.17, using Lemmata 3.36, 3.37, and 4.3. □

Now we consider vertical lifts of sections of the dual bundle.

Theorem 5.21 (Vertical Lift of One-Forms Is a Topological Monomorphism) *Let $\pi_{\mathsf{E}} : \mathsf{E} \to \mathsf{M}$ be a C^{ω}-vector bundle. Then the mapping*

$$\Gamma^{\omega}(\mathsf{E}^*) \ni \lambda \mapsto \lambda^{\mathrm{v}} \in \Gamma^{\omega}(\mathsf{T}^*\mathsf{E})$$

is a topological monomorphism.

Proof This follows in the same manner as Theorem 5.17, using Lemmata 3.41, 3.42, and 4.4. □

Next we consider vertical evaluations of sections of the dual bundle.

Theorem 5.22 (Vertical Evaluations of One-Forms Is a Topological Monomorphism) *Let* $\pi_{\mathsf{E}}\colon \mathsf{E} \to \mathsf{M}$ *be a* C^ω*-vector bundle. Then the mapping*

$$\Gamma^\omega(\mathsf{E}^*) \ni \lambda \mapsto \lambda^{\mathrm{e}} \in \mathrm{C}^\omega(\mathsf{E})$$

is a topological monomorphism.

Proof Since the given map is clearly injective, we focus on its topological properties.

We let $\mathbb{G}_{\mathsf{M}}$ be a C^ω-Riemannian metric on M, $\mathbb{G}_{\pi_{\mathsf{E}}}$ be a C^ω-vector bundle connection in E, ∇^{M} be the Levi-Civita connection for $\mathbb{G}_{\mathsf{M}}$, and $\nabla^{\pi_{\mathsf{E}}}$ be a C^ω-vector bundle connection in E. Corresponding to this, we have a Riemannian metric $\mathbb{G}_{\mathsf{E}}$ on E with its Levi-Civita connection ∇^{E}, as in Sect. 3.2.1. We denote the associated seminorms for $\Gamma^\omega(\mathsf{E}^*)$ and $\mathrm{C}^\omega(\mathsf{E})$ by $p^\omega_{\mathcal{K},\boldsymbol{a}}$ and $q^\omega_{\mathcal{L},\boldsymbol{a}}$ for $\mathcal{K} \subseteq \mathsf{M}$ and $\mathcal{L} \subseteq \mathsf{E}$ compact, and for $\boldsymbol{a} \in \mathrm{c}_0(\mathbb{Z}_{\geq 0}; \mathbb{R}_{>0})$.

Let us make some preliminary computations.

By Lemma 3.51, we have

$$\begin{aligned}
\lambda^{\mathrm{e}}(e) &= \lambda^{\mathrm{e}}(e),\\
\nabla^{\mathsf{E}}\lambda^{\mathrm{e}}(e) &= \widehat{A}^1_1((\nabla^{\pi_{\mathsf{E}}}\lambda)^{\mathrm{e}}(e)) + \widehat{A}^1_0(\lambda^{\mathrm{e}}(e)) + \widehat{C}^1_0(\lambda^{\mathrm{v}}(e)),\\
&\vdots\\
(\mathrm{Sym}_m \otimes \mathrm{id}_{\mathsf{T}^*\mathsf{E}}) \circ \nabla^{\mathsf{E},m}\lambda^{\mathrm{e}}(e) &= \sum_{s=0}^{m} \widehat{A}^m_s((\mathrm{Sym}_s \otimes \mathrm{id}_{\mathsf{T}^*\mathsf{E}}) \circ (\nabla^{\mathsf{M},\pi_{\mathsf{E}},s}\lambda)^{\mathrm{e}}(e))\\
&\quad + \sum_{s=0}^{m-1} \widehat{C}^m_s((\mathrm{Sym}_s \otimes \mathrm{id}_{\mathsf{T}^*\mathsf{E}}) \circ (\nabla^{\mathsf{M},\pi_{\mathsf{E}},s}\lambda)^{\mathrm{v}}(e)).
\end{aligned} \tag{5.7}$$

Just as in the proof of Theorem 5.17, by Lemmata 4.11 and 4.13, and the appropriate analogue of Eq. (3.33) that would appear in a fully fleshed out proof of Lemma 3.49, we have bounds

$$\|\widehat{A}^m_s(\mathrm{Sym}_s(\beta_s))\|_{\mathbb{G}_{\mathsf{E}}} = \|\mathrm{Sym}_m \circ A^m_s(\beta_s)\|_{\mathbb{G}_{\mathsf{E}}} \leq \|A^m_s\|_{\mathbb{G}_{\mathsf{E}}}\|\beta_s\|_{\mathbb{G}_{\mathsf{E}}}$$

and

$$\|\widehat{C}^m_s(\mathrm{Sym}_s(\gamma_s))\|_{\mathbb{G}_{\mathsf{E}}} = \|\mathrm{Sym}_m \circ C^m_s(\gamma_s)\|_{\mathbb{G}_{\mathsf{E}}} \leq \|C^m_s\|_{\mathbb{G}_{\mathsf{E}}}\|\gamma_s\|_{\mathbb{G}_{\mathsf{E}}}.$$

Let $\mathcal{L} \subseteq \mathsf{E}$ be compact. By Lemmata 4.17 and 4.18 with $r = 0$, there exist $C_1, \sigma_1, \rho_1 \in \mathbb{R}_{>0}$ such that

$$\|A^k_s(e)\|_{\mathbb{G}_{\mathsf{E}}} \leq C_1\sigma_1^{-k}\rho_1^{-(k-s)}(k-s)!, \qquad k \in \mathbb{Z}_{\geq 0},\ s \in \{0, 1, \dots, k\},\ e \in \mathcal{L},$$

and

$$\|C_s^k(e)\|_{\mathbb{G}_{\mathsf{E}}} \le C_1\sigma_1^{-k}\rho_1^{-(k-s)}(k-s)!, \qquad k \in \mathbb{Z}_{\ge 0},\ s \in \{0,1,\dots,k-1\},\ e \in \mathcal{L}.$$

Without loss of generality, we assume that $C_1 \ge 1$ and $\sigma_1, \rho_1 \le 1$. Thus, using Lemma 4.6 and abbreviating $\sigma_2 = \sigma_1\rho_1$, we have

$$\begin{aligned}&\|\widehat{A}_s^k((\mathrm{Sym}_s \otimes \mathrm{id}_{\mathsf{T}^*\mathsf{E}}) \circ (\nabla^{\mathsf{M},\pi_{\mathsf{E}},s}\lambda)^{\mathrm{e}}(e))\|_{\mathbb{G}_{\mathsf{M},\pi_{\mathsf{E}}}}\\ &\qquad \le C_1\sigma_2^{-k}(k-s)!\|(\mathrm{Sym}_s \otimes \mathrm{id}_{\mathsf{T}^*\mathsf{E}}) \circ (\nabla^{\mathsf{M},\pi_{\mathsf{E}},s}\lambda)^{\mathrm{e}}(e)\|_{\mathbb{G}_{\mathsf{M},\pi_{\mathsf{E}}}}\end{aligned}$$

for $k \in \mathbb{Z}_{\ge 0}$, $s \in \{0,1,\dots,k\}$, $e \in \mathcal{L}$, and

$$\begin{aligned}&\|\widehat{C}_s^k((\mathrm{Sym}_s \otimes \mathrm{id}_{\mathsf{T}^*\mathsf{E}}) \circ (\nabla^{\mathsf{M},\pi_{\mathsf{E}},s}\lambda)^{\mathrm{e}}(e))\|_{\mathbb{G}_{\mathsf{M},\pi_{\mathsf{E}}}}\\ &\qquad \le C_1\sigma_2^{-k}(k-s)!\|(\mathrm{Sym}_s \otimes \mathrm{id}_{\mathsf{T}^*\mathsf{E}}) \circ (\nabla^{\mathsf{M},\pi_{\mathsf{E}},s}\lambda)^{\mathrm{e}}(e)\|_{\mathbb{G}_{\mathsf{M},\pi_{\mathsf{E}}}}\end{aligned}$$

for $k \in \mathbb{Z}_{\ge 0}$, $s \in \{0,1,\dots,k-1\}$, $e \in \mathcal{L}$. Thus, by (1.4) and (5.7),

$$\begin{aligned}&\|j_m\lambda^{\mathrm{e}}(e)\|_{\mathbb{G}_{\mathsf{E},m}}\\
&\quad \le \sum_{k=0}^{m}\frac{1}{k!}\|\,\mathrm{Sym}_k \circ \nabla^{\mathsf{E},k}\lambda^{\mathrm{e}}(e)\|_{\mathbb{G}_{\mathsf{E}}}\\
&\quad \le \sum_{k=0}^{m}\frac{1}{k!}\left\|\sum_{s=0}^{k}\widehat{A}_s^k((\mathrm{Sym}_s \circ \mathrm{id}_{\mathsf{T}^*\mathsf{E}}) \circ (\nabla^{\mathsf{M},\pi_{\mathsf{E}},s}\lambda)^{\mathrm{e}}(e))\right\|_{\mathbb{G}_{\mathsf{M},\pi_{\mathsf{E}}}}\\
&\qquad + \sum_{k=0}^{m-1}\frac{1}{k!}\left\|\sum_{s=0}^{k}\widehat{C}_s^k((\mathrm{Sym}_s \circ \mathrm{id}_{\mathsf{T}^*\mathsf{E}}) \circ (\nabla^{\mathsf{M},\pi_{\mathsf{E}},s}\lambda)^{\mathrm{v}}(e))\right\|_{\mathbb{G}_{\mathsf{M},\pi_{\mathsf{E}}}}\\
&\quad \le \sum_{k=0}^{m}C_1\sigma_2^{-k}\frac{s!(k-s)!}{k!}\frac{1}{s!}\left\|\sum_{s=0}^{k}(\mathrm{Sym}_s \circ \mathrm{id}_{\mathsf{T}^*\mathsf{E}}) \circ (\nabla^{\mathsf{M},\pi_{\mathsf{E}},s}\lambda)^{\mathrm{e}}(e)\right\|_{\mathbb{G}_{\mathsf{M},\pi_{\mathsf{E}}}}\\
&\qquad + \sum_{k=0}^{m-1}C_1\sigma_2^{-k}\frac{s!(k-s)!}{k!}\frac{1}{s!}\left\|\sum_{s=0}^{k}(\mathrm{Sym}_s \circ \mathrm{id}_{\mathsf{T}^*\mathsf{E}}) \circ (\nabla^{\mathsf{M},\pi_{\mathsf{E}},s}\lambda)^{\mathrm{v}}(e)\right\|_{\mathbb{G}_{\mathsf{M},\pi_{\mathsf{E}}}}\end{aligned}$$

for $e \in \mathcal{L}$ and $m \in \mathbb{Z}_{\ge 0}$. Now note that

$$\frac{s!(k-s)!}{k!} \le 1, \quad C_1\sigma_2^{-k} \le C_1\sigma_2^{-m},$$

for $s \in \{0, 1, \dots, m-1\}$, $k \in \{0, 1, \dots, s\}$, since $\sigma_2 \le 1$. Then

$$\begin{aligned}
&\|j_m\lambda^{\mathrm{e}}(e)\|_{\mathbb{G}_{\mathsf{E},m}} \\
&\le C_1\sigma_2^{-m}\sum_{k=0}^{m}\sum_{s=0}^{m}\frac{1}{s!}\left\|(\mathrm{Sym}_s\circ\mathrm{id}_{\mathsf{T}^*\mathsf{E}})\circ(\nabla^{\mathsf{M},\pi_{\mathsf{E}},s}\lambda)^{\mathrm{e}}(e)\right\|_{\mathbb{G}_{\mathsf{M},\pi_{\mathsf{E}}}} \\
&\quad + C_1\sigma_2^{-m}\sum_{k=0}^{m-1}\sum_{s=0}^{m-1}\frac{1}{s!}\left\|(\mathrm{Sym}_s\circ\mathrm{id}_{\mathsf{T}^*\mathsf{E}})\circ(\nabla^{\mathsf{M},\pi_{\mathsf{E}},s}\lambda)^{\mathrm{v}}(e)\right\|_{\mathbb{G}_{\mathsf{M},\pi_{\mathsf{E}}}} \\
&= (m+1)C_1\sigma_2^{-m}\sum_{s=0}^{m}\frac{1}{s!}\left\|(\mathrm{Sym}_s\circ\mathrm{id}_{\mathsf{T}^*\mathsf{E}})\circ(\nabla^{\mathsf{M},\pi_{\mathsf{E}},s}\lambda)^{\mathrm{e}}(e)\right\|_{\mathbb{G}_{\mathsf{M},\pi_{\mathsf{E}}}} \\
&\quad + (m+1)C_1\sigma_2^{-m}\sum_{s=0}^{m-1}\frac{1}{s!}\left\|(\mathrm{Sym}_s\circ\mathrm{id}_{\mathsf{T}^*\mathsf{E}})\circ(\nabla^{\mathsf{M},\pi_{\mathsf{E}},s}\lambda)^{\mathrm{v}}(e)\right\|_{\mathbb{G}_{\mathsf{M},\pi_{\mathsf{E}}}}.
\end{aligned}$$

Now let $\sigma < \sigma_2$ and note that

$$\lim_{m\to\infty}(m+1)\frac{\sigma_2^{-m}}{\sigma^{-m}} = 0.$$

Thus there exists $N \in \mathbb{Z}_{>0}$ such that

$$(m+1)C_1\sigma_2^{-m} \le C_1\sigma^{-m}, \qquad m \ge N.$$

Let

$$C = \max\left\{C_1, 2C_1\frac{\sigma}{\sigma_2}, 3C_1\left(\frac{\sigma}{\sigma_2}\right)^2, \dots, (N+1)C_1\left(\frac{\sigma}{\sigma_2}\right)^N\right\}.$$

We then immediately have $(m+1)C_1\sigma_2^{-m} \le C\sigma^{-m}$ for all $m \in \mathbb{Z}_{\ge 0}$. We then have, using (1.4),

$$\begin{aligned}
&\|j_m\lambda^{\mathrm{e}}(e)\|_{\mathbb{G}_{\mathsf{E},m}} \\
&\le C\sigma^{-m}\left(\sum_{s=0}^{m}\frac{1}{s!}\left\|(\mathrm{Sym}_s\circ\mathrm{id}_{\mathsf{T}^*\mathsf{E}})\circ(\nabla^{\mathsf{M},\pi_{\mathsf{E}},s}\lambda)^{\mathrm{e}}(e)\right\|_{\mathbb{G}_{\mathsf{M},\pi_{\mathsf{E}}}}\right. \\
&\quad\left. + \sum_{s=0}^{m-1}\frac{1}{s!}\left\|(\mathrm{Sym}_s\circ\mathrm{id}_{\mathsf{T}^*\mathsf{E}})\circ(\nabla^{\mathsf{M},\pi_{\mathsf{E}},s}\lambda)^{\mathrm{v}}(e)\right\|_{\mathbb{G}_{\mathsf{M},\pi_{\mathsf{E}}}}\right) \\
&\le C\sqrt{m+1}\,\sigma^{-m}(\|j_m\lambda(\pi_{\mathsf{E}}(e))(e)\|_{\mathbb{G}_{\mathsf{M},\pi_{\mathsf{E}},m}} + \|j_{m-1}\lambda(\pi_{\mathsf{E}}(e))\|_{\mathbb{G}_{\mathsf{M},\pi_{\mathsf{E}},m-1}}).
\end{aligned}$$

By modifying C and σ just as we did in the preceding, we get

$$\|j_m\lambda^{\mathrm{e}}(e)\|_{\mathbb{G}_{\mathsf{E},m}} \leq C\sigma^{-m}(\|j_m\lambda(\pi_{\mathsf{E}}(e))(e)\|_{\mathbb{G}_{\mathsf{M},\pi_{\mathsf{E}},m}} + \|j_{m-1}\lambda(\pi_{\mathsf{E}}(e))\|_{\mathbb{G}_{\mathsf{M},\pi_{\mathsf{E}},m-1}}).$$

We take

$$\alpha = \max\{1, \sup\{\|e\|_{\mathbb{G}_{\pi_{\mathsf{E}}}} \mid e \in \mathcal{L}\}$$

and then use Lemma 4.11 to arrive at

$$\|j_m\lambda^{\mathrm{e}}(e)\|_{\mathbb{G}_{\mathsf{E},m}} \leq 2\alpha C\sigma^{-m}\|j_m\lambda(\pi_{\mathsf{E}}(e))(e)\|_{\mathbb{G}_{\mathsf{M},\pi_{\mathsf{E}},m}}.$$

Now, given $\boldsymbol{a} \in c_0(\mathbb{Z}_{\geq 0}; \mathbb{R}_{>0})$, we define $\boldsymbol{a}' \in c_0(\mathbb{Z}_{\geq 0}; \mathbb{R}_{>0})$ by $a'_0 = 2\alpha C a_0$ and $a'_j = a_j\sigma^{-1}$, $j \in \mathbb{Z}_{>0}$, we then have

$$q^{\omega}_{\mathcal{L},\boldsymbol{a}}(\lambda^{\mathrm{e}}) \leq p^{\omega}_{\pi_{\mathsf{E}}(\mathcal{L}),\boldsymbol{a}'}(\lambda),$$

and this gives continuity of vertical evaluation.

Now we turn to showing that the mapping of the lemma is open onto its image. By Lemma 3.52, we have

$$\begin{aligned}
\lambda^{\mathrm{e}}(e) &= \lambda^{\mathrm{e}}(e), \\
(\nabla^{\pi_{\mathsf{E}}}\lambda)^{\mathrm{e}}(e) &= \widehat{B}^1_1(\nabla^{\mathsf{E}}\lambda^{\mathrm{e}}(e)) + \widehat{B}^1_0(\lambda^{\mathrm{e}}(e)) + \widehat{D}^1_0(\lambda^{\mathrm{v}}(e)), \\
(\mathrm{Sym}_2 \otimes \mathrm{id}_{\mathsf{TE}}) \circ (\nabla^{\mathsf{M},\pi_{\mathsf{E}},2}\lambda)^{\mathrm{e}}(e) &= \widehat{B}^2_2(\nabla^{\mathsf{E},2}\lambda^{\mathrm{e}}(e)) + \widehat{B}^2_1(\nabla^{\mathsf{E}}\lambda^{\mathrm{e}}(e)) + \widehat{B}^2_0(\lambda^{\mathrm{e}}(e)) \\
&\quad + \widehat{D}^2_1((\nabla^{\mathsf{M},\pi_{\mathsf{E}}}\lambda)^{\mathrm{v}}(e)) + \widehat{D}^1_0(\lambda^{\mathrm{v}}(e)), \\
&\ \ \vdots \\
(\mathrm{Sym}_m \otimes \mathrm{id}_{\mathsf{T}^*\mathsf{E}}) \circ (\nabla^{\mathsf{M},\pi_{\mathsf{E}},m}\lambda)^{\mathrm{e}}(e) &= \sum_{s=0}^{m} \widehat{B}^m_s((\mathrm{Sym}_s \otimes \mathrm{id}_{\mathsf{T}^*\mathsf{E}}) \circ \nabla^{\mathsf{E},s}\lambda^{\mathrm{e}}(e)) \\
&\quad + \sum_{s=0}^{m-1} \widehat{D}^m_s((\mathrm{Sym}_s \otimes \mathrm{id}_{\mathsf{T}^*\mathsf{E}}) \circ \nabla^{\mathsf{E},s}\lambda^{\mathrm{v}}(e)).
\end{aligned} \tag{5.8}$$

Just as in the proof of Theorem 5.17, by Lemmata 4.11 and 4.13, and the appropriate analogue of Eq. (3.33) that would appear in a fully fleshed out proof of Lemma 3.50, we have bounds

$$\begin{aligned}
\|\widehat{B}^m_s(\mathrm{Sym}_s(\beta_s))\|_{\mathbb{G}_{\mathsf{E}}} &= \|\mathrm{Sym}_m \circ B^m_s(\beta_s)\|_{\mathbb{G}_{\mathsf{E}}} \leq \|B^m_s\|_{\mathbb{G}_{\mathsf{E}}}\|\beta_s\|_{\mathbb{G}_{\mathsf{E}}}, \\
\|\widehat{D}^m_s(\mathrm{Sym}_s(\gamma_s))\|_{\mathbb{G}_{\mathsf{E}}} &= \|\mathrm{Sym}_m \circ D^m_s(\gamma_s)\|_{\mathbb{G}_{\mathsf{E}}} \leq \|D^m_s\|_{\mathbb{G}_{\mathsf{E}}}\|\gamma_s\|_{\mathbb{G}_{\mathsf{E}}}.
\end{aligned}$$

Proceeding analogously to the continuity proof above and using Lemma 4.6, we deduce that there exist $C_1, \sigma_1 \in \mathbb{R}_{>0}$ such that

$$\begin{aligned}\|j_m\lambda(\pi_{\mathsf{E}}(e))(e)\|_{\mathbb{G}_{\mathsf{M},\pi_{\mathsf{E}},m}} \\ \leq C_1\sigma_1^{-m}(\|j_m\lambda^{\mathrm{e}}(e)\|_{\mathbb{G}_{\mathsf{E},m}} + \|j_{m-1}\lambda^{\mathrm{v}}(e)\|_{\mathbb{G}_{\mathsf{E},m-1}}), \qquad e \in \mathcal{L}. \qquad (5.9)\end{aligned}$$

Now let $\mathcal{K} \subseteq \mathsf{M}$ be compact and let $\boldsymbol{a} \in c_0(\mathbb{Z}_{\geq 0}, \mathbb{R}_{>0})$. Define

$$\mathcal{L} = \pi_{\mathsf{E}}^{-1}(\mathcal{K}) \cap \{e \in \mathsf{E} \mid \|e\|_{\mathbb{G}_{\pi_{\mathsf{E}}}} = 1\},$$

noting that $\mathcal{L}$ is compact. Let $n = \dim(\mathsf{M})$ and let k be the fibre dimension of E. By Lemma 4.12, and Eqs. (1.3) and (4.10), we have

$$\begin{aligned}\|j_m\lambda(x)\|_{\mathbb{G}_{\mathsf{M},\pi_{\mathsf{E}},m}} &\leq \sum_{j=0}^{m}\sqrt{k\binom{n+j-1}{j}}\sup\{\|j_m\lambda(\pi_{\mathsf{E}}(e))(e)\|_{\mathbb{G}_{\mathsf{M},\pi_{\mathsf{E}},m}} \mid e \in \mathcal{L}\} \\ &\leq \sum_{j=0}^{m}k\binom{n+j-1}{j}\sup\{\|j_m\lambda(\pi_{\mathsf{E}}(e))(e)\|_{\mathbb{G}_{\mathsf{M},\pi_{\mathsf{E}},m}} \mid e \in \mathcal{L}\} \\ &\leq m^2 2^{n+m}\sup\{\|j_m\lambda(\pi_{\mathsf{E}}(e))(e)\|_{\mathbb{G}_{\mathsf{M},\pi_{\mathsf{E}},m}} \mid e \in \mathcal{L}\}\end{aligned}$$

for $x \in \mathcal{K}$. For $\sigma_2 < \frac{1}{2}$,

$$\lim_{m\to\infty} m^2\frac{2^m}{\sigma_2^{-m}} = 0.$$

By by now familiar arguments, one of which the reader can find in the first part of the proof, we can combine this with (5.9) to arrive at $C, \sigma \in \mathbb{R}_{>0}$ for which

$$\begin{aligned}\|j_m\lambda(x)\|_{\mathbb{G}_{\mathsf{M},\pi_{\mathsf{E}},m}} \leq C\sigma^{-m}(&\sup\{\|j_m\lambda^{\mathrm{e}}(e)\|_{\mathbb{G}_{\mathsf{M},\pi_{\mathsf{E}},m}} \mid e \in \mathcal{L}\} \\ &+ \sup\{\|j_m\lambda^{\mathrm{v}}(e)\|_{\mathbb{G}_{\mathsf{E},m}} \mid e \in \mathcal{L}\})\end{aligned}$$

for $x \in \mathcal{K}$. Taking $\boldsymbol{a}' \in c_0(\mathbb{Z}_{\geq 0}; \mathbb{R}_{>0})$ to be defined by $a'_0 = Ca_0$, $a'_j = \sigma^{-1}a_j$, $j \in \mathbb{Z}_{>0}$, we have

$$q^{\omega}_{\mathcal{K},\boldsymbol{a}}(\lambda) \leq p^{\omega}_{\mathcal{L},\boldsymbol{a}'}(\lambda^{\mathrm{e}}) + p^{\omega}_{\mathcal{L},\boldsymbol{a}'}(\lambda^{\mathrm{v}}).$$

By Lemma 1 from the proof of Theorem 5.1, this shows that the mapping

$$\Gamma^{\omega}(\mathsf{E}^*) \ni \lambda \mapsto (\lambda^{\mathrm{e}}, \lambda^{\mathrm{v}}) \in \mathrm{C}^{\omega}(\mathsf{E}) \oplus \Gamma^{\omega}(\mathsf{TE})$$

is open onto its image. This part of the lemma now follows from the following simple fact.

Lemma 1 *Let $\mathcal{S}$, $\mathcal{T}_1$, and $\mathcal{T}_2$ be topological spaces and let $\Phi\colon \mathcal{S} \to \mathcal{T}_1 \times \mathcal{T}_2$ be an open mapping onto its image. Then the mappings $\mathrm{pr}_1 \circ \Phi$ and $\mathrm{pr}_2 \circ \Phi$ are open onto their images.*

Proof Let $\mathcal{O} \subseteq \mathcal{S}$ be open so that $\Phi(\mathcal{O})$ is open in image(Φ). Then, for each $(y_1, y_2) \in \mathcal{O}$, there exists a neighbourhood $\mathcal{N}_1 \subseteq \mathrm{image}(\mathrm{pr}_1 \circ \Phi)$ of y_1 and a neighbourhood $\mathcal{N}_2 \subseteq \mathrm{image}(\mathrm{pr}_2 \circ \Phi)$ of x_2 such that $\mathcal{N}_1 \times \mathcal{N}_2 \subseteq \Phi(\mathcal{O})$. This immediately gives the lemma. ∇

Thus we arrive at the conclusion that the mapping

$$\Gamma^\omega(\mathsf{E}^*) \ni \lambda \mapsto \lambda^{\mathrm{e}} \in C^\omega(\mathsf{E})$$

is open onto its image, as desired. □

Finally, we consider vertical evaluations of sections of the endomorphism bundle.

Theorem 5.23 (Vertical Evaluation of Endomorphisms Is a Topological Monomorphism) *Let $\pi_{\mathsf{E}}\colon \mathsf{E} \to \mathsf{M}$ be a C^ω-vector bundle. Then the mapping*

$$\Gamma^\omega(\mathrm{End}(\mathsf{E})) \ni L \mapsto L^{\mathrm{e}} \in \Gamma^\omega(\mathsf{TE})$$

is a topological monomorphism.

Proof This follows in the same manner as Theorem 5.22, using Lemmata 3.56, 3.57, and 4.7. □

As an illustration of how continuity of these lifts can be helpful, let us consider the continuity of the map that assigns to a vector field on a manifold the tangent lift of that vector field. Precisely, let M be a real analytic manifold and let $X \in \Gamma^\omega(\mathsf{TM})$ be a real analytic vector field. The ***tangent lift*** of X is the vector field $X^T \in \Gamma^\omega(\mathsf{TTM})$ on TM whose flow is the derivative of the flow for X:

$$\Phi_t^{X^T}(v_x) = T_x\Phi_t^X(v_x) \quad \Rightarrow \quad X^T = \left.\frac{\mathrm{d}}{\mathrm{d}t}\right|_{t=0} T_x\Phi_t^X(v_x). \tag{5.10}$$

Let us give a formula for the tangent lift that reduces the continuity of the mapping $X \mapsto X^T$ to continuity of familiar operations.

Lemma 5.24 (Decomposition of the Tangent Lift via an Affine Connection) *Let $r \in \{\infty, \omega\}$ and let M be a C^r-manifold with a C^r-affine connection ∇^{M}. Then*

$$X^T(v_x) = \mathrm{hlft}(v_x, X(x)) + \mathrm{vlft}(v_x, \nabla^{\mathsf{M}}_{v_x} X + T(X(x), v_x)),$$

where T is the torsion of ∇^{M}.

Proof Let $v_x \in \mathsf{TM}$ and let $Y \in \Gamma^r(\mathsf{TM})$ be such that $Y(x) = v_x$. Note that

$$\frac{\mathrm{d}}{\mathrm{d}s}\bigg|_{s=0} \Phi_t^X \circ \Phi_s^Y(x) = T_x\Phi_t^X(Y(x)).$$

Also compute

$$\begin{aligned}\frac{\mathrm{d}}{\mathrm{d}s}\bigg|_{s=0} \Phi_t^X \circ \Phi_s^Y &= \frac{\mathrm{d}}{\mathrm{d}s}\bigg|_{s=0} \Phi_s^Y \circ \Phi_t^X \circ \Phi_{-t}^X \Phi_{-s}^Y \circ \Phi_t^X \circ \Phi_s^Y(x) \\ &= Y(\Phi_t^X(x)) + T_x\Phi_t^X\left(\frac{\mathrm{d}}{\mathrm{d}s}\bigg|_{s=0} \Phi_{-t}^X \circ \Phi_{-s}^Y \circ \Phi_t^X \circ \Phi_s^Y(\Phi_t^X(x))\right).\end{aligned}$$

Note that, for $f \in \mathrm{C}^r(\mathsf{M})$,

$$f \circ \Phi_{-t}^X \circ \Phi_{-s}^Y \circ \Phi_t^X \circ \Phi_s^Y(x) = f(x) + st\mathscr{L}_{[Y,X]}f(x) + o(|st|),$$

by [1, Proposition 4.2.34]. Therefore,

$$\frac{\mathrm{d}}{\mathrm{d}s}\bigg|_{s=0} \Phi_{-t}^X \circ \Phi_{-s}^Y \circ \Phi_t^X \circ \Phi_s^Y(\Phi_t^X(x)) = t[Y,X](\Phi_t^X(x)).$$

Putting the above calculations together gives

$$T_x\Phi_t^X(Y(x)) = Y(\Phi_t^X(x)) - t[X,Y](\Phi_t^X(x)).$$

Thus, making use of (5.10),

$$\Phi_t^{X^{\mathrm{h}}} \circ \Phi_t^{X^T}(Y(x)) = \tau_{\gamma_-}^{(t,0)}(Y(\Phi_t^X(x)) - t[X,Y](\Phi_t^X(x))),$$

where γ_- is the integral curve of $-X$ through $\Phi_t^X(x)$ and τ_{γ_-} is parallel translation along γ_-. If γ is the integral curve of X through x note that $\tau_{\gamma_-}^{(t,0)} = \tau_\gamma^{(0,t)}$. Now we compute

$$\begin{aligned}\frac{\mathrm{d}}{\mathrm{d}t}\bigg|_{t=0} \Phi_t^{-X^{\mathrm{h}}} \circ \Phi_t^{X^T}(Y(x)) &= \frac{\mathrm{d}}{\mathrm{d}t}\bigg|_{t=0} \tau_\gamma^{(0,t)}(Y(\Phi_t^X(x)) - t[X,Y](\Phi_t^X(x))) \\ &= \nabla_X Y(x) - [X,Y](x) = \nabla_Y X(x) + T(X(x),Y(x)).\end{aligned}$$

Note that, since X^T and X^{h} are both vector fields over X, it follows that

$$t \mapsto \tau_\gamma^{(0,t)}(Y(\Phi_t^X(x)))$$

is a curve in $\mathsf{T}_x\mathsf{M}$. Thus the derivative of this curve at $t = 0$ is in $\mathsf{V}_{Y(x)}\mathsf{TM}$. Thus we have shown that

$$\left.\frac{\mathrm{d}}{\mathrm{d}t}\right|_{t=0} \Phi_t^{-X^{\mathrm{h}}} \circ \Phi_t^{X^T}(v_x) = \mathrm{vlft}(v_x, \nabla_{v_x} X(x) + T(X(x), v_x)).$$

Finally, for $f \in \mathrm{C}^r(\mathsf{M})$, using the first terms in the Baker–Campbell–Hausdorff formula as in [1, Corollary 4.1.27], we have

$$f \circ \Phi_t^{-X^{\mathrm{h}}} \circ \Phi_t^{X^T}(v_x) = f \circ \Phi_t^{X^T - X^{\mathrm{h}}} + o(|t|^2).$$

Differentiating with respect to t and evaluating at $t = 0$ gives the result. □

Now we can combine Theorems 5.1(i), 5.19, 5.20, and 5.23, and Corollary 5.14 to give the following result.

Corollary 5.25 (Continuity of Tangent Lift) *If* M *is a* C^ω*-manifold, then the mapping*

$$\Gamma^\omega(\mathsf{TM}) \ni X \mapsto X^T \in \Gamma^\omega(\mathsf{TTM})$$

is continuous.

5.4 Continuity of Composition Operators

In this section we consider continuity of the various sorts of compositions. For real analytic manifolds M and N, there are three sorts of composition operators we can consider:

$$\begin{aligned} C_\Phi \colon \mathrm{C}^\omega(\mathsf{N}) &\to \mathrm{C}^\omega(\mathsf{M}) & S_f \colon \mathrm{C}^\omega(\mathsf{M};\mathsf{N}) &\to \mathrm{C}^\omega(\mathsf{M}) \\ f &\mapsto \Phi^* f, & \Phi &\mapsto f \circ \Phi, \end{aligned}$$

$$\begin{aligned} C_{\mathsf{M},\mathsf{N}} \colon \mathrm{C}^\omega(\mathsf{M};\mathsf{N}) \times \mathrm{C}^\omega(\mathsf{N}) &\to \mathrm{C}^\omega(\mathsf{M}) \\ (\Phi, f) &\mapsto f \circ \Phi, \end{aligned}$$

the first being defined for fixed $\Phi \in \mathrm{C}^\omega(\mathsf{M};\mathsf{N})$ and the second for fixed $f \in \mathrm{C}^\omega(\mathsf{N})$. We call these the ***composition operator*** associated with Φ, the ***superposition operator*** associated with f, and the ***joint composition operator***, respectively. The superposition operator is also known as the "nonlinear composition operator" or the "Nemytskii operator." In general, one studies these mappings for classes of function spaces, e.g., Lebesgue spaces or Hardy spaces. The questions one can ask for such operators include the following.

1. *Well-definedness*: Here, for instance, one wishes to know for which f's or Φ's do the operators S_f or C_Φ maps one function space into another.
2. *Continuity*: The continuity of the linear composition operator C_Φ is often fairly easily established, and also often coincides with the well-definedness of the operator. The continuity of the nonlinear superposition operator S_f, however, is often quite difficult to establish. Moreover, there are important cases where continuity of this operator does not coincide with its well-definedness, a well-known example of this being in the Lipschitz class [21].
3. *Boundedness*: Of course, in the linear case, continuity of C_Φ implies boundedness, although not necessarily the converse if the function spaces in question are not metrisable or, better, not bornological [37, Theorem 13.1.1]. For example, in the real analytic case in which we are interested here, these spaces are not metrisable, but are bornological. The boundedness and continuity of S_f are not generally logically comparable, e.g., this mapping is nonlinear.
4. *Real analyticity*: It is sometimes the case that the superposition operator, though nonlinear, admits a convergent power series expansion, in which case it is said to be "real analytic." We mention this here mostly because this is *not* what we are considering here; here we are considering operators on spaces of real analytic mappings, not operators which are themselves real analytic.

A reader interested in a detailed discussion of superposition operators for various classes of function spaces is referred to [3].

5.4.1 The Real Analytic Composition Operator

We first consider the continuity of the linear composition operator. This continuity is established is a rather *ad hoc* way during the course of the proof of their Lemma 2.5 by Jafarpour and Lewis [36] using a local description of the real analytic topology. Here we give an intrinsic proof using our more systematic analysis.

Theorem 5.26 (Composition Induces a Continuous Map Between Function Spaces) *Let* M *and* N *be* C^ω*-manifolds. If* $\Phi \in \mathrm{C}^\omega(\mathsf{M}; \mathsf{N})$*, then the mapping*

$$\begin{aligned} \Phi^*\colon \mathrm{C}^\omega(\mathsf{N}) &\to \mathrm{C}^\omega(\mathsf{M}) \\ f &\mapsto f \circ \Phi \end{aligned}$$

is continuous. Moreover, if Φ *is a proper surjective submersion or a proper embedding, then* Φ^* *is open onto its image. In case* Φ *is a proper embedding, for any compact* $\mathcal{K} \subseteq \mathsf{M}$ *and any* $\boldsymbol{a} \in c_0(\mathbb{Z}_{\ge 0}; \mathbb{R}_{>0})$*, there exists* $\boldsymbol{a}' \in c_0(\mathbb{Z}_{\ge 0}; \mathbb{R}_{>0})$ *such that*

$$q^\omega_{\Phi(\mathcal{K}),\boldsymbol{a}}(f) \le p^\omega_{\mathcal{K},\boldsymbol{a}'}(\Phi^* f), \qquad f \in \mathrm{C}^\omega(\mathsf{N}),$$

where $p^{\omega}_{\mathcal{K},\boldsymbol{a}}$ *and* $q^{\omega}_{\mathcal{L},\boldsymbol{a}}$ *are the seminorms for* $\mathrm{C}^{\omega}(\mathsf{M})$ *and* $\mathrm{C}^{\omega}(\mathsf{N})$*, respectively, associated to* $\mathcal{K} \subseteq \mathsf{M}$ *and* $\mathcal{L} \subseteq \mathsf{N}$ *compact, and* $\boldsymbol{a} \in c_0(\mathbb{Z}_{\geq 0}; \mathbb{R}_{>0})$.

Proof We let ∇^{M} and ∇^{N} be C^{ω}-affine connections on M and N, respectively, and let $\mathbb{G}_{\mathsf{M}}$ and $\mathbb{G}_{\mathsf{N}}$ be C^{ω}-Riemannian metrics on M and N, respectively.

From Lemma 3.63 we have the formula

$$\mathrm{Sym}_m \circ \nabla^{\mathsf{M},m} \Phi^* f = \sum_{s=0}^{m} \widehat{A}^m_s (\mathrm{Sym}_s \circ \Phi^* \nabla^{\mathsf{N},s} f). \tag{5.11}$$

By Lemma 4.11, we have

$$\|A^m_s(\beta_s)\|_{\mathbb{G}_{\mathsf{M}}} \leq \|A^m_s\|_{\mathbb{G}_{\mathsf{M}},\mathbb{G}_{\mathsf{N}}} \|\beta_s\|_{\mathbb{G}_{\mathsf{N}}}$$

for $\beta_s \in \mathrm{T}^s(\mathrm{T}^*_x\mathsf{M})$, $m \in \mathbb{Z}_{>0}$, and $s \in \{0, 1, \dots, m\}$. By Lemma 4.13,

$$\|\mathrm{Sym}_s(A)\|_{\mathbb{G}_{\mathsf{M}},\mathbb{G}_{\mathsf{N}}} \leq \|A\|_{\mathbb{G}_{\mathsf{M}},\mathbb{G}_{\mathsf{N}}}$$

for $A \in \mathrm{T}^s(\mathrm{T}^*\mathsf{N})$ and $s \in \mathbb{Z}_{>0}$. Thus, recalling (3.29) (and its analogue that would arise in a spelled out proof of Lemma 3.63),

$$\|\widehat{A}^m_s(\mathrm{Sym}_s(\beta_s))\|_{\mathbb{G}_{\mathsf{M}}} = \|\mathrm{Sym}_m \circ A^m_s(\beta_s)\|_{\mathbb{G}_{\mathsf{M}}} \leq \|A^m_s\|_{\mathbb{G}_{\mathsf{M}},\mathbb{G}_{\mathsf{N}}} \|\beta_s\|_{\mathbb{G}_{\mathsf{M}}},$$

for $\beta_s \in \mathrm{T}^s(\Phi^*\mathrm{T}^*\mathsf{N})$, $m \in \mathbb{Z}_{\geq 0}$, $s \in \{1, \dots, m\}$.

Let $\mathcal{K} \subseteq \mathsf{M}$ be compact. By Lemmata 4.17 and 4.18 with $r = 0$, there exist $C_1, \sigma_1, \rho_1 \in \mathbb{R}_{>0}$ such that

$$\|A^k_s(x)\|_{\mathbb{G}_{\mathsf{M}},\mathbb{G}_{\mathsf{N}}} \leq C_1 \sigma_1^{-k} \rho_1^{-(k-s)} (k-s)!, \qquad k \in \mathbb{Z}_{\geq 0},\ s \in \{0, 1, \dots, k\},\ x \in \mathcal{K}.$$

By Lemma 4.8, let $C_2 \in \mathbb{R}_{>0}$ be such that

$$\|\Phi^* \nabla^{\mathsf{N},m} f(x)\|_{\mathbb{G}_{\mathsf{M}}} \leq C_2^m \|\nabla^{\mathsf{N},m} f(\Phi(x))\|_{\mathbb{G}_{\mathsf{N}}}, \qquad x \in \mathcal{K},\ m \in \mathbb{Z}_{\geq 0}.$$

Without loss of generality, we assume that $C_1, C_2 \geq 1$ and $\sigma_1, \rho_1 \leq 1$. Thus, abbreviating $\sigma_2 = \sigma_1 \rho_1$, we have

$$\|\widehat{A}^k_s(\Phi^* \mathrm{Sym}_s \circ \nabla^{\mathsf{N},s} f(x))\|_{\mathbb{G}_{\mathsf{M}}} \leq C_1 C_2^s \sigma_2^{-k} (k-s)! \|\mathrm{Sym}_s \circ \nabla^{\mathsf{N},s} f(\Phi(x))\|_{\mathbb{G}_{\mathsf{N}}}$$

for $k \in \mathbb{Z}_{\geq 0}$, $s \in \{0, 1, \dots, k\}$, $x \in \mathcal{K}$. Thus, by (1.4) and (5.11),

$$\begin{aligned}
\|j_m(\Phi^* f)(x)\|_{\mathbb{G}_{\mathsf{M},m}} &\leq \sum_{k=0}^{m} \frac{1}{k!} \|\mathrm{Sym}_k \circ \nabla^{\mathsf{M},k} \Phi^* f(x)\|_{\mathbb{G}_{\mathsf{M}}} \\
&= \sum_{k=0}^{m} \frac{1}{k!} \left\| \sum_{s=0}^{k} \widehat{A}_s^k (\Phi^* \mathrm{Sym}_s \circ \nabla^{\mathsf{N},s} f(\Phi(x))) \right\|_{\mathbb{G}_{\mathsf{M}}} \\
&\leq \sum_{k=0}^{m} \sum_{s=0}^{k} C_1 \sigma_2^{-k} \frac{s!(k-s)!}{k!} \frac{C_2^s}{s!} \|\mathrm{Sym}_s \circ \nabla^{\mathsf{N},s} f(\Phi(x))\|_{\mathbb{G}_{\mathsf{N}}}
\end{aligned}$$

for $x \in \mathcal{K}$ and $m \in \mathbb{Z}_{\geq 0}$. Now note that

$$\frac{s!(k-s)!}{k!} \leq 1, \quad C_1 \sigma_2^{-k} C_2^s \leq C_1 C_2^m \sigma_2^{-m},$$

for $s \in \{0, 1, \dots, m\}$, $k \in \{0, 1, \dots, s\}$, since $\sigma_2 \leq 1$. Then

$$\begin{aligned}
\|j_m(\Phi^* f)(x)\|_{\mathbb{G}_{\mathsf{M},m}} &\leq C_1 C_2^m \sigma_2^{-m} \sum_{k=0}^{m} \sum_{s=0}^{k} \frac{1}{s!} \|\mathrm{Sym}_s \circ \nabla^{\mathsf{N},s} f(\Phi(x))\|_{\mathbb{G}_{\mathsf{N}}} \\
&\leq C_1 C_2^m \sigma_2^{-m} \sum_{k=0}^{m} \sum_{s=0}^{m} \frac{1}{s!} \|\mathrm{Sym}_s \circ \nabla^{\mathsf{N},s} f(\Phi(x))\|_{\mathbb{G}_{\mathsf{N}}} \\
&= (m+1) C_1 C_2^m \sigma_2^{-m} \sum_{s=0}^{m} \frac{1}{s!} \|\mathrm{Sym}_s \circ \nabla^{\mathsf{N},s} f(\Phi(x))\|_{\mathbb{G}_{\mathsf{N}}}.
\end{aligned}$$

Now let $\sigma < C_2^{-1} \sigma_2$ and note that

$$\lim_{m \to \infty} (m+1) \frac{C_2^m \sigma_2^{-m}}{\sigma^{-m}} = 0.$$

Thus there exists $N \in \mathbb{Z}_{>0}$ such that

$$(m+1) C_1 C_2^m \sigma_2^{-m} \leq C_1 \sigma^{-m}, \qquad m \geq N.$$

Let

$$C = \max \left\{ C_1, 2C_1 C_2 \frac{\sigma}{\sigma_2}, 3C_1 C_2^2 \left(\frac{\sigma}{\sigma_2}\right)^2, \dots, (N+1) C_1 C_2^N \left(\frac{\sigma}{\sigma_2}\right)^N \right\}.$$

We then immediately have $(m+1)C_1C_2^m\sigma_2^{-m} \le C\sigma^{-m}$ for all $m \in \mathbb{Z}_{\ge 0}$. We then have, using (1.4),

$$\begin{aligned}\|j_m(\Phi^* f)(x)\|_{\mathbb{G}_{\mathsf{M},m}} &\le C\sigma^{-m}\left(\sum_{s=0}^{m}\frac{1}{s!}\|\mathrm{Sym}_s \circ \nabla^{\mathsf{N},s} f(\Phi(x))\|_{\mathbb{G}_{\mathsf{N}}}\right)\\ &= C\sqrt{m+1}\sigma^{-m}\|j_m f(\Phi(x))\|_{\mathbb{G}_{\mathsf{N},m}}.\end{aligned}$$

By modifying C and σ guided by what we did just preceding, we get

$$\|j_m(\Phi^* f)(x)\|_{\mathbb{G}_{\mathsf{M},m}} \le C\sigma^{-m}\|j_m f(\Phi(x))\|_{\mathbb{G}_{\mathsf{N},m}}.$$

Now, for $\boldsymbol{a} \in c_0(\mathbb{Z}_{\ge 0};\mathbb{R}_{>0})$, let $\boldsymbol{a}' \in c_0(\mathbb{Z}_{\ge 0};\mathbb{R}_{>0})$ be defined by $a_0' = Ca_0$ and $a_j' = a_j\sigma^{-1}$, $j \in \mathbb{Z}_{>0}$. Then we have

$$p^{\omega}_{\mathcal{K},\boldsymbol{a}}(\Phi^* f) \le q^{\omega}_{\Phi(\mathcal{K}),\boldsymbol{a}'}(f),$$

and this gives continuity of Φ^*.

Now we turn to the final assertion concerning the openness of Φ^* in particular cases. First we note that, by Lemma 3.64, we have

$$\mathrm{Sym}_m \circ \Phi^*\nabla^{\mathsf{N},m} f(x) = \sum_{s=0}^{m}\widehat{B}^m_s(\mathrm{Sym}_s \circ \nabla^{\mathsf{M},s}\Phi^* f(x)).$$

First consider the case where Φ is a proper surjective submersion. For $\mathcal{L} \subseteq \mathsf{N}$ compact and for $y \in \mathcal{L}$, since Φ is surjective, there exists $x \in \mathsf{M}$ such that $\Phi(x) = y$. Also, since Φ is proper, $\Phi^{-1}(\mathcal{L})$ is compact. We can now reproduce the steps from the proof above, now making use of the second part of Lemma 4.8, to prove that

$$q^{\omega}_{\mathcal{L},\boldsymbol{a}}(f) \le p^{\omega}_{\Phi^{-1}(\mathcal{L}),\boldsymbol{a}'}(\Phi^* f),$$

which suffices to prove the openness of Φ^* by Lemma 1 from the proof of Theorem 5.1.

Finally consider the case where Φ is a proper embedding. Here, Lemma 1 from the proof of Theorem 2.25 immediately gives openness of Φ^* in the case that Φ is a proper embedding. For the final assertion, we can follow the same argument as was sketched for the openness of Φ^* when Φ is a proper surjective submersion to give

$$q^{\omega}_{\Phi(\mathcal{K}),\boldsymbol{a}}(f) \le p^{\omega}_{\mathcal{K},\boldsymbol{a}'}(\Phi^* f),$$

as desired. □

The matter of determining general conditions under which Φ^* is an homeomorphism onto its image or has closed image are taken up in [18, 19].

Remarks 5.27 (Adaptation to the Smooth Case) The preceding proof can be adapted to the smooth case. Indeed, much of the elaborate work of the proof can be simplified by not having to pay attention to the exponential growth of m-jet norms as $m \to \infty$. In the smooth case, one works with fixed orders of derivatives. ∘

5.4.2 The Real Analytic Superposition Operator

Note that our definition of the weak-PB topology ensures continuity of the superposition operator $S_f \colon \mathsf{C}^\omega(\mathsf{M}; \mathsf{N}) \to \mathsf{C}^\omega(\mathsf{M})$ for $f \in \mathsf{C}^\omega(\mathsf{N})$. Indeed, the weak-PB topology is defined expressly so that these mappings are continuous. Thus the most meaningful assertions regarding the superposition operator result from the other descriptions of the topology for the space of real analytic mappings as given in Theorems 2.25 and 2.32, and where continuity of superposition is not a tautology; thus these assertions are really about the various equivalent characterisations of the weak-PB topology, rather than about the continuity of the superposition operator. What we shall do in this section is prove continuity of the *joint* composition operator.

In our development of the continuity results in this chapter up to this point, we have made dedicated use of the seminorms for the real analytic topology developed in Sect. 2.4. Our results in this section have to do with continuity involving spaces of mappings, and so one may like to use the semimetrics (2.20) that define the uniformity for this topology. We do not, however, take this approach, instead proving the result in the holomorphic case, and then using the descriptions from Sect. 2.5.4 for the real analytic topology for spaces of mappings derived from germs of holomorphic extensions. The reader may wish to explore using the semimetrics (2.20) to prove Theorem 5.29 below.

The first step, then, is to prove continuity for the joint composition operator in the holomorphic case.

Theorem 5.28 (Continuity of the Holomorphic Joint Composition Operator) *If* M *and* N *are holomorphic manifolds, then the mapping*

$$\begin{aligned} C_{\mathsf{M},\mathsf{N}} \colon \mathsf{C}^{\mathrm{hol}}(\mathsf{M}; \mathsf{N}) \times \mathsf{C}^{\mathrm{hol}}(\mathsf{N}) &\to \mathsf{C}^{\mathrm{hol}}(\mathsf{M}) \\ (\Phi, f) &\mapsto f \circ \Phi \end{aligned}$$

is continuous, using the compact-open topology for $\mathsf{C}^{\mathrm{hol}}(\mathsf{M}; \mathsf{N})$.

Proof We use the semimetrics (2.21) for the topology of $\mathsf{C}^{\mathrm{hol}}(\mathsf{M}; \mathsf{N})$ and the seminorms (2.2) for the topology of $\mathsf{C}^{\mathrm{hol}}(\mathsf{M})$ and $\mathsf{C}^{\mathrm{hol}}(\mathsf{N})$. We use metrics d_{M} and d_{N} whose metric topologies agree with the topologies for M and N, respectively. We denote the seminorms for $\mathsf{C}^{\mathrm{hol}}(\mathsf{M})$ by $p^{\mathrm{hol}}_{\mathcal{K}}$ for $\mathcal{K} \subseteq \mathsf{M}$ compact, and we denote the seminorms for $\mathsf{C}^{\mathrm{hol}}(\mathsf{N})$ by $q^{\mathrm{hol}}_{\mathcal{L}}$ for $\mathcal{L} \subseteq \mathsf{N}$ compact. Let $\mathcal{K} \subseteq \mathsf{M}$ be compact and let

$\epsilon \in \mathbb{R}_{>0}$. Let $f_0 \in \mathrm{C}^{\mathrm{hol}}(\mathsf{N})$ and $\Phi_0 \in \mathrm{C}^{\mathrm{hol}}(\mathsf{M}; \mathsf{N})$. Let $y \in \Phi_0(\mathcal{K})$ and let $r_y \in \mathbb{R}_{>0}$ be such that

1. $\mathsf{B}_{\mathsf{d}_\mathsf{N}}(r_y/2, y)$ and $\mathsf{B}_{\mathsf{d}_\mathsf{N}}(r_y, y)$ are precompact neighbourhoods of y,
2. $\mathrm{cl}(\mathsf{B}_{\mathsf{d}_\mathsf{N}}(r_y/2, y)) \subseteq \mathsf{B}_{\mathsf{d}_\mathsf{N}}(r_y, y)$, and
3. $|f_0(y') - f_0(y)| < \frac{\epsilon}{6}$ for $y' \in \mathsf{B}_{\mathsf{d}_\mathsf{N}}(r_y, y)$.

Let $y_1, \dots, y_k \in \Phi_0(\mathcal{K})$ be such that $\Phi_0(\mathcal{K}) \subseteq \mathcal{V} \triangleq \cup_{j=1}^k \mathsf{B}_{\mathsf{d}_\mathsf{N}}(r_{y_j}/2, y_j)$. Denote $\mathcal{L} = \cup_{j=1}^k \mathrm{cl}(\mathsf{B}_{\mathsf{d}_\mathsf{N}}(r_{y_j}/2, y_j))$, which is compact. Note that $\mathcal{L} \subseteq \cup_{j=1}^k \mathsf{B}_{\mathsf{d}_\mathsf{N}}(r_{y_j}, y_j)$. By the Lebesgue Number Lemma [11, Theorem 1.6.11], let $\delta \in \mathbb{R}_{>0}$ be such that, if $z_1, z_2 \in \mathcal{L}$ satisfy $\mathsf{d}_\mathsf{N}(z_1, z_2) < \delta$, then $z_1, z_2 \in \mathsf{B}_{\mathsf{d}_\mathsf{N}}(r_{y_j}, y_j)$ for some $j \in \{1, \dots, k\}$. Let

$$\mathcal{O} = \mathcal{B}(\mathcal{K}, \mathcal{L}) \cap \{\Phi \in \mathrm{C}^{\mathrm{hol}}(\mathsf{M}; \mathsf{N}) \mid \mathsf{d}_\mathsf{N}(\Phi(x), \Phi_0(x)) < \delta,\ x \in \mathcal{K}\}.$$

Let $\mathcal{P} \subseteq \mathrm{C}^{\mathrm{hol}}(\mathcal{L})$ be a neighbourhood of f_0 such that

$$q_{\mathcal{L}}^{\mathrm{hol}}(f - f_0) < \frac{\epsilon}{6}, \qquad f \in \mathcal{O}.$$

If $\Phi \in \mathcal{O}$ and if $x \in \mathcal{K}$, then

$$\begin{aligned} &\mathsf{d}_\mathsf{N}(\Phi(x), \Phi_0(x)) < \delta \\ \implies &\Phi(x), \Phi_0(x) \in \mathsf{B}_{\mathsf{d}_\mathsf{N}}(r_{y_j}, y_j) \text{ for some } j \in \{1, \dots, k\} \\ \implies &|f_0 \circ \Phi(x) - f_0 \circ \Phi_0(x)| < \frac{\epsilon}{6}. \end{aligned}$$

Also, if $\Phi \in \mathcal{O}$ then $\Phi(\mathcal{K}) \subseteq \mathcal{V} \subseteq \mathcal{L}$. Therefore, for $\Phi \in \mathcal{O}$, $f \in \mathcal{P}$, and $x \in \mathcal{K}$, we have

$$\begin{aligned} |f \circ \Phi(x) - f \circ \Phi_0(x)| &\leq |f \circ \Phi(x) - f_0 \circ \Phi(x)| + |f_0 \circ \Phi(x) - f_0 \circ \Phi_0(x)| \\ &\quad + |f \circ \Phi_0(x) - f_0 \circ \Phi_0(x)| \\ &\leq \frac{\epsilon}{6} + \frac{\epsilon}{6} + \frac{\epsilon}{6} = \frac{\epsilon}{2}. \end{aligned}$$

Then, for $\Phi \in \mathcal{O}$ and $f \in \mathcal{P}$, we calculate

$$\begin{aligned} p_{\mathcal{K}}^{\mathrm{hol}}(f \circ \Phi - f_0 \circ \Phi_0) &= \sup\{|f \circ \Phi(x) - f_0 \circ \Phi_0(x)| \mid x \in \mathcal{K}\} \\ &\leq \sup\{|f \circ \Phi(x) - f \circ \Phi_0(x)| \mid x \in \mathcal{K}\} \\ &\quad + \sup\{|(f - f_0) \circ \Phi_0(x)| \mid x \in \mathcal{K}\} \\ &\leq \frac{\epsilon}{2} + \frac{\epsilon}{6} < \epsilon. \end{aligned}$$

This gives continuity of the joint composition operator. □

The preceding result is a specialisation to the holomorphic case of a well-known result for continuous mappings, e.g., [33, Exercise 2.4.10]. Indeed, this exercise asks the reader to show continuity for C^{ν}-mappings, $\nu \in \mathbb{Z}_{\geq 0} \cup \{\infty\}$.

We can now state the desired result in the real analytic case.

Theorem 5.29 (Continuity of the Real Analytic Joint Composition Operator) *If* M *and* N *are real analytic manifolds, then the joint composition operator* $C_{\mathsf{M},\mathsf{N}}$ *is continuous.*

Proof We let $\overline{\mathsf{M}}$ and $\overline{\mathsf{N}}$ be complexifications of M and N, respectively, and let $\overline{\mathscr{N}}_{\mathsf{M}}$ and $\overline{\mathscr{N}}_{\mathsf{N}}$ be the directed sets of neighbourhoods of M and N in $\overline{\mathsf{M}}$ and $\overline{\mathsf{N}}$. We consider the product $\overline{\mathscr{N}}_{\mathsf{M}} \times \overline{\mathscr{N}}_{\mathsf{N}}$ to be a directed set by the partial order

$$(\overline{\mathcal{U}}_1, \overline{\mathcal{V}}_1) \preceq (\overline{\mathcal{U}}_2, \overline{\mathcal{V}}_2) \iff \overline{\mathcal{U}}_2 \subseteq \overline{\mathcal{U}}_1 \text{ and } \overline{\mathcal{V}}_2 \subseteq \overline{\mathcal{V}}_1.$$

For $\overline{\mathcal{U}} \in \overline{\mathscr{N}}_{\mathsf{M}}$ and $\overline{\mathcal{V}} \in \overline{\mathscr{N}}_{\mathsf{N}}$, we have the diagram

$$\begin{array}{c}
\mathrm{C}^{\mathrm{hol},\mathbb{R}}(\overline{\mathcal{U}}; \overline{\mathcal{V}}) \times \mathrm{C}^{\mathrm{hol},\mathbb{R}}(\overline{\mathcal{V}}) \\
\Big\downarrow {\scriptstyle r_{\overline{\mathcal{U}},\mathsf{M}} \times \mathrm{id}} \\
\varinjlim_{\overline{\mathcal{U}}' \in \overline{\mathscr{N}}_{\mathsf{M}}} \mathrm{C}^{\mathrm{hol},\mathbb{R}}(\overline{\mathcal{U}}'; \overline{\mathcal{V}}) \times \mathrm{C}^{\mathrm{hol},\mathbb{R}}(\overline{\mathcal{V}}) \\
\Big\| {\scriptstyle \simeq_{\mathrm{top}}} \\
\varinjlim_{\overline{\mathcal{U}}' \in \overline{\mathscr{N}}_{\mathsf{M}}} \mathrm{C}^{\mathrm{hol},\mathbb{R}}(\overline{\mathcal{U}}'; \overline{\mathsf{N}}) \times \mathrm{C}^{\mathrm{hol},\mathbb{R}}(\overline{\mathcal{V}}) \\
\Big\downarrow {\scriptstyle \mathrm{id} \times r_{\overline{\mathcal{V}},\mathsf{N}}} \\
\varinjlim_{\overline{\mathcal{U}}' \in \overline{\mathscr{N}}_{\mathsf{M}}} \mathrm{C}^{\mathrm{hol},\mathbb{R}}(\overline{\mathcal{U}}'; \overline{\mathsf{N}}) \times \varinjlim_{\overline{\mathcal{V}}' \in \overline{\mathscr{N}}_{\mathsf{N}}} \mathrm{C}^{\mathrm{hol},\mathbb{R}}(\overline{\mathcal{V}}')
\end{array}$$

The middle homeomorphism follows from Lemma 2.31 and the upper and lower arrows are continuous. If we denote

$$r_{(\overline{\mathcal{U}},\overline{\mathcal{V}}),\mathsf{M}\times\mathsf{N}} = (\mathrm{id} \times r_{\overline{\mathcal{V}},\mathsf{N}}) \circ (r_{\overline{\mathcal{U}},\mathsf{M}} \times \mathrm{id}),$$

we wish to study the final topology associated with the family of mappings

$$r_{(\overline{\mathcal{U}}_2,\overline{\mathcal{V}}_2),\mathsf{M}\times\mathsf{N}} \colon \mathrm{C}^{\mathrm{hol},\mathbb{R}}(\overline{\mathcal{U}}; \overline{\mathcal{V}}) \times \mathrm{C}^{\mathrm{hol},\mathbb{R}}(\overline{\mathcal{V}}) \\ \to \varinjlim_{\overline{\mathcal{U}}' \in \overline{\mathscr{N}}_{\mathsf{M}}} \mathrm{C}^{\mathrm{hol},\mathbb{R}}(\overline{\mathcal{U}}'; \overline{\mathsf{N}}) \times \varinjlim_{\overline{\mathcal{V}}' \in \overline{\mathscr{N}}_{\mathsf{N}}} \mathrm{C}^{\mathrm{hol},\mathbb{R}}(\overline{\mathcal{V}}'),$$

$$(\overline{\mathcal{U}}, \overline{\mathcal{V}}) \in \overline{\mathscr{N}}_{\mathsf{M}} \times \overline{\mathscr{N}}_{\mathsf{N}}.$$

In order to distinguish this final topology from the product topology for the product of the direct limits, we denote the product with the final topology by

$$\varinjlim_{(\overline{\mathcal{U}},\overline{\mathcal{V}})\in\overline{\mathscr{N}}_{\mathsf{M}}\times\overline{\mathscr{N}}_{\mathsf{N}}} \mathsf{C}^{\mathrm{hol},\mathbb{R}}(\overline{\mathcal{U}};\overline{\mathsf{N}})\times\mathsf{C}^{\mathrm{hol},\mathbb{R}}(\overline{\mathcal{V}}).$$

We then have the diagram

$$\begin{array}{ccccc}
\mathsf{C}^{\omega}(\mathsf{M};\mathsf{N}) & \longleftarrow & \mathsf{C}^{\omega}(\mathsf{M};\mathsf{N})\times\mathsf{C}^{\omega}(\mathsf{N}) & \longrightarrow & \mathsf{C}^{\omega}(\mathsf{N})\\
\Vert\,{\simeq_{\mathrm{top}}} & & \Vert\,{\simeq_{\mathrm{set}}} & & \Vert\,{\simeq_{\mathrm{top}}}\\
\varinjlim_{\overline{\mathcal{U}}'\in\overline{\mathscr{N}}_{\mathsf{M}}}\mathsf{C}^{\mathrm{hol},\mathbb{R}}(\overline{\mathcal{U}}';\overline{\mathsf{N}}) & \longleftarrow & \varinjlim_{(\overline{\mathcal{U}}',\overline{\mathcal{V}}')\in\overline{\mathscr{N}}_{\mathsf{M}}\times\overline{\mathscr{N}}_{\mathsf{N}}}\mathsf{C}^{\mathrm{hol},\mathbb{R}}(\overline{\mathcal{U}}';\overline{\mathsf{N}})\times\mathsf{C}^{\mathrm{hol},\mathbb{R}}(\overline{\mathcal{V}}') & \longrightarrow & \varinjlim_{\overline{\mathcal{V}}'\in\overline{\mathscr{N}}_{\mathsf{N}}}\mathsf{C}^{\mathrm{hol},\mathbb{R}}(\overline{\mathcal{V}}')\\
\uparrow r_{\overline{\mathcal{V}},\mathsf{M}} & \nwarrow & \uparrow r_{(\overline{\mathcal{U}},\overline{\mathcal{V}}),\mathsf{M}\times\mathsf{N}} & \nearrow & \uparrow r_{\overline{\mathcal{V}},\mathsf{N}}\\
\mathsf{C}^{\mathrm{hol},\mathbb{R}}(\overline{\mathcal{U}};\overline{\mathcal{V}}) & \longleftarrow & \mathsf{C}^{\mathrm{hol},\mathbb{R}}(\overline{\mathcal{U}};\overline{\mathcal{V}})\times\mathsf{C}^{\mathrm{hol},\mathbb{R}}(\overline{\mathcal{V}}) & \longrightarrow & \mathsf{C}^{\mathrm{hol},\mathbb{R}}(\overline{\mathcal{V}})
\end{array}$$

We regard $\mathsf{C}^{\omega}(\mathsf{M};\mathsf{N})\times\mathsf{C}^{\omega}(\mathsf{N})$ as having the product topology, with the first component having the weak-PB topology (or one of its equivalents) and the second component as having the C^{ω}-topology (in one of its many equivalent forms). The horizontal arrows are the projections. Our first task will be to prove that the upper middle vertical bijection is an homeomorphism with the given topologies.

First, the continuity of the dashed diagonal arrows follows from the continuity of the bottom horizontal arrows and the lower outer vertical arrows. The continuity of these dashed diagonal arrows and the universal property of the outer direct limits gives the continuity of middle projections if the product of direct limits has the final topology. Thus, since the product topology is the initial topology associated with the projections, we conclude that the final topology is finer than the product topology.

We now prove the converse. For this, we refer to the constructions at the beginning of the proof of Theorem 2.32 and assume that N is a Stein manifold which leads to inclusions

$$\mathsf{C}^{\omega}(\mathsf{M};\mathsf{N})\to\mathsf{C}^{\omega}(\mathsf{M};\mathbb{R}^{N}),\quad \mathsf{C}^{\mathrm{hol},\mathbb{R}}(\overline{\mathsf{M}};\overline{\mathsf{N}})\to\mathsf{C}^{\mathrm{hol},\mathbb{R}}(\overline{\mathsf{M}};\mathbb{C}^{N})$$

for which the topologies on the left are induced by the inclusion and one of the many equivalent topologies on the right, by Theorems 2.32 and 2.28. Since initial topologies are transitive, cf. [34, Proposition 2.11.1], the inclusions

$$\mathsf{C}^{\omega}(\mathsf{M};\mathsf{N})\times\mathsf{C}^{\omega}(\mathsf{N})\to\mathsf{C}^{\omega}(\mathsf{M};\mathbb{R}^{N})\times\mathsf{C}^{\omega}(\mathsf{N}),$$
$$\mathsf{C}^{\mathrm{hol},\mathbb{R}}(\overline{\mathcal{U}};\overline{\mathsf{N}})\times\mathsf{C}^{\mathrm{hol},\mathbb{R}}(\overline{\mathcal{V}})\to\mathsf{C}^{\mathrm{hol},\mathbb{R}}(\overline{\mathcal{U}};\mathbb{C}^{N})\times\mathsf{C}^{\mathrm{hol},\mathbb{R}}(\overline{\mathcal{V}}),$$
$$(\overline{\mathcal{U}},\overline{\mathcal{V}})\in\overline{\mathscr{N}}_{\mathsf{M}}\times\overline{\mathscr{N}}_{\mathsf{N}},$$

induce the product topologies on the left from the product topologies on the right. Now we note that we have inclusions

$$\begin{aligned} &\mathrm{C}^{\omega}(\mathsf{M};\mathbb{R}^N)\times \mathrm{C}^{\omega}(\mathsf{N})\to \mathrm{C}^{\omega}(\mathsf{M}\times\mathsf{N};\mathbb{R}^N\oplus\mathbb{R}) \\ &(\boldsymbol{\Phi},f)\mapsto((x,y)\mapsto\boldsymbol{\Phi}(x)\oplus f(y)) \end{aligned} \tag{5.12}$$

and

$$\begin{aligned} &\mathrm{C}^{\mathrm{hol},\mathbb{R}}(\overline{\mathcal{U}};\mathbb{C}^N)\times \mathrm{C}^{\mathrm{hol},\mathbb{R}}(\overline{\mathcal{V}})\to \mathrm{C}^{\mathrm{hol},\mathbb{R}}(\overline{\mathcal{U}}\times\overline{\mathcal{V}};\mathbb{C}^N\oplus\mathbb{C}) \\ &(\boldsymbol{\Phi},f)\mapsto((x,y)\mapsto\boldsymbol{\Phi}(x)\oplus f(y)). \end{aligned} \tag{5.13}$$

We claim that these are both topological monomorphisms. Let $\mathcal{C}\subseteq\mathsf{M}\times\mathsf{N}$ be compact, and let $\boldsymbol{a}\in c_0(\mathbb{Z}_{\ge 0};\mathbb{R}_{>0})$. Let $\mathcal{K}\subseteq\mathsf{M}$ and $\mathcal{L}\subseteq\mathsf{N}$ be compact and such that $\mathcal{C}\subseteq\mathcal{K}\times\mathcal{L}$. Then the inequality

$$p^{\omega}_{\mathcal{C},\boldsymbol{a}}(\boldsymbol{\Phi}\oplus f)\le p^{\omega}_{\mathcal{K},\boldsymbol{a}}(\boldsymbol{\Phi})+p^{\omega}_{\mathcal{L},\boldsymbol{a}}(f)$$

gives continuity of (5.12). Now let $\mathcal{K}\subseteq\mathsf{M}$ and $\mathcal{L}\subseteq\mathsf{N}$ be compact, and let $\boldsymbol{a},\boldsymbol{b}\in c_0(\mathbb{Z}_{\ge 0};\mathbb{R}_{>0})$. Define $\boldsymbol{c}\in c_0(\mathbb{Z}_{\ge 0};\mathbb{R}_{>0})$ by $c_m=\max\{a_m,b_m\}$, $m\in\mathbb{Z}_{\ge 0}$. Then the inequality

$$p^{\omega}_{\mathcal{K},\boldsymbol{a}}(\boldsymbol{\Phi})+p^{\omega}_{\mathcal{L},\boldsymbol{b}}(f)\le p^{\omega}_{\mathcal{K}\times\mathcal{L},\boldsymbol{c}}(\boldsymbol{\Phi}\oplus f)$$

establishes the openness of (5.12) by Lemma 1 from the proof of Theorem 5.1. The continuity and openness of (5.13) is proved similarly.

Since our preceding constructions have led to mappings with domain $\mathsf{M}\times\mathsf{N}$, let us understand an aspect of the direct limit topology for $\mathrm{C}^{\omega}(\mathsf{M}\times\mathsf{N};\mathsf{P})$ for a real analytic manifold P. In particular, let us understand the directed set $\overline{\mathscr{N}}_{\mathsf{M}\times\mathsf{N}}$ of neighbourhoods of $\mathsf{M}\times\mathsf{N}$ in the product $\overline{\mathsf{M}}\times\overline{\mathsf{N}}$ of the complexifications. We claim that the subset of neighbourhoods of the form $\overline{\mathcal{U}}\times\overline{\mathcal{V}}$, $\overline{\mathcal{U}}\in\overline{\mathscr{N}}_{\mathsf{M}}$, $\overline{\mathcal{V}}\in\overline{\mathscr{N}}_{\mathsf{N}}$, is cofinal in $\overline{\mathscr{N}}_{\mathsf{M}\times\mathsf{N}}$. To see this, note that the projections

$$\mathrm{pr}_1\colon\overline{\mathsf{M}}\times\overline{\mathsf{N}}\to\overline{\mathsf{M}},\quad \mathrm{pr}_2\colon\overline{\mathsf{M}}\times\overline{\mathsf{N}}\to\overline{\mathsf{N}}$$

are open. Therefore, if $\overline{\mathcal{W}}\in\overline{\mathscr{N}}_{\mathsf{M}\times\mathsf{N}}$, then $\overline{\mathcal{U}}\triangleq\mathrm{pr}_1(\overline{\mathcal{W}})\in\overline{\mathscr{N}}_{\mathsf{M}}$ and $\overline{\mathcal{V}}\triangleq\mathrm{pr}_2(\overline{\mathcal{W}})\in\overline{\mathscr{N}}_{\mathsf{N}}$. One moreover easily verifies that $\overline{\mathcal{U}}\times\overline{\mathcal{V}}\subseteq\overline{\mathcal{W}}$, giving our claim. It then follows that the direct limit topology for $\mathrm{C}^{\omega}(\mathsf{M}\times\mathsf{N};\mathsf{P})$ is given by the direct limit

$$\varinjlim_{(\overline{\mathcal{U}},\overline{\mathcal{V}})\in\overline{\mathscr{N}}_{\mathsf{M}}\times\overline{\mathscr{N}}_{\mathsf{N}}}\mathrm{C}^{\mathrm{hol},\mathbb{R}}(\overline{\mathcal{U}}\times\overline{\mathcal{V}};\overline{\mathsf{P}}).$$

Now the preceding constructions give rise to the diagram

$$
\begin{array}{ccc}
\mathrm{C}^{\omega}(\mathsf{M};\mathsf{N})\times\mathrm{C}^{\omega}(\mathsf{N}) & \longrightarrow & \mathrm{C}^{\omega}(\mathsf{M}\times\mathsf{N};\mathbb{R}^{N}\oplus\mathbb{R}) \\
\big\downarrow & & \big\downarrow \\
\varinjlim_{(\overline{\mathcal{U}}',\overline{\mathcal{V}}')\in\overline{\mathscr{N}}_{\mathsf{M}}\times\overline{\mathscr{N}}_{\mathsf{N}}}\mathrm{C}^{\mathrm{hol},\mathbb{R}}(\overline{\mathcal{U}}';\overline{\mathsf{N}})\times\mathrm{C}^{\mathrm{hol},\mathbb{R}}(\overline{\mathcal{V}}') & \xrightarrow{\iota_{\mathsf{M},\mathsf{N}}} & \varinjlim_{(\overline{\mathcal{U}}',\overline{\mathcal{V}}')\in\overline{\mathscr{N}}_{\mathsf{M}}\times\overline{\mathscr{N}}_{\mathsf{N}}}\mathrm{C}^{\mathrm{hom},\mathbb{R}}(\overline{\mathcal{U}}'\times\overline{\mathcal{V}}';\mathbb{C}^{N}\oplus\mathbb{C}) \\
\big\uparrow{\scriptstyle r_{(\overline{\mathcal{U}},\overline{\mathcal{V}}),\mathsf{M}\times\mathsf{N}}} & & \big\uparrow{\scriptstyle r_{\overline{\mathcal{U}}\times\overline{\mathcal{V}},\mathsf{M}\times\mathsf{N}}} \\
\mathrm{C}^{\mathrm{hol},\mathbb{R}}(\overline{\mathcal{U}};\overline{\mathcal{V}})\times\mathrm{C}^{\mathrm{hol},\mathbb{R}}(\overline{\mathcal{V}}) & \xrightarrow[\iota_{\overline{\mathcal{U}}\times\overline{\mathcal{V}}}]{} & \mathrm{C}^{\mathrm{hol},\mathbb{R}}(\overline{\mathcal{U}}\times\overline{\mathcal{V}};\mathbb{C}^{N}\oplus\mathbb{C})
\end{array}
$$

The upper left vertical arrow is the bijection whose continuity we wish to establish. The upper right vertical is a topological isomorphism, as we have seen. The lower two vertical mappings are those giving rise to the final topologies, which is the direct limit topology on the right. The upper and lower horizontal arrows are the homeomorphisms onto their image as described above. The middle horizontal arrow is defined so that the upper square commutes, and one readily verifies that the lower square commutes, essentially because the large outer square commutes.

For this diagram, we shall first show that the middle horizontal arrow, $\iota_{\mathsf{M},\mathsf{N}}$, is an homeomorphism onto its image. First of all, given the continuity of the bottom diagonal arrow and the universal property of the final topology, we conclude that $\iota_{\mathsf{M},\mathsf{N}}$ is continuous. For openness, we first note that, for

$$[(\Phi, f)]_{\mathsf{M}\times\mathsf{N}} \in \varinjlim_{(\overline{\mathcal{U}}',\overline{\mathcal{V}}')\in\overline{\mathscr{N}}_{\mathsf{M}}\times\overline{\mathscr{N}}_{\mathsf{N}}} \mathrm{C}^{\mathrm{hol},\mathbb{R}}(\overline{\mathcal{U}}';\overline{\mathsf{N}})\times\mathrm{C}^{\mathrm{hol},\mathbb{R}}(\overline{\mathcal{V}}'),$$

we have

$$\iota_{\overline{\mathcal{U}}\times\overline{\mathcal{V}}}(r^{-1}_{(\overline{\mathcal{U}},\overline{\mathcal{V}}),\mathsf{M}\times\mathsf{N}}([(\Phi, f)]_{\mathsf{M}\times\mathsf{N}})) = r^{-1}_{\overline{\mathcal{U}}\times\overline{\mathcal{V}},\mathsf{M}\times\mathsf{N}}(\iota_{\mathsf{M},\mathsf{N}}([(\Phi, f)]_{\mathsf{M}\times\mathsf{N}}));$$

this is easily directly verified from the definitions. Now let

$$\mathcal{O} \subseteq \varinjlim_{(\overline{\mathcal{U}}',\overline{\mathcal{V}}')\in\overline{\mathscr{N}}_{\mathsf{M}}\times\overline{\mathscr{N}}_{\mathsf{N}}} \mathrm{C}^{\mathrm{hol},\mathbb{R}}(\overline{\mathcal{U}}';\overline{\mathsf{N}})\times\mathrm{C}^{\mathrm{hol},\mathbb{R}}(\overline{\mathcal{V}}')$$

be open. Then

$$r^{-1}_{\overline{\mathcal{U}}\times\overline{\mathcal{V}},\mathsf{M}\times\mathsf{N}}(\iota_{\mathsf{M},\mathsf{N}}(\mathcal{O})) = \iota_{\overline{\mathcal{U}}\times\overline{\mathcal{V}}}(r^{-1}_{(\overline{\mathcal{U}},\overline{\mathcal{V}}),\mathsf{M}\times\mathsf{N}}(\mathcal{O})),$$

and this latter set is open since $\iota_{\overline{\mathcal{U}}\times\overline{\mathcal{V}}}$ is an homeomorphism onto its image and since $r_{(\overline{\mathcal{U}},\overline{\mathcal{V}}),\mathsf{M}\times\mathsf{N}}$ is continuous. Now we note that openness of

$$r^{-1}_{\overline{\mathcal{U}}\times\overline{\mathcal{V}},\mathsf{M}\times\mathsf{N}}(\iota_{\mathsf{M},\mathsf{N}}(\mathcal{O}))$$

for every $\overline{\mathcal{U}} \in \overline{\mathscr{N}}_{\mathsf{M}}$ and $\overline{\mathcal{V}} \in \overline{\mathscr{N}}_{\mathsf{N}}$ is precisely the openness of $\iota_{\mathsf{M},\mathsf{N}}(\mathcal{O})$.

Now, given that $\iota_{\mathsf{M},\mathsf{N}}$ is an homeomorphism onto its image, this is precisely the assertion that the final topology for

$$\varinjlim_{(\overline{\mathcal{U}}',\overline{\mathcal{V}}')\in\overline{\mathscr{N}}_{\mathsf{M}}\times\overline{\mathscr{N}}_{\mathsf{N}}} \mathrm{C}^{\mathrm{hol},\mathbb{R}}(\overline{\mathcal{U}}';\overline{\mathsf{N}}) \times \mathrm{C}^{\mathrm{hol},\mathbb{R}}(\overline{\mathcal{V}}')$$

is the initial topology associated with $\iota_{\mathsf{M},\mathsf{N}}$. Therefore, the continuity of the upper dashed diagonal arrow gives the desired continuity of the upper vertical arrow by the universal property of the initial topology.

This completes the proof of the assertion that the product topology of $\mathrm{C}^{\omega}(\mathsf{M};\mathsf{N}) \times \mathrm{C}^{\omega}(\mathsf{N})$ is the same as the final topology for

$$\varinjlim_{(\overline{\mathcal{U}}',\overline{\mathcal{V}}')\in\overline{\mathscr{N}}_{\mathsf{M}}\times\overline{\mathscr{N}}_{\mathsf{N}}} \mathrm{C}^{\mathrm{hol},\mathbb{R}}(\overline{\mathcal{U}}';\overline{\mathsf{N}}) \times \mathrm{C}^{\mathrm{hol},\mathbb{R}}(\overline{\mathcal{V}}').$$

With this at hand, it is relatively straightforward to prove the continuity of the joint composition map. For $\overline{\mathcal{U}} \in \overline{\mathscr{N}}_{\mathsf{M}}$ and $\overline{\mathcal{V}} \in \overline{\mathscr{N}}_{\mathsf{N}}$, we have the diagram

$$\begin{array}{ccc}
\mathrm{C}^{\mathrm{hol},\mathbb{R}}(\overline{\mathcal{U}};\overline{\mathcal{V}}) \times \mathrm{C}^{\mathrm{hol},\mathbb{R}}(\overline{\mathcal{V}}) & \xrightarrow{r_{(\overline{\mathcal{U}},\overline{\mathcal{V}}),\mathsf{M}\times\mathsf{N}}} & \mathrm{C}^{\omega}(\mathsf{M};\mathsf{N}) \times \mathrm{C}^{\omega}(\mathsf{N}) \\
\downarrow & \searrow & \downarrow \\
\mathrm{C}^{\mathrm{hol},\mathbb{R}}(\overline{\mathcal{U}}) & \xrightarrow[r_{\overline{\mathcal{U}},\mathsf{M}}]{} & \mathrm{C}^{\omega}(\mathsf{M})
\end{array}$$

where the vertical arrows are the joint composition operators. The left vertical arrow is continuous by Theorem 5.28. Since the final topology for $\mathrm{C}^{\omega}(\mathsf{M};\mathsf{N}) \times \mathrm{C}^{\omega}(\mathsf{N})$ is equal to the product topology, the universal property of the final topology, along with the continuity of the dashed diagonal arrow, shows that the right vertical arrow is continuous if its domain has the product topology. □

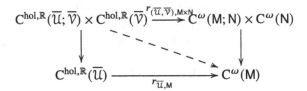

List of Symbols

$(x \mapsto f(x))$	The mapping sending x to $f(x)$, 1
$\Phi^* \pi_{\mathsf{E}} \colon \Phi^* \mathsf{E} \to \mathsf{N}$	Pull-back bundle, 6
A^*	Dual of linear map A, 2, 103
E^*	Dual bundle of E, 6
$\Phi^* \eta$	Pull-back of section, 116
$\Phi^* \nabla^{\pi_{\mathsf{F}}}$	Pull-back connection, 117
V^*	The algebraic dual of V, 2
V'	Topological dual of V, 17
$[\overline{\xi}]_A$	Germ of $\overline{\xi}$, 41
$[X, Y]$	Lie bracket of X and Y, 7
$A \odot B$	Symmetric tensor product of A and B, 3
$\mathsf{U} \otimes_\pi \mathsf{V}$	Projective tensor product, 25
$\Gamma^\omega(\mathsf{E}) \otimes_{\mathsf{C}^\omega(\mathsf{M})}^\pi \Gamma^\omega(\mathsf{F})$	Balanced projective tensor product, 264
$q \otimes_\pi p$	Seminorm for projective tensor product, 25
$\overset{\circ}{\underset{i \in I}{\bigcup}} X_i$	Disjoint union, 1
$\mathfrak{S}_k$	Symmetric group on $\{1, \ldots, k\}$, 2
$\mathfrak{S}_{k,l}$	Ordered permutations, 2
$\mathfrak{S}_{k|l}$	Product of $\mathfrak{S}_k$ and $\mathfrak{S}_l$, 2
$\Delta_{k,m}$	Inclusion for symmetric tensors, 131
$\Gamma^r(\mathsf{E})$	C^r-sections of E, 6
$\Gamma^{\mathrm{hol},\mathbb{R}}(\mathsf{E})$	Real sections of an holomorphic vector bundle, 41
$\iota_{\mathsf{E},k,m}$	Jet bundle inclusion, 129
$\hat{\iota}_{\mathsf{E},k,m}$	Jet bundle inclusion, 135
Φ_t^X	Flow of vector field X, 7
$\pi_{\mathsf{E},k}$	Projection from k-jets to 0-jets, 11
$\pi_{\mathsf{E},l}^k$	Projection from k-jets to l-jets, 11, 63
Π_{S}	Second fundamental form of S, 109
$\pi_{\mathsf{T}^*\mathsf{M}}$	Cotangent bundle projection, 6
$\pi_{\mathsf{T}\mathsf{M}}$	Tangent bundle projection, 6
$\widehat{\Phi}$	Vector bundle mapping induced by Φ, 182

© The Author(s), under exclusive license to Springer Nature Switzerland AG 2023

A. D. Lewis, *Geometric Analysis on Real Analytic Manifolds*, Lecture Notes in Mathematics 2333, https://doi.org/10.1007/978-3-031-37913-0

References

1. R. Abraham, J.E. Marsden, T.S. Ratiu, *Manifolds, Tensor Analysis, and Applications*. Applied Mathematical Sciences, vol. 75, 2nd edn. (Springer, New York, 1988)
2. C.D. Aliprantis, K.C. Border, *Infinite-Dimensional Analysis: A Hitchhiker's Guide*, 3rd edn. (Springer, New York, 2006)
3. J. Appell, P.P. Zabrejko, *Nonlinear Superposition Operators*. Cambridge Tracts in Mathematics, vol. 95 (Cambridge University Press, New York, 1990)
4. S. Axler, *Linear Algebra Done Right*. Undergraduate Texts in Mathematics, 3rd edn. (Springer, New York, 2015)
5. A. Baernstein, II, Representation of holomorphic functions by boundary integrals. Trans. Am. Math. Soc. **160**, 27–37 (1971)
6. R. Bhatia, *Matrix Analysis*. Graduate Texts in Mathematics, vol. 169 (Springer, New York, 1997)
7. K.D. Bierstedt, An introduction to locally convex inductive limits, in *Functional Analysis and its Applications*. ICPAM Lecture Notes (World Scientific, Singapore, 1988), pp. 35–133
8. V.I. Bogachev, *Measure Theory*, vol. 2 (Springer, New York, 2007)
9. E. Borel, Sur quelles points de la théorie des fonctions. Annales Scientifiques de l'École Normale Supérieure. Quatrième Série **12**(3), 44 (1895)
10. N. Bourbaki, *Algebra II*. Elements of Mathematics (Springer, New York, 1990)
11. D. Burago, Y. Burago, S. Ivanov, *A Course in Metric Geometry*. Graduate Studies in Mathematics, vol. 33 (American Mathematical Society, Providence, 2001)
12. H. Cartan, *Séminaire Henri Cartan, Fonctions analytiques de plusieurs variables complexes*, vol. 4 (École Normale Supérieure, 1951–52). Lecture Notes
13. H. Cartan, Variétés analytiques réelles et variétés analytiques complexes. Bulletin de la Société Mathématique de France **85**, 77–99 (1957)
14. K. Cieliebak, Y. Eliashberg, *From Stein to Weinstein and Back: Symplectic Geometry of Affine Complex Manifolds*. American Mathematical Society Colloquium Publications, vol. 59 (American Mathematical Society, Providence, 2012)
15. M. De Wilde, Théorème du graphe fermé et espaces à réseau absorbant. Bulletin Mathématique de la Société des Sciences Mathématiques de Roumanie. Nouvelle Série **59**(2), 225–238 (1967)
16. P. Domański, Notes on real analytic functions and classical operators, in *Proceedings of the Third Winter School in Complex Analysis and Operator Theory*, ed. by O. Blasco, J. Bonet, J. Calabuig, D. Jornet. Contemporary Mathematics, vol. 561 (American Mathematical Society, Providence, 2010), pp. 3–47
17. P. Domański, M. Langenbruch, Composition operators on spaces of real analytic functions. Math. Nachr. **254–255**(1), 68–86 (2003)

A. D. Lewis, *Geometric Analysis on Real Analytic Manifolds*, Lecture Notes in Mathematics 2333, https://doi.org/10.1007/978-3-031-37913-0

18. P. Domański, M. Langenbruch, Composition operators with closed image on spaces of real analytic functions. Bullet. Lond. Math. Soc. **38**(4), 635–646 (2006)
19. P. Domański, D. Vogt, The space of real-analytic functions has no basis. Pol. Akad. Nauk. Inst. Mat. Stud. Math. **142**(2), 187–200 (2000)
20. P. Drábek, Continuity of Nemyckij's operator in Hölder spaces. Comment. Math. Univ. Caro. **16**(1), 37–57 (1975)
21. S.P. Franklin, Spaces in which sequences suffice. Pol. Akademia Nauk Fundamenta Mathematicae **61**, 51–56 (1965)
22. H.L. Goldschmidt, Existence theorems for analytic linear partial differential equations. Ann. Math. **86**(2), 246–270 (1967)
23. H. Grauert, On Levi's problem and the imbedding of real-analytic manifolds. Ann. Math. **68**, 460–472 (1958)
24. H. Grauert, R. Remmert, *Coherent Analytic Sheaves*. Grundlehren der Mathematischen Wissenschaften, vol. 265 (Springer, New York, 1984)
25. H. Grauert, R. Remmert, *Theory of Stein Spaces*. Grundlehren der mathematischen Wissenschaften, vol. 236 (Springer, New York, 1979). Reprint: [26]
26. H. Grauert, R. Remmert, *Theory of Stein Spaces*. Classics in Mathematics (Springer, Berlin, 2004). Original: [25]
27. A. Grothendieck, Résumé de la théorie métrique des produits tensoriels topologiques. Bol. Soc. Mat. São Paulo **8**, 1–59 (1953)
28. A. Groethendieck, *Topological Vector Spaces*. Notes on Mathematics and its Applications (Gordon & Breach Science, New York, 1973)
29. R.C. Gunning, *Introduction to Holomorphic Functions of Several Variables. Volume I: Function Theory*. Wadsworth & Brooks/Cole Mathematics Series (Wadsworth & Brooks/Cole, Belmont, 1990)
30. P.R. Halmos, *Finite-Dimensional Vector Spaces*. Undergraduate Texts in Mathematics, 2nd edn. (Springer, New York, 1986)
31. E. Hewitt, K. Stromberg, *Real and Abstract Analysis*. Graduate Texts in Mathematics, vol. 25 (Springer, New York, 1975)
32. M.W. Hirsch, *Differential Topology*. Graduate Texts in Mathematics, vol. 33 (Springer, New York, 1976)
33. J. Horváth, *Topological Vector Spaces and Distributions*, vol. I (Addison Wesley, Reading, 1966)
34. T.W. Hungerford, *Algebra*. Graduate Texts in Mathematics, vol. 73 (Springer, New York, 1980)
35. S. Jafarpour, A.D. Lewis, *Time-Varying Vector Fields and Their Flows*. Springer Briefs in Mathematics (Springer, New York, 2014)
36. H. Jarchow, *Locally Convex Spaces*. Mathematical Textbooks (Teubner, Leipzig, 1981)
37. J. Jost, *Postmodern Analysis*. Universitext, 3rd edn. (Springer, New York, 2005)
38. L. Kaup, B. Kaup, *Holomorphic Functions of Several Variables*. Studies in Mathematics, vol. 3 (Walter de Gruyter, Berlin, 1983)
39. J.L. Kelley, *General Topology* (Van Nostrand Reinhold, London, 1955). Reprint: [40]
40. J.L. Kelley, *General Topology*. Graduate Texts in Mathematics, vol. 27 (Springer, New York, 1975). Original: [39]
41. S. Kobayashi, K. Nomizu, *Foundations of Differential Geometry, Volume I*. Interscience Tracts in Pure and Applied Mathematics, vol. 15 (Interscience Publishers, New York, 1963)
42. I. Kolář, P.W. Michor, J. Slovák, *Natural Operations in Differential Geometry* (Springer, New York, 1993)
43. H. Komatsu, Projective and injective limits of weakly compact sequences of locally convex spaces. J. Math. Soc. Japan **19**(3), 366–383 (1967)
44. S.G. Krantz, *Function Theory of Several Complex Variables*, 2nd edn. (AMS Chelsea Publishing, Providence, 1992)
45. S.G. Krantz, H.R. Parks, *A Primer of Real Analytic Functions*. Birkhäuser Advanced Texts, 2nd edn. (Birkhäuser, Boston, 2002)

46. A. Kriegl, P.W. Michor, *The Convenient Setting of Global Analysis*. American Mathematical Society Mathematical Surveys and Monographs, vol. 57 (American Mathematical Society, Providence, 1997)
47. A. Martineau, Sur les fontionelles analytiques et la transformation de Fourier–Borel. J. Anal. Math. **11**, 1–164 (1963)
48. A. Martineau, Sur la topologie des espaces de fonctions holomorphes. Math. Ann. **163**, 62–88 (1966)
49. R. Meise, D. Vogt, *Introduction to Functional Analysis*. Oxford Graduate Texts in Mathematics, vol. 2 (Oxford University Press, Oxford, 1997)
50. P.W. Michor, *Manifolds of Differentiable Mappings*. Shiva Mathematics Series, vol. 3 (Shiva Publishing Limited, Orpington, 1980)
51. P.W. Michor, *Topics in Differential Geometry*. Graduate Studies in Mathematics, vol. 93 (American Mathematical Society, Providence, 2008)
52. J. Mujica, A Banach–Dieudonné theorem for germs of holomorphic functions. J. Funct. Anal. **57**(1), 31–48 (1984)
53. E. Nelson, *Tensor Analysis* (Princeton University Press, Princeton, 1967)
54. J.W. Neuberger, Tensor products and successive approximations for partial differential equations. Isr. J. Math. **6**(2), 121–132 (1968)
55. E.A. Nigsch, A nonlinear theory of generalized tensor fields on Riemannian manifolds. Ph.D. Thesis, Universität Wien, Vienna (2010)
56. B. O'Neill, The fundamental equations of a submersion. Michigan Math. J. **13**(4), 459–469 (1968)
57. V.P. Palamodov, The projective limit functor in the category of linear topological spaces. Math. USSR-Sbornik **75**(4), 529–558 (1968)
58. S. Ramanan, *Global Calculus*. Graduate Studies in Mathematics, vol. 65 (American Mathematical Society, Providence, 2005)
59. R. Remmert, Holomorphe und meromorphe Abbildungen analytischer. Ph.D. Thesis, Westfälische Wilhelms–Universität Münster (1954)
60. W. Rudin, *Principles of Mathematical Analysis*. International Series in Pure & Applied Mathematics, 3rd edn. (McGraw-Hill, New York, 1976)
61. S. Sasaki, On the differential geometry of tangent bundles of Riemannian manifolds. Tôhoku Math. J. **10**, 338–354 (1958)
62. D.J. Saunders, *The Geometry of Jet Bundles*. London Mathematical Society Lecture Note Series, vol. 142 (Cambridge University Press, New York, 1989)
63. H.H. Schaefer, M.P. Wolff, *Topological Vector Spaces*. Graduate Texts in Mathematics, vol. 3, 2nd edn. (Springer, New York, 1999)
64. V. Thilliez, Sur les fonctions composées ultradifférentiables. Journal de Mathématiques Pures et Appliquées. Neuvième Sér **76**, 499–524 (1997)
65. D. Vogt, Section spaces of real analytic vector bundles and a theorem of Grothendieck and Poly, in *Linear and Non-Linear Theory of Generalized Functions and its Applications*. Banach Center Publications, vol. 88 (Polish Academy of Sciences, Institute for Mathematics, Warsaw, 2010), pp. 315–321
66. D. Vogt, A fundamental system of seminorms for $A(K)$ (2013). http://arxiv.org/abs/1309.6292v1. ArXiv:1309.6292v1 [math.FA]
67. R.O. Wells, Jr., *Differential Analysis on Complex Manifolds*. Graduate Texts in Mathematics, vol. 65, 3rd edn. (Springer, New York, 2008)
68. J. Wengenroth, *Derived Functors in Functional Analysis*. Lecture Notes in Mathematics, vol. 1810 (Springer, New York, 2003)
69. H. Whitney, Differentiable manifolds. Ann. Math. **37**(3), 645–680 (1936)
70. H. Whitney, F. Bruhat, Quelques propriétés fondamentales des ensembles analytiques-réels. Comment. Math. Helv. **33**, 132–160 (1959)
71. S. Willard, *General Topology* (Addison Wesley, Reading, 1970). Reprint: [72]
72. S. Willard, *General Topology* (Dover Publications, New York, 2004). Original: [71]

Index

A. D. Lewis, *Geometric Analysis on Real Analytic Manifolds*, Lecture Notes in Mathematics 2333, https://doi.org/10.1007/978-3-031-37913-0

LECTURE NOTES IN MATHEMATICS Springer

Editors in Chief: J.-M. Morel, B. Teissier;

Editorial Policy

1. Lecture Notes aim to report new developments in all areas of mathematics and their applications – quickly, informally and at a high level. Mathematical texts analysing new developments in modelling and numerical simulation are welcome.

   Manuscripts should be reasonably self-contained and rounded off. Thus they may, and often will, present not only results of the author but also related work by other people. They may be based on specialised lecture courses. Furthermore, the manuscripts should provide sufficient motivation, examples and applications. This clearly distinguishes Lecture Notes from journal articles or technical reports which normally are very concise. Articles intended for a journal but too long to be accepted by most journals, usually do not have this "lecture notes" character. For similar reasons it is unusual for doctoral theses to be accepted for the Lecture Notes series, though habilitation theses may be appropriate.

2. Besides monographs, multi-author manuscripts resulting from SUMMER SCHOOLS or similar INTENSIVE COURSES are welcome, provided their objective was held to present an active mathematical topic to an audience at the beginning or intermediate graduate level (a list of participants should be provided).

   The resulting manuscript should not be just a collection of course notes, but should require advance planning and coordination among the main lecturers. The subject matter should dictate the structure of the book. This structure should be motivated and explained in a scientific introduction, and the notation, references, index and formulation of results should be, if possible, unified by the editors. Each contribution should have an abstract and an introduction referring to the other contributions. In other words, more preparatory work must go into a multi-authored volume than simply assembling a disparate collection of papers, communicated at the event.

3. Manuscripts should be submitted either online at www.editorialmanager.com/lnm to Springer's mathematics editorial in Heidelberg, or electronically to one of the series editors. Authors should be aware that incomplete or insufficiently close-to-final manuscripts almost always result in longer refereeing times and nevertheless unclear referees' recommendations, making further refereeing of a final draft necessary. The strict minimum amount of material that will be considered should include a detailed outline describing the planned contents of each chapter, a bibliography and several sample chapters. Parallel submission of a manuscript to another publisher while under consideration for LNM is not acceptable and can lead to rejection.

4. In general, **monographs** will be sent out to at least 2 external referees for evaluation.

   A final decision to publish can be made only on the basis of the complete manuscript, however a refereeing process leading to a preliminary decision can be based on a pre-final or incomplete manuscript.

   Volume Editors of **multi-author works** are expected to arrange for the refereeing, to the usual scientific standards, of the individual contributions. If the resulting reports can be forwarded to the LNM Editorial Board, this is very helpful. If no reports are forwarded or if other questions remain unclear in respect of homogeneity etc, the series editors may wish to consult external referees for an overall evaluation of the volume.

5. Manuscripts should in general be submitted in English. Final manuscripts should contain at least 100 pages of mathematical text and should always include

   - a table of contents;
   - an informative introduction, with adequate motivation and perhaps some historical remarks: it should be accessible to a reader not intimately familiar with the topic treated;
   - a subject index: as a rule this is genuinely helpful for the reader.
   - For evaluation purposes, manuscripts should be submitted as pdf files.

6. Careful preparation of the manuscripts will help keep production time short besides ensuring satisfactory appearance of the finished book in print and online. After acceptance of the manuscript authors will be asked to prepare the final LaTeX source files (see LaTeX templates online: https://www.springer.com/gb/authors-editors/book-authors-editors/manuscriptpreparation/5636) plus the corresponding pdf- or zipped ps-file. The LaTeX source files are essential for producing the full-text online version of the book, see http://link.springer.com/bookseries/304 for the existing online volumes of LNM). The technical production of a Lecture Notes volume takes approximately 12 weeks. Additional instructions, if necessary, are available on request from lnm@springer.com.

7. Authors receive a total of 30 free copies of their volume and free access to their book on SpringerLink, but no royalties. They are entitled to a discount of 33.3 % on the price of Springer books purchased for their personal use, if ordering directly from Springer.

8. Commitment to publish is made by a *Publishing Agreement*; contributing authors of multiauthor books are requested to sign a *Consent to Publish form*. Springer-Verlag registers the copyright for each volume. Authors are free to reuse material contained in their LNM volumes in later publications: a brief written (or e-mail) request for formal permission is sufficient.

Addresses:

Professor Jean-Michel Morel, CMLA, École Normale Supérieure de Cachan, France
E-mail: moreljeanmichel@gmail.com

Professor Bernard Teissier, Equipe Géométrie et Dynamique,
Institut de Mathématiques de Jussieu – Paris Rive Gauche, Paris, France
E-mail: bernard.teissier@imj-prg.fr

Springer: Ute McCrory, Mathematics, Heidelberg, Germany,
E-mail: lnm@springer.com

Printed in the United States
by Baker & Taylor Publisher Services

Printed in the United States
by Baker & Taylor Publisher Services